Principles of Electrochemistry

Principles of Electrochemistry
Second Edition

Jiří Koryta

Institute of Physiology,
Czechoslovak Academy of Sciences, Prague

†Jiří Dvořák

Department of Physical Chemistry, Faculty of Science,
Charles University, Prague

Ladislav Kavan

J. Heyrovský Institute of Physical Chemistry and Electrochemistry,
Czechoslovak Academy of Sciences, Prague

JOHN WILEY & SONS

Chichester · New York · Brisbane · Toronto · Singapore

Other Wiley Editorial Offices

John Wiley & Sons, Inc., 605 Third Avenue,
New York, NY 10158-0012, USA

Jacaranda Wiley Ltd, G.P.O. Box 859, Brisbane,
Queensland 4001, Australia

John Wiley & Sons (Canada) Ltd, Worcester Road,
Rexdale, Ontario M9W 1L1, Canada

John Wiley & Sons (SEA) Pte Ltd, 37 Jalan Pemimpin #05-04,
Block B, Union Industrial Building, Singapore 2057

Library of Congress Cataloging-in-Publication Data

Koryta, Jiří.
Principles of electrochemistry.—2nd ed. / Jiří Koryta, Jiří Dvořák, Ladislav Kavan.
p. cm.
Includes bibliographical references and index.
ISBN 0 471 93713 4 : ISBN 0 471 93838 6 (pbk)
1. Electrochemistry. I. Dvořák, Jiří, II. Kavan, Ladislav. III. Title.
QD553.K69 1993
541.3′7—dc20 92-24345
CIP

British Library Cataloguing in Publication Data

A catalogue record for this book is available
from the British Library

ISBN 0 471 93713 4 (cloth)
ISBN 0 471 93838 6 (paper)

Typeset in Times 10/12 pt by The Universities Press (Belfast) Ltd.
Printed and bound in Great Britain by Biddles Ltd, Guildford, Surrey

Contents

Preface to the First Edition

Although electrochemistry has become increasingly important in society and in science the proportion of physical chemistry textbooks devoted to electrochemistry has declined both in extent and in quality (with notable exceptions, e.g. W. J. Moore's *Physical Chemistry*).

As recent books dealing with electrochemistry have mainly been addressed to the specialist it has seemed appropriate to prepare a textbook of electrochemistry which assumes a knowledge of basic physical chemistry at the undergraduate level. Thus, the present text will benefit the more advanced undergraduate and postgraduate students and research workers specializing in physical chemistry, biology, materials science and their applications. An attempt has been made to include as much material as possible so that the book becomes a starting point for the study of monographs and original papers.

Monographs and reviews (mainly published after 1970) pertaining to individual sections of the book are quoted at the end of each section. Many reviews have appeared in monographic series, namely:

Advances in Electrochemistry and Electrochemical Engineering (Eds P. Delahay, H. Gerischer and C. W. Tobias), Wiley–Interscience, New York, published since 1961, abbreviation in References *AE*.

Electroanalytical Chemistry (Ed. A. J. Bard), M. Dekker, New York, published since 1966.

Modern Aspects of Electrochemistry (Eds J. O'M. Bockris, B. E. Conway and coworkers), Butterworths, London, later Plenum Press, New York, published since 1954, abbreviation *MAE*.

Electrochemical compendia include:

The Encyclopedia of Electrochemistry (Ed. C. A. Hempel), Reinhold, New York, 1961.

Comprehensive Treatise of Electrochemistry (Eds J. O'M. Bockris, B. E. Conway, E. Yeager and coworkers), 10 volumes, Plenum Press, 1980–1985, abbreviation *CTE*.

Electrochemistry of Elements (Ed. A. J. Bard), M. Dekker, New York, a multivolume series published since 1973.

Physical Chemistry. An Advanced Treatise (Eds H. Eyring, D. Henderson and W. Jost), Vol. IXA,B, Electrochemistry, Academic Press, New York, 1970, abbreviation *PChAT.*
Hibbert, D. B. and A. M. James, *Dictionary of Electrochemistry,* Macmillan, London, 1984.

There are several more recent textbooks, namely:

Bockris, J. O'M. and A. K. N. Reddy, *Modern Electrochemistry,* Plenum Press, New York, 1970.
Hertz, H. G., *Electrochemistry—A Reformulation of Basic Principles,* Springer-Verlag, Berlin, 1980.
Besson, J., *Precis de Thermodynamique et Cinétique Électrochimique,* Ellipses, Paris, 1984, and an introductory text.
Koryta, J., *Ions, Electrodes, and Membranes,* 2nd Ed., John Wiley & Sons, Chichester, 1991.
Rieger, P. H., *Electrochemistry,* Prentice-Hall, Englewood Cliffs, N.J., 1987.

The more important data compilations are:

Conway, B. E., *Electrochemical Data,* Elsevier, Amsterdam, 1952.
CRC Handbook of Chemistry and Physics (Ed. R. C. Weast), CRC Press, Boca Raton, 1985.
CRC Handbook Series in Inorganic Electrochemistry (Eds L. Meites, P. Zuman, E. B. Rupp and A. Narayanan), CRC Press, Boca Raton, a multivolume series published since 1980.
CRC Handbook Series in Organic Electrochemistry (Eds L. Meites and P. Zuman), CRC Press, Boca Raton, a multivolume series published since 1977.
Horvath, A. L., *Handbook of Aqueous Electrolyte Solutions, Physical Properties, Estimation and Correlation Methods,* Ellis Horwood, Chichester, 1985.
Oxidation–Reduction Potentials in Aqueous Solutions (Eds A. J. Bard, J. Jordan and R. Parsons), Blackwell, Oxford, 1986.
Parsons, R., *Handbook of Electrochemical Data,* Butterworths, London, 1959.
Perrin, D. D., *Dissociation Constants of Inorganic Acids and Bases in Aqueous Solutions,* Butterworths, London, 1969.
Standard Potentials in Aqueous Solutions (Eds A. J. Bard, R. Parsons and J. Jordan), M. Dekker, New York, 1985.

The present authors, together with the late (Miss) Dr V. Boháčková, published their *Electrochemistry,* Methuen, London, in 1970. In spite of the favourable attitude of the readers, reviewers and publishers to that book (German, Russian, Polish, and Czech editions have appeared since then) we now consider it out of date and therefore present a text which has been largely rewritten. In particular we have stressed modern electrochemical materials (electrolytes, electrodes, non-aqueous electrochemistry in general), up-to-date charge transfer theory and biological aspects of electrochemistry. On the other hand, the presentation of electrochemical methods is quite short as the reader has access to excellent monographs on the subject (see page 301).

The Czech manuscript has been kindly translated by Dr M. Hyman-Štulíková. We are much indebted to the late Dr A Ryvolová, Mrs M. Kozlová and Mrs D. Tůmová for their expert help in preparing the manuscript. Professor E. Budevski, Dr J. Ludvík, Dr L. Novotný and Dr J. Weber have supplied excellent photographs and drawings.

Dr K. Janáček, Dr L. Kavan, Dr K. Micka, Dr P. Novák, Dr Z. Samec and Dr J. Weber read individual chapters of the manuscript and made valuable comments and suggestions for improving the book. Dr L. Kavan is the author of the section on non-electrochemical methods (pages 319 to 329).

We are also grateful to Professor V. Pokorný, Vice-president of the Czechoslovak Academy of Sciences and chairman of the Editorial Council of the Academy, for his support.

Lastly we would like to mention with devotion our teachers, the late Professor J. Heyrovský and the late Professor R. Brdička, for the inspiration we received from them for our research and teaching of electrochemistry, and our colleague and friend, the late Dr V. Boháčková, for all her assistance in the past.

Prague, March 1986

Jiří Koryta
Jiří Dvořák

Preface to the Second Edition

The new edition of *Principles of Electrochemistry* has been considerably extended by a number of new sections, particularly dealing with 'electrochemical material science' (ion and electron conducting polymers, chemically modified electrodes), photoelectrochemistry, stochastic processes, new aspects of ion transfer across biological membranes, biosensors, etc. In view of this extension of the book we asked Dr Ladislav Kavan (the author of the section on non-electrochemical methods in the first edition) to contribute as a co-author discussing many of these topics. On the other hand it has been necessary to become less concerned with some of the 'classical' topics the details of which are of limited importance for the reader.

Dr Karel Micka of the J. Heyrovský Institute of Physical Chemistry and Electrochemistry has revised very thoroughly the language of the original text as well as of the new manuscript. He has also made many extremely useful suggestions for amending factual errors and improving the accuracy of many statements throughout the whole text. We are further much indebted to Prof. Michael Grätzel and Dr Nicolas Vlachopoulos, Federal Polytechnics, Lausanne, for valuable suggestions to the manuscript.

During the preparation of the second edition Professor Jiří Dvořák died after a serious illness on 27 February 1992. We shall always remember his scientific effort and his human qualities.

Prague, May 1992 Jiří Koryta

Chapter 1

Equilibrium Properties of Electrolytes

Substances are frequently spoken of as being electro-negative, or electro-positive, according as they go under the supposed influence of direct attraction to the positive or negative pole. But these terms are much too significant for the use for which I should have to put them; for though the meanings are perhaps right, they are only hypothetical, and may be wrong; and then, through a very imperceptible, but still very dangerous, because continual, influence, they do great injury to science, by contracting and limiting the habitual views of those engaged in pursuing it. I propose to distinguish such bodies by calling those *anions* which go to the anode of the decomposing body; and those passing to the cathode, *cations*; and when I have the occasion to speak of these together, I shall call them *ions*. Thus, the chloride of lead is an electrolyte, and when electrolysed evolves two ions, chlorine and lead, the former being an anion, and the latter a cation.

M. Faraday, 1834

1.1 Electrolytes: Elementary Concepts

1.1.1 *Terminology*

A substance present in solution or in a melt which is at least partly in the form of charged species—*ions*—is called an *electrolyte.* The decomposition of electroneutral molecules to form electrically charged ions is termed *electrolytic dissociation.* Ions with a positive charge are called *cations*; those with a negative charge are termed *anions.* Ions move in an electric field as a result of their charge—cations towards the *cathode,* anions to the *anode.* The cathode is considered to be that electrode through which negative charge, i.e. electrons, enters a heterogeneous electrochemical system (electrolytic cell, galvanic cell). Electrons leave the system through the anode. Thus, in the presence of current flow, reduction always occurs at the cathode and oxidation at the anode. In the strictest sense, in the absence of current passage the concepts of anode and cathode lose their meaning. All these terms were introduced in the thirties of the last century by M. Faraday.

R. Clausius (1857) demonstrated the presence of ions in solutions and verified the validity of Ohm's law down to very low voltages (by electrolysis of a solution with direct current and unpolarizable electrodes). Up until that time, it was generally accepted that ions are formed only under the influence of an electric field leading to current flow through the solution.

The electrical conductivity of electrolyte solutions was measured at the very beginning of electrochemistry. The resistance of a conductor R is the proportionality constant between the applied voltage U and the current I passing through the conductor. It is thus the constant in the equation $U = RI$, known as Ohm's law. The reciprocal of the resistance is termed the conductance. The resistance and conductance depend on the material from which the conductor is made and also on the length L and cross-section S of the conductor. If the resistance is recalculated to unit length and unit cross-section of the conductor, the quantity $\rho = RS/L$ is obtained, termed the *resistivity*. For conductors consisting of a solid substance (metals, solid electrolytes) or single component liquids, this quantity is a characteristic of the particular substance. In solutions, however, the resistivity and the *conductivity* $\kappa = 1/\rho$ are also dependent on the electrolyte concentration c. In fact, even the quantity obtained by recalculation of the conductivity to unit concentration, $\Lambda = \kappa/c$, termed the molar conductivity, is not independent of the electrolyte concentration and is thus not a material constant, characterizing the given electrolyte. Only the limiting value at very low concentrations, called the limiting molar conductivity Λ^0, is such a quantity.

A study of the concentration dependence of the molar conductivity, carried out by a number of authors, primarily F. W. G. Kohlrausch and W. Ostwald, revealed that these dependences are of two types (see Fig. 2.5) and thus, apparently, there are two types of electrolytes. Those that are fully dissociated so that their molecules are not present in the solution are called *strong electrolytes*, while those that dissociate incompletely are *weak electrolytes*. Ions as well as molecules are present in solution of a weak electrolyte at finite dilution. However, this distinction is not very accurate as, at higher concentration, the strong electrolytes associate forming *ion pairs* (see Section 1.2.4).

Thus, in weak electrolytes, molecules can exist in a similar way as in non-electrolytes—a molecule is considered to be an electrically neutral species consisting of atoms bonded together so strongly that this species can be studied as an independent entity. In contrast to the molecules of non-electrolytes, the molecules of weak electrolytes contain at least one bond with a partly ionic character. Strong electrolytes do not form molecules in this sense. Here the bond between the cation and the anion is primarily ionic in character and the corresponding chemical formula represents only a formal molecule; nonetheless, this formula correctly describes the composition of the ionic crystal of the given strong electrolyte.

The first theory of solutions of weak electrolytes was formulated in 1887 by S. Arrhenius (see Section 1.1.4). If the molar conductivity is introduced into the equations following from Arrhenius' concepts of weak electrolytes, Eq. (2.4.17) is obtained, known as the Ostwald dilution law; this equation provides a good description of one of these types of concentration dependence of the molar conductivity. The second type was described by Kohlrausch using the empirical equation (2.4.15), which was later theoretically interpreted by P. Debye and E. Hückel on the basis of concepts of the activity coefficients of ions in solutions of completely dissociated electrolytes, and considerably improved by L. Onsager. An electrolyte can be classified as strong or weak according to whether its behaviour can be described by the Ostwald or Kohlrausch equation. Similarly, the 'strength' of an electrolyte can be estimated on the basis of the van't Hoff coefficient (see Section 1.1.4).

1.1.2 *Electroneutrality and mean quantities*

Prior to dissolution, the ion-forming molecules have an overall electric charge of zero. Thus, a homogeneous liquid system also has zero charge even though it contains charged species. In solution, the number of positive elementary charges on the cations equals the number of negative charges of the anions. If a system contains s different ions with molality m_i (concentrations c_i or mole fractions x_i can also be employed), each bearing z_i elementary charges, then the equation

$$\sum_{i=1}^{s} z_i m_i = 0 \tag{1.1.1}$$

called the *electroneutrality condition*, is valid on a macroscopic scale for every homogeneous part of the system but not for the boundary between two phases (see Chapter 4).

From the physical point of view there cannot exist, under equilibrium conditions, a measurable excess of charge in the bulk of an electrolyte solution. By electrostatic repulsion this charge would be dragged to the phase boundary where it would be the source of a strong electric field in the vicinity of the phase. This point will be discussed in Section 3.1.3.

In Eq. (1.1.1), as elsewhere below, z_i is a dimensionless number (the charge of species i related to the charge of a proton, i.e. the *charge number* of the ion) with sign $z_i > 0$ for cations and $z_i < 0$ for anions.

The electroneutrality condition decreases the number of independent variables in the system by one; these variables correspond to components whose concentration can be varied independently. In general, however, a number of further conditions must be maintained (e.g. stoichiometry and the dissociation equilibrium condition). In addition, because of the electroneutrality condition, the contributions of the anion and cation to a number of solution properties of the electrolyte cannot be separated (e.g. electrical conductivity, diffusion coefficient and decrease in vapour pressure) without assumptions about individual particles. Consequently, *mean values* have been defined for a number of cases.

For example, the molality can be expressed for an electrolyte as a whole, m_1; the amount of substance ('number of moles') is expressed in moles of formula units as if the electrolyte were not dissociated. For a strong electrolyte whose formula unit contains ν_+ cations and ν_- anions, i.e. a total of $\nu = \nu_+ + \nu_-$ ions, the molalities of the ions are related to the total molality by a simple relationship, $m_+ = \nu_+ m_1$ and $m_- = \nu_- m_1$. The mean molality is then

$$m_\pm = (m_+^{\nu_+} m_-^{\nu_-})^{1/\nu} = m_1(\nu_+^{\nu_+} \nu_-^{\nu_-})^{1/\nu} \tag{1.1.2}$$

The mean molality values $m_\pm$ (moles per kilogram), mole fractions $x_\pm$ (dimensionless number) and concentrations $c_\pm$ (moles per cubic decimetre) are related by equations similar to those for non-electrolytes (see Appendix A).

1.1.3 *Non-ideal behaviour of electrolyte solutions*

The chemical potential is encountered in electrochemistry in connection with the components of both solutions and gases. The chemical potential μ_i of component i is defined as the partial molar Gibbs energy of the system, i.e. the partial derivative of the Gibbs energy G with respect to the amount of substance n_i of component i at constant pressure, temperature and amounts of all the other components except the ith. Consider that the system does not exchange matter with its environment but only energy in the form of heat and volume work. From this definition it follows for a reversible isothermal change of the pressure of one mole of an ideal gas from the reference value p_{ref} to the actual value p_{act} that

$$\mu_{\text{act}} - \mu_{\text{ref}} = RT \ln \frac{p_{\text{act}}}{p_{\text{ref}}} \tag{1.1.3}$$

which is usually written in the form

$$\mu = \mu^0 + RT \ln p \tag{1.1.4}$$

where p is the dimensionless pressure ratio $p_{\text{act}}/p_{\text{ref}}$. The reference state is taken as the state at the given temperature and at a pressure of 10^5 Pa. The dimensionless pressure p is therefore expressed as multiples of this reference pressure. Term μ^0 has the significance of the chemical potential of the gas at a pressure equal to the standard pressure, $p = 1$, and is termed the *standard chemical potential*. This significance of quantities μ^0 and p should be recalled, e.g. when substituting pressure values into the Nernst equation for gas electrodes (see Section 3.2); if the value of the actual pressure in some arbitrary units were substituted (e.g. in pounds per square inch), this would affect the value of the standard electrode potential.

The chemical potential μ_i of the components of an ideal mixture of liquids (the components of an ideal mixture of liquids obey the Raoult law over the whole range of mole fractions and are completely miscible) is

$$\mu_i = \mu_i^* + RT \ln x_i \tag{1.1.5}$$

The standard term μ_i^* is the chemical potential of the pure component i (i.e. when $x_i = 1$) at the temperature of the system and the corresponding saturated vapour pressure. According to the Raoult law, in an ideal mixture the partial pressure of each component above the liquid is proportional to its mole fraction in the liquid,

$$p_i = p_i^0 x_i \tag{1.1.6}$$

where the proportionality constant p_i^0 is the vapour pressure above the pure substance.

In a general case of a mixture, no component takes preference and the standard state is that of the pure component. In solutions, however, one component, termed the solvent, is treated differently from the others, called solutes. Dilute solutions occupy a special position, as the solvent is present in a large excess. The quantities pertaining to the solvent are denoted by the subscript 0 and those of the solute by the subscript 1. For $x_1 \rightarrow 0$ and $x_0 \rightarrow 1$, $p_0 = p_0$ and $p_1 = k_1 x_1$. Equation (1.1.5) is again valid for the chemical potentials of both components. The standard chemical potential of the solvent is defined in the same way as the standard chemical potential of the component of an ideal mixture, the standard state being that of the pure solvent. The standard chemical potential of the dissolved component μ_1^* is the chemical potential of that pure component in the physically unattainable state corresponding to linear extrapolation of the behaviour of this component according to Henry's law up to point $x_1 = 1$ at the temperature of the mixture T and at pressure $p = k_1$, which is the proportionality constant of Henry's law.

For a solution of a non-volatile substance (e.g. a solid) in a liquid the vapour pressure of the solute can be neglected. The reference state for such a substance is usually its very dilute solution—in the limiting case an infinitely dilute solution—which has identical properties with an ideal solution and is thus useful, especially for introducing activity coefficients (see Sections 1.1.4 and 1.3). The standard chemical potential of such a solute is defined as

$$\mu_1^* = \lim_{x_0 \rightarrow 1} (\mu_1 - RT \ln x_1) \tag{1.1.7}$$

where μ_1 is the chemical potential of the solute, x_1 its mole fraction and x_0 the mole fraction of the solvent.

In the subsequent text, wherever possible, the quantities μ_i^0 and μ_i^* will not be distinguished by separate symbols: only the symbol μ_i^0 will be employed.

In real mixtures and solutions, the chemical potential ceases to be a linear function of the logarithm of the partial pressure or mole fraction. Consequently, a different approach is usually adopted. The simple form of the equations derived for ideal systems is retained for real systems, but a different quantity a, called the *activity* (or fugacity for real gases), is introduced. Imagine that the dissolved species are less 'active' than would correspond to their concentration, as if some sort of 'loss' of the given interaction were involved. The *activity* is related to the chemical potential by the relationship

$$\mu_i = \mu_i^0 + RT \ln a_i \tag{1.1.8}$$

As in electrochemical investigations low pressures are usually employed, the analogy of activity for the gaseous state, the *fugacity,* will not be introduced in the present book.

Electrolyte solutions differ from solutions of uncharged species in their greater tendency to behave non-ideally. This is a result of differences in the forces producing the deviation from ideality, i.e. the forces of interaction between particles of the dissolved substances. In non-electrolytes, these are *short-range forces* (non-bonding interaction forces); in electrolytes, these are *electrostatic forces* whose relatively greater range is given by Coulomb's law. Consider the process of concentrating both electrolyte and non-electrolyte solutions. If the process starts with infinitely dilute solutions, then their initial behaviour will be ideal; with increasing concentration coulombic interactions and at still higher concentrations, van der Waals non-bonding interactions and dipole–dipole interactions will become important. Thus, non-ideal behaviour must be considered for electrolyte solutions at much lower concentrations than for non-electrolyte solutions. 'Respecting non-ideal behaviour' means replacing the mole fractions, molalities and molar concentrations by the corresponding activities in all the thermodynamic relationships. For example, in an aqueous solution with a molar concentration of $10^{-3}\,\text{mol}\cdot\text{dm}^{-3}$, sodium chloride has an activity of 0.967×10^{-3}. Non-electrolyte solutions retain their ideal properties up to concentrations that may be as much as two orders of magnitude higher, as illustrated in Fig. 1.1.

Thus, the deviation in the behaviour of electrolyte solutions from the ideal depends on the composition of the solution, and the activity of the components is a function of their mole fractions. For practical reasons, the form of this function has been defined in the simplest way possible:

$$a_x = \gamma_x x \tag{1.1.9}$$

where the quantity γ_x is termed the *activity coefficient* (the significance of the subscript x will be considered later). However, the complications connected with solution non-ideality have not been removed but only transferred to the activity coefficient, which is also a function of the concentration. The form of this function can be found either theoretically (the theory has been quite successful for electrolyte solutions, see Section 1.3) or empirically. Practical calculations can be based on one of the theoretical or semiempirical equations for the activity coefficient (for the simple ions, the activity coefficient values are tabulated); the activity coefficient is then multiplied by the concentration and the activity thus

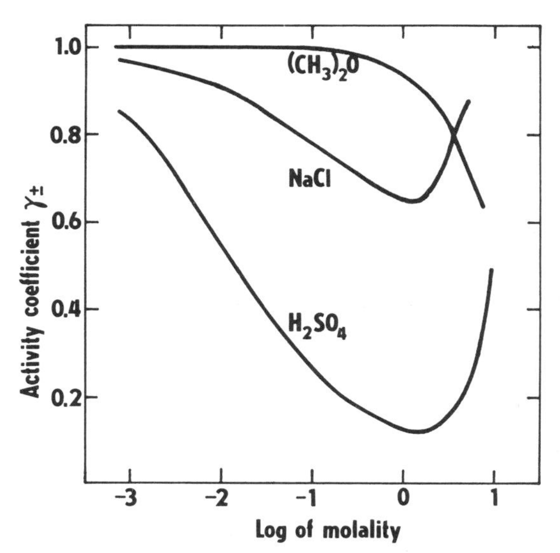

Fig. 1.1 The activity coefficient γ of a non-electrolyte and mean activity coefficients $\gamma_{\pm}$ of electrolytes as functions of molality

obtained is substituted into a simple 'ideal' equation (e.g. the law of mass action for chemical equilibrium).

Activity a_x is termed the rational activity and coefficient γ_x is the *rational activity coefficient.* This activity is not directly given by the ratio of the fugacities, as it is for gases, but appears nonetheless to be the best means from a thermodynamic point of view for description of the behaviour of real solutions. The rational activity corresponds to the mole fraction for ideal solutions (hence the subscript x). Both a_x and γ_x are dimensionless numbers.

In practical electrochemistry, however, the molality m or molar concentration c is used more often than the mole fraction. Thus, the *molal activity* a_m, *molal activity coefficient* γ_m, *molar activity* a_c and *molar activity coefficient* γ_c are introduced. The adjective 'molal' is sometimes replaced by 'practical'.

The following equations provide definitions for these quantities:

$$a_{i,m} = \gamma_{i,m}\frac{m_i}{m_i^0}, \qquad \lim_{m\to 0} \gamma_{i,m} = 1$$
$$a_{i,c} = \gamma_{i,c}\frac{c_i}{c_i^0}, \qquad \lim_{c\to 0} \gamma_{i,c} = 1 \qquad (1.1.10)$$

The standard states are selected as $m_i^0 = 1\ \text{mol}\cdot\text{kg}^{-1}$ and $c_i^0 = 1\ \text{mol}\cdot\text{dm}^{-3}$. In this convention, the ratio m_i/m_i^0 is numerically identical with the actual molality (expressed in units of moles per kilogram). This is, however, the dimensionless relative molality, in the same way that pressure p in Eq. (1.1.4) is the dimensionless relative pressure. The ratio c_i/c_i^0 is analogous. The symbols $m \to 0$ and $c \to 0$ in the last two equations indicate that the molalities or concentrations of all the components except the solvent are small.

Because of the electroneutrality condition, the individual ion activities and activity coefficients cannot be measured without additional extrathermodynamic assumptions (Section 1.3). Thus, *mean quantities* are defined for dissolved electrolytes, for all concentration scales. E.g., for a solution of a single strong binary electrolyte as

$$a_\pm = (a_+^{\nu_+} a_-^{\nu_-})^{1/\nu}, \qquad \gamma_\pm = (\gamma_+^{\nu_+} \gamma_-^{\nu_-})^{1/\nu} \tag{1.1.11}$$

The numerical values of the activity coefficients γ_x, γ_m and γ_c (and also of the activities a_x, a_m and a_c) are different (for the recalculation formulae see Appendix A). Obviously, for the limiting case (for a very dilute solution)

$$\gamma_{\pm,x} = \gamma_{\pm,m} = \gamma_{\pm,c} \approx 1 \tag{1.1.12}$$

The activity coefficient of the solvent remains close to unity up to quite high electrolyte concentrations; e.g. the activity coefficient for water in an aqueous solution of 2 M KCl at 25°C equals $\gamma_{0,x} = 1.004$, while the value for potassium chloride in this solution is $\gamma_{\pm,x} = 0.614$, indicating a quite large deviation from the ideal behaviour. Thus, the activity coefficient of the solvent is not a suitable characteristic of the real behaviour of solutions of electrolytes. If the deviation from ideal behaviour is to be expressed in terms of quantities connected with the solvent, then the *osmotic coefficient* is employed. The osmotic pressure of the system is denoted as π and the hypothetical osmotic pressure of a solution with the same composition that would behave ideally as π^*. The equations for the osmotic pressures π and π^* are obtained from the equilibrium condition of the pure solvent and of the solution. Under equilibrium conditions the chemical potential of the pure solvent, which is equal to the standard chemical potential at the pressure p, is equal to the chemical potential of the solvent in the solution under the osmotic pressure π,

$$\mu_0^0(T, p) = \mu_0(T, p + \pi) = \mu_0^0(T, p + \pi) + RT \ln a_0 \tag{1.1.13}$$

where a_0 is the activity of the solvent (the activity of the pure solvent is unity). As approximately

$$\mu_0^0(T, p + \pi) - \mu_0^0(T, p) = v_0 \pi \tag{1.1.14}$$

we obtain for the osmotic pressure

$$\pi = -\frac{RT}{v_0} \ln a_0 \tag{1.1.15}$$

where v_0 is the molar volume of the solvent. For a dilute solution $\ln a_0 = \ln x_0 = \ln(1 - \sum x_i) \approx -\sum x_i \approx M_0 \sum m_i$, giving for the ideal osmotic pressure (M_0 is the relative molecular mass of the solvent)

$$\pi^* = \frac{RTM_0}{v_0} \sum m_i \tag{1.1.16}$$

and in terms of molar concentration c of a single electrolyte dissociating into ν ions

$$\pi^* = \nu RTc \tag{1.1.17}$$

The ratio π/π^* (which is experimentally measurable) is termed the molal osmotic coefficient

$$\phi_m = \frac{\pi}{\pi^*} = \frac{-\ln a_0}{M_0 \sum m_i} \tag{1.1.18}$$

The rational osmotic coefficient is defined by the equation

$$\ln a_0 = \phi_x \ln x_0 \tag{1.1.19}$$

For a solution of a single electrolyte, the relationship between the mean activity coefficient and the osmotic coefficient is given by the equation

$$\ln \gamma_\pm = -(1 - \phi_m) - \int_0^m (1 - \phi_{m'})\, \mathrm{d} \ln m' \tag{1.1.20}$$

following from the definitions and from the Gibbs–Duhem equation.

In view of the electrostatic nature of forces that primarily lead to deviation of the behaviour of electrolyte solutions from the ideal, the activity coefficient of electrolytes must depend on the electric charge of all the ions present. G. N. Lewis, M. Randall and J. N. Brønsted found experimentally that this dependence for dilute solutions is described quite adequately by the relationship

$$\log \gamma_\pm = Az_+z_-\sqrt{I} \tag{1.1.21}$$

in which the constant A for 25° and water has a value close to 0.5 $\mathrm{dm^{3/2} \cdot mol^{-1/2}}$. Quantity I, called the ionic strength, describes the electrostatic effect of individual ionic species by the equation

$$I = \tfrac{1}{2} \sum_i c_i z_i^2 \tag{1.1.22}$$

(In fact, the symbol I_c should be used, as the molality ionic strength I_m can be defined analogously; in dilute aqueous solutions, however, values of c and m, and thus also I_c and I_m, become identical.) Equation (1.1.21) was later derived theoretically and is called the Debye–Hückel limiting law. It will be discussed in greater detail in Section 1.3.1.

1.1.4 *The Arrhenius theory of electrolytes*

At the end of the last century S. Arrhenius formulated the first quantitative theory describing the behaviour of weak electrolytes. The existence of ions in solution had already been demonstrated at that time, but very little was known of the structure of solutions and the solvent was regarded as an inert medium. Similarly, the concepts of the activity and activity coefficient were not employed. Electrochemistry was limited to aqueous solutions. However, the basis of classical thermodynamics was already formulated (by J. W. Gibbs, W. Thomson and H. v. Helmholtz) and electrolyte solutions had also been investigated thermodynamically especially by means of cryoscopic, osmometric and vapour pressure measurements.

Van't Hoff introduced the correction factor i for electrolyte solutions; the measured quantity (e.g. the osmotic pressure, π) must be divided by this factor to obtain agreement with the theory of dilute solutions of non-electrolytes ($\pi/i = RTc$). For the dilute solutions of some electrolytes (now called strong), this factor approaches small integers. Thus, for a dilute sodium chloride solution with concentration c, an osmotic pressure of $2RTc$ was always measured, which could readily be explained by the fact that the solution, in fact, actually contains twice the number of species corresponding to concentration c calculated in the usual manner from the weighed amount of substance dissolved in the solution. Small deviations from integral numbers were attributed to experimental errors (they are now attributed to the effect of the activity coefficient).

For other electrolytes, now termed weak, factor i has non-integral values depending on the overall electrolyte concentration. This fact was explained by Arrhenius in terms of a reversible dissociation reaction, whose equilibrium state is described by the law of mass action.

A weak electrolyte $B_{\nu_+}A_{\nu_-}$ dissociates in solution to yield ν ions consisting of ν_+ cations B^{z+} and ν_- anions A^{z-},

$$B_{\nu_+}A_{\nu_-} \rightleftarrows \nu_+B^{z+} + \nu_-A^{z-} \tag{1.1.23}$$

Thus the magnitude of the constant called the *thermodynamic* or *real dissociation constant*,

$$K = \frac{a_B^{\nu_+} a_A^{\nu_-}}{a_{BA}} \tag{1.1.24}$$

is a measure of the 'strength of the electrolyte'. The smaller its value, the weaker the electrolyte. The activity can be replaced by the concentrations according to Eq. (1.1.10), yielding

$$K = \frac{[B^{z+}]^{\nu_+}[A^{z-}]^{\nu_-}}{[B_{\nu_+}A_{\nu_-}]} \frac{\gamma_B^{\nu_+}\gamma_A^{\nu_-}}{\gamma_{BZ}} = K' \frac{\gamma_B^{\nu_+}\gamma_A^{\nu_-}}{\gamma_{BA}} \tag{1.1.25}$$

where

$$K' = \frac{[B^{z+}]^{\nu_+}[A^{z-}]^{\nu_-}}{[B_{\nu_+}A_{\nu_-}]} \tag{1.1.26}$$

is called the *apparent dissociation constant.* Constant K depends on the temperature; the dependence on the pressure is usually neglected as equilibria in the condensed phase are involved. Constant K' also depends on the ionic strength and increases with increasing ionic strength, as follows from substitution of the limiting relationship (1.1.21) into Eq. (1.1.25). For simplicity, consider monovalent ions, that is $\nu_+ = \nu_- = 1$, so that $\log \gamma_B = \log \gamma_A = -A\sqrt{I}$ and $\log \gamma_{BA} = 0$. Obviously, then, $\gamma_B = \gamma_A = 10^{0.5\sqrt{I}}$, $\gamma_{BA} = 1$ and substitution and rearrangement yield

$$K' = K10^{\sqrt{I}} \qquad (1.1.27)$$

It should be noted that the activity appearing in the dissociation constant K is the dimensionless relative activity, and constant K' contains the dimensionless relative concentration or molality terms. Constants K and K' are thus also dimensionless. However, their numerical values correspond to the units selected for the standard state, i.e. moles per cubic decimetre or moles per kilogram.

Because the dissociation constants for various electrolytes differ by several order of magnitude, the following definition

$$pK = -\log K; \qquad pK' = -\log K' \qquad (1.1.28)$$

is introduced to characterize the electrolyte strength in terms of a logarithmic quantity. Operator p appears frequently in electrochemistry and is equal to the log operator times -1 (i.e. $px = -\log x$).

The degree of dissociation α is the equilibrium degree of conversion, i.e. the fraction of the number of molecules originally present that dissociated at the given concentration. The degree of dissociation depends directly on the given dissociation constant. Obviously $\alpha = [B^{z+}]/\nu_+ c = [A^{z-}]/\nu_- c$, $[B_{\nu+}A_{\nu-}] = c(1 - \alpha)$ and the dissociation constant is then given as

$$K' = \nu_+^{\nu_+}\nu_-^{\nu_-} \frac{\alpha^\nu c^{\nu-1}}{1-\alpha} \qquad (1.1.29)$$

The most common electrolytes are uni-univalent ($\nu = 2$, $\nu_+ = \nu_- = 1$), for which

$$K' = \frac{[B^+][A^-]}{[BA]} = \frac{\alpha^2 c}{1-\alpha} \qquad (1.1.30)$$

The relationship for α follows:

$$\alpha = \frac{-K' + (K'^2 + 4K'c)^{1/2}}{2c} \qquad (1.1.31)$$

In moderately diluted solutions, i.e. for concentrations fulfilling the condition, $c \gg K'$,

$$\alpha \simeq (K'/c)^{1/2} \ll 1 \qquad (1.1.32)$$

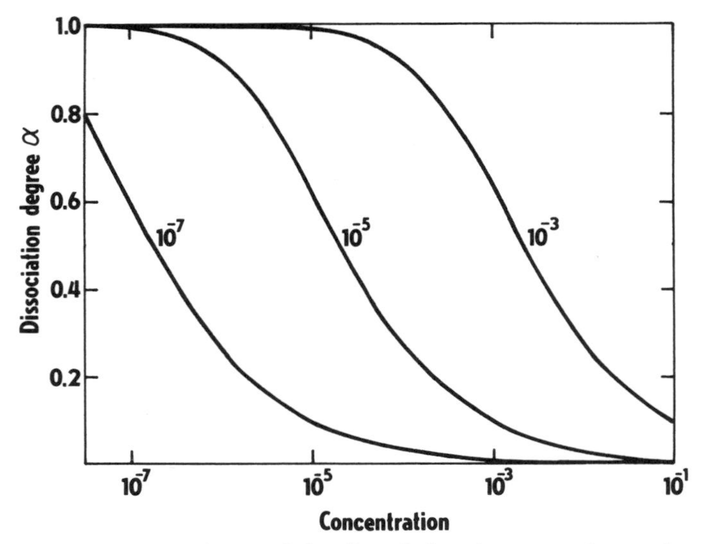

Fig. 1.2 Dependence of the dissociation degree α of a week electrolyte on molar concentration c for different values of the apparent dissociation constant K' (indicated at each curve)

In the limiting case (readily obtained by differentiation of the numerator and denominator with respect to c) it holds that $\alpha \rightarrow 1$ for $c \rightarrow 0$, i.e. each weak electrolyte at sufficient dilution is completely dissociated and, on the other hand, for sufficiently large c, $\alpha \rightarrow 0$, i.e. the highly concentrated electrolyte is dissociated only slightly. The dependence of α on c is given in Fig. 1.2.

For strong electrolytes, the activity of molecules cannot be considered, as no molecules are present, and thus the concept of the dissociation constant loses its meaning. However, the experimentally determined values of the dissociation constant are finite and the values of the degree of dissociation differ from unity. This is not the result of incomplete dissociation, but is rather connected with non-ideal behaviour (Section 1.3) and with ion association occurring in these solutions (see Section 1.2.4).

Arrhenius also formulated the first rational definition of acids and bases:

An acid (HA) is a substance from which hydrogen ions are dissociated in solution:

$$HA \rightleftarrows H^+ + A^-$$

A base (BOH) is a substance splitting off hydroxide ions in solution:

$$BOH \rightleftarrows B^+ + OH^-$$

This approach explained many of the properties of acids and bases and many processes in which acids and bases appear, but not all (e.g. processes in non-aqueous media, some catalytic processes, etc.). It has a drawback coming from the attempt to define acids and bases independently. However, as will be seen later, the acidity or basicity of substances appears only on interaction with the medium with which they are in contact.

References

Dunsch, L., *Geschichte der Elektrochemie,* Deutscher Verlag für Grundstoffindustrie, Leipzig, 1985.
Ostwald, W., *Die Entwicklung der Elektrochemie in gemeinverständlicher Darstellung,* Barth, Leipzig, 1910.

1.2 Structure of Solutions

1.2.1 *Classification of solvents*

The classical period of electrochemistry dealt with aqueous solutions. Gradually, however, other, 'non-aqueous' solvents became important in both chemistry and electrochemistry. For example, some important substances (e.g. the Grignard reagents and other homogeneous catalysts) decompose in water. A number of important biochemical substances (proteins, enzymes, chlorophyll, vitamin B_{12}) are insoluble in water but are soluble, for example, in anhydrous liquid hydrogen fluoride, from which they can be reisolated without loss of biochemical activity. The whole aluminium industry is based on electrolysis of a solution of aluminium oxide in fused cryolite. Many more examples could be given of chemical processes employing solvents other than water. Basically any substance can be used as a solvent at temperatures between its melting and boiling points (provided it is stable in this temperature range). Three types of solvent can be distinguished.

Molecular solvents consist of molecules. The cohesive forces between neighbouring molecules in the liquid phase depend on hydrogen bonds or other 'bridges' (oxygen, halogen), on dipole–dipole interactions or on van der Waals interactions. These solvents act as dielectrics and do not appreciably conduct electric current. *Autoionization* occurs to a slight degree in some of them, leading to low electric conductivity (for example $2H_2O \rightleftarrows H_3O^+ + OH^-$; in the melt, $2HgBr_2 \rightleftarrows HgBr^+ + HgBr_3^-$; in the liquefied state, $2NO_2 \rightleftarrows NO^+ + NO_3^-$).

Ionic solvents consisting of ions are mostly fused salts. However, not all salts yield ions on melting. For example, fused $HgBr_2$, $POCl_3$, BrF_3 and others form molecular liquids. On mixing, however, the molecular solvents H_2O and H_2SO_4 can form ionic solvents that contain only the H_3O^+ and HSO_4^- ions. Ionic solvents have high ionic electric conductivity. Most exist at high temperatures (e.g., at normal pressure, NaCl between 800 and 1465°C) but some salts have low melting points (e.g. ethylpyridinium bromide at −114°C, tetramethylammonium thiocyanate at −50.5°C) and, in addition, there is a number of low-melting-point eutectics (for example $AlCl_3 + KCl + NaCl$ in a ratio of 60:14:26 mol % melts at 94°C). The ions present in these solvents can be either monoatomic (for example Na^+ and Cl^- in fused NaCl) or polyatomic (for example cryolite—Na_3AlF_6—contains Na^+, AlF_6^{3-}, AlF_4^- and F^- ions).

1.2.2 *Liquid structure*

Molecular liquids are not at all amorphous, as would first appear. Methods of structural analysis (X-ray diffraction, NMR, IR and Raman spectroscopy) have demonstrated that the liquid retains the structure of the original crystal to a certain degree. Water is the most ordered solvent (and has been investigated most extensively). Under normal conditions, 70 per cent of the water molecules exist in 'ice floes', clusters of about 50 molecules with a structure similar to that of ice and a mean lifetime of 10^{-11} s. Hydrogen bonds lead to intermolecular spatial association. Hydrogen bonds are also formed in other solvents, but result in the formation of chains (e.g. in alcohols) or rings (e.g. rings containing six molecules are formed in liquid HF). Thus, the degree of organization is lower in these solvents than in water, although the strength of hydrogen bonds increases in the order: HCl, H_2SO_4 (practically monomers) $< NH_3 < H_2O <$ HF. Mixing of a highly organized solvent with a less organized one leads to structure modification. If, for example, ethanol is added to water, ethanol molecules first enter the water structure and strengthen it; at higher concentrations this order is reversed.

It is not the purpose of chemistry, but rather of statistical thermodynamics, to formulate a theory of the structure of water. Such a theory should be able to calculate the properties of water, especially with regard to their dependence on temperature. So far, no theory has been formulated whose equations do not contain adjustable parameters (up to eight in some theories). These include continuum and mixture theories. The continuum theory is based on the concept of a continuous change of the parameters of the water molecule with temperature. Recently, however, theories based on a model of a mixture have become more popular. It is assumed that liquid water is a mixture of structurally different species with various densities. With increasing temperature, there is a decrease in the number of low-density species, compensated by the usual thermal expansion of liquids, leading to the formation of the well-known maximum on the temperature dependence of the density of water ($0.999973\ g \cdot cm^{-3}$ at 3.98°C).

There are various theories on the structure of these species and their size. Some authors have assumed the presence of monomers and oligomers up to pentamers, with the open structure of ice I, while others deny the presence of monomers. Other authors assume the presence of the structure of ice I with loosely arranged six-membered rings and of structures similar to that of ice III with tightly packed rings. Most often, it is assumed that the structure consists of clusters of the ice I type, with various degrees of polymerization, with the maximum of the cluster size distribution in the region of oligomers and with a low concentration of large species.

The following simple concept is sufficient for our purposes. To a given water molecule, two further water molecules can become bonded fairly strongly through their negative 'ends' by electrostatic forces in the direction of the O—H bond (hydrogen bonds). Additional two water molecules are then bound to the original molecule through their positive 'ends'. In the sterically most favourable position these two molecules occupy the positions of the remaining two apices of a tetrahedron whose first two apices lie in the direction of the O—H bonds. This process results in a tetrahedral structure, the effective charge distribution in the water molecule being quadrupolar. In the solid state, each water molecule has four nearest neighbours (Fig. 1.3). This is a very 'open' arrangement that collapses partially on melting (the density of water is larger than that of ice). Thus, liquid water retains clusters of molecules with the structure of ice that constantly collapse and reform. About half of the water molecules are present in these clusters at a given instant. X-ray structural analysis indicates that, in the liquid state, each water molecule has an average of 4.5 nearest neighbours. This is far less than would correspond to the most closely packed arrangement (12 nearest neighbours). The existence of a certain degree of ordering in liquid water can also explain the unusually high value of the heat of vaporization, entropy of vaporization, boiling point, and dielectric constant of water compared with similar simple substances, such as hydrogen sulphide, hydrogen fluoride, and ammonia.

Ionic liquids are also not completely randomly arranged but have a structure similar to that of a crystal. However, in contrast to crystals, the ionic liquid structure contains far more vacancies, interstitial cavities, dislocations, and other perturbations.

1.2.3 *Ion solvation*

If a substance is to be dissolved, its ions or molecules must first move apart and then force their way between the solvent molecules which interact with the solute particles. If an ionic crystal is dissolved, electrostatic interaction forces must be overcome between the ions. The higher the dielectric constant of the solvent, the more effective this process is. The solvent–solute interaction is termed *ion solvation* (*ion hydration* in aqueous solutions). The importance of this phenomenon follows from comparison of the energy changes accompanying solvation of ions and uncharged molecules: for monovalent ions, the enthalpy of hydration is about $400\,\mathrm{kJ \cdot mol^{-1}}$, and equals about $12\,\mathrm{kJ \cdot mol^{-1}}$ for simple non-polar species such as argon or methane.

The simplest theory of interactions between ions and the solvent, proposed by M. Born, assumed that the ions are spheres with a radius of r_i

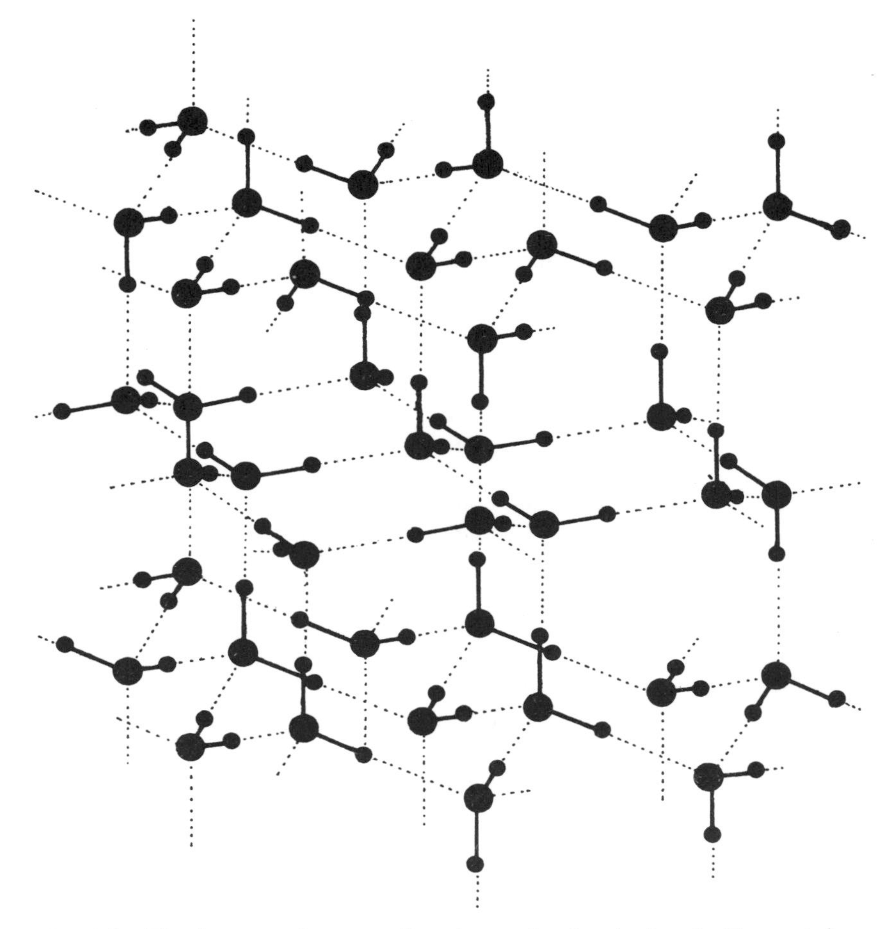

Fig. 1.3 The hexagonal array of water molecules in ice I. Even at low temperatures the hydrogen atoms (smaller circles) are randomly ordered

and charge z_ie (e is the elementary charge, i.e. the charge of a proton), located in a continuous, structureless dielectric with permittivity ε. The balance of the changes in the Gibbs free energy during ion solvation is based on the following hypothetical process. Consider an ion in a vacuum whose charge z_ie is initially discharged, yielding electric work w_1 to the system,

$$w_1 = \int_{z_ie}^{0} \psi_1 \, dq \tag{1.2.1}$$

where ψ_1 is the electric potential at the surface of the ion,

$$\psi_1 = \frac{q}{4\pi\varepsilon_0 r_i} \tag{1.2.2}$$

The quantity ε_0 is the permittivity of a vacuum ($\varepsilon_0 = 8.85418782(7) \times 10^{-12}\,\mathrm{F\cdot m^{-1}}$) and q is the charge of the ion, variable during discharging. Substitution of (1.2.2) into Eq. (1.2.1) and integration yield

$$w_1 = \frac{1}{4\pi\varepsilon_0 r_i}\int_{z_i e}^{0} q\,\mathrm{d}q = -\frac{(z_i e)^2}{8\pi\varepsilon_0 r_i} \tag{1.2.3}$$

The uncharged ion is transferred into a solvent with permittivity $\varepsilon = D\varepsilon_0$, where D is the relative permittivity (dielectric constant) of the medium. No work is gained or lost in this process. In the solvent, the ion is again recharged to the value of the electric potential at its surface,

$$\psi_2 = \frac{q}{4\pi D\varepsilon_0 r_i} \tag{1.2.4}$$

The corresponding electric work is

$$w_2 = \frac{1}{4\pi D\varepsilon_0 r_i}\int_0^{z_i e} q\,\mathrm{d}q = \frac{(z_i e)^2}{8\pi D\varepsilon_0 r_i} \tag{1.2.5}$$

The transfer of one mole of ions from a vacuum into the solution is connected with work $N_A(w_1 + w_2)$. This work is identical with the molar Gibbs energy of solvation $\Delta G_{s,i}$:

$$\Delta G_{s,i} = N_A(w_1 + w_2) \tag{1.2.6}$$

The Born equation for the Gibbs energy of solvation is thus

$$\Delta G_{s,i} = -\frac{z_i^2 e^2 N_A}{8\pi\varepsilon_0 r_i}\left(1 - \frac{1}{D}\right) \tag{1.2.7}$$

Although the Born equation is a rough approximation, it is often used for comparison of the solvation effects of various solvents. The simplification involved in the Born theory is based primarily on the assumption that the permittivity of the solvent is the same in the immediate vicinity of the ion as in the pure solvent, and the work required to compress the solvent around the ion is neglected.

The ionic radii are often difficult to ascertain. Mostly, crystallographic radii r_c corrected by the additive term δ, with a constant common value for cations and a different constant value for all anions, are used:

$$r_i = r_c + \delta \tag{1.2.8}$$

Pauling's ion radii are often used, with values of δ of 0.085 nm for cations and 0.010 nm for anions.

As every solvent has its characteristic structure, its molecules are bonded more or less strongly to the ions in the course of solvation. Again, solvation has a marked effect on the structure of the surrounding solvent. The number of molecules bound in this way to a single ion is termed the *solvation* (*hydration*) *number* of this ion.

The values of the dipole–dipole and ion–dipole interaction energies are required for estimation of the energy conditions in liquid water and in aqueous solutions. The former is given by the sum of the coulombic interaction energies between the individual charges of the two dipoles. Assuming that both dipoles are colinear and identical (i.e. have identical charge q and length l, and the dipole moment equals $\mu = ql$) and that their centres lie a distance r apart, then the *dipole–dipole* interaction energy is given by the relationship

$$U_{dd} = \frac{1}{4\pi\varepsilon_0}\left(\frac{q^2}{r} + \frac{-q^2}{r-l} + \frac{-q^2}{r+l} + \frac{q^2}{r}\right)$$
$$= -\frac{2q^2l^2}{4\pi\varepsilon_0 r(r+l)(r-l)} \qquad (1.2.9)$$

If $r \gg l$ (fulfilled even for the closest approach because the molecular dimensions are large compared with the distance between the positive and negative charges in the dipoles), then

$$U_{dd} = -\frac{\mu^2}{2\pi\varepsilon_0 r^3} \qquad (1.2.10)$$

The expression for the *ion–dipole* interaction energy U_{id} is obtained analogously as the sum of the energies of coulombic interaction of an ion (charge q) with charges q' and $-q'$ on the ends of the dipole, i.e.

$$U_{id} = \frac{1}{4\pi\varepsilon_0}\left(\frac{-q\,|q'|}{r-\frac{1}{2}l} + \frac{q\,|q'|}{r+\frac{1}{2}l}\right) = \frac{q\,|q'|\,l}{\pi\varepsilon_0(4r^2 - l^2)} \qquad (1.2.11)$$

If again $r \gg l$, then

$$U_{id} = -\frac{\mu\,|q'|}{4\pi\varepsilon_0 r^2} \qquad (1.2.12)$$

On approaching to a distance equal to their diameters (0.276 nm), water molecules ($\mu = 6.23 \times 10^{-28}$ C · cm) form a quite stable entity with potential energy $U_{dd} = -3.32 \times 10^{-20}$ J = 0.2 eV. If a comparably large univalent ion approaches a dipolar water molecule ($|q'| = 1.6 \times 10^{-19}$ C), the absolute energy value is almost four times larger, that is $U_{id} = -1.18 \times 10^{-19}$ J = 0.7 eV. If one water molecule in liquid water is replaced in the tetrahedral lattice by an uncharged particle of the same dimensions, then the four closest water molecules suffice to retain the original arrangement. If the new particle has a sufficiently large charge, e.g. positive, then the arrangement must change. Two water molecules are bonded more strongly than previously and the other two must rotate their negative 'end' towards the cation (Fig. 1.4). Thus, depending on the size and charge of the ion, the original arrangement can either be retained or a new, strong structure can be formed, or some state between these two extremes emerges. The

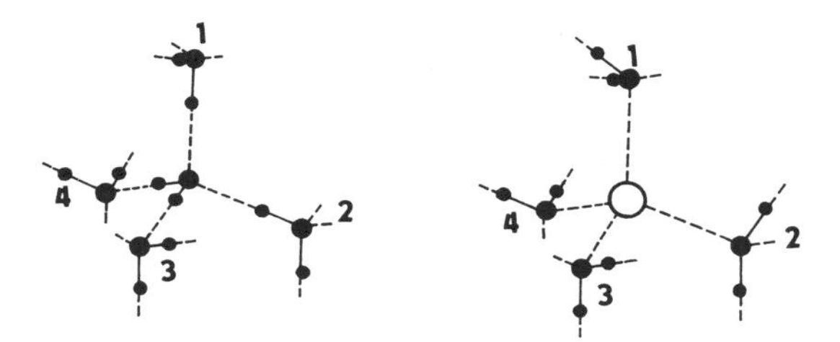

Fig. 1.4 Disturbing effect of a cation (void circle) on water structure. Molecules 1 and 2 show reversed orientation

ordering of the liquid can be completely destroyed if the ion disturbs the original arrangement without forming a new one. Negative hydration numbers are then obtained for some ions, as if these salts produced 'depolymerization' of the water structure. This phenomenon is termed *structure breaking*. The destructive effect increases with increasing ionic radius, e.g. in the order

$$Li^+ < Na^+ < Rb^+ < Cs^+,$$

$$Cl^-, NO_3^- < Br^- < I^- < ClO_3^-$$

and with increasing charge, e.g. in the order

$$Li^+ < Be^{2+} < Al^{3+}$$

Figure 1.5 schematically depicts a partially distorted water structure. A region is formed in the immediate vicinity of the ion where the water molecules are electrostatically bonded so strongly to the ion that they lose the ability to rotate. The value of the permittivity thus decreases sharply (to 6–7 compared to a value of 78.54 at 25°C in pure water). This region is termed the primary hydration sphere. Depending on conditions, regions further from the centre of the ion contain a more or less distorted water structure and regions even further away retain the original structure.

It should be realized that the structure breaking results in an increase in the entropy of the system and thus in a decrease in its Gibbs energy (according to the well-known relationship at constant temperature, $\Delta G = \Delta H - T\Delta S$). If a more complex dissolved particle contains both charged (polar) and uncharged (non-polar) groups, then this entropy factor becomes important. The structure of the solvent remains intact at the 'non-polar surface' of the particle. In order to increase the entropy of the system this 'surface' must be as small as possible to decrease the region where the solvent structure is unbroken. This is achieved by a close approach of two particles with their non-polar regions or by a conformation change resulting in a contact of non-polar regions in the molecule. These, mainly entropic, effects termed *hydrophobic* (in general *solvophobic*)

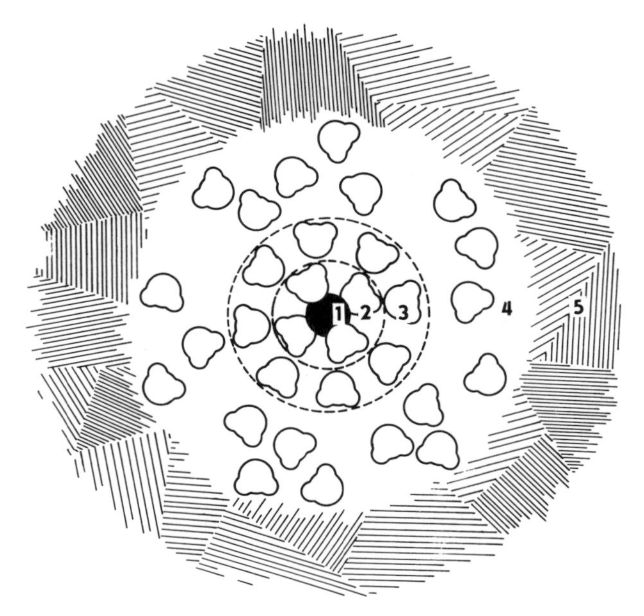

Fig. 1.5 A scheme of hydration: (1) cation, (2) primary hydration sheath (water molecules form a tetrahedron), (3) secondary hydration shell, (4) disorganized water, (5) normal water

interactions are, for example, the important factors affecting the conformational stability of proteins in aqueous solutions.

O. J. Samoilov proposed a statistical approach to hydration. Ions in solution affect the thermal, particularly translational, motion of the solvent molecules in the immediate vicinity of the ion. This translational motion can be identified with the exchange of the solvent molecules in the neighbourhood of the ion. Retardation of this exchange is thus a measure of the ion solvation.

If the water molecule remains in the vicinity of other water molecules for a time τ and in the vicinity of the ion for a time τ_i, then the ratio τ_i/τ is a measure of the solvation. If $\tau_i/\tau \gg 1$, the water molecule is bonded very strongly to the ion. On the other hand, $\tau_i/\tau < 1$ indicates negative solvation—the ion breaks the solvent structure and the solvent molecules in its vicinity can more readily be exchanged with other molecules than those in the solvent alone, where a certain degree of ordering remains. Experimental values of the ratio τ_i/τ for aqueous solutions using $H_2{}^{18}O$ molecules lie in the range from tenths to unities.

Some molecular solvents (such as ammonia, aliphatic amines, hexamethylphosphortriamide) dissolve alkali metals; solutions with molalities of more than $10\,\text{mol} \cdot \text{kg}^{-1}$ are obtained. Ammonia complexes $M(NH_3)_6$ analogous to the amminates or hydrates of salts can be crystallized from these solutions. The solutions are blue (absorb at 7000 cm^{-1}, corresponding to the $s \rightarrow p$ electron transition) and paramagnetic. Paramagnetic resonance measurements indicate that cavities in the liquid with dimensions of about 0.3 nm contain free electrons. Compared with aqueous salt solutions, these metal solutions have about 10^5 times larger molar conductivity, indicating electronic conductivity. Thus, the metals in solution are dissociated to yield solvated cations M^+ and electrons e^- (also solvated). More concentrated solutions also contain species such as M^+e^-, M_2 and Me^-. The stability of these solutions is low because of the formation, for example, of alkali metal amides in solutions of alkali metals in liquid ammonia.

A form of solvation also occurs in *ionic solvents,* but coulombic interactions predominate. The solvent structure can be conceived as resulting from the penetration of two sublattices—cationic and anionic. The cations of the dissolved salt are then scattered throughout the cationic sublattice and the anions in the anionic sublattice. If the dissolved salt and the solvent have a common anion, for example, then the cations are scattered throughout the cationic sublattice and the anionic lattice remains almost unaffected. Let us assume that only one kind of foreign ion, a cation or an anion, is introduced into ionic solvents (while the electroneutrality condition is maintained). If the cation of the dissolved salt has a larger charge than the solvent cation, then vacancies must be scattered throughout the cationic sublattice so that the distribution of electric charge remains neutral. An example of a more complicated situation is the formation of Na^+ and $AlCl_4^-$ ions in an equimolar melt of $AlCl_3 + NaCl$. If the mixture contains a small excess of $AlCl_3$ over the stoichiometric amount, then $Al_2Cl_7^-$ ions are also present. If Cu^+ ions are then dissolved in the melt (as CuCl), they remain free and are scattered throughout the cationic sublattice. However, in the presence of a small excess of NaCl, the anionic sublattice contains a corresponding excess of Cl^- with which the Cu^+ ions form the $CuCl_2^-$ complex which is spread throughout the anionic sublattice.

Three types of methods are used to study solvation in molecular solvents. These are primarily the methods commonly used in studying the structures of molecules. However, *optical spectroscopy* (IR and Raman) yields results that are difficult to interpret from the point of view of solvation and are thus not often used to measure solvation numbers. NMR is more successful, as the chemical shifts are chiefly affected by solvation. Measurement of solvation-dependent kinetic quantities is often used (*electrolytic mobility, diffusion coefficients,* etc). These methods supply data on the region in the immediate vicinity of the ion, i.e. the primary solvation sphere, closely connected to the ion and moving together with it. By means of the third type of methods some static quantities (*entropy* and *compressibility* as well as some non-thermodynamic quantities such as the *dielectric constant*) are measured. These methods also pertain to the secondary solvation sphere, in which the solvent structure is affected by the presence of ions, but the solvent molecules do not move together with the ions. These methods have, understandably, been applied most often to aqueous solutions.

The *ionic mobility* and *diffusion coefficient* are also affected by the ion hydration. The particle dimensions calculated from these values by using Stokes' law (Eq. 2.6.2) do not correspond to the ionic dimensions found, for example, from the crystal structure, and hydration numbers can be calculated from them. In the absence of further assumptions, diffusion measurements again yield only the sum of the hydration numbers of the cation and the anion.

Similarly, concepts of solvation must be employed in the measurement of equilibrium quantities to explain some anomalies, primarily the *salting-out effect.* Addition of an electrolyte to an aqueous solution of a non-electrolyte results in transfer of part of the water to the hydration sheath of the ion, decreasing the amount of 'free' solvent, and the solubility of the non-electrolyte decreases. This effect depends, however, on the electrolyte selected. In addition, the *activity coefficient* values (obtained, for example, by measuring the freezing point) can indicate the magnitude of hydration numbers. Exchange of the open structure of pure water for the more compact structure of the hydration sheath is the cause of lower *compressibility* of the electrolyte solution compared to pure water and of lower *apparent volumes* of the ions in solution in comparison with their effective volumes in the crystals. Again, this method yields the overall hydration number.

The fact that the water molecules forming the hydration sheath have limited mobility, i.e. that the solution is to certain degree ordered, results in lower values of the *ionic entropies*. In special cases, the ionic entropy can be measured (e.g. from the dependence of the standard potential on the temperature for electrodes of the second kind). Otherwise, the heat of solution is the measurable quantity. Knowledge of the lattice energy then permits calculation of the heat of hydration. For a saturated solution, the heat of solution is equal to the product of the temperature and the entropy of solution, from which the entropy of the salt in the solution can be found. However, the absolute value of the entropy of the crystal must be obtained from the dependence of its thermal capacity on the temperature down to very low temperatures. The value of the entropy of the salt can then yield the overall hydration number. It is, however, difficult to separate the contributions of the cation and of the anion.

Various methods have yielded the following average values of the *primary hydration numbers* of the alkali metal cations (the number of methods used is given in brackets and the results are rounded off to the nearest integers): Li^+ 5 (5), Na^+ 5 (5), K^+ 4 (4), Rb^+ 3 (4), and for the halide ions: F^- 4 (3), Cl^- 1 (3), Br^- 1 (3), I^- 1 (2). The error is ± 1 water molecule (except for K^+, where the error is ± 2). For divalent ions the values vary between 10 and 14 (according to G. Kortüm). For illustration of the variability of the results obtained by various methods, the values obtained for the Na^+ ion from mobility measurements were 2–4, from the entropy 4, from the compressibility 6–7, from molar volumes 5, from diffusion 1 and from activity coefficients also 1. For the Cl^- ion, these methods yielded the values (in the same order): 4, 3, 0, −1, 0, 0, 1. Of the divalent ions, for example, solution of Mg^{2+} was measured by the mobility method, yielding a value of 10 to 13, entropy of 13, compressibility of 16 and activity coefficients of 5 (according to B. E. Conway and J. O'M. Bockris).

Remarkable data on primary hydration shells are obtained in non-aqueous solvents containing a definite amount of water. Thus, nitrobenzene saturated with water contains about 0.2 M H_2O. Because of much higher dipole moment of water than of nitrobenzene, the ions will be preferentially solvated by water. Under these conditions the following values of hydration numbers were obtained: Li^+ 6.5, H^+ 5.5, Ag^+ 4.4, Na^+ 3.9, K^+ 1.5, Tl^+ 1.0, Rb^+ 0.8, Cs^+ 0.5, tetraethylammonium ion 0.0, ClO_4^- 0.4, NO_3^- 1.4 and tetraphenylborate anion 0.0 (assumption).

1.2.4 *Ion association*

As already mentioned, the criterion of complete ionization is the fulfilment of the Kohlrausch and Onsager equations (2.4.15) and (2.4.26) stating that the molar conductivity of the solution has to decrease linearly with the square root of its concentration. However, these relationships are valid at moderate concentrations only. At high concentrations, distinct deviations are observed which can partly be ascribed to non-bonding electrostatic and other interaction of more complicated nature (cf. p. 38) and partly to ionic bond formation between ions of opposite charge, i.e. to ion association (ion-pair formation). The separation of these two effects is indeed rather difficult.

The species appearing as strong electrolytes in aqueous solutions lose this property in low-permittivity solvents. The ion-pair formation converts them to a sort of weak electrolyte. In solvents of very low-permittivity (dioxan, benzene) even ion triplets and quadruplets are formed.

The formation of an ion pair

$$A^+ + B^- \rightleftarrows A^+B^- \tag{1.2.13}$$

is described by the association constant K_{ass} and the degree of association $\alpha_{ass} = [A^+B^-]/c$:

$$K_{ass} = \frac{[A^+B^-]}{[A^+][B^-]} = \frac{\alpha_{ass}}{(1-\alpha_{ass})^2 c} \tag{1.2.14}$$

Some values of association constants are listed in Table 1.1.

The dependence of ion-pairing on the dielectric constant is illustrated in the following example. The formation of K^+Cl^- ion pairs in aqueous KCl solutions has not been demonstrated. In methanol ($D = 32.6$), for KCl,

Table 1.1 Values of $\log K_{\mathrm{ass}}$ for some ion pairs in aqueous solutions at 25°C (according to C. W. Davies), where — indicates that ion pair formation could not be proved

	Li^+	Na^+	K^+	Ag^+	Tl^+	Ca^{2+}	Cu^{2+}	Fe^{3+}
OH^-	−0.08	−0.7	—	2.3	0.8	1.30		12.0
F^-				0.4	0.1	1.0	1.23	6.04
Cl^-	—	—	—	3.2	0.5	—	0.4	1.5
Br^-				4.4	1.0		0.0	0.60
NO_3^-	—	−0.6	−0.2	−0.2	0.3	0.28		1.0
SO_4^{2-}	0.6	0.7	1.0	1.3	1.4	2.28	2.36	
$S_2O_3^{2-}$		0.6	0.9	8.8	1.9	1.95		
$P_3O_9^{3-}$		1.16				3.46		
$P_2O_7^{4-}$	3.1	2.4	2.3			6.8		
$P_3O_{10}^{5-}$	3.9	2.7	2.7			8.1		

$\log K_{\mathrm{ass}} = 1.15$, so that at $c = 0.1\ \mathrm{mol\cdot dm^{-3}}$, 32 per cent of the KCl is present as K^+Cl^- pairs. In acetic acid ($D = 6.2$), $\log K_{\mathrm{ass}} = 6.9$, i.e. at the same concentration $\alpha_{\mathrm{ass}} = 99.9$ per cent. A remarkable example is the association of 3×10^{-5} M tetraisopentylammonium nitrate in pure water and in pure dioxan ($D = 2.2$). While in water the concentration of ion pairs is negligible, the concentration of free nitrate ions in pure dioxan is $8 \times 10^{-12}\ \mathrm{mol\cdot dm^{-3}}$. Thus, the free ions are practically absent in low permittivity solvents ($D < 5$).

J. Bjerrum (1926) first developed the theory of ion association. He introduced the concept of a certain critical distance between the cation and the anion at which the electrostatic attractive force is balanced by the mean force corresponding to thermal motion. The energy of the ion is at a minimum at this distance. The method of calculation is analogous to that of Debye and Hückel in the theory of activity coefficients (see Section 1.3.1). The probability $P_i\,\mathrm{d}r$ has to be found for the ith ion species to be present in a volume element in the shape of a spherical shell with thickness $\mathrm{d}r$ at a sufficiently small distance r from the central ion (index k).

When using the Boltzmann distribution, the relationship for the fraction of ions of the ith kind $\mathrm{d}N_i/N_i$ in this spatial element is

$$\frac{\mathrm{d}N_i}{N_i} = 4\pi r^2 \exp\left(-\frac{z_i e\psi}{kT}\right)\mathrm{d}r \tag{1.2.15}$$

where $z_i e\psi$ is the potential energy of the ion (i.e. the electrostatic energy necessary for transferring the ion from infinite distance). For small r values, the potential ψ can be expressed only in terms of the contribution of the central ion, $\psi = z_k e/4\pi\varepsilon r$, so that the probability density P_i is then given by the equation

$$P_i = \frac{\mathrm{d}N_i}{N_i\,\mathrm{d}r} = 4\pi r^2 \exp\left(-\frac{z_i z_k e^2}{4\pi\varepsilon r kT}\right) \tag{1.2.16}$$

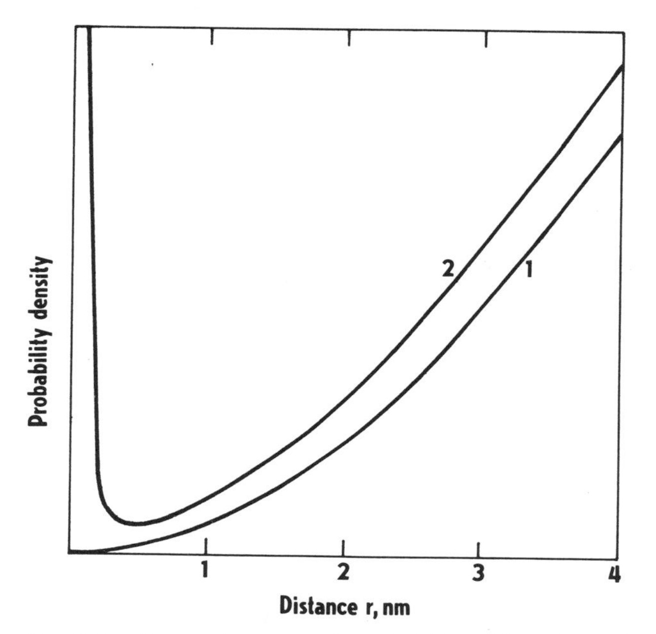

Fig. 1.6 Bjerrum's curves of probability: (1) uncharged particles, (2) two ions with opposite signs (After C. W. Davies)

whose shape is depicted (for ions k and i of opposite sign) in Fig. 1.6. A sharp minimum lies at the critical distance $r_{\min}$, obtained from Eq. (1.2.16) using the condition $\mathrm{d}P_i/\mathrm{d}r = 0$:

$$r_{\min} = -\frac{z_i z_k e^2}{8\pi\varepsilon kT} \tag{1.2.17}$$

For aqueous solutions at 25°C, $r_{\min} = 3.057 \times 10^{-10}\ z_i z_k$ (in metres), i.e. about 0.3 nm, for a uni-univalent electrolyte and about 1.4 nm for a bi-bivalent electrolyte. This distance is considered to be the maximum distance beyond which formation of ion pairs does not occur.

Integration of Eq. (1.2.15) permits calculation of the fraction of ions present in the associated state and thus the degree of association and the association constant

$$K_{\text{ass}} = 4\pi N_{\text{A}} \int_z^2 \frac{\exp y}{y^4}\, \mathrm{d}y \tag{1.2.18}$$

where $y = 2r_{\min}/r$ and $b = 2r_{\min}/a$, a being the distance of closest approach of the ions. J. T. Denison and J. B. Ramsey, W. R. Gilkerson and R. M. Fuoss obtained simplified expressions in the form

$$\ln K_{\text{ass}} = \ln\,(4\pi N_{\text{A}} a^3) - \frac{z_i z_k e^2}{4\pi kTa}\frac{1}{\varepsilon} - \ln\frac{z_i z_k e^2}{4\pi kTa\varepsilon} \tag{1.2.19}$$

Bjerrum's theory includes approximations that are not fully justified: the ions are considered to be spheres, the dielectric constant in the vicinity of the ion is considered to be equal to that in the pure solvent, the possibility of interactions between ions other than pair formation (e.g. the formation of hydrogen bonds) is neglected and the effect of ion solvation during formation of ion pairs is not considered (the effect of the solvation on ion-pair structure is illustrated in Fig. 1.7).

Although the Bjerrum theory is thus not in general quantitatively applicable, the concept of ion association is very useful. It has assisted in an explanation of various phenomena observed in the study of homogeneous

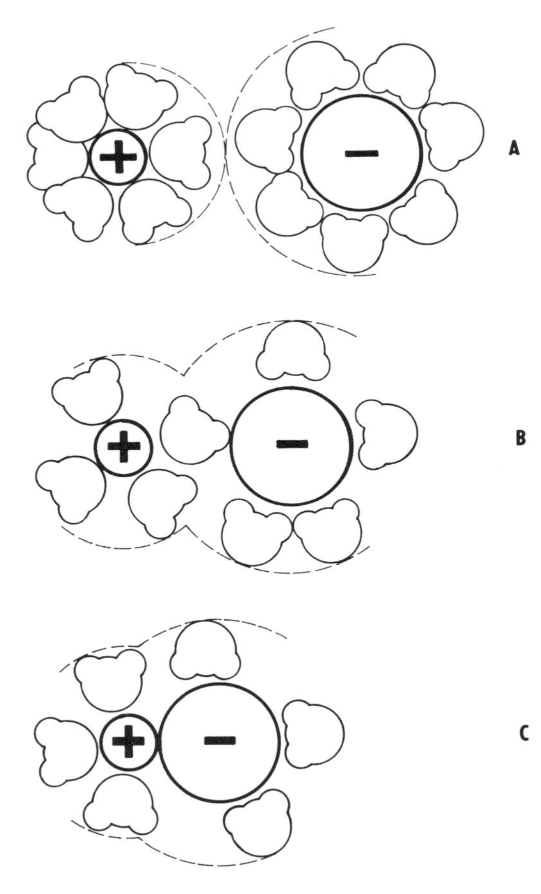

Fig. 1.7 Possible hydration modes of an ion pair: (A) contact of primary hydration shells, (B) sharing of primary hydration shells, (C) direct contact of ions

catalysis, electrical mobility, the behaviour of polyelectrolytes, etc.

Ion pairs (A^+B^-) are formed in fused salts through a process in which the cations or anions of the solvent and of the solute exchange positions in the solvent lattice until the cations and anions occupy neighbouring positions. If the solvent is denoted as XY, then this process can be expressed by the scheme:

$$\begin{array}{ccccc} X^+Y^-X^+ & & Y^-X^+Y^- & & X^+Y^-X^+Y^-X^+Y^- \\ Y^-A^+Y^- & + & X^+B^-X^+ & \rightleftarrows & Y^-A^+B^-X^+Y^-X^+ \\ X^+Y^-X^+ & & Y^-X^+Y^- & & X^+Y^-X^+Y^-X^+Y^- \end{array}$$

According to M. Blander the association constant depends on the change in the Gibbs energy, ΔG, during exchange of a solvent anion Y^-, next to a cation of the dissolved salt A^+, for an anion of the dissolved salt B^- according to the relationship

$$\ln\left(1+\frac{K_{ass}}{n}\right) = -\frac{\Delta G}{RT} \tag{1.2.20}$$

where n is the coordination number of the cation A^+. The energy ΔG is proportional to the combination of the reciprocal values of the distances between the centres of the corresponding pairs of ions (i.e. the sum of the radii of these ions):

$$\Delta G \approx \frac{1}{d_{AB}} + \frac{1}{d_{XY}} - \left(\frac{1}{d_{AY}} + \frac{1}{d_{XB}}\right) \tag{1.2.21}$$

Clearly, it is the size of ions that is decisive in ion-pair formation. Moreover, the coulombic interactions can extend even to more distant neighbours. The Blander equation is then, of course, no longer applicable.

References

Ben-Naim, A., *Hydrophobic Interactions,* Plenum Press, New York, 1980.
Bloom, H., and J. O'M. Bockris, Molten electrolytes, *MAE,* **2,** 262 (1959).
Burger, K., *Solvation, Ionic and Complex Formation Reactions in Non-aqueous Solvents,* Elsevier, Amsterdam, 1983.
Case, B., Ion solvation, Chapter 2 of *Reactions of Molecules at Electrodes* (Ed. N. S. Hush), Wiley–Interscience, London, 1971.
Conway, B. E., *Ionic Hydration in Chemistry and Biophysics,* Elsevier, Amsterdam, 1981.
Davies, C. W., *Ion Association,* Butterworths, London, 1962.
Debye, P., *Polar Molecules,* Dover Publication Co., New York, 1945.
Denison, J. T., and J. B. Ramsey, *J. Am. Chem. Soc.,* **77,** 2615 (1955).
Desnoyers, J. E., and C. Jolicoeur, Ionic solvation, *CTE,* **5,** Chap. 1 (1982).
Dogonadze, R. R., E. Kálmán, A. A. Kornyshev, and J. Ulstrup (Eds), *The Chemical Physics of Solvation,* Elsevier, Amsterdam, 1985.
Eisenberg, D., and W. Kautzmann, *The Structure and Properties of Water,* Oxford University Press, Oxford, 1969.

Fuoss, R. M., and F. Accascina, *Electrolytic Conductance,* Intersceince Publishers, New York, 1959.
Gilkerson, W. R., *J. Chem. Phys.,* **25,** 1199 (1956).
Inmann, D., and D. G. Lovering (Eds), *Ionic Liquids,* Plenum Press, New York, 1981.
Klotz, I. M., Structure of water, in *Membranes and Ion Transport* (Ed. E. E. Bittar), Vol. I, John Wiley & Sons, New York, 1970.
Mamantov, G. (Ed.), *Characterization of Non-aqueous Solvents,* Plenum Press, New York, 1978.
Marcus, Y., *Ion Solvation,* John Wiley & Sons, Chichester, 1986.
Papatheodorou, G. W., Structure and thermodynamics of molten salts, *CTE,* **5,** Chap. 5 (1982).
Samoilov, O. Ya., *Structure of Aqueous Electrolyte Solutions and Hydration of Ions,* Consultants Bureau, New York, 1965.
Škarda, V., J. Rais, and M. Kyrš, *J. Nucl. Inorg. Chem.,* **41,** 1443 (1979).
Sundheim, G., *Fused Salts,* McGraw-Hill, New York, 1964.
Tanaka, N., H. Ohtuki, and R. Tamamushi (Eds), *Ions and Molecules in Solution,* Elsevier, Amsterdam, 1983.
Tanford, C., *The Hydrophobic Effect: Formation of Micelles and Biological Membranes,* 2nd ed., John Wiley & Sons, New York, 1980.
Trémillon, B., *Chemistry in Non-aqueous Solvents,* Reidel, Dordrecht, 1974.

1.3 Interionic Interactions

Thermodynamics describes the behaviour of systems in terms of quantities and functions of state, but cannot express these quantities in terms of model concepts and assumptions on the structure of the system, intermolecular forces, etc. This is also true of the activity coefficients; thermodynamics defines these quantities and gives their dependence on the temperature, pressure and composition, but cannot interpret them from the point of view of intermolecular interactions. Every theoretical expression of the activity coefficients as a function of the composition of the solution is necessarily based on extrathermodynamic, mainly statistical concepts. This approach makes it possible to elaborate quantitatively the theory of individual activity coefficients. Their values are of paramount importance, for example, for operational definition of the pH and its potentiometric determination (Section 3.3.2), for potentiometric measurement with ion-selective electrodes (Section 6.3), in general for all the systems where liquid junctions appear (Section 2.5.3), etc.

The expression for the chemical potential of a component of a real solution can be separated into two terms:

$$\begin{aligned}\mu_i - \mu_i^0 = \Delta\mu_i &= RT\ln x_i + RT\ln\gamma_i \\ &= \Delta\mu_{i,\mathrm{id}} + \Delta\mu_i^E \qquad (1.3.1)\end{aligned}$$

where $\Delta\mu_{i,\mathrm{id}} = RT\ln x_i$ is the difference in the chemical potentials between the standard and actual states, for ideal behaviour, and $\Delta\mu_i^E$ is the correction for the real behaviour, which can be identified with the expression containing the activity coefficient. It can further be written for all

$\Delta\mu_i$'s that

$$\Delta\mu_i^E = \left(\frac{\partial \Delta G^E}{\partial n_i}\right)_{T,p,n_j \neq n_i} \tag{1.3.2}$$

Thus, the activity coefficient can be calculated if ΔG^E is known; this is the difference between the work of the reversible transition of the real system at constant temperature and pressure from the standard to the actual state and the work in the same process in the ideal system:

$$RT \ln \gamma_i = \left(\frac{\partial \Delta G^E}{\partial n_i}\right)_{T,p,n_j \neq n_i} \tag{1.3.3}$$

As has already been mentioned, to carry out such a calculation is not a matter of thermodynamics, but requires adopting certain assumptions on the structure of the system and on interactions between particles.

According to the statistical thermodynamics of solutions the logarithm of the activity coefficient for an electrolyte solution can be expanded in the series

$$\ln \gamma_{\pm} = \alpha x_1^n + \beta x_1 + \gamma x_1^2 + \cdots \tag{1.3.4}$$

where n, α, β, . . . are constants (where $\alpha < 0$, $0 < n < 1$). These relationships have been verified experimentally. The term αx_1^n describes the effect of long-range interactions in electrolyte solutions (i.e. ion–ion interactions in contrast to short-range dipole–dipole and ion–dipole interactions; see Section 1.2.3). At medium concentrations, the value of this term is comparable with that of the term βx_1, and at low concentrations up to $10^{-3}\,\text{mol}\cdot\text{dm}^{-3}$ the term αx_1^n predominates. The aim of the theory of electrolytes is to provide a theoretical interpretation of the coefficients in Eq. (1.3.4). At low concentrations, the Debye–Hückel theory is valid; this theory neglects all types of interaction except for electrostatic, being satisfied with calculation of the term αx_1^n. The Debye–Hückel theory forms the basis and is still the valid nucleus of all the other theories of strong electrolytes, which are more or less elaborations or modifications of the original concepts.

1.3.1 *The Debye–Hückel limiting law*

In infinitely dilute solutions (in the standard state) ions do not interact, their electric field corresponds to that of point charges located at very large distances and the solution behaves ideally. As the solution becomes more concentrated, the ions approach one another, whence their fields become deformed. This process is connected with electrical work depending on the interactions of the ions. Differentiation of this quantity with respect to n_i permits calculation of the activity coefficient; this differentiation is identical with the differentiation $\partial G^E/\partial n_i$ and thus with the term $RT \ln \gamma_i$.

The work connected with the mutual approach of the ions and deformation of their electric fields cannot be calculated directly. However, because this work depends on a change in the thermodynamic functions of state, its value depends only on the initial and final states, and the processes can be separated into several steps. According to P. Debye and E. Hückel this separation is carried out so that the actual concentration change, i.e. the actual approach of the ions, occurs with uncharged species. The ions are supposed to be discharged in the standard state, i.e. they behave at the concentration $1\,\text{mol}\cdot\text{dm}^{-3}$ as if they were at infinite dilution. Work w_1 connected with this process can be readily calculated, as discharging of isolated ions is considered. The ion concentration is then changed from the standard state to concentration c. This transition is connected only with work corresponding to ideal behaviour, and not with electrical work. The action of forces other than electrostatic interionic forces in the electrolyte solution is neglected, as these are far more short-range interactions than electrostatic forces. For recharging of the ions at concentration c electrical work w_2 is expended. The calculation of work w_2 is more complicated, as this work is connected with the formation of a space charge. Debye and Hückel introduced the concept of an *ionic atmosphere*; i.e. an excess charge due to ions with the opposite charge is concentrated around each ion and the charge density is distributed in the ionic atmosphere according to the Maxwell–Boltzmann distribution law. This excess charge is equal to the charge of the ion considered (with opposite sign).

The Debye–Hückel theory yields the coefficient γ_c, but the whole subsequent calculation is accompanied by numerous approximations, valid only at high dilutions, so that in the whole region where the theory is valid it may be assumed that $\gamma_x \approx \gamma_m \approx \gamma_c$.

First, electric work w_1 and w_2 is calculated for a single ion species denoted as k. For w_1 the same procedure as for the quantity w_2 in the Born treatment of solvation Gibbs energy (Eq. (1.2.5) will be used giving ($\varepsilon = D\varepsilon_0$))

$$w_1 = -\frac{(z_k e)^2}{8\pi\varepsilon r} \tag{1.3.5}$$

where r is the distance from the centre of the ion.

At the final concentration c, the potential ψ_k at distance r is given not only by the potential of the ion, ψ_k^0, but also by the potentials of the surrounding ions. Debye and Hückel assumed that a spherical ionic atmosphere of statistically prevailing ions with opposite charge forms around each ion, giving rise to the potential ψ_a. Thus $\psi_k = \psi_k^0 + \psi_a$. The potential of the space charge density ρ is given by the Poisson equation

$$\nabla^2\psi = -\frac{\rho}{\varepsilon} \tag{1.3.6}$$

where ∇^2 is the Laplace operator, whose form depends on the coordinate system used. If the volume element dV contains dN_i ions of sort i with charge z_ie, the space charge density ρ_i formed by ions of sort i is given as

$$\rho_i = \frac{dN_i}{dV} ez_i \tag{1.3.7}$$

If the origin of the coordinate system is located in the centre of the central kth ion, the number of particles dN_i in the volume dV is given by a distribution function, expressed by Debye and Hückel in terms of the Boltzmann distribution law (cf. p. 215)

$$dN_i = \bar{N}_i \exp\left(\frac{-ez_i\psi_k}{kT}\right) dV \tag{1.3.8}$$

Here, k is the Boltzmann constant and $\bar{N}_i = N_i/V$ is the total number of ions of sort i divided by the total volume. Thus

$$\rho = \sum \rho_i = e \sum \bar{N}_i z_i \exp\left(\frac{-ez_i\psi_k}{kT}\right) \tag{1.3.9}$$

The exponential is expanded in a series and, except for unity, only the linear term is retained (this is permissible only when the electrostatic energy $e\,|z_i\psi_k|$ is small compared with the energy of thermal motion kT):

$$\rho = e \sum \bar{N}_i z_i \left(1 - \frac{ez_i\psi_k}{kT}\right) = -\frac{e^2\psi_k}{kT} \sum \bar{N}_i z_i^2 \tag{1.3.10}$$

(Here $\sum N_i z_i = 0$ because of the electroneutrality condition.) Equation (1.3.10) is substituted into Eq. (1.3.6) and the Laplace operator is expressed in polar coordinates (for the spherically symmetric problem):

$$\nabla^2\psi_k = \frac{d^2\psi_k}{dr^2} + \frac{2}{r}\frac{d\psi_k}{dr} = \frac{e^2 \sum \bar{N}_i z_i^2}{kT}\psi_k = \kappa^2\psi_k \tag{1.3.11}$$

(The important quantity κ will be discussed later.) The general integral of Eq. (1.3.11) is

$$\psi_k = \frac{k_1 e^{-\kappa r}}{r} + \frac{k_2 e^{\kappa r}}{r} \tag{1.3.12}$$

where k_1 and k_2 are integration constants determined from the boundary conditions. If $r \to \infty$, then $\psi_k \to 0$, from which it follows that $k_2 = 0$. The constant k_1 is determined by substitution of the expression for ψ_k into Eq. (1.3.10) for ρ. As the electroneutrality of the ion–ionic atmosphere system is required, the integration of the space charge around the central ion over the whole volume of the solution yields a charge of the same magnitude and of the opposite sign to that of the central ion. The volume element dV is expressed in polar coordinates as $dV = 4\pi r^2\, dr$. Thus,

$$\int \rho\, dV = \int_r 4\pi r^2 \rho\, dr = -\int_r 4\pi r^2 \frac{e^2}{kT} \sum \bar{N}_i z_i^2 \psi_k\, dr$$

$$= -\int_r 4\pi r^2 \frac{e^2}{kT} \sum \bar{N}_i z_i^2 \frac{k_1 e^{-\kappa r}}{r}\, dr$$

$$= -4\pi k_1 \varepsilon \kappa^2 \int_r r e^{-\kappa r}\, dr = -z_k e \qquad (1.3.13)$$

The Debye–Hückel limiting law is the least accurate approximation to the actual situation, analogous to the ideal gas law. It is based on the assumption that the ions are material points and that the potential of the ionic atmosphere is distributed from $r = 0$ to $r \to \infty$. Within these limits the last equation is integrated by parts yielding, for constant k_1, the value $ez_k/4\pi\varepsilon$. Potential ψ_k is given by the expression

$$\psi_k = \frac{ez_k}{4\pi\varepsilon r} e^{-\kappa r} \approx \frac{ez_k}{4\pi\varepsilon r} - \frac{ez_k \kappa}{4\pi\varepsilon} = \frac{ez_k}{4\pi\varepsilon r} - \frac{ez_k}{4\pi\varepsilon L_D} \qquad (1.3.14)$$

The final approximate form of Eq. (1.3.14) was again obtained by expanding the exponential in a series and retaining only the linear term. Obviously, the potential ψ_k can be expanded to give two terms, the first of which $(ez_k/4\pi\varepsilon r)$ describes the contribution of the central ion and the second $(ez_k/4\pi\varepsilon L_D)$ the contribution of the ionic atmosphere.

The ionic atmosphere can thus be replaced by the charge at a distance of $L_D = \kappa^{-1}$ from the central ion. The quantity L_D is usually termed the effective radius of the ionic atmosphere or the Debye length. The parameter κ is directly related to the ionic strength I

$$I = \tfrac{1}{2} \sum_{i=1}^{s} \bar{N}_i z_i^2 = \tfrac{1}{2} \sum_{i=1}^{s} c_i z_i^2 \qquad (1.3.15)$$

$$\kappa^2 = (2e^2 N_A / \varepsilon kT) I \qquad (1.3.16)$$

When numerical values are assigned to the constants in Eq. (1.3.16), it gives

$$L_D = 6.2881 \times 10^{-11} \left(\frac{TD}{I}\right)^{1/2} \text{m} \qquad (1.3.17)$$

and, for water at 25°C,

$$L_D = 9.6223 \times 10^{-9} I^{-1/2}\, \text{m}, \quad \text{or}$$

$$L_D = 0.30428 I^{-1/2}\, \text{nm}$$

(The first value is valid for basic SI units, the second is more practical: I in moles per cubic decimetre, L_D in nanometres.) The radii of the ionic atmosphere for various solution concentrations of a single binary electrolyte (for which $I = -\tfrac{1}{2} z_+ z_- \nu c$) are listed in Table 1.2.

Table 1.2 Radii of ionic atmosphere L_D (in nm) at various concentrations of aqueous solutions at 25°C (according to Eq. 1.3.17)

	Charge type of the electrolyte				
$c\ (\mathrm{mol \cdot dm^{-3}})$	1–1	1–2	2–2	1–3	2–3
10^{-5}	96.1	55.5	48.0	39.2	24.8
10^{-4}	30.4	17.5	15.2	12.4	7.8
10^{-3}	9.61	5.55	4.80	3.92	2.48
10^{-2}	3.04	1.75	1.52	1.24	0.78
10^{-1}	0.96	0.56	0.48	0.39	0.25
1	0.30	0.18	0.15	0.12	0.08

The work expended in recharging ion k is

$$\begin{aligned} w_2 &= \int_0^{z_k e} \psi_k\,\mathrm{d}q = \frac{1}{4\pi\varepsilon r}\int_0^{z_k e} q\,\mathrm{d}q - \frac{1}{4\pi\varepsilon}\int_0^{z_k e} \kappa q\,\mathrm{d}q \\ &= -w_1 - \frac{1}{4\pi\varepsilon}\left(\frac{2N_A I}{\varepsilon kT}\right)^{1/2}\int_0^{z_k e} q^2\,\mathrm{d}q \\ &= -w_1 - \frac{z_k^2 e^2 \kappa}{12\pi\varepsilon} \end{aligned} \qquad (1.3.18)$$

The overall electrical work connected with the described process for ion k is then

$$w_k = w_1 + w_2 = \frac{-z_k^2 e^2 \kappa}{12\pi\varepsilon} \qquad (1.3.19)$$

and, for all the ions present in volume of the solution V,

$$W_{el} = -\frac{Ve^2\kappa \sum N_i z_i^2}{12\pi\varepsilon} = -\frac{e^3 N_A^{3/2}}{12\pi\varepsilon(V\varepsilon kT)^{1/2}}\left(\sum_{i=1}^{s} n_i z_i^2\right)^{3/2} \qquad (1.3.20)$$

Work W_{el} is then identified with the correction for non-ideal behaviour ΔG^E and the activity coefficient is obtained from the equation

$$RT \ln \gamma_k = \left(\frac{\partial \Delta G^E}{\partial n_k}\right)_{p,T,n_i \neq n_k} = \left(\frac{\partial W_{el}}{\partial n_k}\right)_{p,T,n_i \neq n_k} \qquad (1.3.21)$$

Differentiation assuming that V is independent of n_i (which is fulfilled for point charges) yields

$$\begin{aligned} -\log \gamma_k &= \frac{\varepsilon e^3 N_A^{3/2}}{RT \ln 10 \times 12\pi\varepsilon(V\varepsilon kT)^{1/2}}\left(\frac{\partial(\sum n_i z_i^2)^{3/2}}{\partial n_k}\right)_{n_i \neq n_k} \\ &= \frac{z_k^2 e^3 N_A^{1/2}\sqrt{I}}{\ln 10 \times 4\pi\sqrt{2}(\varepsilon kT)^{3/2}} \end{aligned} \qquad (1.3.22)$$

It is convenient to introduce

$$A = \frac{e^3 N_A^{1/2}}{\ln 10 \times 4\pi\sqrt{2}(\varepsilon kT)^{3/2}} \quad (1.3.23)$$

yielding the usual form of the equation for the activity coefficient (cf. p. 11):

$$\log \gamma_k = -Az_k^2\sqrt{I} \quad (1.3.24)$$

termed the *Debye–Hückel limiting law.* Coefficient A has the numerical value

$$A = 5.77057 \times 10^4 (DT)^{-3/2} \quad m^{3/2} \cdot mol^{-1/2}$$
$$= 1.82481 \times 10^6 (DT)^{-3/2} \quad dm^{3/2} \cdot mol^{-1/2}$$

and, for water at 25°C,

$$A = 1.61039 \times 10^{-2}\, m^{3/2} \cdot mol^{-1/2}$$
$$= 0.50925\, dm^{3/2} \cdot mol^{-1/2}$$

In these expressions, the first value is valid for basic SI units and the second for I in moles per cubic decimetre substituted into Eq. (1.3.24). Equation (1.3.24) is a very rough approximation and does not involve the individual characteristics of ion k. It is valid for a uni-univalent electrolyte only up to an ionic strength of $10^{-3}\, mol \cdot dm^{-3}$ (see also Fig. 1.8).

In view of the definition of the mean activity coefficient and of the electroneutrality condition, $\nu_+ z_+ = -\nu_- z_-$, the limiting law also has the form

$$\log \gamma_\pm = Az_+ z_- \sqrt{I} \quad (1.3.25)$$

1.3.2 *More rigorous Debye–Hückel treatment of the activity coefficient*

A more rigorous approach requires integration of Eq. (1.3.13) not over the whole volume of the solution but over the effective volume after the volume of the central ion has been excluded, as this region is not accessible for the ionic atmosphere. Thus, integration is carried out from $r = a$, i.e. from the distance of closest approach, equal to the effective ion diameter, which is the smallest mean distance to which the centres of other ions can approach the central ion. This value varies for various electrolytes (Table 1.3). In this refined procedure the integration constant appearing in Eq. (1.3.13) attains the value $k_1 = [z_k e/4\pi\varepsilon(1 + \kappa a)] \exp(\kappa a)$ and the potential $\psi_k = z_k e/4\pi\varepsilon[r(1 + \kappa a)]^{-1} \exp[\kappa(a - r)]$. This expression can again be separated into the contribution of the isolated central ion ψ_k^0 and the contribution of the ionic atmosphere ψ_a. As a result of the principle of linear field superposition, these two quantities can be added algebraically. The contribution of the

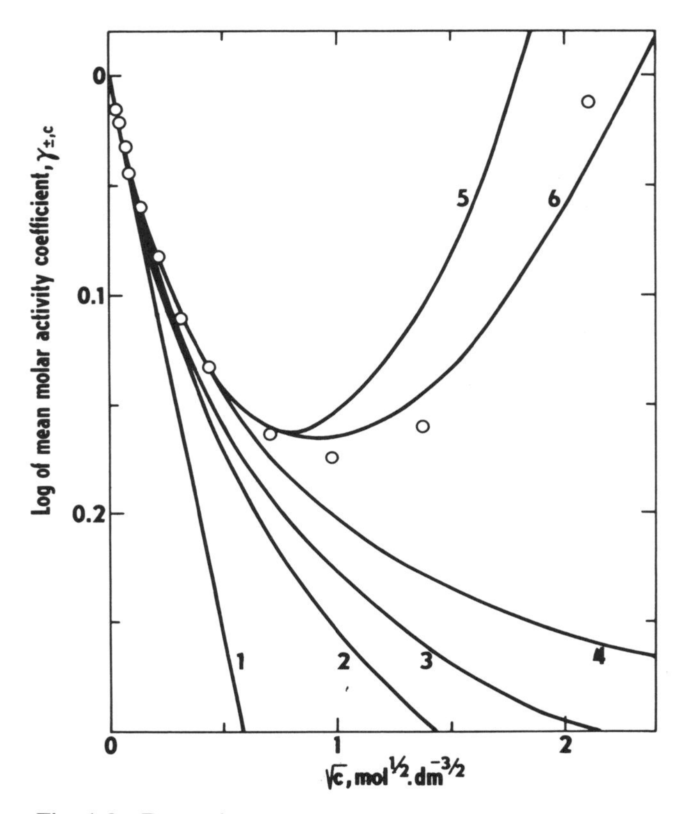

Fig. 1.8 Dependence of the mean activity coefficient $\gamma_{\pm,c}$ of NaCl on the square root of molar concentration c at 25°C. Circles are experimental points. Curve 1 was calculated according to the Debye–Hückel limiting law (1.3.25), curve 2 according to the approximation $aB = 1$ (Eq. 1.3.32); curve 3 according to the Debye–Hückel equation (1.3.31), $a = 325$ nm; curve 4 according to the Bates–Guggenheim approximation (1.3.33); curve 5 according to the Bates–Guggenheim approximation + linear term $0.1\,C$; curve 6 according to Eq. (1.3.38) for $a = 0.4$ nm, $C = 0.055\,\mathrm{dm^3 \cdot mol^{-1}}$

ionic atmosphere is then given by the expressions

$$(\psi_a)_{r>a} = \psi_k - \psi_k^0 = \frac{z_k e}{4\pi\varepsilon r}\left(\frac{e^{\kappa(a-r)}}{1+\kappa a} - 1\right) \tag{1.3.26}$$

$$(\psi_a)_{r\leqq a} = -\frac{z_k e}{4\pi\varepsilon}\frac{\kappa}{1+\kappa a} = -\frac{z_k e}{4\pi\varepsilon}\frac{1}{L_D + a} \tag{1.3.27}$$

because at a distance of $r < a$, no ion of the ionic atmosphere can be present

Table 1.3 Effective ion diameters (According to B. E. Conway)

Ion	a (nm)
Rb^+, Cs^+, NH_4^+, Tl^+, Ag^+	0.25
K^+, Cl^-, Br^-, I^-, CN^-, NO_2^-, NO_3^-	0.3
OH^-, F^-, CNS^-, NCO^-, HS^-, ClO_3^-, ClO_4^-, BrO_3^-, IO_4^-, MnO_4^-, $HCOO^-$, citrate$^-$, $CH_3NH_3^+$	0.35
Hg_2^{2+}, SO_4^{2-}, $S_2O_3^{2-}$, $S_2O_6^{2-}$, $S_2O_8^{2-}$, Se_4^{2-}, CrO_4^{2-}, HPO_4^{2-}, PO_4^{3-}, $Fe(CN)_6^{3-}$, $Cr(NH_3)_6^{3+}$, $Co(NH_3)_6^{3+}$, $Co(NH_3)_5H_2O^{3+}$, $NH_3^+CH_2COOH$, $C_2H_5NH_3^+$	0.4
Na^+, $CdCl^+$, ClO_2^-, IO_3^-, HCO_3^-, $H_2PO_4^-$, HSO_3^-, $H_2AsO_4^-$, Pb^{2+}, CO_3^{2-}, SO_3^{2-}, MoO_4^{2-}, $Co(NH_3)_5Cl^{2+}$, $Fe(CN)_5NO^{2-}$, CH_3COO^-, CH_2ClCOO^-, $(CH_3)_4N^+$, $(C_2H_5)_2NH_2^+$, $NH_2CH_2COO^-$	0.45
Sr^{2+}, Ba^{2+}, Ra^{2+}, Cd^{2+}, Hg^{2+}, S^{2-}, $S_2O_4^{2-}$, WO_4^{2-}, $Fe(CN)_6^{4-}$, $CHCl_2COO^-$, CCl_3COO^+, $(C_2H_5)_3NH^+$, $C_3H_7NH_3^+$	0.5
Li^+, Ca^{2+}, Cu^{2+}, Zn^{2+}, Sn^{2+}, Mn^{2+}, Fe^{2+}, Ni^{2+}, Co^{2+}, Co(ethylendiamine)$_3^{3+}$, $C_6H_5COO^-$, $C_6H_4OHCOO^-$, $(C_2H_5)_4N^+$, $(C_3H_7)_2NH_2^+$	0.6
Mg^{2+}, Be^{2+}	0.8
H^+, Al^{3+}, Fe^{3+}, Cr^{3+}, Sc^{3+}, La^{3+}, In^{3+}, Ce^{3+}, Pr^{3+}, Nd^{3+}, Sm^{3+}	0.9
Th^{4+}, Zr^{4+}, Ce^{4+}, Sn^{4+}	1.1

and therefore the potential remains constant and equal to the value for $r = a$.

In view of this equation the effect of the ionic atmosphere on the potential of the central ion is equivalent to the effect of a charge of the same magnitude (that is $-z_k e$) distributed over the surface of a sphere with a radius of $a + L_D$ around the central ion. In very dilute solutions, $L_D \gg a$; in more concentrated solutions, the Debye length L_D is comparable to or even smaller than a. The radius of the ionic atmosphere calculated from the centre of the central ion is then $L_D + a$.

Work w_2 can now be calculated and, from this value, the total electric work W_{el}, found in an analogous way to that previously used for the case of the limiting law, yielding

$$W_{el} = -\frac{Ve^2 \sum \bar{N}_i z_i^2}{4\pi\varepsilon a^3 \kappa^2}[\ln(1+\kappa a) - \kappa a + \tfrac{1}{2}(\kappa a)] \qquad (1.3.28)$$

(The expression in square brackets follows from the integral $a^3 \int [q^2/(1+\kappa a)]\, dq$ after introducing the limits 0 and $z_k e$.)

The expression for the activity coefficient is calculated from the differential $\partial W_{el}/\partial n_k$ (neglecting the dependence of V on n_k) by substitution into Eq. (1.3.21). Rearrangement yields

$$\log \gamma_k = -\frac{z_k^2 e^2}{\ln 10 \times 8\pi\varepsilon kT}\frac{\kappa}{1+\kappa a} = -\frac{A z_k^2 \sqrt{I}}{1 + Ba\sqrt{I}} \qquad (1.3.29)$$

where the constant A is identical with the constant in the limiting law and

$$B = \left(\frac{2e^2 N_A}{\varepsilon kT}\right)^{1/2} \tag{1.3.30}$$

The numerical value of this constant is

$$\begin{aligned} B &= 1.5903 \times 10^{10} (DT)^{-1/2} \qquad \mathrm{m^{1/2} \cdot mol^{-1/2}} \\ &= 502.90 (DT)^{-1/2} \qquad \mathrm{dm^{3/2} \cdot mol^{-1/2} \cdot nm^{-1}} \end{aligned}$$

and, for water at 25°C,

$$\begin{aligned} B &= 1.0392 \times 10^{8}\ \mathrm{m^{1/2} \cdot mol^{-1/2}} \\ &= 3.2864\ \mathrm{dm^{3/2} \cdot mol^{-1/2} \cdot nm^{-1}} \end{aligned}$$

In the above two equations, the former value is valid for basic SI units and the latter value for I in moles per cubic decimetre and a in nanometres. The parameter a represents one of the difficulties connected with the Debye–Hückel approach as its direct determination is not possible and is, in most cases, found as an adjustable parameter for the best fit of experimental data in the Eq. (1.3.29). For common ions the values of effective ion radii vary from 0.3 to 0.5. Analogous to the limiting law, the mean activity coefficient can be expressed by the equation

$$\log \gamma_{\pm} = \frac{A z_+ z_- \sqrt{I}}{1 + Ba\sqrt{I}} \tag{1.3.31}$$

where a is the average of the effective ion diameters of the cation a_+ and for the anion a_- (this averaging is the source of difficulties in cases of more complicated systems like electrolyte mixtures). Equation (3.1.31) is a definite improvement in comparison of the limiting law (Fig. 1.8). As the constant B in the units $\mathrm{mol^{-3/2} \cdot nm^{-1}}$ is about 3.3 the product aB being not far from unity some authors employ a simplified equation (for aqueous solutions at 25°C)

$$\log \gamma_{\pm} = \frac{A z_+ z_- \sqrt{I}}{1 + \sqrt{I}} \tag{1.3.32}$$

The Bates–Guggenheim equation

$$\log \gamma_i = \frac{A z_i \sqrt{I}}{1 + 1.5\sqrt{I}} \tag{1.3.33}$$

is also often used. Like the limiting law, the latter two equations do not require any specific information on the ions considered; nonetheless, they describe the $\gamma - I$ dependence more closely than the limiting law (see Fig. 1.8).

If the ionic strength in concentrated aqueous or non-aqueous solutions is expressed in terms of the molalities, then the constants $A' = A\sqrt{\rho}$ and $B' = B\sqrt{\rho}$ are used, where ρ is the density of the solvent.

1.3.3 *The osmotic coefficient*

The osmotic pressure of an electrolyte solution π can be considered as the ideal osmotic pressure π^* decreased by the pressure π_{el} resulting from electric cohesion between ions. The work connected with a change in the concentration of the solution is $\pi\,dV = \pi^*\,dV - \pi_{el}\,dV$. The electric part of this work is then $\pi_{el}\,dV = dW_{el}$, and thus $\pi_{el} = (\partial W_{el}/\partial V)_{T,n}$. The osmotic coefficient ϕ is given by the ratio π/π^*, from which it follows that

$$1 - \phi = \frac{\pi_{el}}{\pi^*} = \left(\frac{\partial W_{el}}{\partial V}\right)_{T,n} \frac{1}{kN_A T \sum c_i} \tag{1.3.34}$$

as the ideal osmotic pressure is $\pi^* = kN_A T \sum c_i$. In order to obtain $(\partial W_{el}/\partial V)_{T,n}$, Eq. (1.3.20) is differentiated, resulting in

$$1 - \phi = \frac{e^3 N_A^{1/2}}{3\pi \sum c_i}\left(\frac{I}{2\varepsilon kT}\right)^{3/2} \tag{1.3.35}$$

Consider a dilute solution of a single electrolyte, so that $\sum c_i = c_+ + c_-$, $c_+z_+ + c_-z_- = 0$ and $I = -\frac{1}{2}z_+z_-c$. Comparison of Eq. (1.3.35) with Eq. (1.3.22) written for $\gamma_\pm$ and with Eq. (1.3.23) yields the limiting relationship

$$\begin{aligned} 1 - \phi &= -\frac{e^3 N_A^{1/2}}{6\pi(2\varepsilon kT)^{3/2}} z_+z_-(\tfrac{1}{2}\nu c)^{1/2} \\ &= -\tfrac{1}{3}\ln\gamma_\pm = -\frac{A \ln 10}{3} z_+z_-(\tfrac{1}{2}\nu c)^{1/2} \end{aligned} \tag{1.3.36}$$

For water at 0°C (the osmotic coefficient is mostly determined from cryoscopic measurements),

$$1 - \phi = -0.2658 z_+z_-(\nu c)^{1/2} \tag{1.3.37}$$

1.3.4 *Advanced theory of activity coefficients of electrolytes*

According to Eq. (1.3.31) $\log\gamma$ should be a monotonous function of $\sqrt{I}$ which obviously is not the case. A possible remedy of this situation would be, for example, the introduction of an additional linear term on the right-hand side of that equation,

$$\log\gamma_\pm = -\frac{Az_+z_-\sqrt{I}}{1 + Ba\sqrt{I}} + CI \tag{1.3.38}$$

In this equation, the coefficient C as well as the coefficient a must be taken as adjustable parameters found experimentally. The result is shown in Fig. 1.8, curve 6. Otherwise a quite good approximation for ionic strengths up to 0.3 is achieved by putting $C = -0.1z_+z_-$.

Thus, a suitable refinement of the Debye–Hückel theory must provide a theoretical interpretation of the term CI. Originally this term was qualitatively interpreted as a salting-out effect: during solvation the ions orient solvent molecules so that the electrostatic forces leading to non-ideality of the solution are weakened. The greater the concentration of ions, the greater the portion of the solvent molecules that is oriented; the activity coefficients again increase with increasing ionic strength.

Most authors have accounted for the mutual influence of ions and solvent molecules only by assuming a firmly bound solvent sheath around the ion. The structure of the bulk solvent and the influence of electrolyte concentration on this structure are not taken into consideration.

The necessity of considering the solvent sheath around the ions follows from the fact that the effective ionic diameters a calculated from the Debye–Hückel equation (1.3.31) using the experimental activity coefficient values are too large compared with the corresponding crystallographic quantities. This led to the concept that Eq. (1.3.31) is valid for the activity coefficients of the solvated ions that form a mixture with the remaining 'free' solvent differing from ideal behaviour only in the electrostatic interaction between the solvated ions. If, on the other hand, the activity coefficients are calculated from the experimental data, then the molality or mole fraction is expressed in terms of the amount of anhydrous electrolyte and all the solvent is considered as 'free'. This calculation yields a formal activity coefficient for the unsolvated ions, whose existence is fictitious. Thus, if the theoretical expressions for the activity coefficients are to be compared with the experimental values, the values for the solvated species must be recalculated to those for unsolvated species. This recalculation is particularly important for concentrated solutions, where the decrease in the amount of 'free' solvent is significant.

The method of calculating these values was proposed by R. A. Robinson and R. H. Stokes. The Gibbs energy G of a system containing 1 mole of uni-univalent electrolyte ($n_1 = 1$, total number of moles of ions being 2) and n_0 moles of solvent is expressed for solvated ions (dashed quantities) on one hand and for non-solvated ions (without dash) on the other; it is assumed that $\mu_0 = \mu_0'$ and that the hydration number h (number of moles of the solvent bound to 1 mole of the electrolyte) is independent of concentration:

$$\begin{aligned} G &= n_0\mu_0 + \mu_+ + \mu_- \\ &= (n_0 - h)\mu_0' + \mu_+' + \mu_-' \\ &= (n_0 - h)\mu_0 + \mu_+' + \mu_-' \end{aligned} \qquad (1.3.39)$$

Chemical potentials are split up into standard and variable terms,

$$(\mu_+^0 + \mu_+^{0\prime} + \mu_-^0 + \mu_-^0)/RT + h\mu_0^0/RT + h \ln a_0 + 2 \ln \frac{n_0 + 2 - h}{n_0 + 2} + \ln \gamma_{+,x} + \ln \gamma_{-,x} = \ln \gamma_+' + \ln \gamma_-' \qquad (1.3.40)$$

For $n_0 \rightarrow \infty$, all the logarithmic terms approach zero. Thus the combination of the standard potentials given by the first terms in Eq. (1.3.40) equals zero, so that the mean activity coefficient is given by the equation

$$2 \ln \gamma_{\pm,x} = 2 \ln \gamma'_{\pm,x} - h \ln a_0 - 2 \ln \frac{n_0 + 2 - h}{n_0 + 2} \tag{1.3.41}$$

Robinson and Stokes identified the quantity $0.4343 \ln \gamma'_{\pm,x}$ with the expression on the left-hand side in Eq. (1.3.31). For the system considered here, $m = n_1/n_0M_0$ (where M_0 is the molar mass of the solvent, for water $M_0 = 0.018$ kg/mol), so that

$$\ln[(n_0 + 2 - h)/(n_0 + 2)] = \ln[(1 + 2M_0m - hM_0m)/(1 + 2M_0m)]$$

and, according to Appendix A, $\gamma_{\pm,x} = \gamma_{\pm,m}(1 + 2M_0m)$. Hence,

$$\log \gamma_{\pm,m} = -\frac{A'\sqrt{I}}{1 + B'a\sqrt{I}} - h/2 \log a_0 - \log[1 + (2 - h)M_0m] \tag{1.3.42}$$

As in view of Eq. (1.1.18) $\ln a_0 = -\phi_m M_0 m$ and $1 \gg (2 - h)M_0m$ Eq. (1.3.42) gives

$$\log \gamma_{\pm,m} = DH + 0.4343[\phi_m M_0 h + (2 - h)M_0]m \tag{1.3.43}$$

where DH denotes the Debye–Hückel term (1.3.31) while the second term on the right-hand side corresponds to CI of Eq. (1.3.38).

The Robinson–Stokes equation (1.3.43) can be used to calculate the parameters a and h from the measured values of $\gamma_{\pm,m}$ and ϕ_m. In this way the hydration number can also be found, but only as the sum of the ion hydration numbers of all the ions in the electrolyte 'molecule'. The appropriate calculations have been carried out for a number of systems (see Table 1.4). Equation (1.3.43) describes the experimental data up to molalities of the order of units, so that it is formally very successful. Ionic diameters vary from 0.35 to 0.62 nm. The calculated hydration numbers for salts with a common anion retain the order expected for the cations of these salts ($H^+ > Li^+ > Na^+ > K^+ > Rb^+$), but do not retain the order expected for the anions with a common cation ($I^- > Br^- > Cl^-$). Furthermore, these numbers do not preserve additivity: for example $h_{NaCl} - h_{KCl} = 1.6$, but $h_{NaCl} - h_{KI} = 3.0$. If it is assumed that chloride ions are practically not hydrated and the whole hydration number of chloride salts is attributed to the cation, then the numbers obtained are too large (e.g. 12 for $CaCl_2$ and 13.7 for $MgCl_2$). If the radius of the solvated cation is estimated considering the volume of bound solvent and this is added to the crystallographic radius of the anion, numbers are obtained that exceed the suitable effective diameters to a differing degree, depending on the type of electrolyte (e.g. by 0.07 nm for uni-univalent and by 0.13 nm for uni-divalent electrolytes, etc.). This discrepancy is explained by penetration of the anion solvation sheath into the sheath of the cation (Fig. 1.7).

The contribution of short-range forces to the activity coefficient can be described much better and in greater detail by the methods of the statistical thermodynamics of liquids, which has already created several models of electrolyte solutions. However, the procedures employed in the statistical

Table 1.4 Values of the parameters a and h (Eq. 1.3.42). (According to R. A. Robinson and R. H. Stokes)

	h	a (nm)		h	a (nm)
HCl	8.0	0.447	RbI	0.6	0.356
HBr	8.6	0.518	$MgCl_2$	13.7	0.502
HI	10.6	0.569	$MgBr_2$	17.0	0.546
$HClO_4$	7.4	0.509	MgI_2	19.0	0.618
LiCl	7.1	0.432	$CaCl_2$	12.0	0.473
LiBr	7.6	0.456	$CaBr_2$	14.6	0.502
LiI	9.0	0.560	CaI_2	17.0	0.569
$LiClO_4$	8.7	0.563	$SrCl_2$	10.7	0.461
NaCl	3.5	0.397	$SrBr_2$	12.7	0.489
NaBr	4.2	0.424	SrI_2	15.5	0.558
NaI	5.5	0.447	$BaCl_2$	7.7	0.445
$NaClO_4$	2.1	0.404	$BaBr_2$	10.7	0.468
KCl	1.9	0.363	BaI_2	15.0	0.544
KBr	2.1	0.385	$MnCl_2$	11.0	0.474
KI	2.5	0.416	$FeCl_2$	12.0	0.480
NH_4Cl	1.6	0.375	$CoCl_2$	13.0	0.481
RbCl	1.2	0.349	$NiCl_2$	13.0	0.486
RbBr	0.9	0.348	$Zn(ClO_4)_2$	20.0	0.618

thermodynamics of liquid mixtures are conceptually and mathematically rather complicated so that the reader must be referred to special monographs on the subject. On the other hand, it matters little at this stage that statistical thermodynamic procedures are so far lacking in the present textbook. The derived equations have not yet even been worked out to an experimentally verifiable form. In fact, the theory of ionic solutions should yield an equation permitting calculation of the activity coefficients of ions on the basis of knowledge of the structure of the ion considered and of all the other components of the solution. The methods of statistical thermodynamics that would, in principle, be capable of carrying out this calculation are developing rapidly and it seems that the future looks promising in this area.

1.3.5 *Mixtures of strong electrolytes*

The derivation of the equations of the Debye–Hückel theory did not require differentiation between a solution of a single electrolyte and an electrolyte mixture provided that the limiting law approximation Eq. (1.3.24), was used, which does not contain any specific ionic parameter. If, however, approximation (1.3.29) is to be used, containing the effective ionic diameter a, it must be recalled that this quantity was introduced as the minimal mean distance of approach of both positive and negative ions to the central ion. Thus, this quantity a is in a certain sense an average of effects of all the ions but, at the same time, a characteristic value for the given central ion.

If the validity of Eq. (1.3.31) is assumed for the mean activity coefficient of a given electrolyte even in a mixture of electrolytes, and quantity a is calculated for the same measured electrolyte in various mixtures, then different values are, in fact, obtained which differ for a single total solution molality depending on the relative representation and individual properties of the ionic components.

A number of authors have suggested various mixing rules, according to which the quantity a could be calculated for a measured electrolyte in a mixture, starting from the known individual parameters of the single electrolytes and the known composition of the solution. However, none of the proposed mixing relationships has found broad application. Thus, the question about the dependence of the mean activity coefficients of the individual electrolytes on the relative contents of the various electrolytic components was solved in a different way.

The approach introduced by E. A. Guggenheim and employed by H. S. Harned, G. Åkerlöf, and other authors, especially for a mixture of two electrolytes, is based on the Brønsted assumption of specific ion interactions: in a dilute solution of two electrolytes with constant overall concentration, the interaction between ions with charges of the same sign is non-specific for the type of ion, while interaction between ions with opposite charges is specific.

Guggenheim used this assumption to employ Eq. (1.3.38) for the activity coefficient of the electrolyte, where the product aB was set equal to unity and the specific interaction between oppositely charged ions was accounted for in the term CI. Consider a mixture of two uni-univalent electrolytes $A_I B_I$ and $A_{II} B_{II}$ with overall molality m and individual representations $y_I = m_I/m$ and $y_{II} = m_{II}/m$, where m_I and m_{II} are molalities of individual electrolytes. According to Guggenheim,

$$\begin{aligned} \ln \gamma_I &= -A^* + [2y_I b_{I,I} + (b_{II,I} + b_{I,II})(1 - y_I)]m \\ \ln \gamma_{II} &= -A^* + [2(1 - y_I) b_{II,II} + (b_{II,I} + b_{I,II}) y_I]m \end{aligned} \tag{1.3.44}$$

where the b's are specific interaction constants ($b_{I,II}$ is the parameter of the interaction of A_I with B_{II}, $b_{I,I}$ is the parameter of the interaction of A_I with B_I, etc.) and $A^* = \ln 10\, A\sqrt{I}/(1 + \sqrt{I})$.

Two limiting activity coefficients can be obtained from Eq. (1.3.44) for each electrolyte:

$$\begin{aligned} \lim_{y_I \to 0} \ln \gamma_I &= \ln \gamma_I^0 = -A^* + (b_{II,I} + b_{I,II})m \\ \lim_{y_I \to 1} \ln \gamma_I &= \ln \gamma_I^1 = -A^* + 2b_{I,I}m \\ \lim_{y_{II} \to 0} \ln \gamma_{II} &= \ln \gamma_{II}^0 = -A^* + (b_{II,I} + b_{I,II})m \\ \lim_{y_{II} \to 1} \ln \gamma_{II} &= \ln \gamma_{II}^1 = -A^* + 2b_{II,II}m \end{aligned} \tag{1.3.45}$$

Thus, $\ln \gamma_{\mathrm{I}}^0 = \ln \gamma_{\mathrm{II}}^0$.

Coefficients α_{I} and α_{II} are introduced, describing the combination of the interaction constants:

$$\begin{aligned}\ln 10\, \alpha_{\mathrm{I}} &= 2b_{\mathrm{I,I}} - b_{\mathrm{II,I}} - b_{\mathrm{I,II}} = \frac{\ln \gamma_{\mathrm{I}}^1 - \ln \gamma_{\mathrm{I}}^0}{m} \\ \ln 10\, \alpha_{\mathrm{II}} &= 2b_{\mathrm{II,II}} - b_{\mathrm{II,I}} - b_{\mathrm{I,II}} = \frac{\ln \gamma_{\mathrm{II}}^1 - \ln \gamma_{\mathrm{II}}^0}{m}\end{aligned} \tag{1.3.46}$$

Combination of Eqs (1.3.44) to (1.3.46), introduction of molalities m_{I} and m_{II} and conversion to decadic logarithms yield the equations

$$\begin{aligned}\log \gamma_{\mathrm{I}} &= \log \gamma_{\mathrm{I}}^0 + \alpha_{\mathrm{I}} m_{\mathrm{I}} = \log \gamma_{\mathrm{I}}^1 - \alpha_{\mathrm{I}} m_{\mathrm{II}} \\ \log \gamma_{\mathrm{II}} &= \log \gamma_{\mathrm{II}}^0 + \alpha_{\mathrm{II}} m_{\mathrm{II}} = \log \gamma_{\mathrm{II}}^1 - \alpha_{\mathrm{II}} m_{\mathrm{I}}\end{aligned} \tag{1.3.47}$$

These relationships are termed the *Harned rules* and have been verified experimentally up to high overall molality values (e.g. for a mixture of HCl and KCl up to $2\,\mathrm{mol \cdot kg^{-1}}$). If this linear relationship between the logarithm of the activity coefficient of one electrolyte and the molality of the second electrolyte in a mixture with constant overall molality is not fulfilled, then a further term is added, including the square of the appropriate molality:

$$\begin{aligned}\log \gamma_{\mathrm{I}} &= \log \gamma_{\mathrm{I}}^1 - \alpha_{\mathrm{I}} m_{\mathrm{II}} - \beta_{\mathrm{I}} m_{\mathrm{II}}^2 \\ \log \gamma_{\mathrm{II}} &= \log \gamma_{\mathrm{II}}^1 - \alpha_{\mathrm{II}} m_{\mathrm{I}} - \beta_{\mathrm{II}} m_{\mathrm{I}}^2\end{aligned} \tag{1.3.48}$$

Robinson and Stokes have demonstrated that, for simplifying assumptions, the following relationship is valid between α's and β's:

$$(\alpha_{\mathrm{I}} + \alpha_{\mathrm{II}}) = k - 2m(\beta_{\mathrm{I}} + \beta_{\mathrm{II}}) \tag{1.3.49}$$

where k is a constant independent of the molality.

A particular case of electrolyte mixtures occurs if one electrolyte is present in a large excess over the others, thus determining the value of the ionic strength. In this case the ionic atmospheres of all the ions are formed almost exclusively from these excess ions. Under these conditions, the activities of all the ions present in the solution are proportional to their concentrations, the activity coefficient being a function of the concentration of the excess electrolyte alone.

The excess electrolyte is often termed the *indifferent electrolyte*. From the practical point of view, solutions containing an indifferent electrolyte are very often used in miscellaneous investigations. For example, when determining equilibrium constants (e.g. apparent dissociation constants, Eq. 1.1.26) it is necessary only to indicate the indifferent electrolyte and its concentration, as they do not change when the concentrations of the reactants are changed. Moreover, the indifferent electrolyte is important in the study of diffusion transport (Section 2.5), for elimination of liquid junction potentials (Section 2.5.3) and in the investigations of electrochemical kinetics (Section 5.4).

1.3.6 *Methods of measuring activity coefficients*

Two types of methods are used to measure activity coefficients. *Potentiometric methods* that measure the mean activity coefficient of the dissolved electrolyte directly will be described in Section 3.3.3. However, in galvanic cells with liquid junctions the electrodes respond to individual ion activities (Section 3.2). This is particularly true for pH measurement (Sections 3.3.2 and 6.3). In these cases, extrathermodynamical procedures defining individual ion activities must be employed.

Thermodynamic methods also measure the activity coefficient of the solvent (it should be recalled that the activity coefficient of the solvent is directly related to the osmotic coefficient—Eq. 1.1.19). As the activities of the components of a solution are related by the Gibbs–Duhem equation, the measured activity coefficient of the solvent can readily be used to calculate the activity coefficient of a dissolved electrolyte.

All methods based on osmotic phenomena lead directly to measurement of activities; these include methods measuring the decrease in vapour pressure above the solution, boiling point elevation, freezing point suppression and direct measurement of the osmotic pressure. Of these methods, the most commonly used is the *cryoscopic method*; *ebulioscopic* and *osmometric methods* are technically more complicated and are used less frequently. *Tensiometric methods* are also often used, involving measurement of the vapour pressure of the solvent above the pure solvent p_0^0 and above the solution p_0. The ratio of these two values, p_0/p_0^0, gives the activity of the solvent in the solution. Various methods are used to measure vapour pressure. Special tensiometers can be used to measure vapour pressure directly in tensiometric measurements. Dynamic methods involve passing an inert gas first through the pure solvent, then through a drying adsorbent, further through the solution and then through another adsorbent; the ratio of the amounts of solvent adsorbed on the two adsorbents equals the ratio of the two vapour pressures. The isopiestic method involves placing two solutions of various electrolytes B and C in a single solvent in a special cell and bringing them into equilibrium through the gaseous phase. At osmotic equilibrium, the two solutions have the same vapour pressure and the same solvent activity; the molalities of the two solutions are measured at equilibrium, m_B and m_C. If, for each value of the molality of one of the two electrolytes, for example B, the concentration dependence of the osmotic coefficient ϕ_B is also known, then ϕ_C can be calculated from the equation $\nu_B m_B \phi_B = \nu_C m_C \phi_C$.

Methodically, there is no great difference between measuring the mean activity coefficient in a solution of one electrolyte and measuring this quantity in a mixture of electrolytes. Binary mixtures have been studied most extensively. If osmotic methods are used, then the coefficients α_I and

α_{II} can be calculated and, from these values, the activity coefficients using the above relationships. If the Harned rule is not valid, then similar, more complicated relationships are used that also contain the coefficients β_I and β_{II}. If potentiometric methods are used, then the activity coefficients are obtained directly, i.e. values of γ_I and γ_{II} for the given total molality and relative abundances of the two electrolytes.

References

Conway, B. E., Ionic interactions and activity behaviour, *CTE,* **5,** Chap. 2 (1982).
Friedman, H. L., *Ionic Solution Theory,* Wiley–Interscience, New York, 1962.
Harned, H. S., and B. B. Owen, *The Physical Chemistry of Electrolytic Solutions,* Reinhold, New York, 1950.
Robinson, R. A., and R. H. Stokes, *Electrolyte Solutions,* Butterworths, London, 1959.

1.4 Acids and Bases

The Arrhenius concept was of basic importance because it permitted quantitative treatment of a number of acid–base processes in aqueous solutions, i.e. the behaviour of acids, bases, their salts and mixtures of these substances in aqueous solutions. Nonetheless, when more experimental material was collected, particularly on reaction rates of acid–base catalysed processes, an increasing number of facts was found that was not clearly interpretable on the basis of the Arrhenius theory (e.g. in anhydrous acetone NH_3 reacts with acids in the absence of OH^- and without the formation of water). It gradually became clear that a more general theory was needed. Such a theory was developed in 1923 by J. N. Brønsted and, independently, by T. M. Lowry.

1.4.1 *Definitions*

The generalization was based on the introduction of the concept of donor–acceptor pairs into the theory of acids and bases; this is a fundamental concept in the general interpretation of chemical reactivity. In the same way as a redox reaction depends on the exchange of electrons between the two species forming the redox system, reactions in an acid–base system also depend on the exchange of a chemically simple species—hydrogen cations, i.e. protons. Such a reaction is thus termed *protolytic.* This approach leads to the following definitions:

An acid HA is a substance capable of splitting off a proton:

$$HA \rightarrow H^+ + A^-$$

A base B is a substance capable of bonding a proton:

$$B + H^+ \rightarrow BH^+$$

Substances capable of splitting off a proton, i.e. proton donors, are termed *protogenic* and, on the other hand, substances accepting protons, i.e. proton accceptors, are *protophilic*. Some substances can exhibit both of these properties, i.e. can either accept or donate a proton, and are termed *amphiprotic*. The pairs HA – A and B – HB differ only by the presence or absence of a proton (in the same way as the oxidized form differs from the reduced form in a redox system only by the presence or absence of an electron). The substances HA and A^- or B and BH^+ form a *conjugate pair*. A strong acid is conjugate with a weak base and a strong base is conjugate with a weak acid.

An important characteristic of the definitions should be noted: no mention is made of the charge of species HA, A, B and HB. Acids and bases can be either electrically neutral molecules or ions.

As free proton cannot exist alone in solution, reactions in which a proton is split off from an acid to form a conjugate base cannot occur in an isolated system (in a homogeneous solution although this is possible in electrolysis (Section 5.7.1)). The homogeneous solution must contain another base B_{II} that accepts a proton from the acid HA_I (acid HA_I is, of course, not conjugate with base B_{II}). It will be seen that this second base can even be the solvent molecule, provided it has protophilic properties. Acid–base reactions thus depend on the exchange of a proton between an acid and a base that are not mutually conjugate:†

$$HA_I + B_{II} \rightarrow A_I + HB_{II} \qquad (1.4.1)$$

This scheme describes, in a uniform manner, all the reactions considered in the classical (Arrhenius) theory as different processes, giving them different names. They include, for example,

$HCl + NH_3 \rightarrow Cl^- + NH_4^+$	Salt formation
$HCl + H_2O \rightarrow Cl^- + H_3O^+$	Acid dissociation
$H_2O + NH_3 \rightarrow OH^- + NH_4^+$	Base dissociation
$H_2O + H_2O \rightarrow OH^- + H_3O^+$	Autoionization
$H_3O^+ + OH^- \rightarrow H_2O + H_2O$	Neutralization
$NH_4^+ + H_2O \rightarrow NH_3 + H_3O^+$	Hydrolysis
$H_2O + CH_3COO^- \rightarrow OH^- + CH_3COOH$	Hydrolysis

1.4.2 *Solvents and self-ionization*

To begin with, molecular solvents with high permittivities will be considered. Classification of solvents on the basis of their permittivities agrees roughly with classification as *polar* and *non-polar*, and the borderline between these two categories is usually considered to be a relative dielectric constant of 30–40. Below this value ion pairs are markedly formed. From

† As the acid–base equilibrium is dynamic the reaction $HA + A^{*-} \rightleftarrows A^- + HA^*$, where HA and HA* are identical, unceasingly takes place at a high rate but has no influence on the equilibrium properties of the system (cf. p. 99).

the point of view of acid–base properties, solvents can be classified as *protic*—containing a dissociable proton—and *aprotic*—not able to dissociate a proton. This classification roughly corresponds to division on the basis of whether the solvent molecules are or are not bonded by hydrogen bonds in the liquid state.

Most protic solvents have both *protogenic* and *protophilic* character, i.e. they can split off as well as bind protons. They are called, therefore, *amphiprotic*. These include: water, alcohols, acids (especially carboxylic), ammonia, dimethylsulphoxide and acetonitrile. Solvents that are protogenic and have weak or practically negligible protophilic character include acid solvents, such as sulphuric acid, hydrogen fluoride, hydrogen cyanide, and formic acid.

Aprotic solvents are not protogenic, but can be protophilic, e.g. acetone, 1,4-dioxan, tetrahydrofuran, dimethylformamide, hexamethylphosphortriamide, propylene carbonate and sulpholane. Solvents that do not participate in protolytic reactions, i.e. do not donate or accept a proton, are usually chemically inert, such as benzene, chlorobenzene, chloroform, tetrachloromethane, etc.

The molecules of amphiprotic solvents which are the most important will be designated as SH. *Self-ionization* occurs to a small degree in these solvents according to the equation

$$2\text{HS} \rightarrow \text{H}_2\text{S}^+ + \text{S}^- \qquad (1.4.2)$$

This equilibrium is characterized by the *ion product* of the solvent—thermodynamic

$$K_{\text{HS}} = a(\text{H}_2\text{S}^+) \cdot a(\text{S}^-)$$

or apparent

$$K'_{\text{HS}} = [\text{H}_2\text{S}^+] \cdot [\text{S}^-] \qquad (1.4.3)$$

The activity of the solvent molecule HS in a single-component solvent is constant and is included in K_{HS}. The concentration of ions is mostly quite low. For example, self-ionization occurs in water according to the equation $2H_2O \rightarrow H_3O^+ + OH^-$. The conductivity of pure water at 18°C is only $3.8 \times 10^{-8}\ \Omega^{-1}\,\text{cm}^{-1}$, yielding a degree of self-ionization of 1.4×10^{-19}. Thus, one H_3O^+ or OH^- ion is present for every 7.2×10^8 molecules of water. Some values of K_{HS} are listed in Table 1.5 and the temperature dependence of the ion product of water K_W is given in Table 1.6.

The H_2S^+ ion is generally termed a *lyonium ion* and the S^- ion is termed a *lyate ion*. The symbol H_2S^+ (for example H_3O^+, $CH_3COOH_2{}^+$, etc.) refers only to a 'proton solvated by a suitable solvent' and does not express either the degree of solvation (solvation number) or the structure. For example, two water molecules form the lyonium ion H_3O^+, termed the *oxonium* (formerly hydronium or hydroxonium) *ion*, and the lyate ion OH^-, termed the *hydroxide ion*.

The existence of the species, H_3O^+, was substantiated as early as in 1924

Table 1.5 Self-ionization constants of solvents. (According to B. Trémillon)

Solvent	Cation	Anion	pK_{HS}	°C
Acetamide	$CH_3CONH_3^+$	CH_3CONH^-	14.6	100
Acetonitrile	CH_3CNH^+	CH_2CN^-	19.6	25
Acetic acid	$CH_3CO_2H_2^+$	$CH_3CO_2^-$	14.9	25
Hydrofluoric acid	H_2F^+	F^-	10.7	0
Formic acid	$HCO_2H_2^+$	HCO_2^-	6.2	25
Hydrogen sulphide	H_3S^{2+}	HS^-	32.6	−78
Sulphuric acid	$H_3SO_4^+$	HSO_4^-	2.9	25
Ammonia	NH_4^+	HN_2^-	27.7	25
			32	−60
Dimethylsulphoxide	$C_2H_6SOH^+$	$C_2H_5SO^-$	33.3	25
Water	H_3O^+	HO^-	14.0	25
Ethanol	$C_2H_5OH_2^+$	$C_2H_5O^-$	19.0	25
Ethanolamine	$^+H_3NC_2H_4OH$	$H_2NC_2H_4O^-$	5.5	25
Formamide	$HCONH_3^+$	$HCONH^-$	16.8	25
Hydrazine	$N_2H_5^+$	$N_2H_3^-$	13	25
Methanol	$CH_3OH_2^+$	CH_3O^-	16.7	25

by M. Volmer by the isomorphism of the monohydrate, $HClO_4{\cdot}H_2O$, with NH_4ClO_4. The species present in the crystal obviously was $H_3O^+{\cdot}ClO_4^-$. An exact proof follows from the NMR spectra: the monohydrates of the molecules of nitric, perchloric and sulphuric acids contain protons placed in the apices of equilateral triangles while other structures would give quite different spectra. The H_3O^+ structure is hydrated by further three water molecules, so that the most stable species are the $H_9O_4^+$ and $H_7O_4^-$ ions:

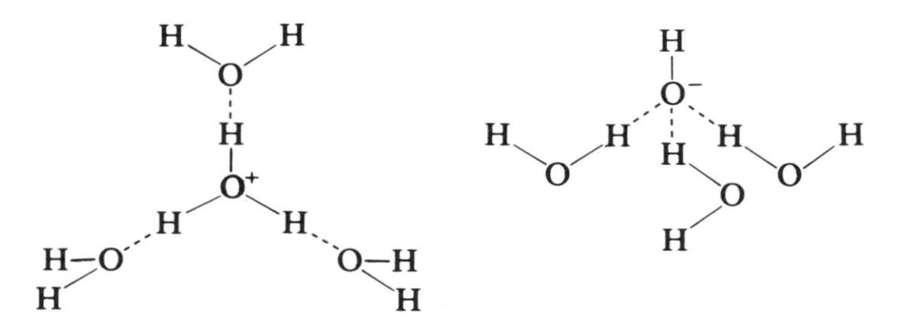

However, the symbols H_3O^+ (often simply H^+) and OH^- will be retained in this book.

Simple self-ionization cannot be assumed *a priori*. The situation is sometimes more complicated, e.g.

$$3HClO_4 \rightarrow Cl_2O_7 + H_3O^+ + ClO_4^-$$

In mixed solvents, the *overall ion product* is defined. For example, for a

Table 1.6 Ion product of water at different temperatures. (According to B. E. Conway)

t (°C)	pK_W	t (°C)	pK_W
0	14.9435	35	13.6801
5	14.7338	40	13.5348
10	14.5346	45	13.3960
15	14.3463	50	13.2617
20	14.1669	55	13.1369
25	13.9965	60	13.0171
30	13.8330		

binary mixture of solvent HS with water, this product is

$$K'_{HS,W} = ([H_2S^+] + [H_3O^+])([S^-] + [OH^-]) \tag{1.4.4}$$

and the equilibria

$$\begin{aligned} HS + OH^- &\rightleftarrows S^- + H_2O && \text{(I)} \\ HS + H_3O^+ &\rightleftarrows H_2S^+ + H_2O && \text{(II)} \\ 2H_2O &\rightleftarrows H_3O^+ + OH^- && \text{(III)} \\ 2HS &\rightleftarrows H_2S^+ + S^- && \text{(IV)} \end{aligned} \tag{1.4.5}$$

are established. The constant of the first reaction K_I indicates whether water is a stronger acid than the solvent HS, constant K_{II} indicates whether water is a stronger base than solvent HS and constants K_{III} and K_{IV} indicate the degree of self-ionization of the two liquids in the mixture. The ratios of the concentrations of the ions

$$\frac{[S^-]}{[OH^-]} = K'_I \frac{[HS]}{[H_2O]}, \qquad \frac{[H_2S^+]}{[H_3O^+]} = K'_{II} \frac{[HS]}{[H_2O]} \tag{1.4.6}$$

thus depend on the relative strengths of HS and water as acids and bases and on the composition of the mixture. Various situations can occur in this system. If the acid–base properties of water and HS are roughly identical, then definition (1.4.4) is valid; however, because the concentrations of the molecules H_2O and HS are given by the composition of the mixture, the dependence of $K_{HS,W}$ on the composition exhibits a maximum. If, on the other hand, solvent HS is almost inert, K_I and K_{II} are small, the concentrations of H_2S^+ and S^- can be neglected in Eq. (1.4.4) and the expression becomes that for K_W of pure water. However, the concentration of water molecules appearing in the equation for K_W decreases with increasing concentration of the solvent HS (which acts as an inert diluent) so that K_W decreases. If water is a sufficiently stronger acid than HS but HS is a stronger base than water, then $K_I < 1$ while $K_{II} > 1$, $[H_3O^+]$ can be neglected compared to $[H_2S^+]$ and $[S^-]$ can be neglected compared to

$[OH^-]$, so that $K_{HS,W} = [H_2S^+][OH^-]$. This situation occurs for a mixture of hydrazine and water, where the equilibrium $H_2O + N_2H_4 \rightleftarrows OH^- + N_2H_5{}^+$ is established. Other combinations of acid and base strengths can be discussed in similar terms.

1.4.3 *Solutions of acids and bases*

In amphiprotic solvents *protolysis* of acids and bases takes place. This reaction is also termed *solvolysis,* classically *dissociation.* The degree of solvolysis depends on the nature of both the dissolved acid or base and on the solvent.

The acid HA reacts with the solvent HS according to the reaction

$$HA + HS \rightleftarrows A^- + H_2S^+, \qquad (1.4.7)$$

with

$$K_A = \frac{a(A^-)a(H_2S^+)}{a(HA)}, \qquad K'_A = \frac{[A^-][H_2S^+]}{[HA]} \qquad (1.4.8)$$

for aqueous solutions

$$K_A = \frac{a(A^-)a(H_3^+O)}{a[HA]}, \qquad K'_A = \frac{[A^-][H_3O^+]}{[HA]} \qquad (1.4.9)$$

where K_A is termed the *true* (*thermodynamic*) *dissociation constant* while K'_A *apparent dissociation constant* of the acid. This constant gives the strength of the acid in the given solvent: the larger the value of K_A or K'_A (the smaller pK_A and pK'_A), the stronger the acid. For the degree of dissociation α [Eq. (1.1.31)] directly applies when setting $[B^+] = [H_2S^+]([H_3O^+]$ in aqueous solutions).

The generally accepted measure of the acidity of any solution is the logarithm of the activity of solvated protons times -1,

$$pH_{HS} = -\log a(H_2S^+) \approx -\log [H_2S^+] \qquad (1.4.10)$$

giving for aqueous solutions the important definition

$$pH = -\log a(H_3O^+) \simeq \log [H_3O^+] \qquad (1.4.11)$$

The subsequent text will be restricted to aqueous solutions and concentrations will be used instead of activities. Thus, the neutral point is given by the condition $[H_2O^+] = [OH^-] = K'^{1/2}_w$,

$$pH = \tfrac{1}{2}pK'_w \qquad (1.4.12)$$

A weak dibasic acid H_2A dissociates in two steps:

$$\begin{aligned} H_2A + H_2O &\rightleftarrows HA^- + H_3O^+ \\ HA^- + H_2O &\rightleftarrows A^{2-} + H_3O^+ \end{aligned} \qquad (1.4.13)$$

with dissociation constants K_1' and K_2' and degrees of dissociation α_1 and α_2, for which

$$K_1' = \frac{[HA^-][H_3O^+]}{[H_2A]} = \frac{c\alpha_1^2(1-\alpha_2^2)}{1-\alpha_1}$$
$$K_2' = \frac{[A^{2-}][H_3O^+]}{[HA^-]} = \frac{c\alpha_1\alpha_2(1+\alpha_2)}{1-\alpha_2} \tag{1.4.14}$$

(simple consideration reveals that $[H_2A] = c(1-\alpha_1)$, $[HA^-] = c\alpha_1(1-\alpha_2)$, $[A^{2-}] = c\alpha_1\alpha_2$, $[H_3O^+] = c\alpha_1 + c\alpha_1\alpha_2$, where $c = [H_2A] + [HA^-] + [A^{2-}]$). For a weak acid ($K_1'$, $K_2' \ll 1$) and not too dilute solution (α_1, $\alpha_2 \ll 1$), the last pair of equations simplifies to yield $K_1' \approx c\alpha_1^2$ and $K_2' \approx c\alpha_1\alpha_2$, so that

$$[H_3O^+] = c\alpha_1 + c\alpha_1\alpha_2 \approx (K_1'c)^{1/2} + K_2' \tag{1.4.15}$$

In most systems, $K_2' < K_1'$. Otherwise (e.g. the first two dissociation steps for ethylenediaminetetraacetic acid—EDTA) the first hydrogen ion capable of dissociation is stabilized by the presence of the second one and can dissociate only when the second begins to dissociate. As a result both hydrogen ions dissociate in one step.

As the second dissociation constant is often much smaller than the first, $\alpha_2 \ll \alpha_1 \ll 1$ and $[H_2S^+] \approx c\alpha_1 = (K_1'c)^{1/2}$. Thus, the acid can be considered as monobasic with the dissociation constant K_1'.

Solvolysis of base B

$$B + HS = HB^+ + S^- \tag{1.4.16}$$

with a dissociation constant

$$K_B' = \frac{[HB^+][S^-]}{[B]} \tag{1.4.17}$$

is usually formulated as dissociation of the cation of the base HB^+ so that Eqs (1.4.8) and (1.4.9) apply for $BH^+ = HA$.

Obviously, for a conjugated acid–base pair

$$K_A'(BH^+)K_B' = K_{SH}, \quad \text{or} \quad pK_A' + pK_B' = pK_{SH} \tag{1.4.18}$$

(for aqueous solutions at 25°C $pK_A' + pK_B' = 14$).

The above treatment of moderately dilute acids and dibasic acids can be used for analogous cases of bases. Table 1.7 lists examples of dissociation constants of bases in aqueous solutions.

The concentration of lyonium or lyate ions cannot be increased arbitrarily. A reasonable limit for a 'dilute solution' can be considered to be a molality of about 1 mol kg^{-1} (e.g. in water, one mole of solute per 55 moles of water). When the ratio is greater than 1:55, it is difficult from a thermodynamic point of view to consider the system as a solution, but it should rather be viewed as a mixture. This situation becomes increasingly

Table 1.7 Dissociation constants of weak acids and bases at 25°C. (From *CRC Handbook of Chemistry and Physics*)

Acid	$pK_A(HA)$	Base	$pK_A(BH^+)$
Acetic acid	4.75	Acetamide	0.63
Benzoic acid	4.19	Ammonia	4.75
Boric acid (20°C)	9.14	Aniline	4.63
Chloroacetic acid	2.85	Dimethylamine	10.73
Formic acid (20°C)	3.75	Hydrazine	5.77
Phenol (20°C)	9.89	Imidazol	6.95
o-Phosphoric acid (first step)	2.12	Methylamine	10.66
		Pyridine	5.25
o-Phosphoric acid (second step)	7.21	Trimethylamine	9.81
o-Phosphoric acid (third step)	12.67		

less favourable when the molecular weight of the solvent increases. The range of reasonably useful values of pH_{HS} in a given solvent is limited on the acid side by the value $pH_{HS} = 0$ ($[H_2S^+]) = 1$) and on the basic side by the value $pH_{HS} = pK'_{HS}$ (when $[S^-] = 1$); the neutral point then lies in the middle where $pH_{HS} = \frac{1}{2}pK'_{HS}$.

The strongest acid that can be introduced into solution is the H_2S^+ ion and the strongest base is the S^- ion. However, these ions can be added to the solution only in the form of a molecule. There are three main possibilities. If a strong acid is dissolved in solution, it quantitatively yields protons that are immediately solvated to form H_2S^+ ions. The corresponding conjugate base cannot recombine with protons in a given solvent (e.g. introduction of HCl into solution yields Cl^- ions) since it has no basic properties in this solvent. The value of pH_{HS} is given directly by the concentration of the dissolved strong acid c_A according to the simple relationship $pH_{HS} = -\log c_A$. If a strong base is dissolved whose conjugate acid form cannot dissociate under the given conditions as bases such as KOH in water or CH_3COOK in glacial acetic acid (the K^+ cation has no dissociable hydrogen), the pH_{HS} value is again given by the concentration of the dissolved strong base, in this case according to the relationship $pH_{HS} = pK'_{HS} + \log c_B$.

After discussing the general properties of an amphiprotic solvent several examples will now be considered. As the Arrhenius concept took only aqueous solutions into account it became a seemingly unambiguous basis for the classification of individual chemical substances as strong or weak acids or bases. However, the designation of a given substance as an acid or base, weak or strong, lacks a logical foundation. Of decisive importance is the behaviour of a given substance with respect to a given solvent, where it can act as a strong or weak acid or base. This generalization follows from the concepts of Brønsted and Lowry.

Water. The pH range that can reasonably be used extends from 0 to 14, with the neutral point at pH 7. A solution with $pH < 7$ is acidic and with $pH > 7$ is basic. The commonest strong acids are $HClO_4$, H_2SO_4, HCl and HNO_3; strong bases include alkali metal and tetraalkylammonium hydroxides.

Methanol and ethanol. The self-ionization reaction is $2ROH \rightleftarrows ROH_2^+ + RO^-$. The pH_{Me} range is about 9.5 and the pH_{Et} about 8.3. $HClO_4$ continues to be a strong acid and HCl becomes a 'medium strong' acid ($pK_{A,Me} = 1.2$, $pK_{A,Et} = 2.1$). Acetic acid has the values of $pK_{A,Me} = 9.7$ and $pK_{A,Et} = 10.4$, i.e. it is much weaker than in water. Strong bases include the alkali metal alcoholates, as they introduce RO^- anions, the conjugate base of alcohol. Water is a weak base in ethanol ($pK_A = 0.3$).

Liquid ammonia. Self-ionization occurs according to the equation $2NH_3 \rightleftarrows NH_4^+ + NH_2^-$ and the pH range at −60°C is 32 units. Acids that are weak in water, such as acetic acid, are strong in ammonia and acids that are very weak in water become medium strong in ammonia. Strong bases include potassium amide, introducing NH_2^- ions, while hydroxides are weak bases (a KOH solution with a concentration of 10^{-2} mol dm^{-3} has a pH_{NH_3} value of about 24.5).

If an amphiprotic solvent contains an acid and base that are neither mutually conjugate nor are conjugated with the solvent, a protolytic reaction occurs between these dissolved components. Four possible situations can arise. If both the acid and base are strong, then neutralization occurs between the lyonium ions and the lyate ions. If the acid is weak and the base strong, the acid reacts with the lyate ions produced by the strong base. The opposite case is analogous. A weak acid and a weak base exchange protons:

Strong acid + strong base	$H_2S^+ + S^- \rightarrow 2HS$	(I)
Weak acid + strong base	$HA + S^- \rightarrow A^- + HS$	(II)
Strong acid + weak base	$H_2S^+ + B \rightarrow HB^+ + HS$	(III)
Weak acid + weak base	$HA + B \rightarrow HB^+ + A^-$	(IV)

(1.4.19)

(Except for the last, these reactions are used in titrimetric neutralization analysis.) Reactions (II) to (IV) can also proceed in the opposite direction. This will be demonstrated on the well-known example of *salt hydrolysis*.

Reaction (II) could be the neutralization of acetic acid by potassium hydroxide, yielding potassium acetate which can be isolated in the crystalline state. On dissolution in water the K^+ cation is only hydrated in solution but does not participate in a protolytic reaction. In this way, the weak base CH_3COO^- is quantitatively introduced into solution in the absence of an equilibrium amount of the conjugate weak acid CH_3COOH. Thus

CH_3COO^- reacts with water to form an equilibrium concentration of CH_3COOH molecules, corresponding to the K'_A value for acetic acid in water:

$$CH_3COO^- + H_2O \rightleftarrows CH_3COOH + OH^-$$

$$K'_H = \frac{[CH_3COOH][OH^-]}{[CH_3COO^-]} = \frac{K'_W}{K'_A(CH_3COOH)} = \frac{\alpha_H^2 c}{1 - \alpha_H} \quad (1.4.20)$$

(the degree of hydrolysis of the salt present at the concentration c is $\alpha_H = [OH^-]/c = [CH_3COOH]/c$, $[CH_3COO^-]/c = 1 - \alpha_H$.) Because of this hydrolysis reaction, the solution becomes alkaline. For small degrees of hydrolysis ($\alpha_H \ll 1$),

$$[H_3O^+] = \frac{K'_W}{[OH^-]} \approx \left[\frac{K'_W K'_A(CH_3COOH)}{c}\right]^{1/2} \quad (1.4.21)$$

The concentration of oxonium ions increases with increasing strength of the acid and the pH decreases.

The case of the salt of a weak base and strong acid is treated in an analogous way. With decreasing strength of the base, the concentration of oxonium ions increases and the pH decreases.

In the case of the salt of a weak acid and a weak base HAB, e.g. ammonium acetate, hydrolysis of both the ions occurs:

$$2HAB + 2H_2O \rightleftarrows A^- + HA + BH^+ + B + H_3O^+ + OH^-$$

For a high degree of hydrolysis $[HA] \approx c$ and $[B] \approx c$, where c is the overall concentration of the salt. The electroneutrality condition gives $[A^-] + [OH^-] = [H_3O^+] + [BH^+]$. Here the concentrations $[A^-]$ and $[BH^+]$ are expressed by means of the acid dissociation constants $K'_A(HA)$ and $K'_A(BH^+)$ and the oxonium concentration. Finally, we have

$$[H_3O^+] = \sqrt{\frac{[K'_A(HA)c + K'_W]K'_A(BH^+)}{K'_A(BH^+) + c}}$$

$$\approx \sqrt{K'_A(HA) \cdot K'_A(BH^+)}$$

$$pH = \tfrac{1}{2}[pK'_A(HA) + pK'_A(BH^+)] \quad (1.4.22)$$

Thus, the relative strengths of the acid (small $pK'_A(HA)$) and of the base (large $pK'_A(BH^+)$) decide whether the solution becomes acidic or alkaline.

Sometimes, hydrolysis leads to complete transformation of the dissolved substance, for example $BCl_3 + H_2O \rightleftarrows B(OH)_3 + 3H_3O^+ + 3Cl^-$ (the solution is acidified by this reaction). Similarly, ammonolysis can occur, for example $BCl_3 + 6NH_3 \rightleftarrows B(NH_2)_3 + 3NH_4{}^+ + 3Cl^-$ (the solution is again acidified and excess lyonium ions $NH_4{}^+$ are formed); fluorolysis can also take place, for example $UO_2F_2 + 6HF \rightleftarrows UF_6 + 3H_3O^+ + 2F^-$ (the solution becomes alkaline as a result of the formation of lyate ions, F^-).

Not only addition of acidic species supplying the ions, H_2S^+, but also of acceptors of S^-, typically of metal ions M^{n+}, results in a shift of the protolytic equilibrium towards the H_2S^+ ions, connected with *consecutive complex formation,*

$$\begin{aligned} M^{n+} + 2HS &\rightleftarrows MS^{(n-1)+} + H_2S^+ \\ MS^{(n-1)+} + 2HS &\rightleftarrows MS_2^{(n-2)+} + H_2S^+ \\ &\text{etc.} \end{aligned} \tag{1.4.23}$$

Thus, the cation acts as a polybasic acid. Another case of consecutive complex formation occurs if the solution contains a complexing agent, X,

$$\begin{aligned} M^{n+} + X &\rightleftarrows MX^{n+} \\ MX^{n+} + X &\rightleftarrows MX_2^{n+} \\ &\cdots\cdots \\ MX_{r-1}^{n+} + X &\rightleftarrows MX_r^{n+} \end{aligned} \tag{1.4.24}$$

with consecutive stability constants defined as

$$K'_{MX_i} = \frac{[MX_i]}{[MX_{i-1}][X]} \tag{1.4.25}$$

for $i = 1, \ldots, r$ (the charge of the species is not indicated). The more stable the complex, the larger is the stability constant. Some examples are given in Table 1.8.

Solutions formed by incomplete neutralization of a weak acid by a strong base or a weak base by a strong acid (cases (II) and (III) in scheme 1.4.19) are called *buffers* and have interesting properties which are particularly important in many fields of chemistry and biology. Aqueous solutions will

Table 1.8 Consecutive stability constants, expressed as $\log K_{MX_i}$, of complexes of ammonia (A), ethylene diamine (B), diethylenetriamine (C) and the anion of ethylenediaminetetraacetic acid (D^{4-}) at 20°C and 0.1 M KNO_3 as indifferent electrolyte. (According to J. Bjerrum and G. Schwarzenbach)

	Mn^{2+}	Fe^{2+}	Co^{2+}	Ni^{2+}	Cu^{2+}	Zn^{2+}	Cd^{2+}
MA	—	—	2.05	2.75	4.13	2.27	2.60
MA_2	—	—	1.57	2.20	3.48	2.34	2.05
MA_3	—	—	0.99	1.69	2.87	2.40	1.39
MA_4	—	—	0.70	1.15	2.11	2.05	0.88
MA_5	—	—	0.12	0.71	—	—	−0.32
MA_6	—	—	−0.14	−0.01	—	—	−1.66
MB	2.8	4.4	6.0	7.9	10.8	6.0	5.7
MB_2	2.1	3.3	4.9	6.6	9.4	5.2	4.6
MB_3	0.9	2.0	3.2	4.7	0.1	1.8	2.1
MC	—	—	8.1	10.7	16.0	8.9	8.4
MC_2	—	—	6.0	8.2	5.3	5.5	5.4
MD^{2-}	14.02	14.33	16.31	18.62	18.80	16.50	16.46

be considered by way of illustration. Consider an aqueous solution of a weak acid HA of concentration $s = [HA] + [A^-]$, to which potassium hydroxide is added so that its final concentration in solution is $b = [K^+]$. The K^+ ions do not participate in the protolytic process and OH^- ions present react with the acid HA to form water molecules and the conjugate base A^- (classically viewed, the salt KA). This reaction is practically quantitative, because OH^- ions are a strong base in aqueous medium. Thus we can set $[A^-] \approx b$, i.e. the A^- anions formed by dissociation of the acid HA can be neglected. Further, $[HA] = s - [A^-] \approx s - b$, so that $K'_A = [H_3O^+]b/(s - b)$, which yields

$$\mathrm{pH} = \mathrm{p}K'_A(\mathrm{HA}) + \log \frac{b}{s - b} \tag{1.4.26}$$

(see Fig. 1.9).

The case of a buffer consisting of a weak base and its acidic form (for example, NH_3 and NH_4^+) is treated in an analogous way. Equations of the type of (1.4.26) are sometimes called the *Henderson–Hasselbalch equations*.

Buffering properties are exhibited by mixtures of weak acids (or bases) and their salts with a composition such that the concentration of the weak component and the salt are not too far apart. Thus, for $b = \frac{1}{2}s$, $\mathrm{pH} = \mathrm{p}K'_A(\mathrm{HA})$. The addition of a quite large amount of strong hydroxide or acid to such a mixture results in a small pH change in the solution (see Fig. 1.9).

The buffering ability of a buffer solution is characterized by its buffering capacity β, which is a differential quotient indicating the change of

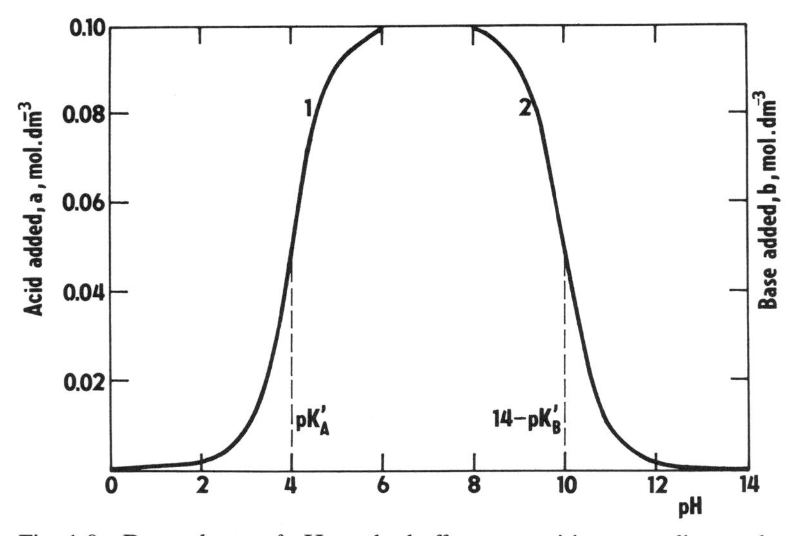

Fig. 1.9 Dependence of pH on the buffer composition according to the Henderson–Hasselbalch equation: (1) acidic buffer (Eq. 1.4.26); (2) basic buffer. Calculation for $K'_A = K'_B = 10^{-4}$, $s = 0.1\ \mathrm{mol \cdot dm^{-3}}$

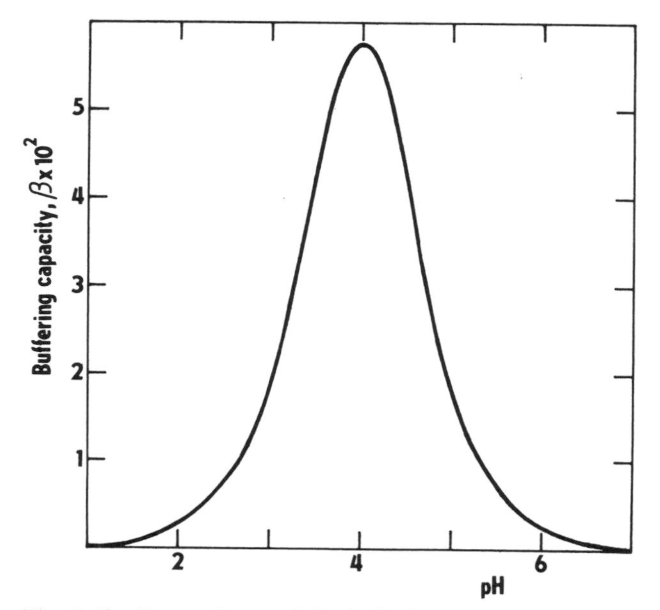

Fig. 1.10 Dependence of the buffering capacity β on the pH of an acid buffer. Calculation according to Eqs (1.4.27) for $K'_A = 10^{-4}$, $s = 0.01$ mol · dm^{-3}

concentration of added strong acid or base in the solution, which is necessary for a given change of pH (Fig. 1.10). On differentiation of (1.4.26) we obtain

$$\beta = \frac{db}{dpH} = b\left(1 - \frac{b}{s}\right) \ln 10 \qquad (1.4.27)$$

Thus, the buffering capacity depends on the composition of the buffer, i.e. on the concentration of the salt a or b. The maximum value found by differentiation of Eq. (1.4.27) with respect to a corresponds, for an acidic buffer, to $b = \frac{1}{2}s$.

For practical purposes, a number of buffer mixtures have been proposed that are useful for various pH ranges. Procedures for their preparation and requirements on purity and definition of the chemicals used are given in laboratory manuals and tables.

If the solvent is not protogenic but protophilic (acetone, dioxan, tetrahydrofuran, dimethylformamide, etc.), self-ionization obviously does not occur. Consequently, the dissolved acids are dissociated to a greater or lesser degree but dissolved bases do not undergo protolysis. Thus, there can exist only strong acids but no strong bases in these solvents. The pH is not defined for a solution that does not contain a dissolved acid (i.e. in the pure solvent or in the solution of a base). The pK_A value can be defined but not the pK_B value, and there is no neutral point. If the solvent is neither protogenic nor protophilic, then neither dissolved acids nor dissolved bases are protolysed. Protolytic reactions can occur in such a solvent, but only between dissolved acids and bases (that are not mutually conjugate).

If the dielectric constant of an amphiprotic solvent is small, protolytic reactions are complicated by the formation of ion pairs. Acetic acid is often given as an example (denoted here as AcOH, with a relative dielectric constant of 6.2). In this solvent, a dissolved strong acid, perchloric acid, is completely dissociated but the ions produced partly form ion pairs, so that the concentration of solvated protons $AcOH_2^+$ and perchlorate anions is smaller than would correspond to a strong acid (their concentrations correspond to an acid with a pK'_A of about 4.85). A weak acid in acetic acid medium, for example HCl, is even less dissociated than would correspond to its dissociation constant in the absence of ion-pair formation. The equilibrium

$$HCl + AcOH \rightleftarrows AcOH_2^+Cl^- \rightleftarrows AcOH_2^+ + Cl^- \qquad (1.4.28)$$

is established. If the constants

$$\begin{aligned} K'_i &= \frac{[AcOH_2^+Cl^-]}{[HCl]} \\ K'_{ass} &= \frac{[AcOH_2^+Cl^-]}{[AcOH_2^+][Cl^-]} \\ K'_A &= \frac{[AcOH_2^+][Cl^-]}{[HCl] + [AcOH_2^+Cl^-]} \end{aligned} \qquad (1.4.29)$$

are introduced, then

$$pK'_A = -pK'_{ass} + \log\left(1 + \frac{1}{K'_i}\right) \qquad (1.4.30)$$

The protolytic reaction between two molecules must be formulated as the formation of the pair

$$B + HA \rightleftarrows BH^+A^- \qquad (1.4.31)$$

that can react as a base with another acid HA_{II} or as an acid with another base B_{II}:

$$\begin{aligned} B_IH^+A_I^- + HA_{II} &\rightleftarrows B_IH^+A_{II}^- + HA_I \\ B_IH^+A_I^- + B_{II} &\rightleftarrows B_{II}H^+A_I^- + B_I \end{aligned} \qquad (1.4.32)$$

Protolytic reactions can also occur in fused salts. The solvent participates in these reactions provided that at least one of its ions has protogenic and/or protophilic character. An example of a solvent in which the cation is aprotic and the anion protophilic is ethylpyridinium bromide (m.p. 114°C). The acid HA is protolysed in this solvent ($HA + Br^- \rightleftarrows HBr + A^-$). Hydrogen bromide acts as a solvated proton and the acidity is expressed as pHBr = −log [HBr] and can be measured with a hydrogen electrode. In ethylammonium chloride (m.p. 108°C) the cation is protogenic and self-ionization occurs, $C_2H_5NH_3^+ + Cl^- \rightleftarrows C_2H_5NH_2 + HCl$, with the ion product $10^{-8.1}$. Fused alkali-metal hydroxides with amphiprotic OH^- anion are the most important of these solvents; the autoprotolysis follows the equation $2OH^- \rightleftarrows H_2O + O^{2-}$. Here, water is the strongest acid, corresponding to solvated protons, and O^{2-} is the strongest base. Metal cations (added as hydroxides) behave as acids as they remove O^{2-} anions, forming compounds that often precipitate.

So far acid–base reactions of molecules in the ground state of energy have been discussed. The properties of these molecules can change considerably in the electronically excited state (K. Weber, T. Förster, A. Weller, *et al.*). For example, in aqueous solution in the ground state, 2-naphthol has $pK'_A = 9.5$, while the value for the excited state is 2.5. Similarly for phenol, in the ground state, $pK'_A = 10.0$, while the value for the excited state is 3.6. Thus, the strength of acids in these and a number of other cases increases by several orders on excitation. Simultaneously, the pH of a solution calculated from the concentrations of species corresponding to the conjugate pair decreases by several pH units. This effect is termed the 'laser pH jump'. The effect is studied by irradiating the solution with a short intense pulse (usually provided by a laser—hence the name) and fast fluorescence or absorption spectroscopy is employed to determine the concentrations of the corresponding excited acid species and the conjugated base. If the spectroscopy is sufficiently resolved in time (with resolution of the order of nanoseconds to picoseconds), the kinetics of the exchange of a proton between the dissolved conjugate pair and the solvent can be studied. It has been found that this is a very fast reaction (with a rate constant of the order of $10^9\ s^{-1}$).

1.4.4 *Generalization of the concept of acids and bases*

The most extensive generalization of the concept of acids and bases was that proposed in 1938 by G. N. Lewis, who concluded that the Brønsted concept of limiting the group of acids to substances with a dissociable hydrogen prevents the general understanding of the chemistry of acids, in the same way as the early definition of oxidants as oxygen-containing substances impeded the general understanding of oxidation–reduction reactions. In view of this, he related the acidic and basic character of substances to their electron structure by the following definition:

A base is a substance capable of acting as an electron-pair donor.
An acid is a substance capable of acting as an electron-pair acceptor.

According to this definition the group of bases practically includes just about the same substances as the Brønsted conception, but with a few additions such as the inert gases (e.g. argon reacts as a base with boron trifluoride). On the other hand, the category of acids is much wider than under the Brønsted definition, as it includes species such as H^+, HCl, $AlCl_3$, $SnCl_3$, SO_2, SO_3, O, Ag^+, Be^{2+}, Cu^{2+}, etc.

The Lewis acids that also fulfil the Brønsted definition form a special group of protonic acids and can be formally considered to be the products of neutralization of a proton by a base (for example Cl^-).

The reaction between an acid and a base produces a coordination compound (complex). In contrast to former theories, this is more an addition than an exchange reaction. On the other hand, an exchange reaction in the Lewis sense is the result of competition between two acceptors for a single donor or between two donors for a single acceptor. Thus the typical Brønsted acid–base reaction $HA + B \rightleftarrows A + HB$ is the result of competition between two bases :A and :B for the acceptor H^+. The reactions $(AlCl_3 \leftarrow A) + :B \rightleftarrows (AlCl_3 \leftarrow B) + :A$ are analogous. The classical reactions depend on replacement of one base (B_I) by another (B_{II}):

$$\begin{array}{rll} AB_I + B_{II} \rightarrow AB_{II} + B_I & & \\ HCl + NH_3 \rightarrow NH_4^+ + Cl^- & \text{Salt formation} & \\ HCl + H_2O \rightarrow H_3O^+ + Cl^- & \text{Dissociation} & (1.4.33) \\ H_2O + NH_3 \rightarrow NH_4^+ + OH^- & \text{Dissociation} & \\ H_2O + CH_3COO^- \rightarrow CH_3COOH + OH^- & \text{Hydrolysis} & \end{array}$$

The Lewis concept permits inclusion of acids and oxidants in a single group of *electrophilic* substances and bases and reductants in a single group of *nucleophilic* substances. The only difference between acid–base and redox reactions lies in the nature of the bond formed.

However, a wide generalization usually brings a number of complications. The reactions between acids and bases have become so specific that they evade mutual comparison and arrangement in a series according to their strengths. Consequently, Lewis complemented his original definition by further, purely phenomenological definitions that are not in any way related to the theory of the structure of molecules and are based only on the behaviour of substances during chemical reactions:

Bases are substances that (similar to OH^- ions) neutralize hydrogen ions or some other acid.

Acids are substances that (similar to H^+ ions) neutralize hydroxide ions or other bases.

Acids and bases have the following properties:

(a) The mutual neutralization reaction is fast and has low activation energy.
(b) Species of the same type can exchange with one another (e.g. one base for another one).
(c) They can be titrated with the help of acid–base indicators.
(d) They catalyse chemical reactions.

These complementary definitions, however, do not solve the problems, as some of substances exhibit only one of the features required, whilst lacking others, and thus their placing in the acid or base group is not certain.

The concept of 'hard' and 'soft' acids and bases ('HSAB') should also be mentioned here. This is not a new theory of acids and bases but represents a useful classification of Lewis acids and bases from the point of view of their reactivity, as introduced by R. G. Pearson.

Hard acids (HA) have low polarizability and small dimensions, higher oxidation numbers and the 'hardness' increases with increasing oxidation number; they bind bases primarily through ionic bonds. Typical examples are H^+, Na^+, Hg^{2+}, Ca^{2+}, Sn^{2+}, VO^{2+}, VO_2^{2+}, $(CH_3)_2Sn^{2+}$, $Al(CH_3)_3$, I^{7+}, I^{5+}, Cl^{7+}, CO_2, R_3C^+, SO_3.

Soft acids (SA) are strongly polarizable, have low or zero oxidation numbers, the 'softness' decreases with increasing oxidation number and if groups with a large negative charge are also bonded to the central atom, then the 'softness' increases; they bind bases primarily through covalent bonds. Typical examples are Cu^+, Ag^+, Hg^+, Cs^+, Pd^{2+}, Cd^{2+}, Hg^{2+}, Tl^{3+}, $Tl(CH_3)_3$, BH_3, I^+, Br^+, HO_2^+, I_2, Br_2, O, Cl, Br, I, metal atoms, metals in the solid phase.

Hard bases (HB) have low polarizability and strong Brønsted basicity; in the V, VI and VII groups of the periodic table, the first atom is always the hardest and the hardness decreases with increasing atomic number in the group. Typical examples are H_2O, OH^-, F^-, O^{2-}, CH_3COO^-, SO_4^{2-}, Cl^-, CO_3^{2-}, NO_3^-, ROH, RO^-, R_2O, NH_3, RNH_2, N_2H_4.

Soft bases(SB) are strongly polarizable and generally have low Brønsted basicity. Typical examples are R_2S, RSH, RS^-, I^-, R_3P, CN^-, CO, C_2H_4, C_6H_6, H^-, R^-.

Obviously, a number of substances cannot be classified exactly. Borderline acids include Fe^{2+}, Sb^{3+}, Ir^{3+}, $B(CH_3)_3$, SO_2, NO^+, R_3C^+ and $C_6H_5^+$ and borderline bases include $C_6H_5NH_2$, N_3^-, Br^-, NO_2^-, and SO^{2-}.

Simple laws govern the reactions of these substances: hard acids react preferentially with hard bases and soft acids with soft bases. Mixing of 'unsymmetrical' complexes results in the reaction

$$\text{HA—SB} + \text{SA—HB} \rightarrow \text{HA—HB} + \text{SA—SB}. \qquad (1.4.34)$$

For example, the ether RO—C$\lessdot$ (of the HB—SA type) is split by hydrogen iodide (of the HA—SB type) to yield ROH (the HB—HA type) and I—C$\lessdot$ (the SB—SA type).

It is obvious that the Lewis theory and the HSAB concept are very important for the description of the mechanisms of chemical reactions; however, for electrochemistry the Brønsted theory is quite adequate.

1.4.5 *Correlation of the properties of electrolytes in various solvents*

It has already been mentioned several times that the strength of acids and bases is expressed by their dissociation constants K_A and K_B. This is actually true if acids or bases are compared in a single solvent. For example, in aqueous solutions values of K_A of 1.38×10^{-3}, 1.74×10^{-5} and 4.27×10^{-7} have been found for chloroacetic acid, acetic acid and carbonic acid, respectively; consequently, in (dilute) aqueous solutions chloroacetic acid is two orders of magnitude stronger than acetic acid which is again two orders stronger than carbonic acid. These values have no absolute significance. They change on transfer to another solvent and even their order can change. If, for example, the protolytic reaction $CH_3COOH + H_2O \rightleftarrows CH_3COO^- + H_3O^+$ occurred in acetic acid medium, with water as a dissolved base, then the standard state to which the activities and equilibrium constants are related would be changed. Comparison of the acidity or basicity of solvents is just as difficult. For example, alcohols react according to the equation $ROH + H_3O^+ \rightleftarrows ROH_2{}^+ + H_2O$. In dilute solutions (water concentration of less than $1\ \text{mol} \cdot \text{dm}^{-3}$), the constant K' for this reaction in methanol is equal to 8.6×10^{-3} and in ethanol, 3.9×10^{-3}. If these values were also valid in the vicinity of composition $[ROH]/[H_2O] = 1$, then the ratio $[ROH_2{}^+]/[H_3O^+] = K'[ROH]/[H_2O]$ describing the proton distribution between the alcohol and water would be small. Consequently, water would have a greater affinity for protons than alcohol, i.e. would be more basic. The opposite is true.

The effect of the medium (solvent) on the dissolved substance can best be expressed thermodynamically. Consider a solution of a given substance (subscript i) in solvent s and in another solvent r taken as a reference. Water (w) is usually used as a reference solvent. The two solutions are brought to equilibrium (saturated solutions are in equilibrium when each is in equilibrium with the same solid phase—the crystals of the dissolved substance; solutions in completely immiscible solvents are simply brought into contact and distribution equilibrium is established). The thermodynamic equilibrium condition is expressed in terms of equality of the chemical potentials of the dissolved substance in both solutions, $\mu_i(\text{w}) = \mu_i(s)$, whence

$$RT \ln \frac{a_i(\text{w})}{a_i(\text{s})} = \mu_i^0(\text{s}) - \mu_i^0(\text{w}) = \Delta G_{\text{tr},i}^{0,\text{w}\to\text{s}} \tag{1.4.35}$$

where $\mu_i^0(\text{s})$ and $\mu_i^0(\text{w})$ are the standard chemical potentials, identical with the standard Gibbs energies of solvation. The quantity $\Delta G_{\text{tr},i}^{0,\text{w}\to\text{s}}$ is the change in the partial molar Gibbs energy of transfer of the dissolved substance from the standard state in the reference solvent, w, to the standard state in the solvent considered, s, and is termed the *standard Gibbs transfer energy*. A negative value of the standard Gibbs energy of transfer for a proton, $\Delta G_{\text{tr},\text{H}^+}^{0,\text{w}\to\text{s}}$ indicates that a proton is more stable in solvent s and a positive value indicates that it is more stable in water.

Some authors prefer the coefficient $\gamma_{\text{tr},i}^{\text{w}\to\text{s}}$ to the transfer Gibbs energy; this coefficient is defined by the relationship

$$\gamma_{tr,i}^{w \to s} = \exp\left(\frac{\Delta G_{tr,i}^{0,w \to s}}{RT}\right) \tag{1.4.36}$$

This coefficient has various names (medium effect, solvation activity coefficient, etc.); the name recommended by the responsible IUPAC commission is the transfer activity coefficient. In this book the effect of solvation in various solvents will be expressed exclusively in terms of standard Gibbs transfer energies.

In electrolytes, the ionic and the overall Gibbs transfer energies must be distinguished. These quantities are defined in the usual manner. For example, for the most usual type of electrolyte $AB \to A + B$,

$$\Delta G_{tr,AB}^{0,w \to s} = \Delta G_{tr,A}^{0,w \to s} + \Delta G_{tr,B}^{0,w \to s} \tag{1.4.37}$$

The Gibbs energy of an ion changes on transfer from one solvent to another primarily because the electrostatic interaction between ions and the medium changes as a result of the varying dielecric constant of the solvent. This can be expressed roughly by the Born equation (see Eq. 1.2.7),

$$\Delta G_{tr,i}^{0,w \to s} = \frac{z_i^2 e^2 N_A}{8\pi\varepsilon_0 r_i}\left(\frac{1}{D_S} - \frac{1}{D_W}\right) \tag{1.4.38}$$

Mostly, $D_S < D_W$ so that the Gibbs energy increases on transfer from water to some other solvent.

The determination of the standard Gibbs energies of transfer and their importance for potential differences at the boundary between two immiscible electrolyte solutions are described in Sections 3.2.7 and 3.2.8.

1.4.6 *The acidity scale*

An exact description of the acidity of solutions and correlation of the acidity in various solvents is one of the most important problems in the theory of electrolyte solutions. In 1909, S. P. L. Sørensen suggested the logarithmic definition of acidity for aqueous solutions considering, at that time, of course, hydrogen instead of oxonium ions (cf. Eq. (1.4.11))

$$\mathrm{pH} = -\log a(H_3O^+) = -\log [H_3O^+] - \log \gamma(H_3O^+) \tag{1.4.39}$$

This *notional definition of the pH scale* can, however, not be used for practical measurements, as it contains the activity coefficients of the individual ions, $\gamma(H_3O^+)$.

Under these conditions, an *operational definition* must be used, i.e. a definition based on the method by which the quantity pH is measured. Obviously, the operational definition must be formulated as closely as possible to the 'absolute' equation (1.4.39). In this way, a practical scale of pH is obtained that is very similar to the absolute scale but is not identical

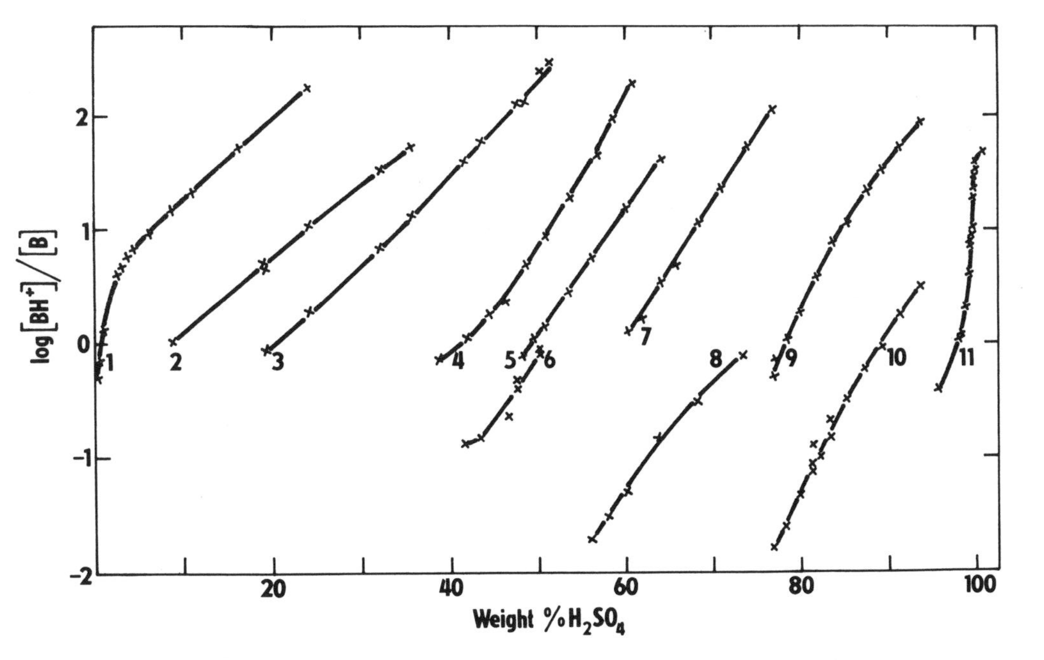

Fig. 1.11 Dissociation ranges of colour indicators for determination of the acidity function H_0 in H_2SO_4–H_2O mixtures: (1) *p*-nitroaniline, (2) *o*-nitroaniline, (3) *p*-chloro-*o*-nitroaniline, (4) *p*-nitrodiphenylamine, (5) 2,4-dichloro-6-nitroaniline, (6) *p*-nitroazobenzene, (7) 2,6-dinitro-4-methylaniline, (8) benzalacetophenone, (9) 6-bromo-2,4-dinitroaniline, (10) anthraquinone, (11) 2,4,6-trinitroaniline. (According to L. P. Hammett and A. J. Deyrup)

with it. This definition has changed somewhat over the years and the definitions have now been unified (*Pure Applied Chemistry,* **57,** 531, 1985). As these definitions are based on pH measurement with the hydrogen electrode they will be discussed in Section 3.3.2. For the same reason the pH scales in non-aqueous solutions will be dealt with in Section 3.2.6.

For solutions of extreme acidity (e.g. the H_2O–H_2SO_4 mixture) the acidity function H_0 was introduced by L. P. Hammett as

$$H_0 = \mathrm{p}K_\mathrm{A}(\mathrm{BH}^+) + \log\frac{c_\mathrm{B}}{c_{\mathrm{BH}^+}} = -\log a_{\mathrm{H}^+} - \log\frac{\gamma_\mathrm{B}}{\gamma_{\mathrm{BH}^+}} \tag{1.4.40}$$

Here, $K_\mathrm{A}(\mathrm{BH}^+)$ denotes the acid dissociation constant of a basic coloured indicator B (for the theory, see Section 1.4.7).

If a given indicator is half dissociated, the value of H_0 is equal to the decadic logarithm of its acidity constant times −1. For practical measurements an indicator may be used only in the range of its colour transition, i.e. at most in the range of the ratio $c_\mathrm{B}/c_{\mathrm{BH}^+}$ from 10 to 10^{-1}. For a larger interval of H_0 values several indicators have to be used.

In view of the term containing activity coefficients, the acidity function depends on the ionic type of the indicator. The definition of H_0 is combined with the assumption that the ratio $\gamma_\mathrm{B}/\gamma_{\mathrm{BH}^+}$ is constant for all indicators of the same charge type (in the present case the base is electroneutral; hence the index 0 in H_0). Thus, the acidity function does not depend on each individual indicator but on the series of indicators.

The acidity function is determined by successive use of a range of indicators. Hammett and Deyrup started with *p*-nitroaniline, the $\mathrm{p}K_{\mathrm{A}_1}$ of which in a dilute aqueous solution is 1.11 (the solvolysis constant has been identified with the acidity constant). Since in a dilute aqueous solution, $\gamma_\mathrm{B} = \gamma_{\mathrm{BH}^+} \approx 1$, the acidity function for the aqueous media goes over to the pH scale. By means of *p*-nitroaniline, the acidity constant of another somewhat more acidic indicator is obtained under conditions such that both forms of each indicator are present at measurable concentrations. Then H_0 as well as the $\mathrm{p}K_{\mathrm{A}_2}$ of the other indicator is determined by using Eq. (1.4.40). By means of this indicator, values of H_0 not accessible with *p*-nitroaniline may be reached. The H_0 scale is further extended by using a third indicator and its $\mathrm{p}K_{\mathrm{A}_3}$ is determined in the same way (see Fig. 1.11). The concentration ratios are determined photometrically in the visual or ultraviolet region. Figure 1.12 shows the dependence of H_0 on the composition of the H_2SO_4–H_2O mixture and was obtained as indicated above.

1.4.7 *Acid–base indicators*

The interest in colour indicators has recently increased as they are used for the direct determination of pH (acid–base indicators) and free calcium ions (fluorescent derivatives based on the calcium chelator EGTA as metallochromic indicators) in biological systems at cellular level.

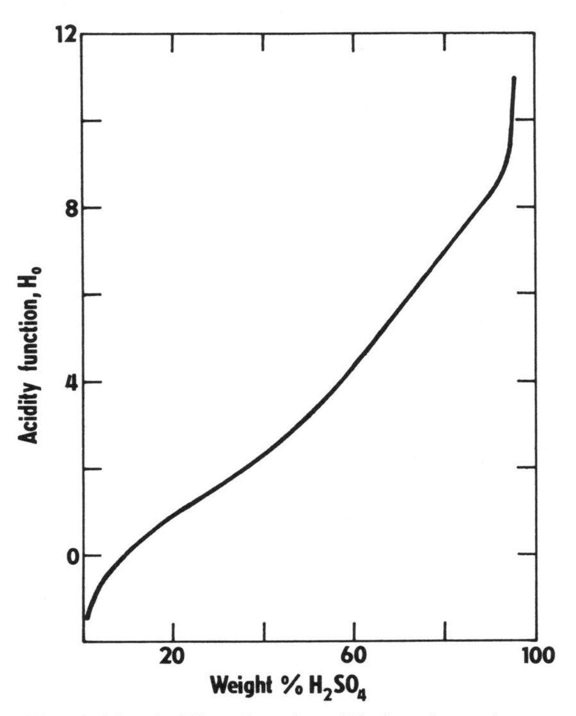

Fig. 1.12 Acidity function H_0 for the mixture H_2SO_4–H_2O as a function of the composition. (According to L. P. Hammett and M. A. Paul)

While the role of acid–base indicators in acidimetric titrations has been surpassed by more advanced, automatic, potentiometric methods various metallochromic indicators are still in use in complexometric analysis. Here we shall deal only with *acid–base indicators*.

According to the early view of Ostwald, acid–base indicators are weak acids or bases, the undissociated form of which differs in colour from the ionic form. For example, the molecule of an indicator HI dissociates in water according to the equation

$$HI + H_2O = H_3O^+ + I^- \tag{1.4.41}$$

The equilibrium constant of this reaction in a dilute solution is

$$K_I = \frac{[H_3O^+][I^-]}{[HI]} \tag{1.4.42}$$

whence

$$pH = pK_I + \log \frac{[I^+]}{[HI]} \tag{1.4.43}$$

showing the dependence of the ratio $[I^-]/[HI]$ on pH.

In a later conception due to Hantzsch, an indicator possesses two tautomeric forms differing in colour, and at least one form functions as an acid or base. If both tautomeric forms are capable of dissociation their ions also differ in colour. However, the ions and molecules of the same form show identical light absorption because such a small variation of the structure of a molecule as the dissociation of a proton cannot cause a large change in molecular property such as light absorption. For example, phenolphthalein does not absorb visible light in the acid and neutral region, but at pH > 8.2 one of its hydroxyl groups becomes ionized and, at the same time, a tautomeric change takes place:

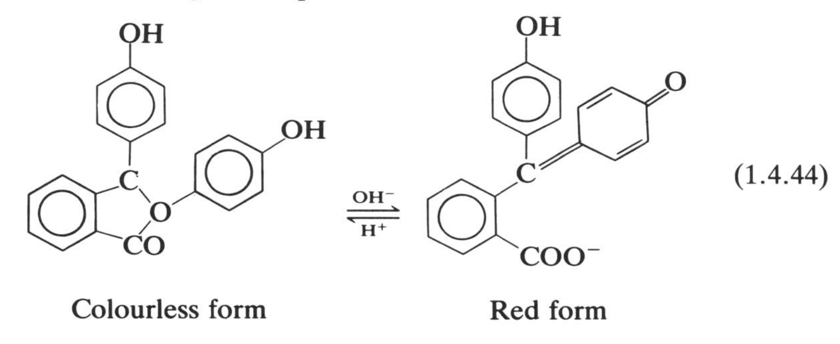

(1.4.44)

Colourless form   Red form

The quinoid group

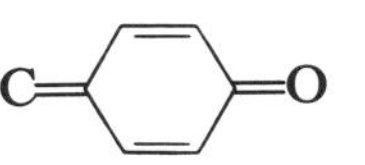

is a chromophore which absorbs light radiation in the whole visible region of the spectrum except for the long wavelength range (red light). On absorption by the quinoid system the light energy is dissipated in the form of the energy of thermal movement of solvent molecules. However, in the case of fluorescence indicators this energy is converted to light of different wavelength which is recorded with a fluorescence spectrometer.

In the case of only one form of the indicator undergoing the tautomeric change to a coloured form (this is the case of phenolphthalein anion) the equilibrium between the original form of the indicator anion and its tautomeric form is characterized by the tautomeric constant

$$K_T = \frac{[I_q^-]}{[I^-]} \qquad (1.4.45)$$

where I^- is the coloured (in the case of phenolphthalein the quinoid) form of the anion of the indicator. For $K \gg 1$

$$[HI] + [I_q^-] = s \qquad (1.4.46)$$

where s is the overall concentration of the indicator. Thus, when substituting for $[HI] = s - [I_q^-]$ into Eq. (1.4.41) an analogy of the Henderson–Hasselbalch equation (1.4.26) is obtained, which is the basis of the determination of pH by means of acid–base indicators.

The determination of intercellular pH is based on measurement of fluorescence emitted by the fluorescence indicator, fluorescein. The starting compound used for this purpose is fluorescein diacetate:

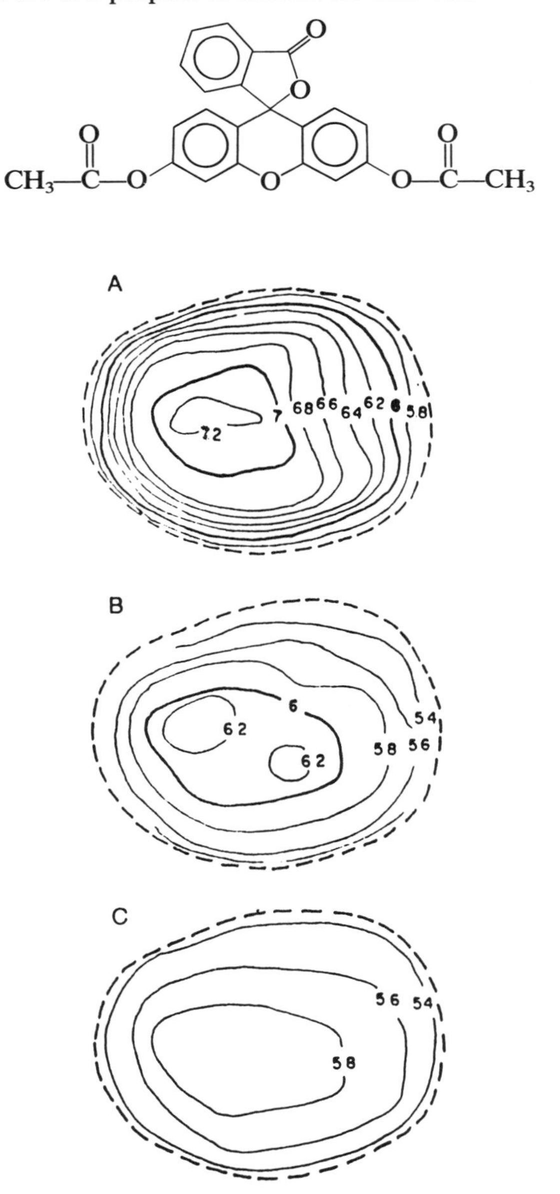

Fig. 1.13 pH maps of *Saccharomyces cerevisiae* cell transferred from water to a 0.2 M buffer of pH 3: (a) after 2 min; (b) after 5 min; (c) after 20 min. (Reproduced by permission of J. Slavík and A. Kotyk)

This substance penetrates into the cell where it is hydrolysed to fluorescein through the action of the enzyme's esterases. The quantity of the fluorescent quinoid form then depends on the local value of pH in the same way as in the case of phenolphthalein. The intensity of the fluorescein radiation is measured with a fluorescence microscope and then processed to a digital image which is the basis of a map of pH distribution in the cell (Fig. 1.13).

References

Albert, A., and E. P. Sergeant, *The Determination of Ionization Constants. A Laboratory Manual,* 3rd ed., Chapman & Hall, London, 1983.

Bates, R. G., *Determination of pH. Theory, Practice,* John Wiley & Sons, New York, 1973.

Burger, K., see page 27.

King, E. J., Acid-base behaviour, in *Physical Chemistry of Organic Solvent Systems* (Eds A. K. Covington and T. Dickinson), Chap. 3, Plenum Press, New York, 1973.

Lengyel, S., and B. E. Conway, Proton solvation and transfer in chemical and electrochemical processes, *CTE,* **5,** Chap. 4.

Rochester, C. R., *Acidity Functions,* John Wiley & Sons, New York, 1970.

Slavík, J., in press.

Trémillon, B., see page 27.

1.5 Special Cases of Electrolytic Systems

1.5.1 *Sparingly soluble electrolytes*

A great many electrolytes have only limited solubility, which can be very low. If a solid electrolyte is added to a pure solvent in an amount greater than corresponds to its solubility, a heterogeneous system is formed in which equilibrium is established between the electrolyte ions in solution and in the solid phase. At constant temperature, this equilibrium can be described by the thermodynamic condition for equality of the chemical potentials of ions in the liquid and solid phases (under these conditions, cations and anions enter and leave the solid phase simultaneously, fulfilling the electroneutrality condition). In the liquid phase, the chemical potential of the ion is a function of its activity, while it is constant in the solid phase. If the formula unit of the electrolyte considered consists of ν_+ cations and ν_- anions, then

$$\nu_+\mu_+(\mathrm{s}) + \nu_-\mu_-(\mathrm{s}) = \nu_+\mu_+^0(\mathrm{l}) + \nu_-\mu_-^0(\mathrm{l}) + \nu_+RT\ln a_+ + \nu_-RT\ln a_- \tag{1.5.1}$$

from which it follows that

$$a_+^{\nu_+}a_-^{\nu_-} = P = \exp\left\{\frac{\nu_+[\mu_+^0(\mathrm{s}) - \mu_+^0(\mathrm{l})] + \nu_-[\mu_-^0(\mathrm{s}) - \mu_-^0(\mathrm{l})]}{RT}\right\} \tag{1.5.2}$$

The product P, termed the (real or thermodynamic) solubility product, is a constant dependent only on the temperature. The analogous (apparent) solubility product P', defined in terms of concentrations or molalities, depends on the ionic strength. Obviously

$$P = P'\gamma_{\pm}^{\nu} \tag{1.5.3}$$

At ionic strengths where the activity coefficient decreases with increasing ionic strength, the product P' increases because the product P is constant.

The solubility, c_{sat}, is the concentration of the saturated solution and thus depends on the product P'. As $c_+ = \nu_+ c_{sat}$ and $c_- = \nu_- c_{sat}$, then

$$c_{sat} = \left(\frac{P'}{\nu_+^{\nu_+}\nu_-^{\nu_-}}\right)^{1/\nu} \tag{1.5.4}$$

If the concentration of one of the ions is increased (for example c_-) by addition of its soluble salt (for example KCl is added to a saturated AgCl solution), then the solubility of the electrolyte, AgCl, as a whole is given by the concentration of the ion present in the lower concentration ($c_{sat} = c_+/\nu_+$). It can readily be seen that $\log c_{sat} = -(\nu_-/\nu_+) \log c_- + (1/\nu_+) \log P - \log \nu_+$. The logarithm of the solubility thus decreases linearly with the logarithm of the increasing concentration of the added ion. The separations by precipitation in analytical chemistry is based on this principle.

Examples of solubility products are listed in Table 1.9. Similarly to the dissociation constants, the solubility products are also dimensionless quantities. However, because of the choice of the standard state, their values numerically correspond to units of moles per cubic decimetre.

The solubility product is measured by determining the ion concentrations by a suitable analytical method and the results are extrapolated to zero ionic strength, where $P' = P$.

1.5.2 *Ampholytes*

Ampholytes are substances which, according to the acidity of the medium, act either as acids or as bases. Such a case has already been dis-

Table 1.9 Solubility products P of some sparingly soluble salts. (According to B. E. Conway)

Salt	t (°C)	P	Salt	t (°C)	P
CdS	25	1.14×10^{-28}	Hg_2I_2	19.2	1.05×10^{-19}
$Ca(COO)_2$	25	1.78×10^{-9}	Hg_2CrO_4	25	2.0×10^{-9}
CoS	20	3×10^{-26}	AgCl	25	1.8×10^{-10}
CuS	25	3×10^{-38}	AgBr	25	6.5×10^{-13}
Cu_2S	25	8.5×10^{-45}	AgI	25	1.0×10^{-16}
Cu_2I_2	18	5.0×10^{-12}	Ag_2S	25	3×10^{-52}
$Mg(OH)_2$	25	4.6×10^{-24}	AgCNS	25	1.44×10^{-12}
Hg_2Cl_2	19.2	5.42×10^{-19}	TlCl	25	2.25×10^{-4}
Hg_2Br_2	19.2	3.9×10^{-23}	TlI	25	6.47×10^{-8}

cussed as the anions of a two-protonic acid possess amphiprotic properties. For example, the hydrogen sulphite anion reacts in two ways: in alkaline medium as an acid, $HSO_3^- + H_2O \rightleftarrows SO_3^{2-} + H_3O^+$, while in acidic medium as a base, $HSO_3^- + H_2O \rightleftarrows H_2SO_3 + OH^-$. As an ampholyte, however, such a substance is usually considered whose neutral molecule HP has both the acidic and basic functions. In an aqueous solution the following equilibria are established:

$$\begin{aligned} HP + H_2O &\rightleftarrows H_3O^+ + P^- \\ H_2P^+ + H_2O &\rightleftarrows H_3O^+ + HP \end{aligned} \tag{1.5.5}$$

with acid dissociation constants

$$K_1 = \frac{a_{H_3O^+} a_{P^-}}{a_{HP}} \quad \text{and} \quad K_2 = \frac{a_{H_3O^+} a_{HP}}{a_{H_2P^+}} \tag{1.5.6}$$

This simple treatment is applicable to cases where HP is a real neutral molecule. Although typical examples of ampholytes are amino acids, in which case the equilibrium (1.5.5) would be written

$$\begin{aligned} NH_2RCOOH + H_2O &\rightleftarrows H_3O^+ + NH_2RCOO^- \\ NH_3^+RCOOH + H_2O &\rightleftarrows NH_2RCOOH + H_3O^+ \end{aligned} \tag{1.5.7}$$

However, in aqueous solutions the amino acids will have undergone internal ionization:

$$NH_2RCOOH \rightleftarrows NH_3^+RCOO^- \tag{1.5.8}$$

This equilibrium lies almost completely to the right-hand side, so that the NH_2RCOOH molecules are practically absent from the solution. A particle of the type $NH_3^+RCOO^-$ is called an amphion (zwitterion). The internal ionization is proven by several empirical facts:

(a) The amino acids have a large dipole moment.
(b) The acidic dissociation reactions described by (1.5.6) and (1.5.7) have the constants $K_1 \approx 10^{-10}$ and $K_2 \approx 10^{-2}$, whereas based upon values for acetic acid and ammonia, the constants should be of the order of 10^{-5} and 10^{-9} for glycine; this difference cannot be attributed to the replacement of hydrogen in the acetic acid molecule by the amine group or, to replacement of hydrogen in the NH_3 molecule by the residue of acetic acid, CH_2COOH.
(c) By blocking the basic functional group (e.g. in the reaction of glycine with formaldehyde) the acidity increases strongly and approaches that of the fatty acids.

In view of this, the protolytic equilibria may be formulated as

$$\begin{aligned} NH_3^+RCOO^- + H_2O &\rightleftarrows NH_2RCOO^- + H_3O^+ \\ NH_3^+RCOOH + H_2O &\rightleftarrows NH_3^+RCOO^- + H_3O^+ \end{aligned} \tag{1.5.9}$$

with acid dissociation constants

$$K_1 = \frac{a_{H_3O^+} a_{NH_2RCOO^-}}{a_{NH_3^+RCOO^-}} \quad \text{and} \quad K_2 = \frac{a_{H_3O^+} a_{NH_3^+RCOO^-}}{a_{NH_3^+RCOOH}} \tag{1.5.10}$$

From (1.5.10) we may conclude that the constant K_1 does not characterize the acidity of the carboxyl group but of the NH_3^+ group, and *vice versa* for K_2. The deviations from the accepted values of the acid dissociation constants of acetic acid and ammonium ion may be attributed to the influence of the constitution of the ionic forms of the amino acid. The increase in acidity of the carboxyl group is caused by the neighbouring positive charge of $—NH_3^+$, while the negative charge of $—COO^-$ is the cause of the increase in basicity of the amine group. The values of the dissociation constants K_1' and K_2' are listed in Table 1.10.

An important quantity characterizing the amino acids, peptides and proteins is the isoelectric point, which is the value of pH at which the ampholytes do not migrate in an electrical field. This occurs when

$$m_{NH_3^+RCOOH} = m_{NH_2RCOO^-} \tag{1.5.11}$$

From (1.5.5) we obtain for a dilute solution

$$K_1 K_2 = a^2_{H_3O^+} \frac{m_{NH_2RCOO^-} \gamma_{NH_2RCOO^-}}{m_{NH_3^+RCOOH} \gamma_{NH_3^+RCOOH}} \approx \frac{a^2_{H_3O^+} m_{NH_2RCOO^-}}{m_{NH_3^+RCOOH}} \tag{1.5.12}$$

Using the condition (1.5.11) for the isoelectric point we have (subscript I

Table 1.10 Dissociation constants of amino acids at 25°C. (According to B. E. Conway)

Amino acid	pK_1'	pK_2'
Glycine	2.34	9.60
Alanine	2.34	9.69
α-Amino-*n*-butyric acid	2.55	9.60
Valine	2.32	9.62
α-Amino-*n*-valeric acid	2.36	9.72
Leucine	2.36	9.60
Isoleucine	2.36	9.68
Norleucine	2.39	9.76
Serine	2.21	9.15
Proline	1.99	10.60
Phenylalanine	1.83	9.13
Tryptophane	2.38	9.39
Methionine	2.28	9.21
Isoserine	2.78	9.27
Hydroxyvaline	2.61	9.71
Taurine[a]	1.5	8.74
β-Alanine	3.60	10.19

[a] For $—SO_3H$.

denoting the isoelectric point)

$$(a_{H_3O^+})_I = \sqrt{K_1 K_2}$$
$$pI = pH_{iso} = \tfrac{1}{2}(pK_1 + pK_2) \tag{1.5.13}$$

At the isoelectric point, the acid dissociation equals the base dissociation. The degrees of dissociation α_1 and α_2 are defined in such a way that

$$m_{NH_3^+RCOO^-} = s(1 - \alpha_1 - \alpha_2)$$
$$m_{NH_3^+RCOOH} = s\alpha_1 \tag{1.5.14}$$
$$m_{NH_2RCOO^-} = s\alpha_2$$

where s is the overall concentration of the ampholyte. From the definitions of K_1 and K_2 and from (1.5.14) we obtain

$$\frac{1}{K_2} = \frac{\alpha_1}{a_{H_3O^+}(1 - \alpha_1 - \alpha_2)} \quad \text{and} \quad K_1 = \frac{a_{H_3O^+}\alpha_2}{1 - \alpha_1 - \alpha_2} \tag{1.5.15}$$

At the isoelectric point $\alpha_1 = \alpha_2 = (\alpha)_I$, so that on eliminating α from Eqs (1.5.15) we obtain Eqs (1.5.13).

Furthermore, at the isoelectric point the total dissociation, $\alpha_1 + \alpha_2$, is at its minimum. This sum may be obtained from Eqs (1.5.15) so that on differentiating with respect to $a_{H_3O^+}$ we have, for activity coefficients except that of H_3O^+ equal to unity,

$$\frac{d(\alpha_1 + \alpha_2)}{da_{H_3O^+}} = \frac{2a_{H_3O^+}(K_2 a_{H_3O^+} + a^2_{H_3O^+} + K_1K_2) - (a^2_{H_3O^+} + K_1K_2)(K_2 + 2a_{H_3O^+})}{(K_2 a_{H_3O^+} + a^2_{H_3O^+} + K_1K_2)^2}$$
$$= 0 \tag{1.5.16}$$

whence, after some rearrangement, Eq. (1.5.13) is again obtained. The course of dissociation is shown schematically in Fig. 1.14.

The degree of dissociation at the isoelectric point can be found by calculating $a_{H_3O^+}$ from Eq. (1.5.15), putting $\alpha_1 = \alpha_2 = (\alpha)_I \ll 1$ and substituting into (1.5.13). In a dilute solution

$$(\alpha)_I = \sqrt{\frac{K_1}{K_2}} \tag{1.5.17}$$

Assuming the sensitivity of the determination of α at the isoelectric point as 1 per cent, the isoelectric point will appear as a real 'point', if $(\alpha)_I > 10^{-2}$, that is $K_1/K_2 > 10^{-4}$, whereas for $(\alpha)_I < 10^{-2}$, that is $K_1/K_2 < 10^{-4}$, a definite interval of pH obeys the isoelectric condition (see Fig. 1.14).

1.5.3 *Polyelectrolytes*

Polyelectrolytes are macromolecular substances whose molecules have a large number of groups that are ionized in solution. They are termed macroions or polyions and are studied most often in aqueous solutions.

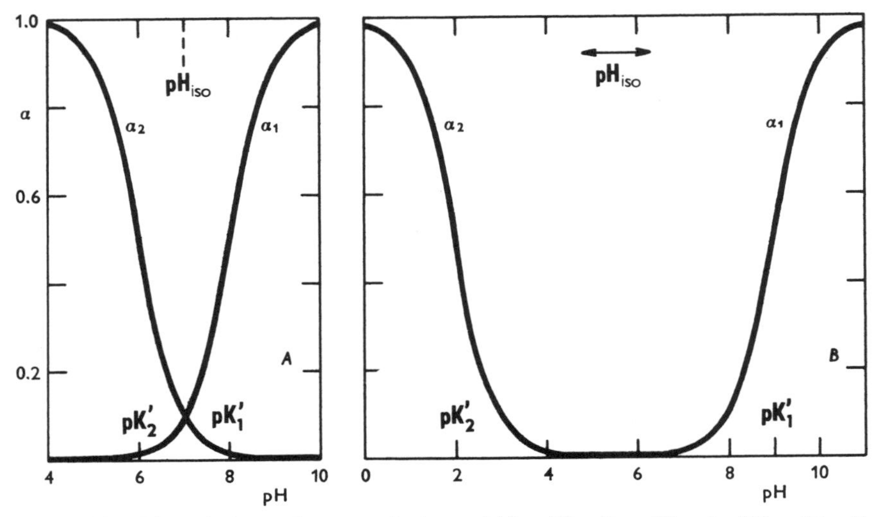

Fig. 1.14 Dissociation of an ampholyte. (A) $pK_1' = 8$, $pK_2' = 6$; (B) $pK_1' = 9$, $pK_2' = 2$. Calculation according to Eq. (1.5.15)

Small counterions (gegenions) ensure electroneutrality. For example, the sodium salt of polyacrylic acid is ionized in solution to form polyanions and the corresponding number of sodium cations:

$$\left(\begin{array}{c} \text{—CH—CH—} \\ \quad\ \ | \\ \quad\ \ \text{C=O} \\ \quad\ \ | \\ \quad\ \ \text{O}^- \end{array} \right)_n + n\text{Na}^+$$

The fact that one of the ions has large dimensions affects the electrochemical properties of the system compared with the corresponding low-molecular system. On the other hand, the presence of an electric charge affects the behaviour in solution of the macromolecular system compared to the corresponding system without electric charge.

Two extreme types of polyanions can be distinguished. In the first, the main chain is flexible and can assume a large number of different conformations so that the overall shape of the ion depends on the solvent. In some solvents (called 'bad'), they are rolled up to form a relatively rigid 'random coil', while in others ('good') the coil is more or less expanded. This type includes most synthetic polyelectrolytes such as the following acids:

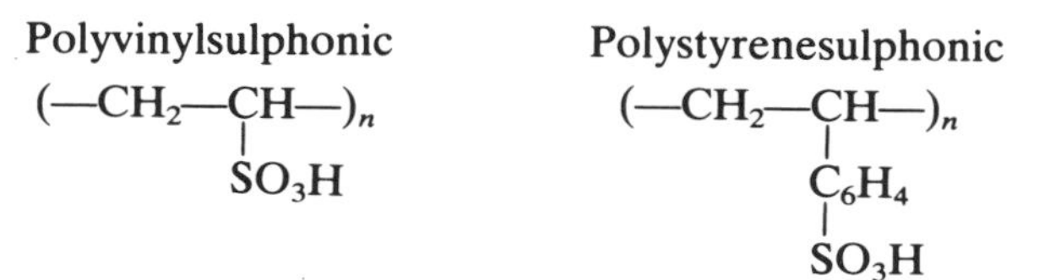

(the first two are weak in aqueous solutions while the other two are strong). In copolymers of maleic acid with styrene, two acid groups are always separated by one or more styrene units, so that a dibasic polyacid is formed:

$$(\text{—}CH_2\text{—}CH(C_6H_5)\text{—}CH(COOH)\text{—}CH(COOH)\text{—})_n$$

Analogously, a tribasic polyacid can be represented by the copolymer of maleic acid with acrylic acid:

$$(\text{—}CH_2\text{—}CH(COOH)\text{—}CH(COOH)\text{—}CH(COOH)\text{—})_n$$

Polymetaphosphoric acid is an example of an inorganic polyacid:

$$(\text{—}O\text{—}P(=O)(OH)\text{—})_n$$

while polyvinylamine is an example of a weak polybase:

$$(\text{—}CH_2\text{—}CH(NH_2)\text{—})_n$$

A second type of polyion assumes stable, specific conformation. This group includes many natural polyelectrolytes. For example, the side chains of proteins have a large number of acidic and basic ionizable groups. The resultant sign and amount of charge of the polyion depends on the pH (the species being a polyampholyte). The tertiary structure of the globular protein is so stable that it is not denatured even at a high value of the overall charge. Another example is the native form of deoxyribonucleic acid, whose double helix is quite stable in solution; the polyion acts in solution as a rigid rod. The difference in behaviour between polyion with a fixed shape and that of low-molecular-weight ions in solution depends only on their size and their overall charge.

The behaviour in solution of the first type of polyions compared with that of uncharged molecules is more involved in that the coil expansion is affected not only by the solvent but also by the electric field formed by the polyion itself, by the counterions and by the ions of other low-molecular-weight electrolytes, if present in the solution. Infinite charge dilution cannot be achieved by diluting the polyelectrolyte solution, as a local high-intensity

Fig. 1.15 A schematic representation of a polyacid solution including polyanions ◎, anions of an added low molecular electrolyte ○ and positive counterions ●

electric field remains inside the macroion, attracting a corresponding number of counterions (see the scheme in Fig. 1.15). Thus, the concept of the ionic strength of the solution lacks sense as an infinitely dilute solution of a polyelectrolyte does not behave ideally at any conditions and cannot be considered to be in the standard state. On the other hand, mutual repulsion of charges on the chain can lead to considerable chain expansion, which could not be achieved by transfer of the uncharged macromolecules from a 'bad' to a 'good' solvent.

The shape of any macromolecule with a flexible chain, both uncharged and charged, i.e. the degree of expanding or rolling-up the random coil, is characterized by the parameter h, corresponding to the distance between the ends of the original chain from which the coil is formed. The probability that the ends will lie at a distance h is characterized by the probability density $w(h)$ derived in textbooks on macromolecular chemistry by statistical methods on the basis of various models with various simplifying assumptions. In the resultant equations the number of basic segments in the macromolecule, Z, and their length, b, is taken into account and the actual system is described more or less accurately (and the equations are more or less complex) depending on the accuracy of the model employed. For polyelectrolytes, $w(h)$ consists of the *a priori* probability $w_0(h)$, corresponding to the uncharged macromolecule, and the Boltzmann factor, containing the excess electrostatic Gibbs energy $\Delta G_{el}^E(h)$:

$$w(h)\,dh = w_0(h)\exp\left(\frac{-\Delta G_{el}^E(h)}{kT}\right)dh \tag{1.5.18}$$

The distribution of distance h is characterized either by the most probable value h_{max}, defined by the condition $dw(h)/dh = 0$, or by the mean square of distance h, defined in the usual manner, $\overline{h^2} = \int h^2 w(h)\,dh / \int w(h)\,dh$. The ratio $(h_{max})^2/h^2$ gives the width of the distribution. The closer this value is to unity, the narrower the distribution. This ratio is equal to $\frac{2}{3}$ for the simplest model in statistics, the 'random walk'.

The problem is then reduced to determining the energy ΔG_{el}^E, which again depends on the model. The simplest concept (W. Kuhn, A. Katchalsky, *et al.*) is to place half of the total charge Q at one end of the polyion and the other half at the other end, from which it follows that $\Delta G_{el}^E(h) = Q^2/4\varepsilon h$, where ε is taken at a first approximation to be the macroscopic permittivity of water. Another greatly simplified model, corresponding to a certain degree to globular proteins, considers the polyion as a spherically symmetrical charge cloud. The electrostatic contribution to the excess Gibbs energy affects the thermodynamic, rheological, transport, and other properties of polyelectrolytes. Here, only dissociation equilibria will be considered briefly.

Consider a molecule of a carboxylic polyacid with degree of polymerization N, bearing N ionizable groups, n of which are actually ionized. K_n' is the dissociation constant for the ionization of an acidic group bound to an n-times ionized molecule; K_0' is the dissociation constant of the group that splits off the first proton and therefore has not yet been affected by the electric field of the other already ionized groups. The expression $-kT \ln K_0'$ gives the Gibbs energy change due to ionization of a neutral molecule. The value of $-kT \ln K_n'$ differs from this expression by the electric work necessary for removal of a further proton from the field of an n-fold ionized molecule. If the electric charge of an n-times ionized molecule is ΔG_{el}^E, this work is equal to $\partial G_{el}^E/\partial n$. Then we have

$$-kT \ln K_n' = -kT \ln K_0' + \frac{\partial G_{el}^E}{\partial n} \tag{1.5.19}$$

Except at the first stage in neutralization titration, i.e. for small values of n, all the molecules in solution are found experimentally to be ionized to the same degree $\alpha = n/N$. Then

$$K_n' = [H_3O^+]\frac{n}{N-n} = [H_3O^+]\frac{\alpha}{1-\alpha} \tag{1.5.20}$$

$$pH = pK_0' - \log\frac{1-\alpha}{\alpha} + \frac{1}{\ln 10 \times kT}\frac{\partial G_{el}^E}{\partial n} \tag{1.5.21}$$

Thus, the dissociation equilibrium is affected by the ionic strength, temperature and dielectric constant of the solvent as well as by the parameter h (involved in ΔG_{el}^{E}). On the other hand, the term $\partial G_{el}^{E}/\partial n$ does not depend on the degree of polymerization (except for very small values of n). The degree of polymerization does not affect, for example, the course of the potentiometric titration of a polyacid.

For a dibasic polyacid, where, of the total of N units, n_1 are ionized to the first stage and n_2 to the second stage, then simultaneously

$$\begin{aligned} \mathrm{pH} + \log \frac{2(N - n_1 - n_2)}{n_1 + 1} &= \mathrm{p}K_0' + \frac{1}{\ln 10 \times kT} \frac{\partial G_{el}^{E}}{\partial (n_1 + n_2)} \\ \mathrm{pH} + \log \frac{n_1}{2(n_2 + 1)} - \frac{e^2}{\ln 10 \times R\varepsilon kT} &= \mathrm{p}K_0' + \frac{1}{\ln 10 \times kT} \frac{\partial G_{el}^{E}}{\partial (n_1 + n_2)} \end{aligned} \tag{1.5.22}$$

Here, R is the average distance between two neighbouring doubly ionized monomeric units and ε is the effective dielectric constant.

For a polyampholyte we have

$$\begin{aligned} \mathrm{pH} &= \mathrm{p}K_{0,\mathrm{A}}' - \log \frac{1 - \alpha_{\mathrm{A}}}{\alpha_{\mathrm{A}}} + \frac{1}{\ln 10 \times kT} \frac{\partial G_{el}^{E}}{\partial n_{\mathrm{A}}} \\ \mathrm{pH} &= \mathrm{p}K_{0,\mathrm{B}}' + \log \frac{1 - \alpha_{\mathrm{B}}}{\alpha_{\mathrm{B}}} - \frac{1}{\ln 10 \times kT} \frac{\partial G_{el}^{E}}{\partial n_{\mathrm{B}}} \end{aligned} \tag{1.5.23}$$

where $K_{0,\mathrm{A}}'$ and $K_{0,\mathrm{B}}'$ are the acid dissociation constants of the monomeric unit of the ampholyte, $\alpha_{\mathrm{A}} = n_{\mathrm{A}}/N_{\mathrm{A}}$ and $\alpha_{\mathrm{B}} = n_{\mathrm{B}}/N_{\mathrm{B}}$, where n_{A} and n_{B} are numbers of ionized acid and basic groups and N_{A} and N_{B} are total numbers of ionizable acid and basic groups, respectively. These equations are, however, invalid in the vicinity of the isoelectric point where large attractive forces act inside the molecules. A result of this is a decrease of the solubility and coagulation of the polyampholyte.

References

Fasman, G. D. (Ed.), *Prediction of Protein Structure and the Principles of Protein Conformation,* Plenum Press, New York 1989

Jurnak, F., and A. McPherson (Eds), *Biological Macromolecules and Asemblies,* Vol. 2, *Nucleic Acids and Interactive Proteins,* John Wiley & Sons, New York, 1985.

Manning, G., *Quart. Rev. Biophys.,* **11,** 179 (1978).

Morawetz, H., *Macromolecules in Solution,* John Wiley & Sons, New York, 1965.

Oosawa, F., *Polyelectrolytes,* M. Dekker, New York, 1971.

Chapter 2

Transport Processes in Electrolyte Systems

Since water as solvent plays the role of a medium where electrolytic displacements take place we shall be able to state for sure, together with Wiedemann, Beetz and Quincke, that the electrical resistance of a solution consists of resistances to movement enforced upon the components of the solution by water particles, by the components themselves and perhaps by the undecomposed molecules of the electrolyte. To separate these various hindrances will be no easy task, particularly because, as stressed by Quincke, they are not necessarily constant but can, for example, depend on the condition of the solution. Even from this standpoint the process of conduction will be, in general, still very complicated.

The discussion becomes different when restricted to *dilute solutions*. In this case one must arrive at the conclusion which essentially helps to simplify the phenomena: In a dilute solution the conductivity depends (besides on the number of dissolved molecules) only on the transported components being independent of their mutual associations. Thus, the more the number of water molecules prevails over those of the electrolyte the more pronounced is the influence exerted by the water molecules on the ions and the less their mutual friction.

F. Kohlrausch, 1879

2.1 Irreversible Processes

In a homogeneous medium of an electrolyte solution, an ionic liquid or a solid electrolyte under conditions of constant pressure and temperature, mechanical, electrostatic and short-range forces act on the individual particles in solution, but these forces average out in time. The effect of these forces is reflected in the activity values of the individual components of the system.

Let us consider another situation where a force (or forces) is not compensated on a time average. Then the particles upon which the force is exerted become transported in the medium. This 'translocation' phenomenon changes with time. Particle transport, of course, also occurs under equilibrium conditions in homogeneous media. Self-diffusion is a process that can be observed and its velocity can be measured, provided that a gradient of isotopically labelled species is formed in the system at constant composition.

The forces leading to transport of matter in the system are connected with the transfer of impulse during collisions between particles, with the effect of external mechanical forces and gravitation on the whole system or part thereof and with the effect of the electrostatic field on the impulse of the particles. The manifestation of forces defined in this manner in the course of transport phenomena can be treated quantitatively by statistical mechanics. However, a different approach is usually employed. These uncompensated mechanical and electrical forces are expressed in terms of 'phenomenological driving forces'. The response of the system to these forces is then described in terms of the fluxes of the corresponding thermodynamic quantities. The flux (or flux intensity by some authors) is defined as the amount of substance, energy or charge passing through a unit area in a unit time. If a single driving force $\mathbf{X}$ acts on one species of particles the corresponding flux is

$$\mathbf{J} = L\mathbf{X} \tag{2.1.1}$$

where L is the phenomenological coefficient, a scalar quantity corresponding to a generalized conductance or to the reciprocal of a generalized resistance. Onsager broadened this simple formulation to include the case where the fluxes of n thermodynamic quantities are the result of n different driving forces. Assuming that the system is not far from equilibrium, the fluxes $\mathbf{J}_i$ are linear functions of the driving forces (*linear irreversible thermodynamics*):

$$\mathbf{J}_i = \sum_k L_{ik}\mathbf{X}_k \tag{2.1.2}$$

According to the Onsager assumption, the square matrix of the phenomenological coefficients

$$\begin{Vmatrix} L_{11} & L_{12} & \cdots & L_{1,n-1} & L_{1n} \\ L_{21} & L_{22} & \cdots & L_{2,n-1} & L_{2n} \\ \hdashline \\ \hdashline L_{n-1,1} & L_{n-1,2} & \cdots & L_{n-1,n-1} & L_{n-1,n} \\ L_{n1} & L_{n2} & \cdots & L_{n,n-1} & L_{nn} \end{Vmatrix} \tag{2.1.3}$$

is symmetrical. Thus, the reciprocity relationship

$$L_{ik} = L_{ki} \tag{2.1.4}$$

is valid.

Four basic transport phenomena can be distinguished:

1. *Diffusion*—transport of matter as a result of differing values of the chemical potential of a given component at various sites within the system, or in the system and its surroundings. Obviously, the particles

present in a given volume at a site of lower concentration have a lower probability to come to the same volume at a site of higher concentration than in the opposite case, where the particles at a site of higher concentration come to the site of lower concentration.

2. *Heat conduction*—the transport of energy resulting from temperature gradients in the system or difference in temperature, between the system and its surroundings. Heat conduction arises from the fact that, while the number of particles per volume unit is identical at various sites in the system, they have different impulses.
3. *Convection*—the transport of mass or energy as a result of streaming in the system produced by the action of external forces. These include mechanical forces (*forced convection*) or gravitation, if there are density gradients in the system (*natural or free convection*).
4. *Conduction of electric current*—transport of charge and mass as the result of the motion of elecrically charged particles in an electric field.

Chemical reactions in the system are irreversible processes, affecting transport processes, as they result in the formation and disappearance of components of the system and in the release or consumption of thermal energy.

The rate of a chemical reaction (the 'chemical flux'), J_{ch}, in contrast to the above processes, is a scalar quantity and, according to the Curie principle, cannot be coupled with vector fluxes corresponding to transport phenomena, provided that the chemical reaction occurs in an isotropic medium. Otherwise (see Chapter 6, page 450), chemical flux can be treated in the same way as the other fluxes.

References

De Groot, S. R., and P. Mazur, *Non-equilibrium Thermodynamics,* Interscience, New York, 1962.

Denbigh, K. G., *Thermodynamics of the Steady State,* Methuen, London, 1951.

Glansdorff, P., and I. Prigogine, *Structure, Stability and Fluctuations,* John Wiley & Sons, New York, 1971.

Onsager, L., *Phys. Rev.,* **31,** 405, 2265 (1931).

Prigogine, I., *Introduction to Thermodynamics of Irreversible Processes,* John Wiley & Sons, New York, 1967.

Yao, Y. L., *Irreversible Thermodynamics,* Science Press, Beijing and Van Nostrand-Reinhold, New York, 1981.

2.2 Common Properties of the Fluxes of Thermodynamic Quantities

Consider a linear tube with cross-section A and length l, filled with a material, and assume that changes in the considered thermodynamic quantity can occur only along the length of the tube (i.e. the 'mantle' of the tube is insulated, the ends being permeable for the thermodynamic quantities). Flux J is a function of coordinate x (see Fig. 2.1).

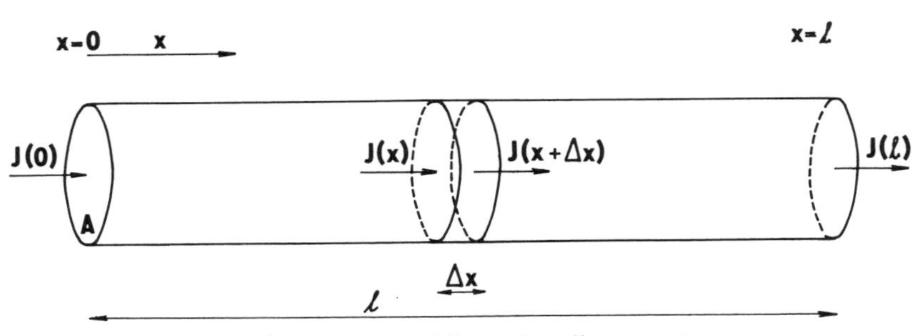

Fig. 2.1 Balance of fluxes in a linear system

The tube contains the thermodynamic quantity in an amount M (amount of a substance, thermal energy, etc.), which has a density (concentration, energy density, etc.) $\psi(x)$ at each point in the tube defined by the relationship

$$\psi(x) = \frac{\partial M}{\partial V} \tag{2.2.1}$$

where V is the volume of the system. It thus holds for M that

$$M = A \int_0^l \psi \, \mathrm{d}x \tag{2.2.2}$$

The change in amount M in the whole tube per unit time is given by the relationship

$$\frac{\partial M}{\partial t} = A \int_0^l \frac{\partial \psi}{\partial t} \mathrm{d}x = -[J(l) - J(0)]A \tag{2.2.3}$$

Consider a section of the tube with length Δx; inside this section and in its vicinity, J is a continuous function of x. If an amount $J(x)A\Delta t$ flows into this section at point x in time Δt and an amount $J(x + \Delta x)A\Delta t$ at point $x + \Delta x$, then its change in volume $A\Delta x$ is given as

$$A \int_x^{x+\Delta x} \frac{\partial \psi}{\partial t} \mathrm{d}x \, \Delta t = -A[J(x + \Delta x) - J(x)]\Delta t \tag{2.2.4}$$

Dividing both sides of the equation by $A\Delta x \Delta t$ and taking the limit yields the continuity relation

$$\frac{\partial \psi}{\partial t} = -\frac{\partial J}{\partial x} \tag{2.2.5}$$

The general case of a three-dimensional body enclosed by a surface S will be treated using vector analysis. The time change in the amount M is given by the relationship

$$\frac{\partial M}{\partial t} = -\oint \mathbf{J} \, \mathrm{d}\mathbf{S} \tag{2.2.6}$$

The normal vector $\mathbf{dS}$ points out of the body and $\mathbf{J} > 0$ for a quantity flowing out of the body; thus the right-hand side of Eq. (2.2.2) must be negative in sign. Obviously,

$$\frac{\partial M}{\partial t} = \int \frac{\partial \psi}{\partial t} \mathrm{d}V \tag{2.2.7}$$

In view of the Gauss–Ostrogradsky theorem

$$\oint \mathbf{J} \, \mathrm{d}\mathbf{S} = \int \operatorname{div} \mathbf{J} \, \mathrm{d}V \tag{2.2.8}$$

substitution into Eq. (2.2.6) gives

$$\int \frac{\partial \psi}{\partial t} \mathrm{d}V = -\int \operatorname{div} \mathbf{J} \, \mathrm{d}V \tag{2.2.9}$$

and differentiation with respect to V yields the continuity equation

$$\frac{\partial \psi}{\partial t} = -\operatorname{div} \mathbf{J} \tag{2.2.10}$$

If in the system the thermodynamic quantity is formed (e.g. by chemical reaction) at a rate of μ units of quantity per unit volume and unit time, then Eq. (2.2.10) must be completed to yield

$$\frac{\partial \psi}{\partial t} = -\operatorname{div} \mathbf{J} + \mu \tag{2.2.11}$$

For thermodynamic quantities other than the amount of substance, the quantity ψ can be expressed by the equation

$$\psi = \frac{\partial M}{\partial V} = \frac{\partial M}{\partial w} \frac{\partial w}{\partial V} = m\rho \tag{2.2.12}$$

where m is the amount of the quantity M per unit of mass (e.g. energy per kilogram), w denotes mass and ρ is the density of the substance present in the system. Equation (2.2.11) then takes the form

$$\rho \frac{\partial m}{\partial t} = -\operatorname{div} \mathbf{J} + \mu \tag{2.2.13}$$

References

De Groot, S. R., and P. Mazur, see page 81.
Denbigh, K. G., see page 81.
Ibl, N., Fundamentals of transport phenomena in electrolytic systems, *CTE,* **6,** 1 (1983).
Levich, V. G., *Physico-Chemical Hydrodynamics,* Prentice-Hall, Englewood Cliffs, 1962.

2.3 Production of Entropy, the Driving Forces of Transport Phenomena

If the system considered is closed and thermally insulated, then its entropy increases during a transport phenomenon

$$\frac{dS}{dt} > 0 \tag{2.3.1}$$

If the system is neither closed nor thermally insulated, then the change in the entropy with time consists of two quantities: of the time change in the entropy as a result of processes occurring within the system S_i and of entropy changes in the surroundings, caused by transfer of the entropy from the system in the reversible process S_e

$$\frac{dS}{dt} = -\frac{dS_e}{dt} + \frac{dS_i}{dt} \tag{2.3.2}$$

If we consider a system of infinitesimal dimensions, then the entropy balance can be described directly by Eq. (2.2.13):

$$\rho \frac{\partial s}{\partial t} = -\mathrm{div}\, \mathbf{J}_s + \sigma \tag{2.3.3}$$

where s is the entropy of the system, related to 1 kg of substance, $\mathbf{J}_s$ is the entropy flux and σ is the rate of entropy production per unit volume. The first term on the right-hand side of this equation corresponds to $-dS_e/dt$ and the second to dS_i/dt in Eq. (2.3.2). The rate of entropy formation for a single transport process (Eq. 2.1.1) is given as

$$\sigma = \frac{1}{T}\mathbf{J}\mathbf{X} = \frac{1}{T} L X^2 \tag{2.3.4}$$

(unit $\mathrm{J\,m^{-3}s^{-1}\,K^{-1}}$). X is the absolute value of the vector of the driving force $\mathbf{X}$. Generally,

$$\sigma = \frac{1}{T}\sum_i \mathbf{J}_i \mathbf{X}_i = \frac{1}{T}\sum_i \sum_k L_{ik} \mathbf{X}_i \mathbf{X}_k \tag{2.3.5}$$

Some authors use the dissipation function Φ (unit $\mathrm{J\,m^{-3}\,s^{-1}}$):

$$\Phi = -T\sigma = -\sum_i \sum_k L_{ik} \mathbf{X}_i \mathbf{X}_k \tag{2.3.6}$$

The rate of entropy production is always positive in the present case, since transport processes are irreversible in nature, i.e. always connected with irreversible losses (dissipation) of energy.

The simplest case of a flux and a driving force is shown in the *conduction of electric current.* This process is governed by Ohm's law for the current density:

$$\mathbf{j} = -\kappa\, \mathrm{grad}\, \phi \tag{2.3.7}$$

where κ is the conductivity and ϕ is the electric potential, or for a system with a single coordinate x,

$$j = -\kappa \frac{\partial \phi}{\partial x} \tag{2.3.8}$$

In a unit volume, the passage of current produces heat:

$$\frac{dq}{dt} = \kappa (\text{grad } \phi)^2 = T\sigma \tag{2.3.9}$$

the rate of entropy production having always a positive value:

$$\sigma = \frac{\kappa (\text{grad } \phi)^2}{T} \tag{2.3.10}$$

In addition to the transport of charge, the current flow in an electrolyte is also accompanied by mass transport. The *migration flux* of species i is given by the equation

$$\mathbf{J}_{i,\text{migr}} = -u_i z_i F c_i \text{ grad } \phi \tag{2.3.11}$$

where u_i is the *mobility*† of the particle (the velocity of the particle under the influence of unit force, in units of $N^{-1} \cdot mol \cdot m \cdot s^{-1}$), z_i its charge number and c_i its concentration. For the migration flux, the driving force is the molar electric energy gradient multiplied by -1, that is

$$\mathbf{X}_{\text{migr}} = -z_i F \text{ grad } \phi \tag{2.3.12}$$

The corresponding phenomenological coefficient is then given by the relationship

$$L_{\text{migr}} = u_i c_i \tag{2.3.13}$$

The flux of charge, connected with the mass flux of the electrically charged species, is given by Faraday's law for the equivalence of the current density and the material fluxes:

$$\mathbf{j} = \sum_i z_i F \mathbf{J}_i \tag{2.3.14}$$

Thus, in the case of migration material fluxes (Eq. 2.3.11) it holds for the partial current density (the contribution of the ith ion to the overall current density) that

$$\begin{aligned} \mathbf{j}_i &= -u_i z_i^2 F^2 c_i \text{ grad } \phi \\ &= -U_i \, |z_i| \, F c_i \text{ grad } \phi \end{aligned} \tag{2.3.15}$$

† The term *mobility* is used to describe the influence of the drag of the medium on the movement of a particle caused by an unspecified force (the unit of u_i is, for example, mol. $m \, s^{-1} \, N^{-1}$) which may be *diffusivity* (diffusion coefficient) with a chemical potential gradient as the driving force and *electrolytic mobility* connected with the electric field.

where U_i is the electrolytic mobility of the ith ion

$$U_i = |z_i| \, Fu_i \tag{2.3.16}$$

The total current density is

$$\begin{aligned}\mathbf{j} &= \sum_i \mathbf{j}_i = \sum_i z_i F \mathbf{J}_{i,\mathrm{migr}} \\ &= -\sum_i U_i \, |z_i| F c_i \, \mathrm{grad}\, \phi = -\kappa \, \mathrm{grad}\, \phi\end{aligned} \tag{2.3.17}$$

where κ is the conductivity of the system, i.e. this result is identical with Eq. (2.3.7).

Fick's first law,

$$\mathbf{J}_{i,\mathrm{diff}} = -D_i \, \mathrm{grad}\, c_i \tag{2.3.18}$$

or

$$J_{i,\mathrm{diff}} = -D_i \frac{\mathrm{d}c_i}{\mathrm{d}x} \tag{2.3.19}$$

is an empirical relationship for *diffusion*. Here, D_i is the diffusion coefficient of the ith component of the system. The concentration gradient cannot be used as a driving force for formulation of the rate of entropy production according to Eq. (2.3.5). Similarly to migration, the gradient of the partial molar Gibbs energy (chemical potential) multiplied by -1 is selected as the driving force. Equation (2.3.13) is again used for the phenomenological coefficient. The same value of u_i can be used for migration and diffusion only in dilute solutions. We set

$$\mathbf{J}_{i,\mathrm{diff}} = -u_i c_i \, \mathrm{grad}\, \mu_i = -RTu_i \, \mathrm{grad}\, c_i \tag{2.3.20}$$

Comparison with Eq. (2.3.18) yields the relationship for the diffusion coefficient

$$D_i = RTu_i \tag{2.3.21}$$

The ratio of the diffusion coefficient and the electrolytic mobility is given by the Nernst–Einstein equation (valid for dilute solutions)

$$\frac{D_i}{U_i} = \frac{RT}{|z_i| \, F} \tag{2.3.22}$$

The *convection flux* along the x coordinate,

$$J_i = c_i v_x \tag{2.3.23}$$

is the amount of substance contained in a column of height v_x (the velocity of the medium) and unit base. The general form of Eq. (2.3.23) is

$$\mathbf{J}_{i,\mathrm{conv}} = c_i \mathbf{v} \tag{2.3.24}$$

For the sake of completeness, the equation for the *heat conduction* (the

Fourier equation) can be introduced:

$$\mathbf{J}_{th} = -\lambda \operatorname{grad} T \tag{2.3.25}$$

expressing the dependence of the heat energy flux ($J \cdot cm^{-2} \cdot s^{-1}$) as a function of the thermal conductivity λ and temperature gradient.

Finally, the scalar flux of the *chemical reaction* is

$$J_{ch} = \frac{d\xi}{dt} = LA \tag{2.3.26}$$

where ξ is the extent of reaction, the phenomenological coefficient L is a function of the rate constant of the reaction and concentrations of the reactants and A is the affinity of the reaction. The affinity of the chemical reaction A is the driving force for the chemical flux.

References

See page 81.

2.4 Conduction of Electricity in Electrolytes

2.4.1 *Classification of conductors*

According to the nomenclature introduced by Faraday, two basic types of conductors can be distinguished, called *first* and *second class conductors.* According to contemporary concepts, electrons carry the electric current in first class conductors; in conductors of the second class, electric current is carried by ions. (The species carrying the charge in a given system are called *charge carriers.*)

The properties of electronic conductors follow from the band theory of the solid state. The energy levels of isolated atoms have definite values and electrons fill these levels according to the laws of quantum mechanics. However, when atoms approach one another, their electron shells interact and the positions of the individual energy levels change. When the set of atoms finally forms a crystal lattice, the original energy levels combine to form energy bands; each of these bands corresponds to an energy level in the isolated atom. Each energy level in a given band can contain a maximum of two electrons.

When some of the energy bands at a given temperature are completely occupied by electrons and bands with higher energy are empty, and a large amount of energy is required to transfer an electron from the highest occupied energy band to the lowest unoccupied band, the substance is an *insulator*.

Metals are examples of the opposite type of solid substance. The conductivity band, corresponding to the highest, partially occupied energy levels of the metal atom in the ground state, contains a sufficient number of

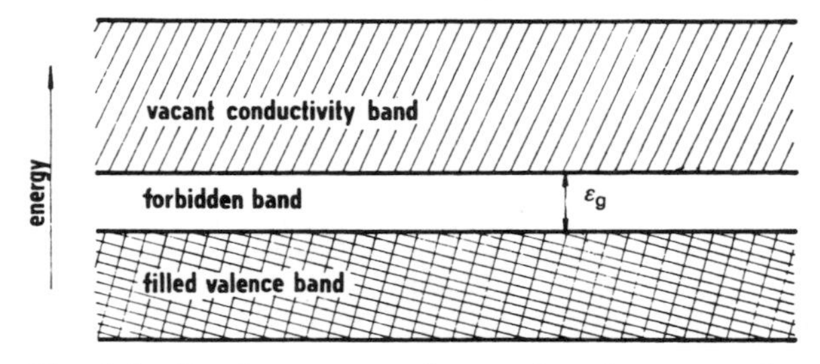

Fig. 2.2 Band structure of a semiconductor. ε_g denotes the energy gap (width of the forbidden band)

electrons even at absolute zero of temperature to produce the typical high conductivity of metals. These loosely bound electrons form an 'electron gas'. Under the influence of an external electric field, their originally random motion becomes oriented. It should be pointed out that the increase in the electron velocity in the direction of the electric field is much smaller than the average velocity of their random motion.

Semiconductors form a special group of electronic conductors. These are substances with chemically bonded valence electrons (forming a 'valence' band); however, when energy is supplied externally (e.g. by irradiation), these electrons can be excited to an energetically higher conduction band (see Fig. 2.2). A 'forbidden' band lies between the valence and conductivity bands. The energy difference between the lowest energy level of the conductivity band and the highest level of the valence band is termed the band gap ε_g (the width of the forbidden band).

Electric current is conducted either by these excited electrons in the conduction band or by 'holes' remaining in place of excited electrons in the original valence energy band. These holes have a positive effective charge. If an electron from a neighbouring atom jumps over into a free site (hole), then this process is equivalent to movement of the hole in the opposite direction. In the valence band, the electric current is thus conducted by these positive charge carriers. Semiconductors are divided into *intrinsic semiconductors,* where electrons are thermally excited to the conduction band, and semiconductors with intentionally introduced impurities, called *doped semiconductors,* where the traces of impurities account for most of the conductivity.

If the impurity is an *electron donor* (for germanium and silicon, group V elements, e.g. arsenic or antimony), then a new energy level is formed below the conduction band (see Fig. 2.3A). The energy difference between the lowest level of the conductivity band and this new level is denoted as ε_d. This quantity is much smaller than the energy gap ε_d. Thus electrons from this donor band pass readily into the conductivity band and represent the main contribution to the semiconductor conductivity. As the charge carriers are electrons, i.e. negatively charged species, these materials are termed *n-type semiconductors*.

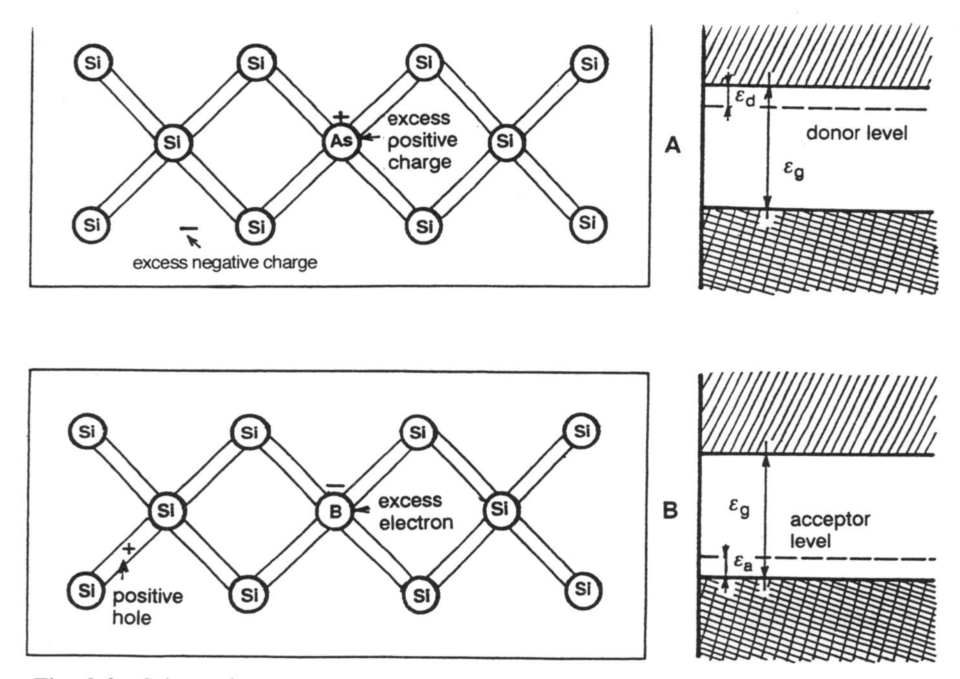

Fig. 2.3 Schematic structure of a silicon semiconductor of (A) *n* type and (B) *p* type together with the energy level array. (According to C. Kittel)

If, on the other hand, the impurity is an *electron acceptor* (here, elements of group III), then the new acceptor band lies closely above the valence band (see Fig. 2.3B). The electrons from the valence band pass readily into this new band and leave holes behind. These holes are the main charge carriers in *p-type semiconductors.*

Compared with metals, semiconductors have quite high resistivity, as conduction of current requires a supply of activation energy. The conductivity of semiconductors increases with increasing temperature.

Germanium provides a good illustration of conductivity conditions in semiconductors. In pure germanium, the concentration of charge carriers is $n_e^0 = 2 \times 10^{13}\ \text{cm}^{-3}$ (in metals, the concentration of free electrons is of the order of $10^{22}\ \text{cm}^{-3}$). The intrinsic conductivity of germanium is about the same as that of pure (conductivity) water, $\kappa = 5.5 \times 10^{8}\ \Omega^{-1}\ \text{cm}^{-1}$ at 25°C. In strongly doped semiconductors, the concentration of charge carriers can increase by up to four orders of magnitude, but is nonetheless still comparable with the concentration of dilute electrolytes.

Ionic (*electrolytic*) *conduction* of electric current is exhibited by electrolyte solutions, melts, solid electrolytes, colloidal systems and ionized gases. Their conductivity is small compared to that of metal conductors and increases with increasing temperature, as the resistance of a viscous medium acts against ion movement and decreases with increasing temperature.

A special class of conductors are *ionically and electronically conducting polymers* (Sections 2.6.4 and 5.5.5).

2.4.2 *Conductivity of electrolytes*

This part will be concerned with the properties of electrolytes (liquid or solid) under ordinary laboratory conditions (i.e. in the absence of strong external electric fields). The electroneutrality condition (Eq. 1.1.1) holds with sufficient accuracy for current flow under these conditions:

$$\sum_i z_i c_i = 0 \tag{2.4.1}$$

where c_i are the concentrations of ions i and z_i are their charge numbers. The basic equations for the current density (Faraday's law), electrolytic mobility and conductivity are (2.3.14), (2.3.15) and (2.3.17).

The *conductivity*

$$\kappa = \sum_i z_i^2 F^2 u_i c_i = \sum_i |z_i| \, F U_i c_i \tag{2.4.2}$$

in dilute solutions is thus a linear function of the concentrations of the components and the proportionality constants are termed the (individual) *ionic conductivities*:

$$\lambda_i = z_i^2 F^2 u_i = |z_i| \, F U_i \tag{2.4.3}$$

Consider a solution in which a single strong electrolyte of a concentration c is dissolved; this electrolyte consists of ν_+ cations B^{z+} in concentration c_+ and ν_- anions A^{z-} in concentration c_-. Obviously

$$\nu_+ z_+ = -\nu_- z_- = \nu_- \, |z_-| \tag{2.4.4}$$

and

$$c = \frac{c_+}{\nu_+} = \frac{c_-}{\nu_-} \tag{2.4.5}$$

Substitution into Eq. (2.4.2) yields

$$\begin{aligned} \kappa &= z_+^2 F^2 u_+ \nu_+ c + z_-^2 F^2 u_- \nu_- c \\ &= (U_+ + U_-) z_+ \nu_+ F c = (U_+ + U_-) \, |z_-| \, \nu_- F c \end{aligned} \tag{2.4.6}$$

The quantity

$$(U_+ + U_-) z_+ \nu_+ F = (U_+ + U_-) \, |z_-| \, \nu_- F = \frac{\kappa}{c} = \Lambda \tag{2.4.7}$$

is called the *molar conductivity,* which is as shown below, a concentration-dependent quantity except in an ideal solution (in practice at high dilution).

For first class conductors, the conductivity is a constant characterizing the ability of a given material at a given temperature to conduct electric current. However, for electrolyte solutions, it depends on the concentration and is not a material constant. Thus the fraction $\Lambda = \kappa/c$ is introduced; however, it will be seen below that the constant characterizing the ability of a given electrolyte to conduct electric current in solution is given by the limiting value of the molar conductivity at zero concentration. The main unit of molar conductivity is $\Omega^{-1} \cdot m^2 \cdot mol^{-1}$, corresponding to κ in $\Omega^{-1} \cdot m^{-1}$ and c in $mol \cdot m^{-3}$. However, units of $\Omega^{-1} \cdot cm^2 \cdot mol^{-1}$ are often used. If units of $\Omega^{-1} \cdot cm^{-1}$ are simultaneously used for κ and the usual units of $mol \cdot dm^{-3}$ for the concentration, then Eq. (2.4.7) becomes

$$\Lambda = \frac{1000\kappa}{c} \tag{2.4.8}$$

When reporting the molar conductivity data, the species whose amount is given in moles should be indicated. Often, a fractional molar conductivity corresponding to one mole of chemical equivalents (called a val) is reported. For example, for sulphuric acid, the concentration c can be expressed as the 'normality', i.e. the species $\frac{1}{2}H_2SO_4$ is considered. Obviously, $\Lambda(H_2SO_4) = 2\Lambda(\frac{1}{2}H_2SO_4)$. Consequently, the concept of the 'equivalent conductivity' is often used, defined by the relationship

$$\Lambda^* = \frac{\Lambda}{z_+\nu_+} = \frac{\Lambda}{|z_-|\,\nu_-} = (U_+ + U_-)F = \lambda_+^* + \lambda_-^* \tag{2.4.9}$$

where $\lambda_+^* = \lambda_+/z_+$ and $\lambda_-^* = \lambda_-/|z_-|$ (cf. Eq. 2.4.3). It follows that, for our example of sulphuric acid, $\Lambda^*(H_2SO_4) = \Lambda(\frac{1}{2}H_2SO_4) = \frac{1}{2}\Lambda(H_2SO_4)$.

Combination of Eqs (2.3.15) and (2.3.17) yields

$$\mathbf{j}_i = -t_i\kappa \operatorname{grad} \phi \tag{2.4.10}$$

where t_i is the *transport* (transference) *number*, giving the contribution of the ith ion to the total conductivity κ, that is

$$t_i = \frac{z_i^2F^2u_ic_i}{\kappa} = \frac{z_i^2F^2u_ic_i}{\sum_j z_j^2F^2u_jc_j} = \frac{|z_i|\,FU_ic_i}{\sum_j |z_j|\,FU_jc_j} \tag{2.4.11}$$

for a single (binary) electrolyte,

$$\begin{aligned} t_+ &= \frac{U_+}{U_+ + U_-} = \frac{\lambda_+^*}{\lambda_+^* + \lambda_-^*} \\ t_- &= \frac{U_-}{U_+ + U_-} = \frac{\lambda_-^*}{\lambda_+^* + \lambda_-^*} \end{aligned} \tag{2.4.12}$$

For a solution of a weak electrolyte of total concentration c, dissociating to form ν_+ cations and ν_- anions and with a degree of dissociation α, we have, considering Eqs (2.4.6) and (2.4.9),

$$\begin{aligned}\kappa &= \alpha(U_+ + U_-)z_+\nu_+Fc = \alpha(U_+ + U_-)\,|z_-|\,\nu_-Fc\\ \Lambda &= \alpha(U_+ + U_-)z_+\nu_+F = \alpha(U_+ + U_-)\,|z_-|\,\nu_-F\\ \Lambda^* &= \alpha(U_+ + U_-)F\end{aligned} \tag{2.4.13}$$

The behaviour of real solutions approaches that of ideal solutions at high dilution. The molar conductivity at limiting dilution, denoted Λ^0, is

$$\begin{aligned}\Lambda^0 &= z_+\nu_+F(U_+^0 + U_-^0) = |z_-|\,\nu_-F(U_+^0 + U_-^0) = \nu_+\lambda_+^0 + \nu_-\lambda_-^0\\ \Lambda^{0*} &= \lambda_+^{*0} + \lambda_-^{*0}\end{aligned} \tag{2.4.14}$$

This equation is valid for both strong and weak electrolytes, as $\alpha = 1$ at the limiting dilution. The quantities $\lambda_i^0 = |z_i|\,FU_i^0$ have the significance of ionic conductivities at infinite dilution. The *Kohlrausch law of independent ionic conductivities* holds for a solution containing an arbitrary number of ion species. At limiting dilution, all the ions conduct electric current independently; the total conductivity of the solution is the sum of the contributions of the individual ions.

Because of the interionic forces, the conductivity is directly proportional to the concentration only at low concentrations. At higher concentrations, the conductivity is lower than expected from direct proportionality. This decelerated growth of the conductivity corresponds to a decrease of the molar conductivity. Figure 2.4 gives some examples of the dependence of

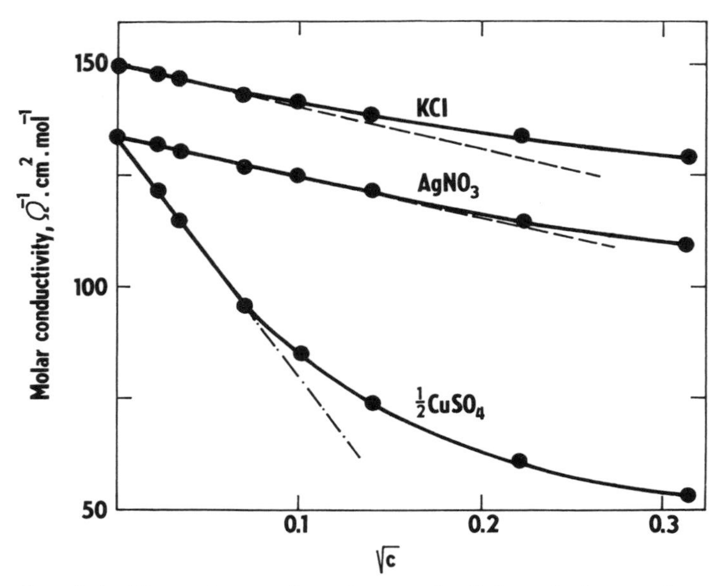

Fig. 2.4 Dependence of molar conductivity of strong electrolytes on the square root of concentration c. The dashed lines demonstrate the Kohlrausch law (Eq. 2.4.15)

the molar conductivity of strong electrolytes on the concentration. It can be seen that even at rather low concentrations, the limiting molar conductivity is still not attained.

A special branch of the theory of strong electrolytes deals with the dependence of the electrical conductivity of electrolytes on concentration (see Section 2.4.3). For very low concentrations, Kohlrausch found empirically that

$$\Lambda = \Lambda^0 - kc^{1/2} \tag{2.4.15}$$

(k is an empirical constant), i.e. the dependence of Λ on $c^{1/2}$ is linear.

Interionic forces are relatively less important for weak electrolytes because the concentrations of ions are relatively rather low as a result of incomplete dissociation. Thus, in agreement with the classical (Arrhenius) theory of weak electrolytes, the concentration dependence of the molar conductivity can be attributed approximately to the dependence of the degree of dissociation α on the concentration. If the degree of dissociation

$$\alpha \approx \frac{\Lambda}{\Lambda^0} \tag{2.4.16}$$

is substituted into the equation for the apparent dissociation constant K', then the result is

$$K' = \frac{\alpha^2 c}{1-\alpha} \approx \frac{\Lambda^2 c}{\Lambda^0(\Lambda^0 - \Lambda)} \tag{2.4.17}$$

sometimes called the Ostwald dilution law. In a suitably linearized form, this equation can be used to calculate the quantities K' and Λ^0 from measured values of Λ and c. However, the resulting values of K' are not thermodynamic dissociation constants; the latter can be found from conductivity measurements by a more complicated procedure (see Section 2.4.5). Figure 2.5 illustrates, as an example, the dependence of the molar conductivity on the concentration for acetic acid compared with hydrochloric acid.

While the molar conductivity of strong electrolytes Λ^0 can be measured directly, for determination of the ionic conductivities the measurable transport numbers must be used (cf. Eq. (2.4.12)). Table 2.1 lists the values of the limiting conductivities of some ions in aqueous solutions.

2.4.3 *Interionic forces and conductivity*

The influence of interionic fores on ion mobilities is twofold. The *electrophoretic effect* (occurring also in the case of the electrophoretic motion of charged colloidal particles in an electric field, cf. p. 242) is caused by the simultaneous movement of the ion in the direction of the applied

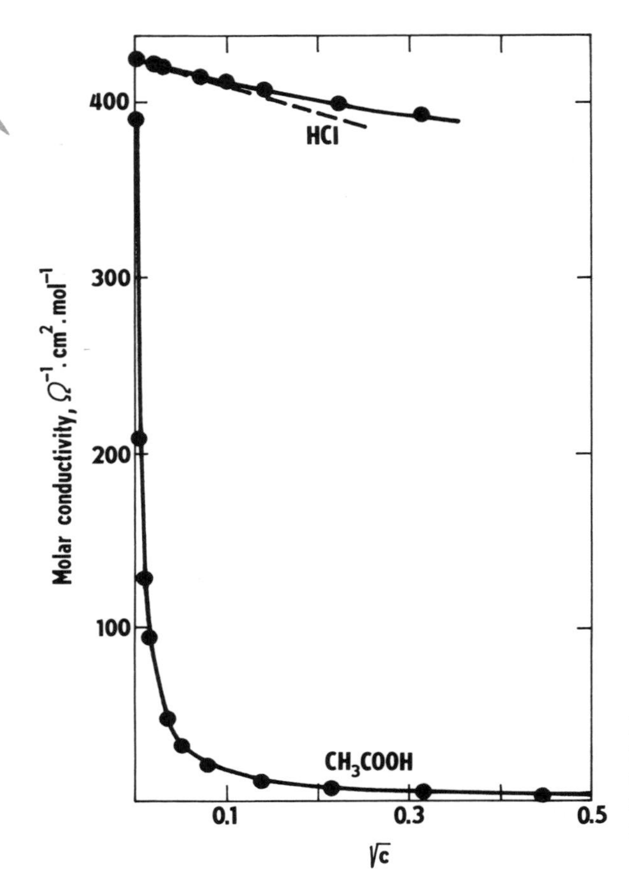

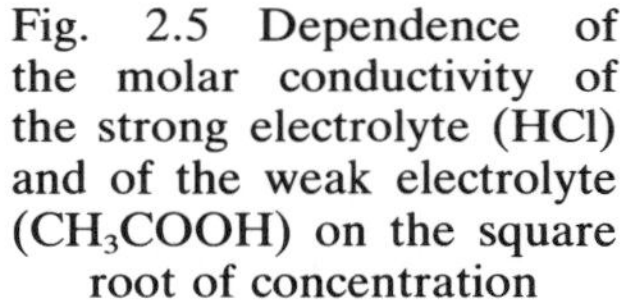
Fig. 2.5 Dependence of the molar conductivity of the strong electrolyte (HCl) and of the weak electrolyte (CH_3COOH) on the square root of concentration

Table 2.1 Ionic conductivities at infinite dilution ($\Omega^{-1} \cdot cm^2 \cdot mol^{-1}$) at various temperatures. (According to R. A. Robinson and R. H. Stokes)

	Temperature (°C)								
Ion	0	5	15	18	25	35	45	55	100
H^+	225.0	250.10	300.60	315	349.81	397.0	441.4	483.1	630
OH^-	105.0	—	—	171	198.30	—	—	—	450
Li^+	19.4	22.76	30.20	32.8	38.68	48.00	58.04	68.74	115
Na^+	26.5	30.30	39.77	42.8	50.10	61.54	73.73	86.88	145
K^+	40.7	46.75	59.66	63.9	73.50	88.21	103.49	119.29	195
Rb^+	43.9	50.13	63.44	66.5	77.81	92.91	108.55	124.25	—
Cs^+	44.0	50.03	63.16	67	77.26	92.10	107.53	123.66	—
Cl^-	41.0	47.51	61.41	66.0	76.35	92.21	108.92	126.40	212
Br^-	42.6	49.25	63.15	68.0	78.14	94.03	110.68	127.86	—
I^-	41.4	48.57	62.17	66.5	76.84	92.39	108.64	125.44	—

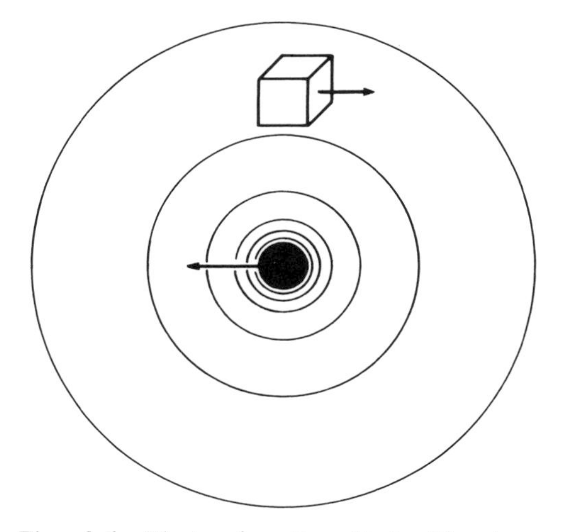

Fig. 2.6 Electrophoretic effect. The ion moves in the opposite direction to the ionic atmosphere

electric field and of the ionic atmosphere in the opposite direction (Fig. 2.6). Both the central ion and the ions of the ionic atmosphere take the neighbouring solvent molecules with them, which results in a retardation of the movement of the central ion.

For very dilute solutions, the motion of the ionic atmosphere in the direction of the coordinates can be represented by the movement of a sphere with a radius equal to the Debye length $L_D = \kappa^{-1}$ (see Eq. 1.3.15) through a medium of viscosity η under the influence of an electric force $z_i e E_x$, where E_x is the electric field strength and z_i is the charge of the ion that the ionic atmosphere surrounds. Under these conditions, the velocity of the ionic atmosphere can be expressed in terms of the Stokes' law (2.6.2) by the equation

$$v_x = \frac{z_i e E_x}{6\pi\eta L_D} \tag{2.4.18}$$

The electrolytic mobility of the ionic atmosphere around the ith ion can then be defined by the expression

$$U_{ai} = \frac{z_i e}{6\pi\eta L_D} = \frac{z_i F}{6\pi N_A \eta L_D} \tag{2.4.19}$$

This quantity can be identified with deceleration of the ion as a result of the motion of the ionic atmosphere in the opposite direction, i.e.

$$-\Delta U_i = \frac{z_i F}{6\pi N_A \eta L_D} \tag{2.4.20}$$

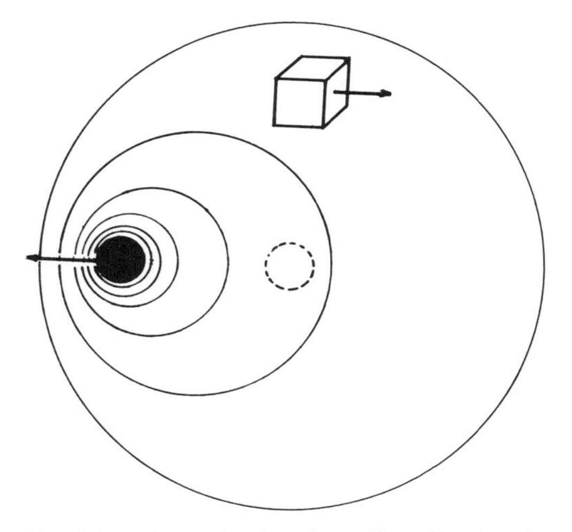

Fig. 2.7 Time-of-relaxation effect. During the movement of the ion the ionic atmosphere is renewed in a finite time so that the position of the ion does not coincide with the centre of the ionic atmosphere

The *time-of-relaxation effect* (see Fig. 2.7) originates in a certain time delay (relaxation) required for the renewal of the spherical symmetry of the ionic atmosphere around the central ion moving under the influence of an applied electric field. The disappearance of the ionic atmosphere after removal of the central ion, similar to its formation, is an exponential function of time; in fact, both of these processes are complete after twice the relaxation time, which is of the order of 10^{-7} to 10^{-9} s, depending on the electrolyte concentration. If the central ion moves under the influence of an external electric field, it becomes asymmetrically located with respect to the centre of the ionic atmosphere. Thus the time average of the forces of interaction of the ionic atmosphere with the central ion is not equal to zero. The external electric field is decreased by the relaxation electric field, as it is oriented in the opposite direction to the external force. Although the relaxation time is several orders of magnitude smaller than the time required for the central ion to pass through the ionic atmosphere (about 10^{-3} s), its effect is important because the strength of the electric field formed by the ionic atmosphere ($\approx 10^5$ V · cm^{-1}) is greater than the strength of the external electric field. Thus, even small changes in the symmetry of the ionic atmosphere have a measurable effect acting against that of the external electric field.

The mathematical theory of the time-of-relaxation effect is based on the interionic electrostatics and the hydrodynamic equation of flow continuity. It is the most involved part of the theory of strong electrolytes. Only the main conclusions will be given here.

The first approximate calculation was carried out by Debye and Hückel and later by Onsager, who obtained the following relationship for the relative strength of the relaxation field $\Delta E/E$ in a very dilute solution of a single uni-univalent electrolyte

$$-\frac{\Delta E}{E}=\frac{F^2}{12(2+\sqrt{2})\pi N_A \varepsilon RT} \tag{2.4.21}$$

In the ideal case, the ionic conductivity is given by the product $z_i F U_i^0$. Because of the electrophoretic effect, the real ionic mobility differs from the ideal by ΔU_i and equals $U^0 + \Delta U_i$. Further, in real systems the electric field is not given by the external field E alone, but also by the relaxation field ΔE, and thus equals $E + \Delta E$. Thus the conductivity (related to the unit external field E) is increased by the factor $(E + \Delta E)/E$. Consideration of both these effects leads to the following expressions for the equivalent ionic conductivity (cf. Eq. 2.4.9):

$$\begin{aligned} \lambda_+^* &= F(U_+^0 + \Delta U_+)\left(1+\frac{\Delta E}{E}\right) \\ \lambda_-^* &= F(U_-^0 + \Delta U_-)\left(1+\frac{\Delta E}{E}\right) \end{aligned} \tag{2.4.22}$$

and for the overall equivalent conductivity of the electrolyte

$$\begin{aligned} \Lambda^* &= F(U_+^0 + \Delta U_+ + U_-^0 + \Delta U_-)\left(1+\frac{\Delta E}{E}\right) \\ &\approx F\left[(U_+^0 + U_-^0)\left(1+\frac{\Delta E}{E}\right) + \Delta U_+ + \Delta U_-\right] \end{aligned} \tag{2.4.23}$$

We shall introduce $FU_+^0 = \lambda_+^0$, $FU_-^0 = \lambda_-^0$ and $F(U_+^0 + U_-^0) = \Lambda^0$. The resulting Onsager's expression (2.4.21) completed by the term $1 + Ba\sqrt{c}$ in the denominator introduced by Falkenhagen (cf. Eq. (1.3.31)) takes both effects of interionic forces at higher concentrations into account:

$$\Lambda^* = \Lambda^{*0} - (B_1\Lambda^0 + B_2)\frac{\sqrt{c}}{1+Ba\sqrt{c}} \tag{2.4.24}$$

where

$$\begin{aligned} B_1 &= \frac{F^2\kappa}{12\pi(2+\sqrt{2})N_A\varepsilon RT\sqrt{c}} = \frac{F^3}{12\pi(1+\sqrt{2})(\varepsilon RT)^{3/2}} \\ B_2 &= \frac{F^2\kappa}{6\pi N_A \eta\sqrt{c}} = \frac{\sqrt{2}F^3}{3\pi N_A\eta(\varepsilon RT)^{1/2}} \end{aligned} \tag{2.4.25}$$

The validity of Eq. (2.4.24) for uni-univalent electrolytes has been verified up to a concentration of 0.1 mol · dm^{-3}. If the ionic radius is not involved in

Eq. (2.4.24) (i.e. the product $Ba\sqrt{c}$ is cancelled), then the *Onsager limiting law* is obtained

$$\Lambda = \Lambda^0 - (B_1\Lambda^0 + B_2)\sqrt{c} \tag{2.4.26}$$

which is an analogy of the Debye–Hückel limiting law. In the same way it is valid for rather dilute solutions (see Fig. 2.4, dashed line). The ratio of the molar conductivities at moderate and limiting dilutions is the conductivity coefficient given for rather dilute solutions by the relation

$$\gamma_\Lambda = 1 - (B_1 + B_2/\Lambda^0)\sqrt{c} \tag{2.4.27}$$

In an analogous way, for a weak uni-univalent electrolyte the Onsager limiting law has the form

$$\Lambda = \alpha[\Lambda^0 - (B_1\Lambda^0 + B_2)\sqrt{\alpha c}] \tag{2.4.28}$$

The ratio of the equivalent conductivity at a given concentration to the limiting equivalent conductivity then is

$$\Lambda/\Lambda^0 = \alpha\gamma_\Lambda \tag{2.4.29}$$

These relations are used in the precise form of Ostwald's dilution law (2.4.17).

Equations (2.4.27) and (2.4.28) are employed for conductometric determination of dissociation constants and solubility products.

2.4.4 *The Wien and Debye–Falkenhagen effects*

The conductivities of strong electrolytes do not depend on the strength of the electric field for weak fields (of the order of $10^4\,\mathrm{V\cdot m^{-1}}$). At high electric field strengths (of about $10^7\,\mathrm{V\cdot m^{-1}}$), Wien observed a significant increase in the conductivity (Fig. 2.8). This effect increases at higher concentrations and at a higher charge number of the electrolyte ions and approaches a limiting value with increasing electric field. This phenomenon is a result of the high ion velocities, preventing rearrangement of the ionic atmospheres during motion. Thus an ionic atmosphere is not formed at all and both the electrophoretic and relaxation effects disappear.

The conductivity also increases in solutions of weak electrolytes. This 'second Wien effect' (or field dissociation effect) is a result of the effect of the electric field on the dissociation equilibria in weak electrolytes. For example, from a kinetic point of view, the equilibrium between a weak acid HA, its anion A^- and the oxonium ion H_3O^+ has a dynamic character:

$$HA + H_2O \underset{k_r}{\overset{k_d}{\rightleftharpoons}} H_3O^+ + A^-$$

where k_d is the rate constant for dissociation and k_r is the rate constant for recombination of the anion with a hydronium ion. Their ratio yields the

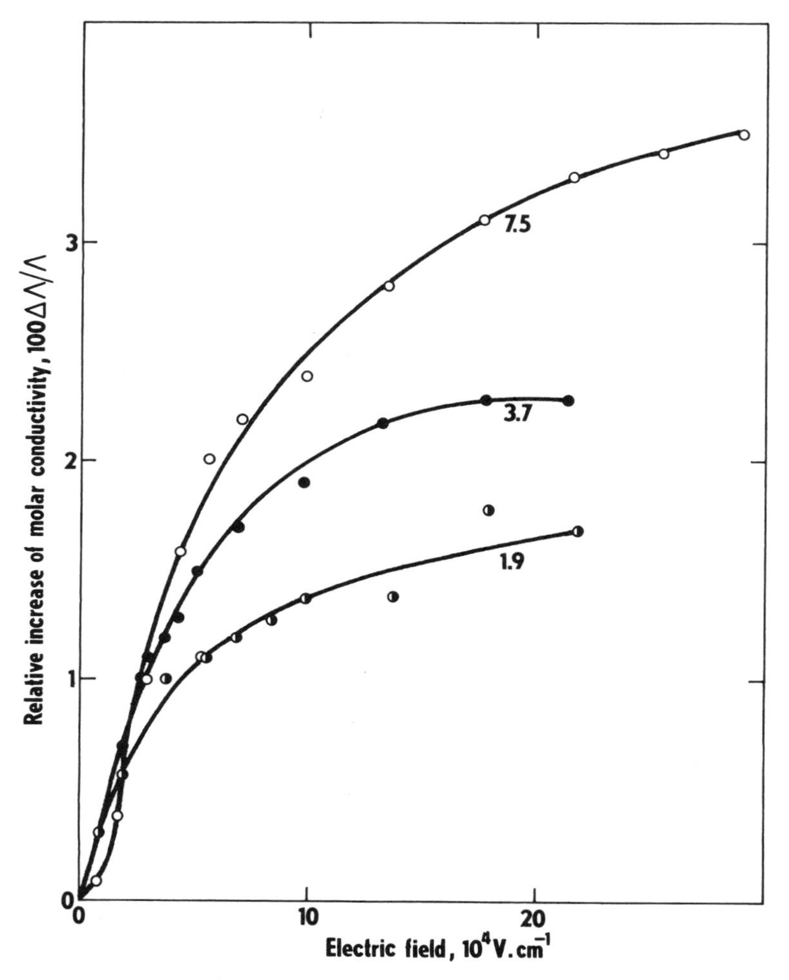

Fig. 2.8 The Wien effect shown by the percentage increase of equivalent conductivity in dependence on the electric field in $Li_3Fe(CN)_6$ solutions in water. Concentrations in mmol · dm^{-3} are indicated at each curve

dissociation constant for a weak acid

$$K_A = \frac{k_d}{k_r} \quad (2.4.30)$$

The bond between the hydrogen atom and the anion is primarily electrostatic in character, so that the acid molecule has the character of an ion pair to a certain extent. As demonstrated by Onsager, the rate of dissociation of an ion pair increases in the presence of an external electric field, while the rate of formation of this ion pair is not affected by the presence of an external field. The value of k_d thus increases in the presence of a field, as does K_A. A weak electrolyte is therefore dissociated to a greater degree in strong electric fields and its conductivity increases.

Debye and Falkenhagen predicted that the ionic atmosphere would not be able to adopt an asymmetric configuration corresponding to a moving central ion if the ion were oscillating in response to an applied electrical field and if the frequency of the applied field were comparable to the reciprocal of the relaxation time of the ionic atmosphere. This was found to be the case at frequencies over 5 MHz where the molar conductivity approaches a value somewhat higher than Λ^0. This increase of conductivity is caused by the disappearance of the time-of-relaxation effect, while the electrophoretic effect remains in full force.

2.4.5 *Conductometry*

Determination of the conductivity of electrolyte solutions. The resistance of second class conductors is measured by the bridge method, similar to measurement of the resistance of first class conductors. Usually, alternating current is used, to avoid polarization of the electrodes (see Chapter 5). This involves certain technical difficulties, connected with the necessity of compensating the capacity and inductive components of the resistors and also with the fact that current can pass through the capacity coupling between the circuit and ground; similarly, strong magnetic fields produce induction effects. Direct current can only be used if a suitable unpolarizable electrode can be found for the given system (see page 209). Therefore, most often an a.c. supply of audio-frequency and an amplitude of a few millivolts is used.

The bridge method is based on various modifications of the well-known Wheatstone bridge.

Instruments for exact measurements usually have a sinusoidal current source and an electronic balance detector. The circuit is made as symmetrical as possible to avoid stray coupling.

During measurement, the conductivity cell is filled with an electrolyte solution; this cell is usually made of glass with sealed platinum electrodes. Various shapes are used, depending on the purpose that it is to serve. Figure 2.9 depicts examples of suitable cell arrangements. The electrodes are covered with platinum black, to avoid electrode polarization. The electrodes are placed close to one another in poorly conductive solutions and further apart in more conductive solutions.

For an evaluation of measurements the conductivity cell is calibrated with a solution of a known conductivity. Both the electrolyte and the solvent must be carefully purified.

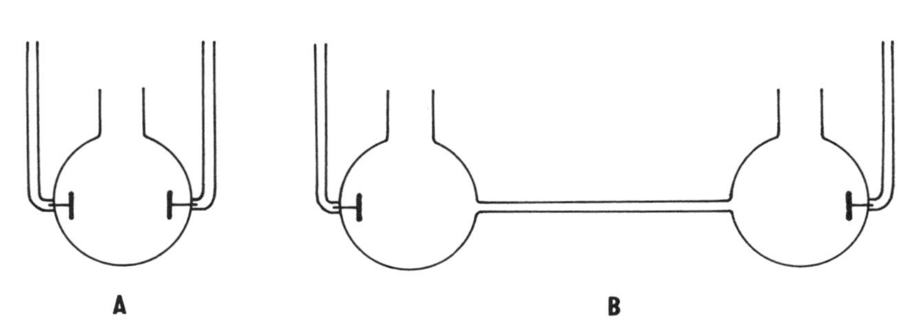

Fig. 2.9 Conductometric cells for (A) low and (B) high conductivity solutions

2.4.6 *Transport numbers*

Transport (transference) numbers are defined by Eqs (2.4.11) and (2.4.12). The experimental arrangement shown in Fig. 2.10 can be used to demonstrate this concept. In the cathode and anode compartments the electrolyte concentration is maintained homogeneous by stirring. The connecting tube is narrow, so that it contains a negligible amount of electrolyte compared with that in the electrode compartments. The electric field is homogeneously distributed and mass transfer occurs only through migration at sufficiently high field strengths. If the charge Q passes through this electrolysis cell, then Q/z_+F moles of cations are discharged on the cathode and $Q/|z_-|F$ moles of anions on the anode. An amount of $Qt_-/|z_-|F$ moles of anions migrates from the cathode space to the anode space and Qt_+/z_+F moles of cations in the opposite direction. Thus the cathode compartment will be depleted by a total of $\Delta n_{+,C} = Qt_-/z_+F$ moles of cations and $\Delta n_{-,C} = Qt_-/|z_-|F$ moles of anions, and the anode compartment by $\Delta n_{+,A} = Qt_+/z_+F$ moles of cations and $\Delta n_{-,A} = Qt_+/|z_-|F$ moles

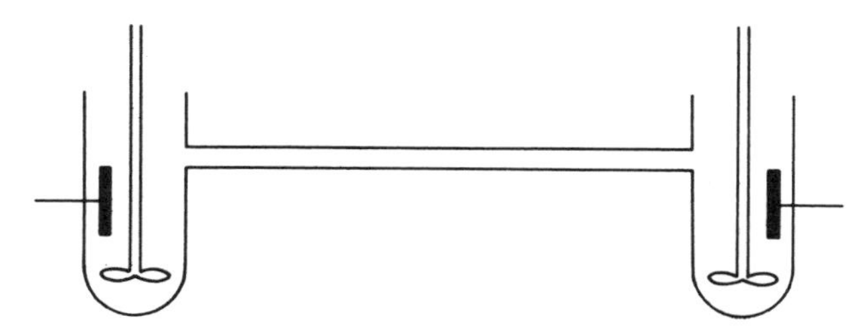

Fig. 2.10 Schematic design of a cell for the determination of transport numbers from measurements of the concentration decrease in electrode compartments (Hittorf's method)

of anions. As it obviously holds that $n_+ = |z_-| n$ and $n_- = z_+ n$ for the amounts of cations n_+, anions n_- and salt n, respectively, then the decrease in the amount of the salt is $\Delta n_C = Qt_-/z_+ |z_-| F$ and $\Delta n_A = Qt_+/z_+ |z_-| F$. Thus the decrease in the amount of salt in the cathode compartment is proportional to the transport number of the anion and the decrease in the amount of salt in the anode compartment is proportional to the transport number of the cation,

$$t_+ = \frac{\Delta n_A}{\Delta n_C + \Delta n_A} \quad \text{and} \quad t_- = \frac{\Delta n_C}{\Delta n_C + \Delta n_A} \tag{2.4.31}$$

Thus, the transport number can be found from measurement of the decrease in the amount of the salt in the electrode compartments.

In view of Eq. (2.4.9), for a strong electrolyte,

$$t_+ = \frac{\lambda_+^*}{\lambda_+^* + \lambda_-^*} = \frac{\lambda_+^*}{\Lambda^*} \quad \text{and} \quad t_- = \frac{\lambda_-^*}{\lambda_+^* + \lambda_-^*} = \frac{\lambda_-^*}{\Lambda^*} \tag{2.4.32}$$

and, for a weak electrolyte,

$$t_+ = \frac{\alpha\lambda_+^*}{\Lambda^*} \quad \text{and} \quad t_- = \frac{\alpha\lambda_-^*}{\Lambda^*} \tag{2.4.33}$$

These equations are used to determine ionic conductivities.

The transport numbers thus depend on the mobilities of both the ions of the electrolyte (or of all the ions present). This quantity is therefore not a characteristic of an isolated ion, but of an ion in a given electrolyte. Table 2.2 lists examples of transport numbers. It can be seen from the table that the transport numbers also depend on the electrolyte concentration. The following rules can be derived from experimental data:

(a) If the transport number is close to 0.5, it depends only very slightly on the concentration.
(b) If the transport number for the cation is less than 0.5, then it decreases with increasing concentration (and, simultaneously, $t_- > 0.5$ increases with increasing concentration); and vice versa.

These rules are based on the theory of conductivity of strong electrolytes accounting for the electrophoretic effect only (the relaxation effect terms outbalance each other).

The methods for determination of transport numbers include the Hittorf method and the concentration cell method (p. 121).

The Hittorf method is based on measuring the concentration changes at the anode and cathode during electrolysis. These changes can be found by a sensitive analytical method, e.g. conductometrically for a suitable cell

Table 2.2 Transport numbers of cations at various concentrations ($mol \cdot dm^{-3}$). Relative accuracy 0.02 per cent. (According to B. E. Conway)

	c					
Electrolyte	0	0.01	0.02	0.05	0.10	0.20
HCl	0.8209	0.8251	0.8266	0.8292	0.8314	0.8337
CH_3COONa	0.5507	0.5537	0.5550	0.5573	0.5594	0.5610
CH_3COOK	0.6427	0.6498	0.6523	0.6569	0.6609	—
KNO_3	0.5072	0.5084	0.5087	0.5093	0.5103	0.5120
NH_4Cl	0.4909	0.4907	0.4906	0.4905	0.4907	0.4911
KCl	0.4906	0.4902	0.4901	0.4899	0.4898	0.4894
KI	0.4892	0.4884	0.4883	0.4882	0.4883	0.4887
KBr	0.4849	0.4833	0.4832	0.4831	0.4833	0.4841
$AgNO_3$	0.4643	0.4648	0.4652	0.4664	0.4682	—
NaCl	0.3963	0.3918	0.3902	0.3876	0.3854	0.3821
LiCl	0.3364	0.3289	0.3261	0.3211	0.3168	0.3112
$CaCl_2$	0.4380	0.4264	0.4220	0.4140	0.4060	0.3953
$\frac{1}{2}Na_2SO_4$	0.386	0.3848	0.3836	0.3829	0.3828	0.3828
$\frac{1}{2}K_2SO_4$	0.479	0.4829	0.4848	0.4870	0.4890	0.4910
$\frac{1}{3}LaCl_3$	0.477	0.4625	0.4576	0.4482	0.4375	0.4233
$\frac{1}{4}K_4Fe(CN)_6$	—	0.515	0.555	0.604	0.647	—
$\frac{1}{3}K_3Fe(CN)_6$	—	—	—	0.475	0.491	—

arrangement. Transport numbers can be found by using Eq. (2.4.31) from the overall change in electrolyte contents in the anode and cathode compartments.

Correct transport numbers could be obtained from concentration changes in the electrode compartments only if the ions were not hydrated in the aqueous solution. When they are hydrated the accompanying water molecules remain in the electrode compartment during discharge of the ions. Accordingly, the measured decrease in electrolyte concentration at the electrode is greater (or smaller) than would correspond to simple charge transport. If n_+ molecules of water are bound to each cation n_- and to each anion, then W moles of water are transported by a charge passage corresponding to 1 F:

$$W = n_+t_+ - n_-t_- \tag{2.4.34}$$

If the cation is more hydrated, then W is a positive number; if the anion is more hydrated, then W is a negative number and water is transported to the anode. Transport numbers calculated from measured concentration changes involving transport of water by solvated ions are sometimes called Hittorf (t_i') numbers; those corrected for the transport of water are called true transport numbers (t_i). These two types of transport numbers are related by the following expressions:

$$t_+ = t'_+ + \frac{cW}{55.5 z_+ \nu_+} \qquad \text{and} \qquad t_- = t'_- - \frac{cW}{55.5\,|z_-|\,\nu_-} \tag{2.4.35}$$

If water is transported to the cathode, then $W > 0$, $t_+ > t'_+$ and $t_- < t'_-$. If water is transported to the anode, then $W < 0$, $t_+ < t'_+$ and $t_- > t'_-$. Ion solvation can be determined from measured true and Hittorf transport numbers. The value of W can be found experimentally by adding a non-electrolyte (e.g. sucrose) to the solution. As it is assumed that this substance does not migrate in the electric field, W is determined by measuring changes in its concentration in the electrode compartment after electrolysis. It is evident from Eq. (2.4.34) that, even when the true transport numbers and the values of W are known, it is not possible to determine the absolute values of hydration numbers, but only relative values under certain assumptions (e.g. that the bromide ion is not hydrated).

References

Barthel, J., Transport properties of electrolytes from infinite dilution to saturation, *Pure Appl. Chem.,* **57,** 355 (1985).

Falkenhagen, H., *Theorie der Elektrolyte,* Hirzel, Leipzig, 1971.

Fuoss, R. M., and F. Accascina, *Electrolytic Conductance,* Interscience, New York, 1959.

Justice, J.-C., Conductance of electrolyte solutions, *CTE,* **5,** Chap. 3 (1982).

Kittel, C., *Introduction to Solid State Physics,* John Wiley & Sons, New York, 1976.

Robbins, J., *Ions in Solution,* Vol. 2, *An Introduction to Electrochemistry,* Oxford University Press, Oxford, 1972.

Shedlovsky, T., and L. Shedlovsky, Conductometry, in *Physical Methods of Chemistry* (Eds. A. Weissberger and B. W. Rossiter), Part IIA, Wiley–Interscience, New York, 1971, p. 163.

2.5 Diffusion and Migration in Electrolyte Solutions

The passage of current through electrolytes as treated in Section 2.4 did not involve any transient phenomena. During migration the current and the potential difference were assumed constant. During the measurement, transient phenomena appear only in connection with electrode charging or with reactions at the electrodes from which electric charge is transferred into the electrolyte. One of the aims of experimental methods is to eliminate this effect during measurements.

On the other hand, it is characteristic for diffusion processes that they usually involve time changes of the concentrations of the diffusing substances, so that they have a transient character. Steady state is a result of a prior transient phenomenon that is inherent in the diffusion process itself.

It is characteristic for the actual diffusion in electrolyte solutions that the individual species are not transported independently. The diffusion of the faster ions forms an electric field that accelerates the diffusion of the slower ions, so that the electroneutrality condition is practically maintained in solution. Diffusion in a two-component solution is relatively simple (i.e. diffusion of a binary salt—see Section 2.5.4). In contrast, diffusion in a three-component electrolyte solution is quite complicated and requires the use of equations such as (2.1.2), taking into account that the flux of one electrically charged component affects the others.

A simple case is the diffusion of a single type of ion in a solution containing a sufficient excess of an indifferent electrolyte (see page 116), which then occurs in the same way as in the case of a non-electrolyte. Isotope (tracer) diffusion has the same character, where a concentration gradient of the radioactive isotope of an ion, present in a much lower concentration, is formed in a solution with a much larger, constant salt concentration.

2.5.1 *The time dependence of diffusion*

We will first consider the simple case of diffusion of a non-electrolyte. The course of the diffusion (i.e. the dependence of the concentration of the diffusing substance on time and spatial coordinates) cannot be derived directly from Eq. (2.3.18) or Eq. (2.3.19); it is necessary to obtain a differential equation where the dependent variable is the concentration c while the time and the spatial coordinates are independent variables. The derivation is thus based on Eq. (2.2.10) or Eq. (2.2.5), where we set $\psi = c$ and substitute from Eq. (2.3.18) or Eq. (2.3.19) for the fluxes. This yields Fick's second law (in fact, this is only a consequence of Fick's first law respecting the material balance—Eq. 2.2.10), which has the form of a partial differential equation

$$\frac{\partial c}{\partial t} = D \operatorname{div} \operatorname{grad} c = D\nabla^2 c \tag{2.5.1}$$

where ∇^2 is the Laplace operator (for its form in Cartesian and spherical coordinates, see footnote p. 109). When the concentration is a function of coordinate x alone, Fick's first law has the most common form:

$$J_d = -D\frac{dc}{dx} \tag{2.5.2}$$

and the second law is then

$$\frac{\partial c}{\partial t} = D\frac{\partial^2 c}{\partial x^2} \tag{2.5.3}$$

In order to integrate the partial differential equations (2.5.1) or (2.5.3) it is necessary to choose suitable initial and boundary conditions, i.e. to define a model of a certain experiment. The initial condition describes the concentration conditions of the system at the beginning of the diffusion process. The values of the concentrations or material fluxes (which are proportional to concentration gradients) at the boundary surfaces of the system or at concentration discontinuities inside the system are described by the boundary conditions.

For illustration, consider the simplest type of diffusion, described by the partial differential equation (2.5.3), also called linear diffusion. The system will be represented by an infinite tube closed at one end (for $x = 0$) and initially filled with a solution with concentration c^0. Diffusion is produced by very fast removal (e.g. by precipitation or an electrode reaction) of the dissolved substance at the $x = 0$ plane (the reference plane). The initial concentration c^0 is retained at large distances from this reference plane ($x \to \infty$). The initial condition is thus

$$x > 0, \qquad t = 0, \qquad c = c^0 \tag{2.5.4}$$

and the boundary conditions are

$$\begin{aligned} x &= 0, \qquad t > 0, \qquad c = 0 \\ x &\to \infty, \qquad t = 0, \qquad c = c^0 \end{aligned} \tag{2.5.5}$$

Solution of the differential equation (2.5.3) together with the initial and boundary conditions (2.5.4) and (2.5.5) yields the relationship

$$c(x, t) = c^0 \operatorname{erf}\left(\frac{x}{2D^{1/2}t^{1/2}}\right) \tag{2.5.6}$$

where the error function erf y is defined by the equation

$$\operatorname{erf} y = \frac{2}{\sqrt{\pi}} \int_0^y \mathrm{e}^{-z^2}\, \mathrm{d}z$$

Table 2.3 Error function $\operatorname{erf}(x) = (2/\sqrt{\pi}) \int_0^x \mathrm{e}^{-y^2}\, \mathrm{d}y$

x	erf (x)	x	erf (x)
0	0.000	1.1	0.880
0.1	0.113	1.2	0.910
0.2	0.223	1.3	0.914
0.3	0.329	1.4	0.952
0.4	0.428	1.5	0.966
0.5	0.521	1.6	0.976
0.6	0.604	1.7	0.984
0.7	0.678	1.8	0.989
0.8	0.742	1.9	0.993
0.9	0.797	2.0	0.995
1.0	0.843		

For $x < 0.1$ in sufficient approximation $\operatorname{erf}(x) = 1.128x - 0.376x^{-3}$.

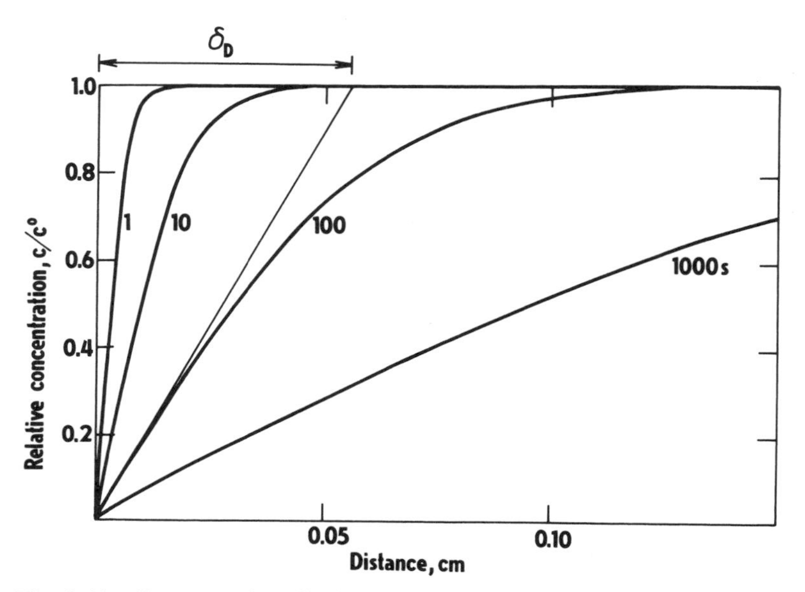

Fig. 2.11 Concentration distribution in the case of linear diffusion with $c = 0$ for $x = 0$ (see Eq. 2.5.6), where $D = 10^{-5}$ cm$^2 \cdot$ s^{-1}, time in seconds is indicated at each curve and the effective diffusion layer thickness is shown for $t = 100$ s

(see Table 2.3). Figure 2.11 depicts the concentration distribution. It can be seen from Eq. (2.5.6) that the ratio c/c^0 is independent of the concentration but only depends on the distance from the reference plane, on the time and on the diffusion coefficient. Assuming that the solution of the diffusing substance is dilute, the diffusion coefficient does not depend on its concentration.

The material flux through the reference plane is given by the concentration gradient for $x = 0$. In the present case the concentration gradient is

$$\frac{\partial c}{\partial x} = \frac{c^0}{(\pi D t)^{1/2}} \exp\left(-\frac{x^2}{4Dt}\right) \tag{2.5.7}$$

and the concentration gradient at the reference plane is then

$$\left(\frac{\partial c}{\partial x}\right)_{x=0} = c^0(\pi D t)^{-1/2} \tag{2.5.8}$$

Obviously, the material flux decreases with increasing time and, finally, for $t \to \infty$ approaches zero.

If a tangent to the curves in Fig. 2.11 is drawn at the origin, it intersects the straight line $c = c^0$ at a distance of

$$\delta_D = (\pi D t)^{1/2} \tag{2.5.9}$$

termed the diffusion layer thickness. It is a measure of the region that is depleted by the diffusion process.

Another example is linear diffusion, with a prescribed concentration gradient at the reference plane, i.e. a prescribed material flux through the reference plane. This type of diffusion transport is important mainly for electrode processes (see Section 5.4). The point of interest in this case is the concentration at the reference plane. In the simplest case, the material flux is constant, so that the boundary condition for $x = 0$ (Eq. 2.5.5) can be replaced by

$$x = 0, \qquad t > 0, \qquad D\frac{\partial c}{\partial x} = K \tag{2.5.10}$$

The resultant concentration distribution (Fig. 2.12) is given by the equation

$$c = c^0 - \frac{2Kt^{1/2}}{D^{1/2}\pi^{1/2}}\exp\left(-\frac{x^2}{4Dt}\right) + x\frac{K}{D}\operatorname{erfc}\left(\frac{x}{2(Dt)^{1/2}}\right) \tag{2.5.11}$$

The error function complement $\operatorname{erfc} y$ is defined by the relationship $\operatorname{erfc} y = 1 - \operatorname{erf} y$. The concentration at the reference plane c^* is

$$c^* = c^0 - 2Kt^{1/2}(\pi D)^{-1/2} \tag{2.5.12}$$

Clearly, the concentration at the reference plane decreases to zero after

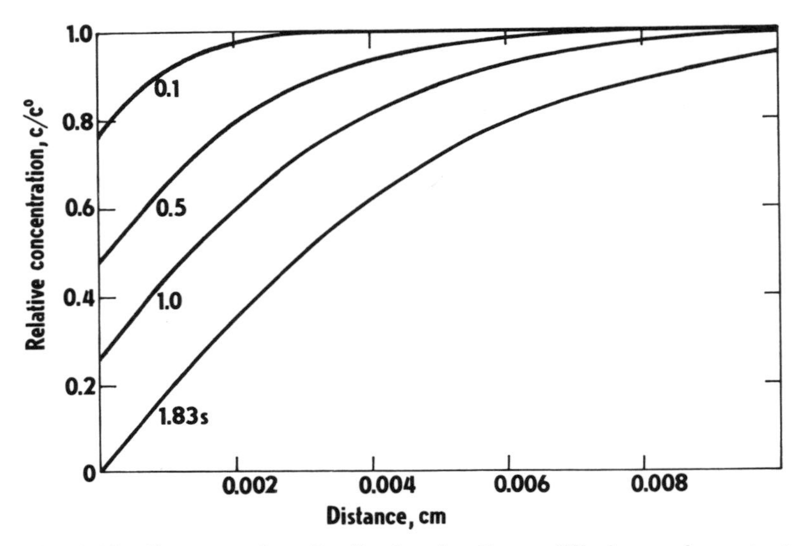

Fig. 2.12 Concentration distribution for linear diffusion and constant concentration gradient at the reference plane $x = 0$ (Eq. 2.5.11), where $K = 10^{-7}\ \mathrm{mol\cdot cm^{-2}\cdot s^{-1}}$, $D = 10^{-5}\ \mathrm{cm^2\cdot s^{-1}}$, $c^0 = 5\times 10^{-5}\ \mathrm{mol\cdot cm^{-3}}$ ($\tau = 1.83$ s)

time

$$t \equiv \tau = \pi D\left(\frac{c^0}{2K}\right)^2 \tag{2.5.13}$$

This case is an example of a time-restricted process as it cannot proceed after the instant $t = \tau$ (the concentration would assume negative values). Time τ is termed the *transition time.*

Consider a tube of length l containing a solution of concentration c_2, with one end ($x = 0$) rinsed with a solution of concentration c_1 and the other ($x = l$) with a solution of concentration c_2. In contrast to the previous example, after a period of time *steady state* will be reached, characterized by the relationship

$$\frac{\partial c}{\partial t} = 0 \tag{2.5.14}$$

for $t \to \infty$. The boundary conditions for the ordinary differential equation

$$D\frac{\mathrm{d}^2 c}{\mathrm{d}x^2} = 0 \tag{2.5.15}$$

are

$$\begin{aligned} x &= 0, \qquad c = c_1 \\ x &= l, \qquad c = c_2 \end{aligned} \tag{2.5.16}$$

The solution of this system is then

$$c = c_1 + (c_2 - c_1)\frac{x}{l} \tag{2.5.17}$$

The material flux at the steady state J_{st} is constant at every point in the system:

$$J_{st} = \frac{D(c_1 - c_2)}{l} \tag{2.5.18}$$

As a last example in this section, let us consider a sphere situated in a solution extending to infinity in all directions. If the concentration at the surface of the sphere is maintained constant (for example $c = 0$) while the initial concentration of the solution is different (for example $c = c^0$), then this represents a model of *spherical diffusion.* It is preferable to express the Laplace operator in the diffusion equation (2.5.1) in spherical coordinates for the centro-symmetrical case.† The resulting partial differential equation

† $\nabla^2 = (\partial^2/\partial x^2) + (\partial^2/\partial y^2) + (\partial^2/\partial z^2) = (\partial^2/\partial r^2) + (\partial/\partial r)(2/r)$, where r is the distance from the origin (located at the centre of the sphere).

is then

$$\frac{\partial c}{\partial t} = D\left(\frac{\partial^2 c}{\partial r^2} + \frac{2}{r}\frac{\partial c}{\partial r}\right)$$

$$\begin{aligned} &r > r_0, \quad &t = 0, \quad &c = c^0 \\ &r = r_0, \quad &t > 0, \quad &c = 0 \\ &r \to \infty, \quad &t > 0, \quad &c = c^0 \end{aligned} \tag{2.5.19}$$

where r_0 is the radius of the sphere.

The concentration gradient for $r = r_0$ is given by the equation

$$\left(\frac{\partial c}{\partial r}\right)_{r=r_0} = \frac{c^0}{\sqrt{\pi D t}} + \frac{c^0}{r_0} \tag{2.5.20}$$

The concentration gradient will obviously assume the steady-state value for $t \gg r_0^2/\pi D$,

$$\left(\frac{\partial c}{\partial r}\right)_{r=r_0} = \frac{c^0}{r_0} \tag{2.5.21}$$

Thus the time during which the transport process attains the steady state depends strongly on the radius of the sphere r_0. The steady state is connected with the dimensions of the surface to which diffusion transport takes place and does, in fact, not depend much on its shape. Diffusion to a semispherical surface located on an impermeable planar surface occurs in the same way as to a spherical surface in infinite space. The properties of diffusion to a disk-shaped surface located in an impermeable plane are not very different. The material flux is inversely proportional to the radius of the surface and the time during which stationary concentration distribution is attained decreases with the square of the disk radius. This is especially important for application of microelectrodes (see page 292).

Further examples of diffusion processes characterized by boundary conditions connected with specific electrode processes will be considered in Section 5.4.

2.5.2 *Simultaneous diffusion and migration*

Consider a dilute electrolyte solution containing s components (non-electrolytes and various ionic species) in which concentration gradients of the components and an electric field are present. The material flux of the ith component is then given by a combination of Eqs (2.3.11) to (2.3.20):

$$\mathbf{J}_i = -u_i(RT \operatorname{grad} c_i + c_i z_i F \operatorname{grad} \phi) \tag{2.5.22}$$

This relationship can be expressed in the form

$$\mathbf{J}_i = -u_i c_i \operatorname{grad} \tilde{\mu}_i \tag{2.5.23}$$

where the driving force for the simultaneous diffusion and migration is the gradient of the electrochemical potential $\tilde{\mu}_i = \mu_i + z_i F\phi$. The concept of the electrochemical potential is discussed in detail in Section 3.1.1.

Equations (2.3.11), (2.3.16) and (2.3.18) can be combined to give

$$\mathbf{J}_i = -D_i \operatorname{grad} c_i - \frac{z_i}{|z_i|} U_i c_i \operatorname{grad} \phi \tag{2.5.24}$$

or, for transport with respect to coordinate x alone,

$$\begin{aligned} J_i &= -D_i \frac{\mathrm{d}c_i}{\mathrm{d}x} - \frac{z_i}{|z_i|} U_i c_i \frac{\mathrm{d}\phi}{\mathrm{d}x} \\ &= -RTu_i \frac{\mathrm{d}c_i}{\mathrm{d}x} - z_i F u_i c_i \frac{\mathrm{d}\phi}{\mathrm{d}x} \end{aligned} \tag{2.5.25}$$

Equations (2.5.22) and (2.5.25) are called the *Nernst–Planck equations.*

Substitution for $\mathbf{J}_i$ from Eq. (2.5.24) into Faraday's law (2.3.14) yields the equation

$$\mathbf{j} = -\sum_{i=1}^{s} z_i F D_i \operatorname{grad} c_i - \sum_{i=1}^{s} |z_i| F U_i c_i \operatorname{grad} \phi \tag{2.5.26}$$

which, combined with the conductivity relationship (2.4.2), yields

$$\mathbf{j} = -\sum_{i=1}^{s} z_i F D_i \operatorname{grad} c_i - \kappa \operatorname{grad} \phi \tag{2.5.27}$$

The electric potential gradient is thus given by the sum of two expressions:

$$\begin{aligned} \operatorname{grad} \phi &= \frac{-\mathbf{j}}{\kappa} - \frac{1}{\kappa} \sum_{i=1}^{s} z_i F D_i \operatorname{grad} c_i \\ &= \operatorname{grad} \phi_{\text{ohm}} + \operatorname{grad} \phi_{\text{diff}} \end{aligned} \tag{2.5.28}$$

The first of these is the *ohmic potential* gradient, characteristic for charge transfer in an arbitrary medium. It is formed only when an electric current passes through the medium. The second expression is that for the *diffusion potential* gradient, formed when various charged species in the electrolyte have different mobilities. If their mobilities were identical, the diffusion electric potential would not be formed. In contrast to the ohmic electric potential, the diffusion electric potential does not depend directly on the passage of electric current through the electrolyte (it does not disappear in the absence of current flow).

2.5.3 *The diffusion potential and the liquid junction potential*

If all the changes taking place in the system occur along the direction of the x axis, then Eqs (2.5.28), (2.4.2), and (2.4.11) yield a relationship for the diffusion potential gradient

$$\frac{d\phi}{dx} = -\frac{RT}{F}\frac{\sum_{i=1}^{s} z_i u_i \, dc_i/dx}{\sum_{j=1}^{s} z_j^2 u_j c_j} = -\frac{RT}{F}\frac{\sum_{i=1}^{s} z_i u_i c_i \, d\ln c_i/dx}{\sum_{j=1}^{s} z_j^2 u_j c_j}$$

$$= -\frac{RT}{F}\sum_{i=1}^{s}\frac{t_i}{z_i}\frac{d\ln c_i}{dx} \tag{2.5.29}$$

The electric potential difference between two points (2 and 1) in the electrolyte during diffusion is thus given by the equation

$$\Delta\phi_{\text{diff}} = \phi_2 - \phi_1 = -\frac{RT}{F}\int_1^2 \sum_{i=1}^{s}\frac{t_i}{z_i}\, d\ln c_i \tag{2.5.30}$$

where the integration limits refer to the solution composition at points 1 and 2. In general the value of $\Delta\phi_{\text{diff}}$ is influenced by the way in which the c_i's and t_i's depend on x. The transport number for a dilute solution is not a function of the concentration only in the case of a single binary salt (see Eqs 2.4.12), so that in this case the diffusion potential is given by the equation

$$\Delta\phi_{\text{diff}} = \frac{z_+ - (z_+ - z_-)t_+}{z_+ z_-}\frac{RT}{F}\ln\frac{c(1)}{c(2)} \tag{2.5.31}$$

where $c(1)$ and $c(2)$ are the concentrations at points 1 and 2.

The potential difference formed at the junction of two electrolytes with different but constant composition and dissolved in the same solvent is very important in electrochemical measurements. Solutions can come into contact either directly (a suitable experimental arrangement must be used to prevent mixing through convection—see below) or can be separated by a diaphragm with sufficiently large pores. This diaphragm can be a glass frit, ceramic disk, etc., and is permeable for all the components of the system and simply prevents mechanical mixing. Thus, the system consists of two solutions, where the concentration is constant everywhere, and of a transition region at their contact, where diffusion occurs. This region has a thickness of d and is either free or enclosed in the pores of a diaphragm:

$$\begin{array}{c} \text{Solution 1} \mid \text{diaphragm} \mid \text{solution 2} \\ x = p \qquad\qquad x = q \end{array} \tag{2.5.32}$$

Here, x is the coordinate normal to the diaphragm, so that $d = q - p$. The *liquid junction potential* $\Delta\phi_L$ is the diffusion potential difference between solutions 2 and 1. The liquid junction potential can be calculated for more complex systems than that leading to Eq. (2.5.31) by several methods. A general calculation of the integral in Eq. (2.5.30) is not possible and thus assumptions must be made for the dependence of the ion concentration on x in the liquid junction. The approximate calculation of L. J. Henderson is often used, assuming that the ion concentration in the liquid junction varies linearly with x between the values in the two solutions:

$$c_i(x) = c_i(p) + [c_i(q) - c_i(p)]\frac{x}{d} \tag{2.5.33}$$

Substitution into Eq. (2.5.30) yields

$$\Delta\phi_L = -\frac{RT}{F}\int_0^d \frac{\sum_{i=1}^{s} z_i u_i\{[c_i(q) - c_i(p)]x/d + c_i(p)\}}{\sum_{j=1}^{s} z_j^2 u_j\{[c_j(q) - c_j(p)]x/d + c_j(p)\}}$$

$$\times d \ln \sum_{i=1}^{s} \{[c_i(q) - c_i(p)]x/d + c_i(p)\}$$

$$= -\frac{RT}{F}\frac{\sum_{i=1}^{s} z_i u_i[c_i(q) - c_i(p)]}{\sum_{i=1}^{s} z_i^2 u_i[c_i(q) - c_i(p)]} \ln \frac{\sum_{i=1}^{s} z_i^2 u_i c_i(q)}{\sum_{i=1}^{s} z_i^2 u_i c_i(p)} \tag{2.5.34}$$

whre $d = q - p$ is the thickness of the liquid junction.

Although the *Henderson formula* depends on a number of simplifications and also employs ion concentrations rather than activities, it can nonetheless be used for estimation of the liquid junction potential up to moderate electrolyte concentrations, apparently as a result of compensation of errors.

If there is only a single uni-univalent electrolyte with a common anion or cation on both sides of the liquid junction and the concentration of both electrolytes is the same, e.g.

0.1 M KCl	0.1 M HCl
solution 1	solution 2

then Eq. (2.5.34) is reduced to the form (Lewis–Sargent equation)

$$\Delta\phi_L = \pm\frac{RT}{F}\ln\frac{U_+(2) + U_-(2)}{U_+(1) + U_-(1)} = \pm\frac{RT}{F}\ln\frac{\Lambda(2)}{\Lambda(1)} \tag{2.5.35}$$

where the symbols 1 and 2 designate the electrolytes in solutions 1 and 2 and the Λ's are the molar conductivities at the given electrolyte concentrations. The minus sign holds when the electrolytes have a common anion and the plus sign holds for a common cation. The Henderson equation for a constantly varying electrolyte mixture approximately approaches conditions for a liquid junction with free diffusion (see Fig. 2.13).

The general *Planck's solution* of the liquid-junction potential is based on the assumption that the fluxes of ions in the liquid junction are in a steady state. As the mathematical treatment is rather space-consuming the reader is recommended to inspect a more recent treatment of this by W. E. Morf

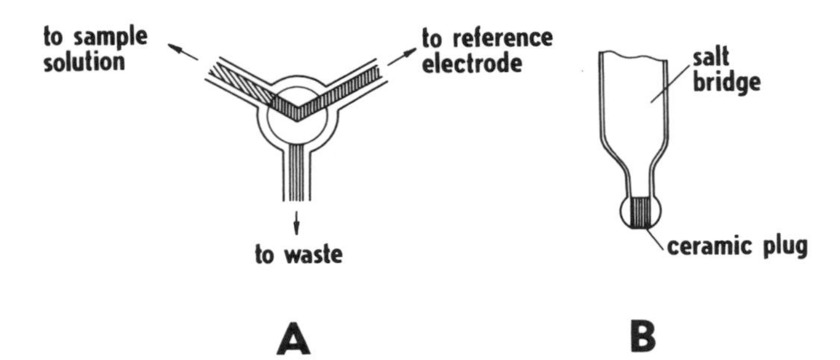

Fig. 2.13 Examples of liquid junctions: (A) liquid junction with 'free' diffusion is formed in a three-way cock which connects the solution under investigation with the salt bridge solution; (B) liquid junction with restrained diffusion is formed in a ceramic plug which connects the salt bridge with the investigated solution

(*Anal. Chem.*, **49,** 810, 1977). Planck's model corresponds to the liquid junction formed inside rather thin-pore diaphragms (Fig. 2.13B).

However, useful arrangements are made to practically eliminate the diffusion potential, i.e. reduce it to the smallest possible value, in potentiometric measurements not concerned directly with measuring the liquid junction potential, e.g. pH measurements, potentiometric titrations, etc. This is achieved by using a salt bridge connecting the two solutions, consisting, for example, of a tube closed at both ends by a porous material and usually filled with a saturated solution of potassium chloride, ammonium nitrate, or sodium formate (if the presence of chloride or potassium ions is undesirable). The theoretical basis for the use of these salt bridges is that the ions of the bridge solutions are present in a large excess compared with the ions in the electrode compartments, and thus are almost solely responsible for the charge transport across the two boundaries. As the transport numbers of the cations and anions of these excess electrolytes are almost identical, the two newly formed diffusion potentials are small and act in opposite directions, so that the final diffusion potential is negligible. For example, the diffusion potential at the contact between 0.1 and 0.01 M HCl is about 40 mV. If a salt bridge containing a saturated potassium chloride solution is placed between the two solutions, the diffusion potential is about 3 mV on the side with the more dilute solution and about 5 mV on the side with the more concentrated solution. The final potential difference is about 2 mV, which is insignificant in many applications.

In potentiometric measurements the simplest approach to the liquid-junction problem is to use a reference electrode containing a saturated solution of potassium chloride, for example the saturated calomel electrode (p. 177). The effect of the diffusion potential is completely suppressed if the solutions in contact contain the same indifferent electrolyte in a sufficient excess, which increases the overall conductivity of the solutions (cf. Eq. 2.5.28).

2.5.4 *The diffusion coefficient in electrolyte solutions*

As demonstrated in the preceding section, an electric potential gradient is formed in electrolyte solutions as a result of diffusion alone. Let us assume that no electric current passes through the solution and convection is absent. The Nernst–Planck equation (2.5.24) then has the form:

$$\mathbf{J}_i = -D_i \operatorname{grad} c_i - \frac{z_i}{|z_i|} U_i c_i \operatorname{grad} \phi_{\mathrm{dif}} \tag{2.5.36}$$

Substitution from Eq. (2.3.22) into Eq. (2.5.36) yields the following relationship for dilute solutions:

$$J_i = -D_i \operatorname{grad} c_i - \frac{D_i z_i F}{RT} c_i \operatorname{grad} \phi_{\mathrm{dif}} \tag{2.5.37}$$

For a single electrolyte, Eqs (2.5.37) and (2.3.14) for $j = 0$ and Eqs (2.4.4) and (2.4.5) yield the equation

$$-\operatorname{grad} \phi_{\mathrm{dif}} = \frac{RT(D_+ - D_-)}{Fc(z_+ D_+ - z_- D_-)} \operatorname{grad} c \tag{2.5.38}$$

where D_+ and D_- are the diffusion coefficients of the cation and anion.

The overall material flux of the electrolyte

$$\mathbf{J} = \frac{\mathbf{J}_+}{\nu_+} = \frac{\mathbf{J}_-}{\nu_-} \tag{2.5.39}$$

where ν_+ and ν_- are the stoichiometric coefficients of the anion and of the cation in the salt molecule. From Eqs (2.5.37) to (2.5.39) we then obtain

$$\mathbf{J} = -\frac{D_+ D_- (\nu_+ + \nu_-)}{\nu_- D_+ + \nu_+ D_-} \operatorname{grad} c \tag{2.5.40}$$

Diffusion of a single electrolyte ('salt') is thus characterized by an effective diffusion coefficient

$$D_{\mathrm{eff}} = \frac{D_+ D_- (\nu_+ + \nu_-)}{\nu_- D_+ + \nu_+ D_-} \tag{2.5.41}$$

Replacement of the diffusion coefficients by the electrolytic mobilities according to Eq. (2.3.22) yields the *Nernst–Hartley equation*:

$$D_{\mathrm{eff}} = \frac{RT}{F} \frac{U_+ U_- (\nu_+ + \nu_-)}{\nu_+ z_+ (U_+ + U_-)} \tag{2.5.42}$$

It could be expected for less dilute solutions that an expression for the effective diffusion coefficient could be derived theoretically from Eq. (2.5.42) by using correction terms for U_+ and U_- taken from the Debye–Hückel–Onsager theory of electrolyte conductivities. However, it has been demonstrated experimentally that this approach does not lead to a correct description of the dependence of the diffusion coefficient on the concentration. The mobility of ions in the diffusion process varies far less with changes in the concentration than when charge is transported in an external electric field, and the effect of the concentration on the mobility can be either retarding, zero or accelerating, depending on the type of salt (increasing concentration always reduces ion mobility in an external electric field). This difference between the two transport phenomena is a result of the fact that diffusion is connected with the movement of cations and anions in the same direction, so that faster species are retarded by slower species, and vice versa, whereas during electric current flow, oppositely charged ions move in opposite directions and the two types of ions retard one another. The electrophoretic effect caused by the retardation of the movement of the central ion by the moving ionic atmosphere is thus of a different magnitude in diffusion. The time-of-relaxation effect is completely absent, as the symmetry of the ionic atmosphere is not disturbed during diffusion.

As mentioned, the gradient of the diffusion electric potential is suppressed in the case of diffusion of ions present in a low concentration in an excess of *indifferent electrolyte* ('base electrolyte'). Under these conditions, the simple form of Fick's law (2.3.18) holds for the diffusion of the given ion. The

Table 2.4 Diffusion coefficients $D \times 10^6$ ($cm^2 \cdot s^{-1}$) determined by means of polarography or chronopotentiometry at various indifferent electrolyte concentrations c ($mol \cdot dm^{-3}$) at 25°C. The composition of the indifferent electrolyte is indicated for each ion. (According to J. Heyrovský and J. Kůta)

	Ag^+	Tl^+			Pb^{2+}		Cd^{2+}	
c	KNO_3	KNO_3	KCl	NaCl	KNO_3	KCl	KNO_3	KCl
0.01	15.85				8.76	8.99		8.15
0.1	15.32	18.2	17.4	17.7	8.28	8.67	6.90	7.15
1.0	15.46	16.5	15.7	15.0	8.02	9.20	6.81	7.90
3.0			13.5	9.2		8.14		7.90

	Zn^{2+}			IO_3^-		$Fe(CN)_6^{4-}$	$Fe(CN)_6^{3-}$
c	KNO_3	KCl	NaOH	KCl	NaCl	KCl	KCl
0.01	6.60	6.76					7.84
0.1	6.38	6.73	6.54	10.15	10.01	6.50	7.62
1.0	6.20	7.23	5.13	9.89	8.92	6.50	7.63
3.0		7.69	4.18	9.36	7.24	6.20	7.36

Table 2.5 Diffusion coefficients ($cm^2 \cdot s^{-1} \times 10^{-5}$) of electrolytes in aqueous solutions at various concentrations c ($mol \cdot dm^{-3}$). (According to H. A. Robinson and R. H. Stokes)

Electrolyte	Temperature (°C)	c = 0	0.001	0.002	0.003	0.005	0.007	0.010
LiCl	25	1.366	1.345	1.337	1.331	1.323	1.318	1.312
NaCl	25	1.610	1.585	1.576	1.570	1.560	1.555	1.545
KCl	20	1.763	1.739	1.729	1.722	1.708	—	1.692
KCl	25	1.993	1.964	1.954	1.945	1.934	1.925	1.917
KCl	30	2.230	—	—	2.174	2.161	2.152	2.144
RbCl	25	2.051	—	2.011	2.007	1.995	1.984	1.973
CsCl	25	2.004	2.013	2.000	1.992	1.978	1.969	1.958
$LiNO_3$	25	1.336	—	—	1.296	1.289	1.283	1.276
$NaNO_3$	25	1.568	—	1.535	—	1.516	1.513	1.503
$KClO_4$	25	1.871	1.845	1.841	1.835	1.829	1.821	1.790
KNO_3	25	1.928	1.899	1.884	1.879	1.866	1.857	1.846
$AgNO_3$	25	1.765	—	—	1.719	1.708	1.698	—
$MgCl_2$	25	1.249	1.187	1.169	1.158	—	—	—
$CaCl_2$	25	1.335	1.263	1.243	1.230	1.213	1.201	1.188
$SrCl_2$	25	1.334	1.269	1.248	1.236	1.219	1.209	—
$BaCl_2$	25	1.385	1.320	1.298	1.283	1.265	—	—
Li_2SO_4	25	1.041	0.990	0.974	0.965	0.950	—	—
Na_2SO_4	25	1.230	1.175	1.160	1.147	1.123	—	—
Cs_2SO_4	25	1.569	1.489	1.454	1.437	1.420	—	—
$MgSO_4$	25	0.849	0.768	0.740	0.727	0.710	—	—
$ZnSO_4$	25	0.846	0.748	0.733	0.724	0.705	—	—
$LaCl_3$	25	1.293	1.175	1.145	1.126	1.105	1.084	—
$K_4Fe(CN)_6$	25	1.468	—	—	1.213	1.184	—	—

diffusion coefficient, of course, depends on the concentration of the indifferent electrolyte.

A similar situation occurs in *tracer diffusion*. This type of diffusion occurs for different abundances of an isotope in a component of the electrolyte at various sites in the solution, although the overall concentration of the electrolyte is identical at all points. Since the labelled and the original ions have the same diffusion coefficient, diffusion of the individual isotopes proceeds without formation of the diffusion potential gradient, so that the diffusion can again be described by the simple form of Fick's law.

Experimental methods for determining diffusion coefficients are described in the following section. The diffusion coefficients of the individual ions at infinite dilution can be calculated from the ionic conductivities by using Eqs (2.3.22), (2.4.2) and (2.4.3). The individual diffusion coefficients of the ions in the presence of an excess of indifferent electrolyte are usually found by electrochemical methods such as polarography or chronopotentiometry (see Section 5.4). Examples of diffusion coefficients determined in this way are listed in Table 2.4. Table 2.5 gives examples of the diffusion coefficients of various salts in aqueous solutions in dependence on the concentration.

2.5.5 *Methods of measurement of diffusion coefficients*

Both steady-state and non-steady-state methods are used in the determination of diffusion coefficients. In the steady-state methods ($\partial c/\partial t = 0$) the flux J and the concentration gradient $\partial c/\partial x$ are measured directly and the diffusion coefficient D is calculated according to Fick's first law, often without assuming that D is constant. In non-steady-state methods, all the parameters of diffusion flux change with position and time and thus equations resulting from integration of Fick's second law must be employed with various boundary conditions, depending on the experimental arrangement. In integration, when necessary, a suitable form of the function $D(x, t)$ must be assumed. Usually the diffusion coefficient is assumed to be constant and thus the final integrated equations can be used only for small concentration ranges in which the change in the diffusion coefficient can be neglected.

A steady-state method. In the diaphragm method the diffusion process is restricted to a porous diaphragm (a glass frit with a pore size of about 10 μm, separating compartments 1 and 2 of volumes V_1 and V_2 (cf. Eq. (2.5.18). The diaphragm is the actual diffusion space with an effective cross-section of pores A and an effective length l. The compartments are filled with solutions of concentrations c_1^0 and c_2^0 at the start of the experiment, and are thoroughly stirred during the experiment. After a short time the solutions penetrate into the pores of the diaphragm and the solution diffuses from 1 to 2 if $c_1^0 > c_2^0$. In the steady state there is no accumulation of the solute in the pores of the diaphragm, i.e. the same amount of solute which leaves the compartment 1 in a certain time interval passes into the compartment 2. The flux of the substance is constant through the entire thickness of the diaphragm but it decreases with time because the concentration difference between compartments 1 and 2 decreases. After the experiment has been running for a certain time, the contents of both compartments are analysed and the concentrations c_1' and c_2' are determined. From these data we may compute the average value of the diffusion coefficients between the concentrations $(c_1^0 + c_1')/2$ and $(c_2^0 + c_2')/2$.

The time dependence of the concentration in 1 and 2 is obtained by solving the differential equations

$$-V_1 \frac{dc_1}{dt} = V_2 \frac{dc_2}{dt} = -A\frac{D}{l}(c_2 - c_1) \qquad (2.5.43)$$

Integration using the relationship $V_1c_1^0 + V_2c_2^0 = V_1c_1 + V_2c_2$ yields

$$\ln \frac{c_1^0 - c_2^0}{c_1' - c_2'} = \frac{AD}{l}\left(\frac{1}{V_1} + \frac{1}{V_2}\right)\Delta t \qquad (2.5.44)$$

where Δt is the time interval during which concentrations c_1 and c_2 assume values c_1' and c_2', respectively.

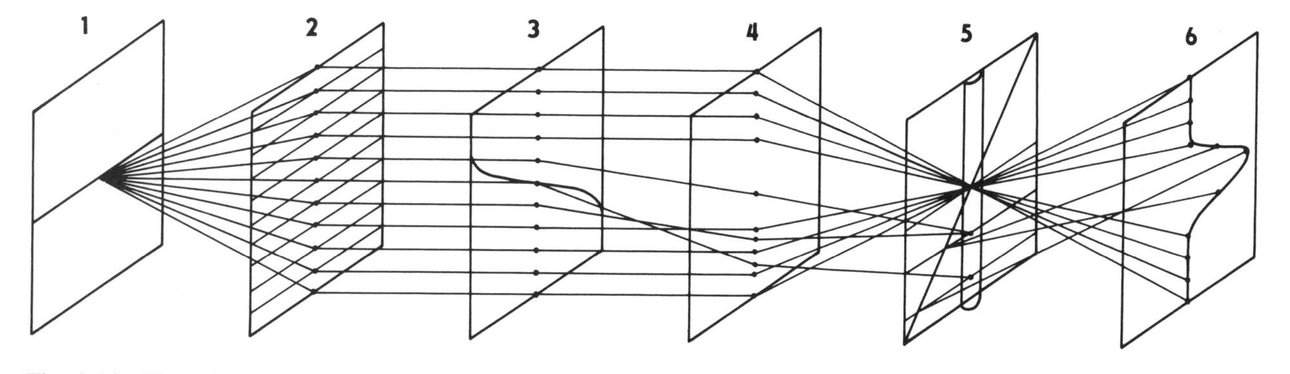

Fig. 2.14 The scheme of the cylindrical lens method for diffusion coefficient measurement: (1) the source with the horizontal slit; (2) the condenser supplying a bandle of parallel beams; (3) the cuvette with a refraction index gradient where the beams are deflected; (4) the objective lens focusing the parallel beams to a single point; (5) the optical member with an oblique slit and a cylindrical lens; (6) the photosensitive material

Non-steady-state methods. The most frequently used methods belonging to this category are the optical methods. The experiment is usually arranged so that the boundary and initial conditions are fulfilled for linear diffusion in a limited space (tube) with one end closed and impermeable for the diffusing substance ($D\partial c/\partial x = 0$) and the other end in contact with constantly stirred pure solution. After an initial delay the gradient of the refractive index $\mathrm{d}n/\mathrm{d}x$ which is directly proportional to the concentration gradient is measured at several points of the cuvette. A parallel beam of light incident upon the cuvette at a right angle to the direction of the concentration gradient will be deflected in a direction towards the region of larger refractive index and the angle of deflection will be directly proportional to the concentration gradient. The resulting dependence of this deviation on the position in the cuvette (Fig. 2.14) is then analysed on the basis of theoretical relationships.

References

Covington, A. K., and M. J. F. Rebello, Reference electrodes and liquid junction effect in ion-selective electrode potentiometry, *Ion-Sel. Electrode Revs.*, **5,** 93 (1983)

Crank, J., *The Mathematics of Diffusion,* Clarendon Press, Oxford, 1964.

Cussler, E. L., *Multicomponent Diffusion,* Elsevier, Amsterdam, 1976.

Cussler, E. L., *Diffusion: Mass Transfer in Fluid Systems,* Cambridge University Press, Cambridge, 1984.

Delahay, P., *New Instrumental Methods in Electrochemistry,* Interscience, New York, 1954.

Geddes, A. L., and R. B. Pontius, Determination of diffusivity, in *Technique of Organic Chemistry* (Ed. A. Weissberger), Vol. 1, Part 2, 3rd ed., Interscience, New York, 1960, p. 895.

Marchiano, S. L., and A. J. Arvía, Diffusion in absence of convection. Steady state and nonsteady state, *CTE,* **6,** 65 (1983).

Morf, W. E., *The Principles of Ion-Selective Electrodes and of Membrane Transport,* Akadémiai Kiadó, Budapest, and Elsevier, Amsterdam, 1981.

Newman, J., *Electrochemical Systems,* Prentice-Hall, Englewood Cliffs, 1973.

Overbeek, J. Th. G., The Donnan equilibrium, in *Progress in Biophysics and Biophysical Chemistry,* Vol. 6, Pergamon Press, London 1956.

2.6 The Mechanism of Ion Transport in Solutions, Solids, Melts and Polymers

Diffusion is one of the most important natural processes. The varying rate of diffusion yields an excellent characteristic of individual states of aggregation. While diffusion coefficients in gases are of the order of $10^{-1}\,\mathrm{cm}^2\cdot\mathrm{s}^{-1}$, those in solutions are of the order of $10^{-5}\,\mathrm{cm}^2\cdot\mathrm{s}^{-1}$ and, in solids, of $10^{-10}\,\mathrm{cm}^2\cdot\mathrm{s}^{-1}$.

The mobility of a species u_i is a parameter connecting diffusion and migration processes. This section will be concerned primarily with the qualitative theory of mobility, which has characteristic properties for transport in dilute electrolyte solutions in electrically neutral liquids, in melts (ionic liquids) and in solid substances.

2.6.1 *Transport in solution*

The transport of dissolved species in a solvent occurs randomly through movement of the Brownian type. The particles of the dissolved substance and of the solvent continuously collide and thus move stochastically with various velocities in various directions. The relationship between the mobility of a particle, the observation time τ and the mean shift $\langle x^2 \rangle$ is given by the *Einstein–Smoluchowski equation* (in three-dimensional case)

$$u_i = \frac{\langle r^2 \rangle}{6RT} \tag{2.6.1}$$

The value of the mean shift of an ion $\langle r^2 \rangle$ does not change markedly, even when a not exceedingly large electric field ($<10^5\ \mathrm{V \cdot m^{-1}}$) is applied to the studied system.

The dissolved species are located within the quasicrystalline solvent structure and vibrate around equilibrium positions. After several hundred vibrations they attain a sufficiently great energy to permit a jump to a neighbouring position in the quasicrystalline lattice if a sufficiently large free space is available. Thus, the frequency of jumps depends on the energy required to release a species from its equilibrium position and on the energy required to form a 'hole' in a neighbouring position in the solvent. For simple ions, the frequency of these jumps in common solvents is about 10^{10}–$10^{11}\ \mathrm{s^{-1}}$ at normal temperatures.

When an electric field is applied, jumps of the ions in the direction of the field are somewhat preferred over those in other directions. This leads to migration. It should be noted that the absolute effect of the field on the ionic motion is small but constant. For example, an external field of $1\ \mathrm{V \cdot m^{-1}}$ in water leads to ionic motion with a velocity of the order of $50\ \mathrm{nm \cdot s^{-1}}$, while the instantaneous velocity of ions as a result of thermal motion is of the order of $100\ \mathrm{m \cdot s^{-1}}$.

The velocity of motion of the particles depends on their dimensions and shape, on the interaction (e.g. association) between the solvent molecules and finally on the interaction between particles of the dissolved substance and solvent molecules. Consider the simplest case, where the molecule of the dissolved substance is much larger than the solvent molecule, is spherical and the interaction between the solute molecules and the solvent is negligible. Then the motion of the particles of the solute can be considered as the motion of spherical particles with radius r_i through a viscous medium with viscosity coefficient η. The velocity $\mathbf{v}$ is then described by the *Stokes law*:

$$\mathbf{v} = \frac{\mathbf{f}_i}{6\pi\eta r_i} \tag{2.6.2}$$

where $\mathbf{f}_i$ is the force acting on the particle. For diffusion, this velocity corresponds to the material flux; thus similar considerations as those employed for convection mass flux (Eq. (2.3.24)) lead to the relationship

$$(\mathbf{J}_i)_{\mathrm{dif}} = c_i \mathbf{v}_{\mathrm{dif}} \tag{2.6.3}$$

where $\mathbf{v}_{\mathrm{dif}}$ is the diffusion velocity.

In Eq. (2.6.2), $\mathbf{f}_i$ is the force acting on a single particle, whereas in Eq. (2.3.20) grad μ is the force acting on 1 mole of particles. Thus,

$$\mathbf{f}_i = -\frac{\mathrm{grad}\,\mu_i}{N_{\mathrm{A}}} \tag{2.6.4}$$

Substitution of Eq. (2.6.4) into (2.6.2) and then into (2.6.3) yields

$$(\mathbf{J}_i)_{\mathrm{dif}} = -\frac{\mathrm{grad}\,\mu_i}{6\pi\eta r_i N_{\mathrm{A}}} \tag{2.6.5}$$

leading to the *Stokes–Einstein equation* for the diffusion coefficient,

$$D_i = \frac{kT}{6\pi\eta r_i} \tag{2.6.6}$$

and to the relationship for the electrical mobility

$$U_i = \frac{|z_i|\,e}{6\pi\eta r_i} \tag{2.6.7}$$

The values derived in this way for the diffusion coefficients exhibit surprising agreement with the experimental values, even for small ions, better than a coincidence in the order of magnitude. More detailed theoretical analysis indicates that the formation of ‘holes’ required for particle jump is analogous to the formation of holes necessary for viscous flow of a liquid. Consequently, the activation energy for diffusion is similar to that for viscous flow.

Wirtz *et al.* obtained better agreement for the diffusion coefficients than that yielded by Eq. (2.6.6), by considering the change in viscosity in the vicinity of the diffusing particle. The semiempirical equation has the form

$$D = \frac{kT}{6\pi\eta\zeta r_i} \tag{2.6.8}$$

where ζ is the microviscosity factor

$$\zeta = 0.16 + 0.4\frac{r_i}{r_{\mathrm{L}}} \tag{2.6.9}$$

with r_{L} as the effective radius of a solvent molecule, obtained from the molar volume.

Using the statistical mechanical theory, Longuet-Higgins and Pople obtained a correlation function for the velocities in a liquid consisting of rigid spheres. McGall, Douglas and Anderson modified this correlation function to yield an expression for the diffusion coefficient

$$D = \frac{3kT\xi}{10\pi\eta r_i} \tag{2.6.10}$$

where ξ is the fraction of the total volume occupied by solvent molecules.

It follows from Eqs. (2.6.6), (2.6.8) and (2.6.10) that the presence of the solvent has two effects on the ionic mobility: the effect of changing viscosity and that of changing the ionic radius as a result of various degrees of solvation of the diffusing particles. If the effective ionic radius does not change in a number of solutions with various viscosities and if ion association does not occur, then the *Walden rule* is valid for these solutions:

$$D_i\eta = \text{constant} \tag{2.6.11}$$

or

$$\Lambda\eta = \text{constant} \tag{2.6.12}$$

As the radii of solvated ions show no striking difference, the diffusion coefficients for individual ions in water vary in the range 5×10^{-10} to $2 \times 10^{-9}\,\text{m}^2 \cdot \text{s}^{-1}$ ($5 \times 10^{-6}\,\text{cm}^2\,\text{s}^{-1}$ to $2 \times 10^{-5}\,\text{cm}^2\,\text{s}^{-1}$, cf. Table 2.4) and their electrolytic mobilities U_i in the range 2×10^{-8} to $10^{-7}\,\text{m}^2 \cdot \text{s}^{-1} \cdot \text{V}^{-1}$.

The oxonium ion H_3O^+ in aqueous solutions has a very different diffusion coefficient value, about five times higher than would be expected on the basis of its dimensions. Similarly, the hydroxide ion also has a much larger diffusion coefficient value than predicted. Analogously, the mobilities of the $H_3SO_4{}^+$ and $HSO_4{}^-$ ions in concentrated sulphuric acid are 50 and 100 times higher than those of the other ions. This effect is explained by the fact that the lyonium and lyate ions need not migrate through a protic solvent, but 'move' through exchange of a proton between neighbouring solvent molecules (the so-called *Grotthus mechanism* of charge transport). It should be recalled that Grotthus knew nothing about ions when he developed his theory of electrolytic decomposition (1809), but assumed that the transport of hydrogen and oxygen in solution during the electrolysis of water occurs through alternate splitting of the water molecule, shift of the atoms of hydrogen and oxygen produced and reformation of a water molecule.

According to contemporary concepts, a proton acts as a quantum mechanical species during transfer between two neighbouring molecules (see Fig. 2.15) and the actual transfer occurs as a result of tunnelling through a potential energy barrier between the initial and final states of the system: $H_3O^+ + H_2O \rightarrow H_2O + H_3O^+$. However, tunnelling can occur only when both species are favourably oriented. The activation energy for proton transfer depends on the rotation energy of a water molecule. This concept of proton transfer between two water molecules is analogous to the transfer of an electron between the oxidized and reduced forms of an ion, for example $Fe^{3+} + e \rightleftarrows Fe^{2+}$, or to the transfer of an electron between an

Fig. 2.15 Transport of protons in water. Proton tunnelling is a fast process but water molecules must first rotate to the position where the transfer is possible

electrode and these species (see Section 5.3.1). The energy required for an orientation of the water molecule suitable for proton transfer plays a role similar to the reorganization energy of the solvation shell involved in these processes.

Eigen pointed out that the mobility of a proton in ice at 0°C is about 50 times larger than in water. In ice, H_2O molecules already occupy fixed positions suitable for accepting a proton, so that the proton mobility is directly proportional to the rate of tunnelling.

2.6.2 *Transport in solids*

Diffusion and migration in solid crystalline electrolytes depend on the presence of defects in the crystal lattice (Fig. 2.16). *Frenkel defects* originate from some ions leaving the regular lattice positions and coming to *interstitial positions*. In this way empty sites (holes or vacancies) are formed, somewhat analogous to the holes appearing in the band theory of electronic conductors (see Section 2.4.1).

Ions are transferred in the lattice in three ways:

1. By a shift of an ion from one interstitial position to another

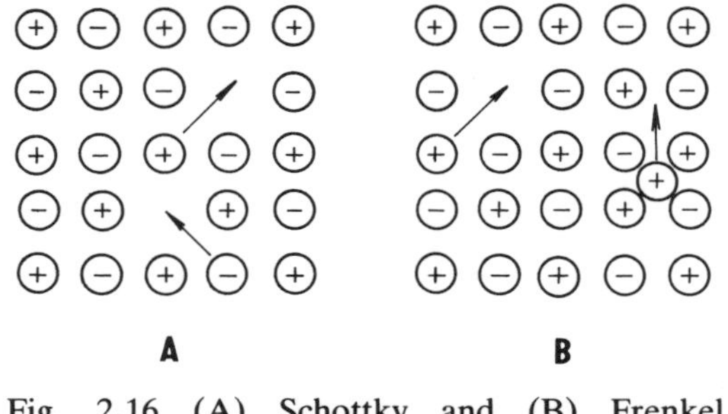

Fig. 2.16 (A) Schottky and (B) Frenkel defects in a crystal

2. By a shift of an ion from a stable position into an interstitial position and of another ion from an interstitial position into the hole formed
3. By a shift of an ion from a stable position into a neighbouring hole, forming a new vacancy

Current is conducted by the Frenkel mechanism, e.g. in silver halides, where the charge carrier is the silver ion (transport number $t_{Ag^+} = 1$). Impurities in crystals favour the Frenkel mechanism. Important materials with high conductivity depending on the Frenkel mechanism include β-alumina, a ceramic material with the composition $Na_2O \cdot 11Al_2O_3$. The structure of β-alumina (see Fig. 2.17) consists of a compact arrangement of blocks of the γ-phase of aluminium oxide of a thickness of four oxygen layers and of a spinel structure. These blocks are separated by bridging layers, containing only oxygen and sodium ions. Each bridging layer contains only one-fourth as many oxygen atoms as the compact layer in blocks. These are arranged so that the sodium ions are located in relatively small concentrations in tunnels between the oxygen ions, where they can migrate relatively freely, leading to the unusually high conductivity of β-alumina. The conductivity mechanism depends on the Frenkel defects in the bridging layer (interstitial oxygen ions close to interstitial aluminium ions in the compact block).

Other ceramic ion conductors similar to β-alumina were introduced by Goodenough *et al.* in 1976; they belong to the so-called Nasicon family. Nasicon is a solid solution of the composition $Na_3Zr_2Si_2PO_{12}$, exhibiting a Na^+ conductivity of 0.2 S/cm at 300°C. Numerous Nasicon derivatives,

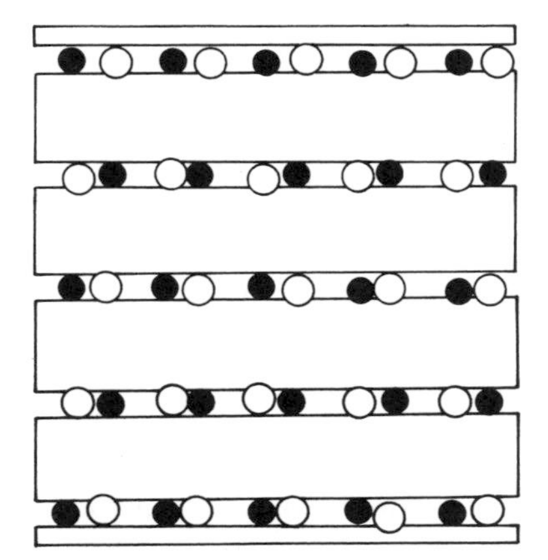

Fig. 2.17 A scheme of β-alumina structure. The gaps between blocks of β-phase Al_2O_3 function as bridging layers sparsely populated by oxygen (○) and sodium (●) ions. (According to R. A. Huggins)

containing Sc, Cr, Fe, In, Yb, Ti, V . . . , etc. were also prepared and characterized as promising solid conductors for high-temperature batteries. The Nasicon structure is composed of two ZrO_6 octahedra and three $(P/Si)O_4$ tetrahedra in a rather loose packing with opened conduction channels. Sodium cations are distributed inside these channels in two symmetrically non-equivalent positions, Na(1) and Na(2), but only some of the available positions are occupied. The migration of Na^+ ions through these channels is therefore rather easy and the observed ionic conductivities are higher by orders of magnitude in comparison with those of other ceramic conductors.

In some ionic crystals (primarily in halides of the alkali metals), there are vacancies in both the cationic and anionic positions (called *Schottky defects*—see Fig. 2.16). During transport, the ions (mostly of one sort) are shifted from a stable position to a neighbouring hole. The Schottky mechanism characterizes transport in important solid electrolytes such as Nernst mass (ZrO_2 doped with Y_2O_3 or with CaO). Thus, in the presence of 10 mol.% CaO, 5 per cent of the oxygen atoms in the lattice are replaced by vacancies. The presence of impurities also leads to the formation of Schottky defects. Most substances contain Frenkel and Schottky defects simultaneously, both influencing ion transport.

For Schottky defects in an ionic crystal with a cubic lattice, the diffusion coefficient is given by the relationship, e.g. for a cation,

$$D = \tfrac{1}{3}a^2\beta p \tag{2.6.13}$$

where a is the distance between the closest cation and anion, β is the fraction of vacancies in the total number of cations and p is the probability of the jump of a vacancy per second, given as

$$p = \nu \exp\left(\frac{-\varepsilon_j}{kT}\right) \tag{2.6.14}$$

where ν is the vibrational frequency of the cations and ε_j is the activation energy for a jump.

Crystals with Frenkel or Schottky defects are reasonably ion-conducting only at rather high temperatures. On the other hand, there exist several crystals (sometimes called 'soft framework crystals'), which show surprisingly high ionic conductivities even at the room or slightly elevated temperatures. This effect was revealed by G. Bruni in 1913; two well known examples are AgI and CuI. For instance, the α-modification of AgI (stable above 146°C, sometimes denoted also as 'γ-modification') exhibits at this temperature an Ag^+ conductivity ($t_+ = 1$) comparable to that of a 0.1 M aqueous solution. (The solid-state Ag^+ conductivity of α-AgI at the melting point is actually higher than that of the melt.) This unusual behaviour can hardly be explained by the above-discussed defect mechanism. It has been anticipated that the conductivity of α-AgI and similar crystals is described by a qualitatively different transport model, the so-called *disordered sub-lattice motion.*

Every ionic crystal can formally be regarded as a mutually interconnected composite of two distinct structures: cationic sublattice and anionic sub-lattice, which may or may not have identical symmetry. Silver iodide exhibits two structures thermodynamically stable below 146°C: sphalerite (below 137°C) and wurtzite (137–146°C), with a plane-centred I^- sub-lattice. This changes into a body-centred one at 146°C, and it persists up to the melting point of AgI (555°C). On the other hand, the Ag^+ sub-lattice is much less stable; it collapses at the phase transition temperature (146°C) into a highly disordered, liquid-like system, in which the Ag^+ ions are easily mobile over all the 42 theoretically available interstitial sites in the I^- sub-lattice. This system shows an Ag^+ conductivity of 1.31 S/cm at 146°C (the regular wurtzite modification of AgI has an ionic conductivity of about 10^{-3} S/cm at this temperature).

Attempts have been made to lower the temperature of appearance of the sub-lattice motions. It was found that substitution in the I^- sub-lattice of AgI, e.g. by WO_4^{2-}, stabilizes this structure up to rather low temperatures: crystals of $(AgI)_{1-x}(Ag_2WO_4)_x$ show, for $x = 0.18$, an Ag^+ conductivity of 0.065 S/cm at 20°C. Addition of cationic species, for instance in Ag_2HgI_4, Ag_4RbI_6, and $Ag_7[N(CH_3)_4]I_8$ has a similar effect.

Lanthanum fluoride (and fluorides of some other lanthanides) has an unusual type of defect (see Section 6.3.2), namely Schottky defects of the molecular hole type (whole LaF_3 molecules are missing at certain sites). Charge carriers (F^-) are formed as the result of interaction of LaF_3 with this hole, leading to dissociation with formation of $LaF_2{}^+$ and F^-.

Only the alkali metal ions are mobile in *oxygen-containing glasses* such as the silicates or borates of the alkali metals. The relatively large free space in the glass permits a jump from one oxygen atom to another with simultaneous charge transfer between the oxygens forming bridges between the silicon or boron atoms.

Similarly, as in the transport of ions in electrolyte solutions, random ion motion predominates over ordered motion in the direction of the field during the passage of electric current through solid substances.

2.6.3 *Transport in melts*

The *mobility of ions in melts* (ionic liquids) has not been clearly elucidated. A very strong, constant electric field results in the ionic motion being affected primarily by short-range forces between ions. It would seem that the ionic motion is affected most strongly either by fluctuations in the liquid density (on a molecular level) as a result of the thermal motion of ions or directly by the formation of cavities in the liquid. Both of these possibilities would allow ion transport in a melt.

2.6.4 *Ion transport in polymers*

Ion conducting polymers (polymer electrolytes) attract increasing attention not only from the academic point of view, but also for their prospective applications in batteries, sensors, electrolysers and other practical devices. Although the ion conducting polymers can formally be regarded as 'solid electrolytes', the mechanism of ion transport is different from that in the above-discussed inorganic crystals with lattice defects, and it resembles the ion transport in liquid media. This follows from the fact that ions are transported in a polymeric host material, which is essentially not as rigid as the defect crystal of the classical inorganic solid electrolyte, i.e. the host motions or rearrangements virtually contribute to the ion transport as well. The ion conducting polymers therefore present a special class of electrolytes with features intermediate between those of solid (defect crystals) and liquid (solutions, melts) electrolytes. The ion conducting polymers contain, besides the organic polymeric backbone, ions in more or less strong interactions with the polymer, and in some cases also solvent molecules.

Ion solvating polymers. Polymers that are conducting even in the dry state, i.e. without attached solvent molecules, are termed *ion solvating* (*or ion-coordinating*) *polymers.* These contain electronegative heteroatoms (prevailingly oxygen, sometimes also nitrogen, sulphur, or phosphorus), which interact with cations by donor–acceptor bonds similar to those in solvated ions in solutions. The complexes of ions with polymers are simply formed by dissolving inorganic salts in suitable polymeric hosts. The most important solvating polymeric hosts are two polyethers: poly(ethylene oxide) (PEO) and poly(propylene oxide) (PPO):

$$\text{—CH}_2\text{—CH}_2\text{—O—} \qquad\qquad \text{—}\underset{\displaystyle \text{CH}_3}{\underset{|}{\text{CH}}}\text{—CH}_2\text{—O—}$$

PEO PPO

Lithium perchlorate is dissolved in PEO owing to the coordination of Li^+ cation by oxygen atoms in the polymeric chain. The complex thus formed has the helical structure (Fig. 2.18), and exhibits ionic conductivities of up to 10^{-4} S/cm at 60°C.

The mechanism of ion transport in such systems is not fully elucidated, but it is presumably dependent on the degree of crystallinity of the polymeric complex (which further depends on the temperature and the salt type). The ionic conductivity was initially attributed to cation hopping between fixed coordination sites in the depicted helical tunnel, i.e. in the crystalline part of the polymer.

A more detailed analysis revealed, however, that the cation hopping

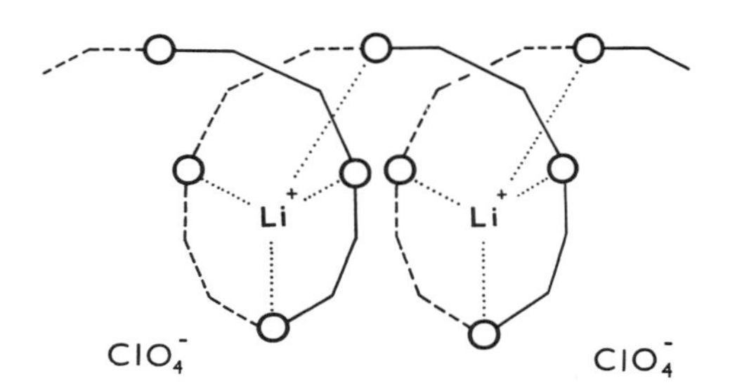

Fig. 2.18 Structure of a poly(ethylene) oxide–$LiClO_4$ complex

between vacant sites along crystalline regions in the polymer has only marginal significance for ion transport. The main arguments are: (1) completely amorphous polymers are also conducting, usually even better than the crystalline ones, and (ii) not only cations, but also anions are the mobile species in ion-solvating polymers. (For instance, the Li^+ transport number in the PEO–$LiClO_4$ complex is in the region of only 0.2–0.35, and drops, moreover, with increasing Li/PEO ratio.) The preconception of an immobile polymer host skeleton is also far from reality, especially for amorphous polymers.

The transition between crystalline and amorphous polymers is characterized by the so-called glass transition temperature, T_g. This important quantity is defined as the temperature above which the polymer chains have acquired sufficient thermal energy for rotational or torsional oscillations to occur about the majority of bonds in the chain. Below T_g, the polymer chain has a more or less fixed conformation. On heating through the temperature T_g, there is an abrupt change of the coefficient of thermal expansion (α), compressibility, specific heat, diffusion coefficient, solubility of gases, refractive index, and many other properties including the chemical reactivity.

In polymer electrolytes (even prevailingly crystalline), most of ions are transported via the mobile amorphous regions. The ion conduction should therefore be related to viscoelastic properties of the polymeric host and described by models analogous to that for ion transport in liquids. These include either the 'free volume model' or the 'configurational entropy model'. The former is based on the assumption that thermal fluctuations of the polymer skeleton open occasionally 'free volumes' into which the ionic (or other) species can migrate. For classical liquid electrolytes, the free volume per molecule, v_f, is defined as:

$$v_f = v - v_0 \tag{2.6.15}$$

where v is the average volume per molecule in the liquid and v_0 is the van der Waals' volume of the molecule. For polymers, the free volume, v_f, can also be expressed as:

$$v_f = v_g[k + \alpha(T - T_g)] \quad (2.6.16)$$

where v_g is the volume per molecule at the glass transition temperature, T_g, α is the thermal expansion coefficient of the polymer, and k is a non-dimensional constant. The ion transport in the polymer is conditioned by the formation of sufficiently large voids in the polymeric skeleton. The probability of this process, p, decreases exponentially with the total-to-free volume ratio, v/v_f:

$$p = \exp(-v/v_f) \quad (2.6.17)$$

Equations (2.6.16) and (2.6.17) can be rearranged to the form:

$$p = \exp[-B/(T - T_0)] \quad (2.6.18)$$

where B and T_0 are constants (T_0 is usually 20–50°C lower than T_g).

Using Eq. (2.6.18) the temperature dependence of various transport properties of polymers, such as diffusion coefficient D, ionic conductivity σ and fluidity (reciprocal viscosity) $1/\eta$ are described, since all these quantities are proportional to p. Except for fluidity, the proportionality constant (pre-exponential factor) also depends, however, on temperature,

$$D \sim T^{1/2}; \qquad \sigma \sim T^{-1/2} \cdot p; \qquad 1/\eta \sim p \quad (2.6.19)$$

Another quasi-thermodynamical model of ion transport in polymers is based on the concept of minimum configurational entropy required for rearrangement of the polymer, giving practically identical $\sigma - T$ and $D - T$ dependences as the preceding model.

The common disadvantage of both the free volume and configuration entropy models is their quasi-thermodynamic approach. The ion transport is better described on a microscopic level in terms of ion size, charge, and interactions with other ions and the host matrix. This makes a basis of the percolation theory, which describes formally the ion conductor as a random mixture of conductive islands (concentration c) interconnected by an essentially non-conductive matrix. (The mentioned formalism is applicable not only for ion conductors, but also for any insulator/conductor mixtures.)

The main conclusion of the percolation theory is that there exists a critical concentration of the conductive fraction (percolation threshold, c_0), below which the ion (charge) transport is very difficult because of a lack of pathways between conductive islands. Above and near the threshold, the conductivity can be expressed as:

$$\sigma = \sigma_0(c - c_0)^n \quad (2.6.20)$$

where the exponent n is about 1.2–1.5, depending on the spatial dimensionality of the system under study.

Besides the polyether-based polymer electrolytes, the nitrogen analogues (polyimines) were also extensively studied. Various other oxygen-containing polymers were further used, e.g. polyesters, polymers containing oxygen heterocycles, crown ethers, etc. Most of the studied polymer electrolytes are reasonably conducting at elevated temperatures, typically 60–100°C. Recently, proton conductors based on PEO complexes with anhydrous acids or acid salts, such as H_3PO_4 or NH_4HSO_4 were prepared; they show conductivities in the range 10^{-5}–10^{-4} S/cm even at the room temperature. Other highly conducting ion-solvating polymers contain etheric solvating groups on an inorganic polymeric skeleton, either phosphazene or siloxane. The former have been prepared by chlorine substitution in poly(dichlorophosphazene), $(\text{—N=PCl}_2\text{—})_n$. Some phosphazene derivatives show very low glass transition temperatures, and are thus regarded as promising ion-solvating polymers. The best known representative is poly[bis(methoxyethoxyethoxide)] phosphazene:

$$
\begin{array}{l}
\quad\quad\quad\text{O—(CH}_2)_2\text{—O—(CH}_2)_2\text{—O—CH}_3 \\
\quad\quad\quad | \\
(\text{—N=P—})_n \\
\quad\quad\quad | \\
\quad\quad\quad\text{O—(CH}_2)_2\text{—O—(CH}_2)_2\text{—O—CH}_3
\end{array}
$$

also abbreviated as MEEP. This polymer forms complexes with $Li^+CF_3SO_3^-$, exhibiting a room-temperature ionic conductivity of up to 10^{-4} S/cm.

The ion solvating polymers have found application mainly in power sources (all-solid lithium batteries, see Fig. 2.19), where polymer electrolytes offer various advantages over liquid electrolyte solutions.

Polymer gels and ionomers. Another class of polymer electrolytes are those in which the ion transport is conditioned by the presence of a low-molecular-weight solvent in the polymer. The most simple case is the so-called *gel polymer electrolyte,* in which the intrinsically insulating polymer (agar, poly(vinylchloride), poly(vinylidene fluoride), etc.) is swollen with an aqueous or aprotic liquid electrolyte solution. The polymer host acts here only as a passive support of the liquid electrolyte solution, i.e. ions are transported essentially in a liquid medium. Swelling of the polymer by the solvent is described by the volume fraction of the pure polymer in the gel (V_p). The diffusion coefficient of ions in the gel (D_p) is related to that in the pure solvent (D_0) according to the equation:

$$D_p = D_0 \exp\left[-kV_p/(1 - V_p)\right] \qquad (2.6.21)$$

where k is a constant.

Another example of ion conducting polymer/ion/solvent systems are polyelectrolytes based on ion-exchange polymers, also called *ionomers*. The ionic conductivity of ion-exchange polymers is usually very low in the dry state, but increases abruptly by orders of magnitude upon addition of a small amount of a liquid solvent (several molecules per one immobilized ionic group). The most important representatives of ionomers are sulphonated fluoropolymers, such as Nafion:

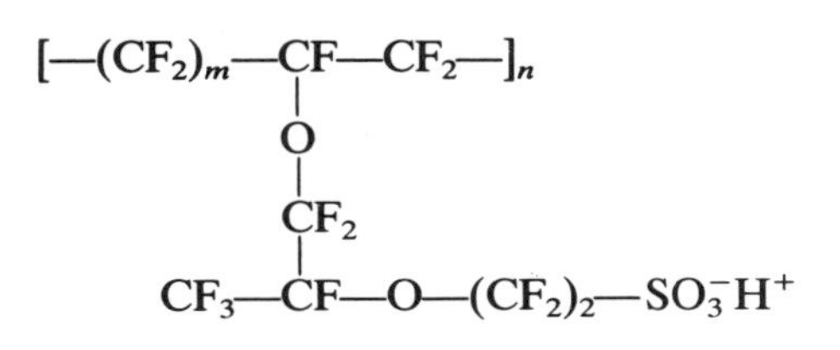

where m is about 13.

This material was first synthesized in the middle 1960s by E.I. Du Pont de Nemours and Co., and was soon recognized as an outstanding ion conductor for laboratory as well as for industrial electrochemistry. The perfluorinated polymeric backbone is responsible for the good chemical and thermal stability of the polymer. Nafion membrane swollen with an electrolyte solution shows high cation conductivity, whereas the transport of anions is almost entirely suppressed. This so-called permselectivity (cf. Section 6.2.1) is a characteristic advantage of Nafion in comparison with classical ion-exchange polymers, in which the selective ion transport is usually not so pronounced.

Thanks to its high stability and permselectivity, Nafion has been used as a Na^+ conductor in membrane electrolysis of brine in the chlor-alkali industry. This application was introduced in the early 1980s and is by far the most important use of ionomer membranes.

Materials similar to Nafion containing immobilized $—COO^-$ or $—NR_3^+$ groups on a perfluorinated skeleton were also synthesized. These are available in the form of solid membranes or solutions in organic solvents; the former can readily be used as solid electrolytes in the so called *solid polymer electrolyte* (SPE) cells, the latter are suitable for preparing ion-exchange polymeric films on electrodes simply by evaporating the polymer solution in a suitable solvent.

The first Nafion–SPE cells were constructed in the later 1960s, i.e. long before the industrial membrane electrolysis of brine. These cells have a typical thin-layer arrangement with swollen membrane functioning as an electrolyte without additional separate liquid phases. A Nafion membrane acts here as a separator of hydrogen and oxygen catalytic electrodes and simultaneously as a solid electrolyte conducting H^+ ions formed at the anode. The interelectrode spacing is very narrow, typically about 0.3 mm; minimizing ohmic losses and permitting relatively high current densities to be attained.

Inhibition of anion transport in Nafion was attributed to the inhomogeneous structure of the ion exchange sites in the polymer network (Gierke cluster network model). It was found that Nafion contains (even in

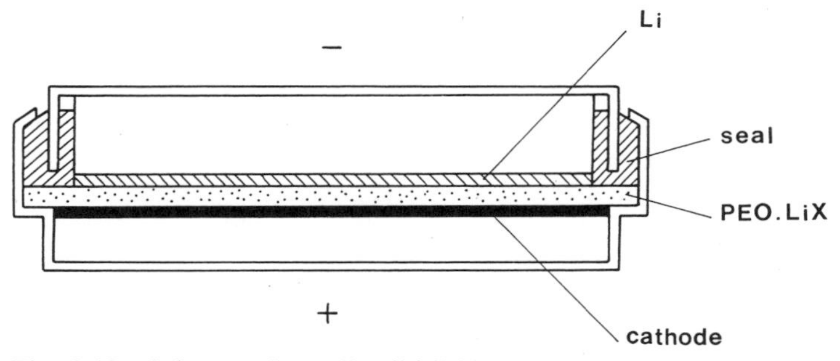

Fig. 2.19 Scheme of an all-solid lithium battery with PEO based electrolyte

the dry state) discrete hydrophilic regions, separated by narrow hydrophobic channels. The hydrophilic regions are formed by a specific aggregation of the ionogenic side chains on the perfluorinated backbone, whereas the channels are composed of hydrophobic fluorocarbon chains (see the polymer formula above). The individual macromolecular chains are thus interconnected by the embedded clusters into a more complicated supramolecular network (Fig. 2.20). This, moreover, explains why Nafion and related materials are practically insoluble in water, in spite of the absence of regular interchain bonds as in classical cross-linked ion-exchange polymers (e.g. polystyrene sulphonate).

By swelling with aqueous electrolyte, cations (and, to lesser extent, also anions) penetrate together with water into the hydrophilic regions and form spherical electrolyte clusters with micellar morphology. The inner surface of clusters and channels is composed of a double layer of the immobilized —SO_3^- groups and the equivalent number of counterions, M^+. Anions in the interior of the clusters are shielded from the —SO_3^- groups by hydrated cations and water molecules. On the other hand, anions are thus

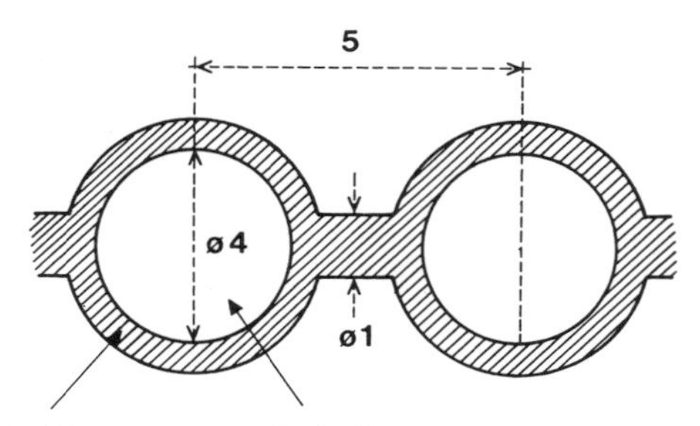

Fig. 2.20 The Gierke model of a cluster network in Nafion. Dimensions are expressed in nm. The shaded area is the double layer region, containing the immobilized —SO_3^- groups with corresponding number of counterions M^+. Anions are expelled from this region electrostatically

'trapped' inside the clusters owing to the high electrostatic barrier in the hydrophobic channel where the shielding effect is suppressed. This makes it clear why the channels are preferentially conductive for cations, i.e. the polymer membrane (despite the presence of free anions) is not anion-conducting.

Diffusion of cations in a Nafion membrane can formally be treated as in other polymers swollen with an electrolyte solution (Eq. (2.6.21). Particularly illustrative here is the percolation theory, since the conductive sites can easily be identified with the electrolyte clusters, dispersed in the non-conductive environment of hydrophobic fluorocarbon chains (cf. Eq. (2.6.20)). The experimental diffusion coefficients of cations in a Nafion membrane are typically 2–4 orders of magnitude lower than in aqueous solution.

References

Conway, B. E., Proton solvation and proton transfer in solution, *MAE,* **3,** 43 (1964).

Dekker, A. J., *Solid State Physics,* Prentice-Hall, Englewood Cliffs, 1957.

Eisenberg, A. E., and H. L. Yeager (Eds), *Perfluorinated Ionomer Membranes,* ACS Symposium Series 180, ACS, Washington, 1982.

Glasstone, S., H. Eyring, and K. J. Laidler, *Theory of Rate Processes,* McGraw-Hill, New York, 1942.

Hertz, H. G., Self-diffusion in solids, *Ber. Bunsenges.*, **75,** 183 (1971).

Hladik, J. (Ed.), *Physics of Electrolytes,* Academic Press, New York, a multivolume series, published since 1972.

Huggins, R. A., Ionically conducting solid state membranes, *AE,* **10,** 323 (1977).

Inman, D., and D. C. Lovering (Eds), *Ionic Liquids,* Plenum Press, New York, 1981.

Kittel, C., see page 104.

Laity, R. W., Electrochemistry of fused salts, *J. Chem. Educ.*, **39,** 67 (1962).

Linford, R. G., *Electrochemical Science and Technology of Polymers,* Vol. 1 & 2, Elsevier, London, 1987 and 1990.

MacCallum, J. R., and C. A. Vincent (Eds), *Polymer Electrolyte Reviews,* Elsevier, London, 1987.

Vashista, P., J. N. Mundy, and G. K. Shenoy (Eds), *Fast Ion Transport in Solids,* North-Holland, Amsterdam, 1979.

Vincent, C. A., The motion of ions in solution under the influence of an electric current, *J. Chem. Educ.*, **53,** 490 (1976).

2.7 Transport in a Flowing Liquid

2.7.1 *Basic concepts*

Processes involving transport of a substance and charge in a streaming electrolyte are very important in electrochemistry, particularly in the study of the kinetics of electrode processes and in technology. If no concentration gradient is formed, transport is controlled by migration alone; convection

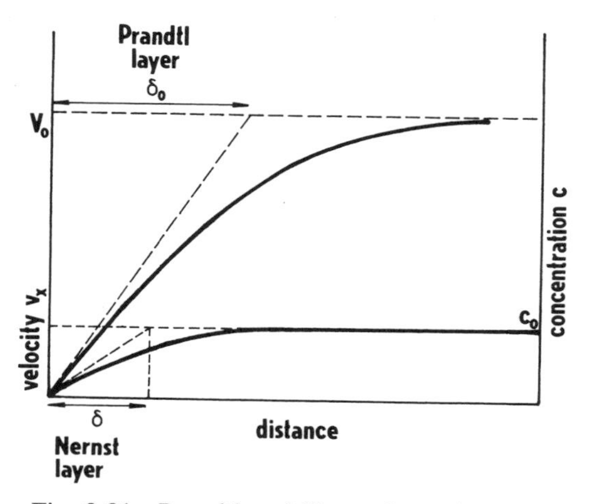

Fig. 2.21 Prandtl and Nernst layer formation at the boundary between a solid plate and flowing liquid ($c^* = 0$)

has no effect. The opposite situation where concentration gradients are formed during transport has interesting characteristics; if transport through migration can be neglected (e.g. at a sufficiently high concentration of an indifferent electrolyte), transport is controlled by diffusion in the streaming liquid, i.e. *convective diffusion* is involved.

To a major part this chapter will consider processes occurring in a steady state, i.e. where $\partial c_i/\partial t = 0$. It will be simultaneously assumed that the liquid is incompressible. A change in its velocity occurs close to the boundary with a solid phase or with another liquid. This is because the molecules of a liquid in direct contact with the solid phase have the same velocity as the solid phase. In other words, their velocity relative to the solid phase is zero. It is therefore convenient to use a reference frame that is fixed with respect to the solid phase. Viscosity forces act on the layers of the liquid that are further apart from the phase boundary, forming a velocity distribution in the liquid, depicted in Fig. 2.21 for liquid flowing along a solid plate. A tangent drawn from the origin to the curve of the dependence of the tangential velocity component, v_x, on the distance y from the plate intercepts the straight line $v_x = V_0$ at a distance δ_0. The quantity δ_0 is termed the thickness of the *hydrodynamic* or *Prandtl layer*. It should be pointed out that δ_0 is a function of the distance from the leading edge of the plate, l. As l increases, the quantity δ_0 also increases; the liquid also moves in a direction perpendicular to the plate, away from the plate with velocity v_y.

If diffusion occurs towards the surface of the plate, the motion of the liquid at a distance from the surface inside the hydrodynamic layer is sufficient to maintain the original concentration in the bulk of the solution, c^0. A decrease of the concentration to a value c^* at the phase boundary (in Fig. 2.21 $c^* = 0$) occurs only in a thin layer within the Prandtl layer, with thickness $\delta < \delta_0$. The concentration c^* is attained in the immediate vicinity of the interface and is determined by its properties (e.g. for transport to an electrode, the electrode potential is decisive—see Section 5.3). The quantity δ is termed the thickness of the *Nernst layer*. For a steady-state concentration gradient in the immediate vicinity of the phase boundary we have

$$\left(\frac{\partial c}{\partial y}\right)_{y=0} = \frac{c^0 - c^*}{\delta} \tag{2.7.1}$$

The steady-state distribution of the concentration of the diffusing substance at the phase boundary is termed the Nernst layer.†

For the material flux (cf. Eq. 2.5.18) we have

$$J_y = \frac{D}{\delta}(c^0 - c^*) = \kappa(c^0 - c^*) \tag{2.7.2}$$

where D is the diffusion coefficient and $\kappa = D/\delta$ is the *mass transfer coefficient* (this term is employed in chemical and electrochemical engineering).

2.7.2 *The theory of convective diffusion*

The material flux during convective diffusion in an indifferent electrolyte (grad $\phi \approx 0$) can be described by a relationship obtained by combination of Eqs (2.3.18) and (2.3.23):

$$\mathbf{J} = \mathbf{J}_{\text{dif}} + \mathbf{J}_{\text{conv}} = -D \operatorname{grad} c + c\mathbf{v} \tag{2.7.3}$$

where $\mathbf{J}_{\text{dif}}$ is the diffusion and $\mathbf{J}_{\text{conv}}$ the convection material flux.

Equation (2.2.10) for $\psi = c$ and the incompressibility condition

$$\operatorname{div} \mathbf{v} = 0 \tag{2.7.4}$$

yield

$$\frac{\partial c}{\partial t} = D\nabla^2 c - \mathbf{v} \operatorname{grad} c \tag{2.7.5}$$

or, for the steady state,

$$D\nabla^2 c - \mathbf{v} \operatorname{grad} c = 0 \tag{2.7.6}$$

In Cartesian coordinates, this equation becomes

$$D\left(\frac{\partial^2 c}{\partial x^2} + \frac{\partial^2 c}{\partial y^2} + \frac{\partial^2 c}{\partial z^2}\right) - \left(v_x \frac{\partial c}{\partial x} + v_y \frac{\partial c}{\partial y} + v_z \frac{\partial c}{\partial z}\right) = 0 \tag{2.7.7}$$

† Nernst studied the theory of heterogeneous reactions around the year 1900 and assumed that the flow of a liquid along a phase boundary with a solid phase involves formation of an immobile liquid layer at this surface. A steady-state concentration gradient is then formed within this layer as a result of diffusion in the limited space (see Eq. 2.5.18). In fact, the diffusion process has the character of convective diffusion everywhere as a result of the velocity distribution in the hydrodynamic layer.

The assumption of a steady state corresponds to laminar flow of the liquid. This is fulfilled only under certain conditions (limited velocity of the liquid, smooth phase boundary between the flowing liquid and the other phase, etc.). Otherwise, turbulent flow occurs, where the local velocity depends on time, pulsation of the system, etc. Mathematically the turbulent flow problem is a very difficult task and it is often doubtful whether, in specific cases, it is possible to obtain any solution at all.

Liquid flow originates from the effect of various forces. In *forced convection,* these are mechanical forces acting, e.g. on a piston forcing the liquid into a vessel, or a stirrer transferring the impulse to a liquid in a vessel, etc. In this manner, pressure gradients are formed in the solution, resulting in motion of the liquid. *Natural* (*free*) *convection* results when density changes are produced in the solution as a result of concentration changes produced by transport processes. It is the force of gravity that causes natural convection and produces hydrostatic pressure gradients in solution that are different from conditions in a liquid of constant density that is at rest.

The relationship between the motion of a viscous liquid and the mentioned forces is given by the *Navier–Stokes equation*

$$\frac{\partial \mathbf{v}}{\partial t} + (\mathbf{v}\,\mathrm{grad})\mathbf{v} = -\frac{1}{\rho}\,\mathrm{grad}\,p + \nu\nabla^2\mathbf{v} + \frac{\mathbf{g}(\rho - \rho_0)}{\rho} \qquad (2.7.8)$$

where ρ is the variable liquid density, ρ_0 is the density of the liquid at rest, ν is its kinematic viscosity ($\nu = \eta/\rho$), $\mathbf{g}$ is the acceleration of gravity and p is the difference between the total pressure and the hydrostatic pressure. The right-hand side of this equation describes the pressure gradient for forced convection, the effect of viscosity forces and the effect of gravitation during natural convection. For sufficiently intense forced convection, the effect of natural convection can be neglected. These conditions are assumed in the subsequent discussion. As it is assumed that the liquid is incompressible, the continuity equation (2.7.4) is valid.

In forced convection, the velocity of the liquid must be characterized by a suitable characteristic value V_0, e.g. the mean velocity of the liquid flow through a tube or the velocity of the edge of a disk rotating in the liquid, etc. For natural convection, this characteristic velocity can be set equal to zero. The dimension of the system in which liquid flow occurs has a certain characteristic value l, e.g. the length of a tube or the longitudinal dimension of the plate along which the liquid flows or the radius of a disk rotating in the liquid, etc. Solution of the differential equations (2.7.5), (2.7.7) and (2.7.8) should yield the value of the material flux at the phase boundary of the liquid with another phase, where the concentration equals c^*.

If a rigorous solution to this problem is at all possible, it consists of two parts:

(a) Solution of the hydrodynamic part of the problem in order to obtain a

relationship for the flow velocity as a function of the spatial coordinates (i.e. solution of Eq. (2.7.8) with suitable boundary conditions) and
(b) Solution of Eq. (2.7.6) or Eq. (2.7.7) on the basis of a known relationship for $\mathbf{v}$, or for v_x, v_y and v_z

This approach is possible only if density gradients are not formed in the solution as a result of transport processes (e.g. in dilute solutions). Otherwise, both differential equations must be solved simultaneously—a very difficult task.

Two examples will now be given of solution of the convective diffusion problem, transport to a rotating disk as a stationary case and transport to a growing sphere as a transient case. Finally, an engineering approach will be mentioned in which the solution is expressed as a function of dimensionless quantities characterizing the properties of the system.

As a *rotating disk* is a very useful device for many types of electrochemical research, convective diffusion to a rotating disk, treated theoretically by V. G. Levich, will be used here as an example of this type of transport process. Consider a disk in the xz plane, rotating around the y axis with radial velocity ω (see Fig. 2.22). If the radius of the disk is sufficiently larger

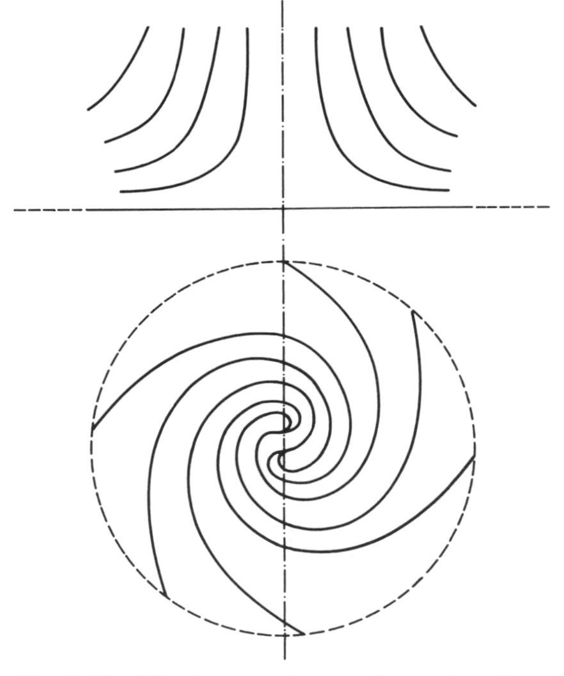

Fig. 2.22 Distribution of streamlines of the liquid at the surface of a rotating disk. Above: a side view; below: a view from the direction of the disk axis. (According to V. G. Levich)

than the thickness of the hydrodynamic layer, then the change of flow at the edge of the disk can be neglected. Solution of the hydrodynamic part of the problem, carried out by von Kármán and Cochrane, indicates that the thickness of the hydrodynamic layer is not a function of the distance from the centre of the disk y; its value in centimetres is given by the relationship

$$\delta_0 = 3.6\left(\frac{\nu}{\omega}\right)^{1/2} \tag{2.7.9}$$

The velocity of the liquid in the direction perpendicular to the surface of the disk in centimetres per second is described by the equation

$$v_y = -0.51\omega^{3/2}\nu^{-1/2}y^2 \tag{2.7.10}$$

It will be assumed that the concentration gradient in the direction perpendicular to the surface of the disk is much greater than in the radial direction. Then Eq. (2.7.7) reduces to the form

$$v_y\frac{\mathrm{d}c}{\mathrm{d}y} = D\frac{\mathrm{d}^2c}{\mathrm{d}y^2} \tag{2.7.11}$$

The boundary conditions are

$$y = 0, \qquad c = c^*$$

$$y \to \infty, \qquad \mathrm{d}c/\mathrm{d}y = 0, \qquad c = c^0$$

For c in moles per cubic centimetre and D in square centimetres per second, the solution to this problem has the form

$$\left(\frac{\mathrm{d}c}{\mathrm{d}y}\right)_{y=0} = 0.62D^{-1/3}\nu^{-1/6}\omega^{1/2}(c^0 - c^*) \tag{2.7.12}$$

and

$$J = -D\left(\frac{\mathrm{d}c}{\mathrm{d}y}\right)_{y=0} = 0.62D^{2/3}\nu^{-1/6}\omega^{1/2}(c^0 - c^*) \tag{2.7.13}$$

The thickness of the Nernst layer (see Eq. 2.7.1) is

$$\delta = 1.61D^{1/3}\nu^{1/6}\omega^{-1/2}$$

Convective diffusion to a growing sphere. In the polarographic method (see Section 5.5) a dropping mercury electrode is most often used. Transport to this electrode has the character of convective diffusion, which, however, does not proceed under steady-state conditions. Convection results from growth of the electrode, producing radial motion of the solution towards the electrode surface. It will be assumed that the thickness of the diffusion layer formed around the spherical surface is much smaller than the radius of the sphere (the drop is approximated as an ideal spherical surface). The spherical surface can then be replaced by a planar surface

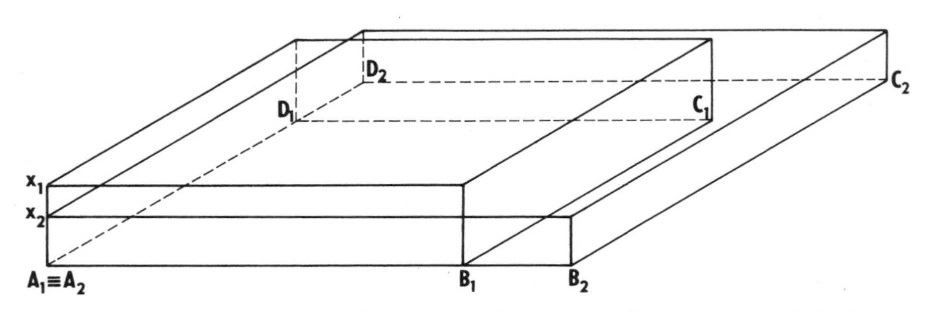

Fig. 2.23 Movement of the solution towards an expanding sphere modelled by a prism with an expanding plane as a base. When the base expands from the original area $A_1B_1C_1D_1$ (equal to $A_0t_1^{2/3}$) to the area $A_2B_2C_2D_2$ ($A_0t_2^{2/3}$) then the point in the solution originally at a distance x_1 must move to the distance x_2 in order to keep a constant volume of the prism V_0 (i.e. to preserve the condition of the incompresssibility of the solution, see Eq. 2.7.4). This model fits the actual situation only when the distance x_1 is small ($x_1 \ll A_1B_1$)

increasing in size according to the same laws as those governing the growth of the surface of a sphere, while the solution phase is modelled by a prism with an expanding plane as base. Let us first determine the velocity of a point in the solution at the distance x from the surface (Fig. 2.23).

The original spherical surface increases as a result of an increase in its volume at a constant velocity u. The area of the surface of the sphere at time t is given by the relationship $A = (4\pi)^{1/3}(3ut)^{2/3} = A_0t^{2/3}$, where A_0 is the area at time $t = 1$. From the incompressibility of the liquid it follows for volume $A_0xt^{2/3}$ that

$$A_0xt^{2/3} = \text{constant} \tag{2.7.14}$$

Differentiation of Eq. (2.7.14) with respect to time gives

$$v = \frac{dx}{dt} = -\frac{2x}{3t} \tag{2.7.15}$$

Substitution into Eq. (2.7.5) respecting only changes along the x coordinate yields the *Ilkovič differential equation*

$$\frac{\partial c}{\partial t} = D\frac{\partial^2 c}{\partial x^2} + \frac{2}{3}\frac{x}{t}\frac{\partial c}{\partial x} \tag{2.7.16}$$

with the initial condition $c = c^0$ and the boundary condition for $x = 0$, $c = c^*$. Solution of this equation yields

$$-D\left(\frac{\partial c}{\partial x}\right)_{x=0} = -(c^0 - c^*)\left(\frac{7D}{3\pi t}\right)^{1/2} = \sqrt{\frac{7}{3}}J_{\text{lin}} \tag{2.7.17}$$

where J_{lin} is the material flux for linear diffusion (cf. Eq. 2.5.8). The increase in the rate of transport through convection is expressed by the coefficient $(7/3)^{1/2}$. This result is only an approximation based on simplifications in the derivation of Eq. (2.7.16). J. Koutecký gave a rigorous solution for diffusion to a growing sphere in the form of a power series. His solution has the form

$$J = \sqrt{\frac{7}{3}} J_{\text{lin}}(1 + 1.04a^{-1}D^{1/2}t^{1/6}) \tag{2.7.18}$$

where $a = (A_0/4\pi)^{1/2}$ in centimetres is the radius of the sphere at time $t = 1$ s.

2.7.3 *The mass transfer approach to convective diffusion*

In hydrodynamic problems it is an advantage to transform all variables into dimensionless parameters by 'scaling' each variable, using constants characterizing the dimensions and other properties of a given system. Thus, in the differential equations describing the systems, these constants are combined into dimensionless numbers or criteria. This formulation of the problem is very convenient because a process may be characterized in a much more general way by using the dimensionless criteria and, furthermore, in more complicated cases it is possible to find the properties of a system using a small model and then to apply the results to a large device keeping the criteria constant.

In consideration of the hydrodynamic problem alone, it is usually attempted to characterize the studied system by three quantities, the characteristic length l (e.g. the length of the plate in the direction of the flowing liquid or the radius of a rotating disk), the velocity of the flowing liquid outside the Prandtl layer V_0 and the kinematic viscosity ν.

The Cartesian coordinates of the system, x, y and z, are referred to the characteristic length l and the components of the velocity v_x, v_y and v_z within the Prandtl layer are referred to the characteristic velocity V_0,

$$\frac{x}{l} = X, \qquad \frac{y}{l} = Y, \qquad \frac{z}{l} = Z \tag{2.7.19}$$

$$\frac{v_u}{V_0} = V_u \tag{2.7.20}$$

where $u = x, y, z$. When the first and third terms on the right-hand side of Eq. (2.7.8) are neglected, the Navier–Stokes equation is obtained in Cartesian coordinates as three equations:

$$V_x \frac{\partial V_u}{\partial X} + V_y \frac{\partial V_u}{\partial Y} + V_z \frac{\partial V_u}{\partial Z} = \frac{\nu}{V_0 l} \nabla^2 V_u \tag{2.7.21}$$

The dimensionless number appearing in Eq. (2.7.21),

$$\frac{V_0 l}{\nu} = \text{Re} \tag{2.7.22}$$

is called the Reynolds number.

Its value, among others, characterizes the flow of the liquid in the system. For not very large Reynolds numbers the flow is laminar, which means that the velocity is stationary and its change monotonic in a perpendicular direction to the phase boundary between the phase boundary and the region of maximum relative velocity of the liquid. For very large values of Re the flow becomes turbulent. The value of the Reynolds number characterizing the start of turbulent flow depends strongly on the properties of the system. Thus, for example, in the case of a liquid flowing along a smooth plate the critical value is $\mathrm{Re} \approx 1.5 \times 10^3$. However, this limiting value may be considerably decreased by attaching various obstacles to the surface or by roughening the surface.

It holds approximately for the thickness of the Prandtl layer that

$$\delta_0 \approx l\sqrt{\mathrm{Re}} \tag{2.7.23}$$

In order for the flow along the boundary to come to a steady state, it is necessary that $\mathrm{Re} \gg 1$. It follows from the form of Eq. (2.7.21) that the solution for the flow rate can be expressed as a function of dimensionless quantities:

$$V_u = f_u(X, Y, Z, \mathrm{Re}) \tag{2.7.24}$$

The concentration of the transported substance c is converted to a dimensionless form, $C = c/(c^0 - c^*)$. Then the equation for convective diffusion at a steady state (2.7.7) can be converted by using Eqs (2.7.19) and (2.7.20) to the form

$$V_x \frac{\partial C}{\partial X} + V_y \frac{\partial C}{\partial Y} + V_z \frac{\partial C}{\partial Z} = \frac{D}{V_0 l}\left(\frac{\partial^2 C}{\partial X^2} + \frac{\partial^2 C}{\partial Y^2} + \frac{\partial^2 C}{\partial Z^2}\right) \tag{2.7.25}$$

The quantity

$$\frac{V_0 l}{D} = \mathrm{Pe} \tag{2.7.26}$$

is the Peclet number. The ratio of the Peclet and Reynolds numbers is called the Schmidt number,

$$\mathrm{Sc} = \frac{\mathrm{Pe}}{\mathrm{Re}} = \frac{\nu}{D} \tag{2.7.27}$$

The solution of Eq. (2.7.25) as the gradient of the dimensionless concentration C with respect to the dimensionless coordinate Y (perpendicular to the phase boundary) for $Y = 0$ is then given by the relationship

$$\left(\frac{\partial C}{\partial Y}\right)_{Y=0} = \phi(X, Y, \mathrm{Re}, \mathrm{Sc}) \tag{2.7.28}$$

Experimental measurements yield the mean value of the material flux $\bar{J}_Y$, independent of the coordinates X and Y. If a further dimensionless parameter, the Sherwood number, is introduced,

$$\mathrm{Sh} = \frac{\bar{J}_Y l}{(c^0 - c^*)D} \tag{2.7.29}$$

then Eqs (2.7.1) and (2.7.28) yield the criterial equation

$$\mathrm{Sh} = F(\mathrm{Sc}, \mathrm{Re}) \tag{2.7.30}$$

For the thickness of the Nernst layer we have (cf. Eq. 2.7.2)

$$\delta = \frac{l}{\mathrm{Sh}} \tag{2.7.31}$$

In convective diffusion to a rotating disk, the characteristic velocity V_0 is given by the product of the disk radius r, as a characteristic dimension of the system, and the radial velocity ω, so that the Reynolds number is given by the equation

$$\mathrm{Re} = \frac{\omega r^2}{\nu} \tag{2.7.32}$$

It follows from Eq. (2.7.29) for $l = r$, and Eqs (2.7.13), (2.7.27) and (2.7.32) that

$$\mathrm{Sh} = 0.62\,\mathrm{Re}^{1/2}\mathrm{Sc}^{1/3} \tag{2.7.33}$$

When the relationship between the material flux and the parameters of the system can be calculated directly by solution of the appropriate differential equations, the criterion equation (2.7.30) has little significance. However, this is not possible in the great majority of practical systems, and thus the empirically determined criterial equation is of general validity for physically similar systems. It can form a basis for designing larger equipment on the basis of experiments with model systems.

References

Agar, J. N., Diffusion and convection at electrodes, *Faraday Soc. Discussions,* **1,** 26 (1947).

Cussler, E. L., see page 20.

Heyrovský, J., and J. Kůta, *Principles of Polarography,* Academic Press, New York, 1966.

Ibl, N., and O. Dossenbach, Convective mass transport, *CTE,* **6,** 133 (1983).

Levich, V. G., see page 20.

Roušar, I., K. Micka, and A. Kimla, *Electrochemical Engineering,* Elsevier, Amsterdam, Vol. 1, 1985, Vol. 2, 1986.

Chapter 3
Equilibria of Charge Transfer in Heterogeneous Electrochemical Systems

Again, the consideration of the electrical potential in the electrolyte, and especially the consideration of the difference of potential in electrolyte and electrode, involve the consideration of quantities of which we have no apparent means of physical measurement, while the difference of potential in pieces of metal of the same kind attached to the electrodes is exactly one of the things which we can and do measure.

J. W. Gibbs, 1899

3.1 Structure and Electrical Properties of Interfacial Regions

In contrast to the previous chapters, dealing with the electrochemical properties of a single phase, this and the next three chapters will be concerned with electrochemical systems consisting of two or more phases in contact, at least one of which is an electronic or electrolytic conductor. The second phase may be either another electrolytic conductor (this case will be considered the most extensively), or another electronic conductor, a dielectric or a vacuum.

Two phases in contact are separated by a surface called a phase boundary or *interface* (Fig. 3.1). The interface is unambiguously defined, for example, in the case of a metal–electrolyte solution interface. Surprisingly enough, even in the case of a liquid–liquid interface where the liquids are partially miscible the transition from one phase to the other is rather sharp.

The term 'interface' will be distinguished from a related term, the interfacial region or *interphase*. This term denotes the region between the two phases where the properties vary markedly in contrast to those in the bulk of the phases. In the case of electrically conducting phases, charge distribution occurs in this region.

The main criterion of the electrochemical properties of the interface is whether electric charge can pass sufficiently rapidly across it in both directions, or whether it is impermeable to charge transfer and thus has the

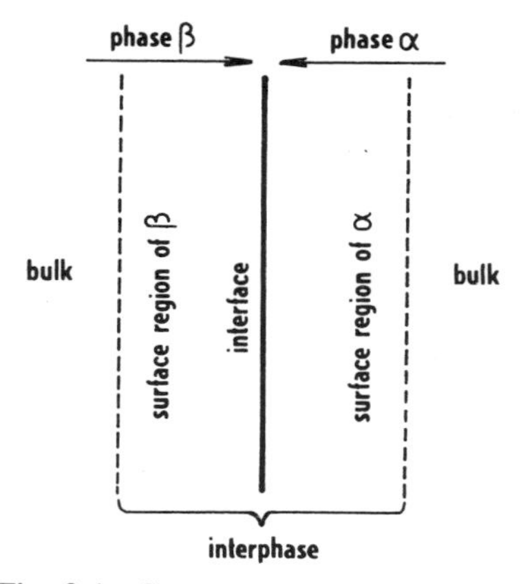

Fig. 3.1 Structure of the interphase

properties of a capacitor without leakage. For the first type of interface, the electrical potential difference between the two phases is a result of charge transfer across the interface; these interfaces are termed *non-polarizable*. In this case, charge transfer involves the transfer of electrons from a metal conductor to an electron acceptor in the second phase, the emission of electrons into a liquid phase with formation of solvated electrons, the transfer of ions from one phase into the other and similar processes occurring in the opposite direction. In the second situation, charge distribution occurs through charging of the two phases with opposite charges from an external source, unequal adsorption of ions with opposite charges on the two sides of the interface (including purely electrostatic attraction and repulsion of ions at the boundary between oppositely charged phases), adsorption and orientation of molecular dipoles and polarization of species in the inhomogeneous force field in the interfacial region. Such an interface is termed *polarizable*. The term *electrical* (electrochemical) *double layer* is often used for the corresponding interface. An electrical double layer is, of course, present even at non-polarizable interfaces. In this case, however, it cannot be charged (i.e. charge separation cannot be attained) by introduction of charge from an external source, but only through charge transfer across the interface. The electrical double layer will be dealt with in Chapter 4.

This chapter will include equilibria at non-polarizable interfaces for a metal or semiconductor phase–electrolyte system (a galvanic cell in the broadest sense) and for two electrolytes (the solid electrolyte–electrolyte solution interface, or that between two immiscible electrolyte solutions).

3.1.1 *Classification of electrical potentials at interfaces*

The distribution equilibrium for uncharged species distributed between two phases is determined by the equality of the chemical potentials of the components in the two phases. The *chemical potential* of the ith component in phase α is defined by the relationship (cf. Section 1.1.3)

$$\mu_i(\alpha) = \left(\frac{\partial G'}{\partial n_i}\right)_{n_{j\neq i},p,T} = \left(\frac{\partial U'}{\partial n_i}\right)_{n_{j\neq i},V,S} \tag{3.1.1}$$

where G' is the Gibbs energy, U' is the internal energy of phase α and n_i is the amount of the given component (in moles). According to an alternative definition, the chemical potential is a measure of the work that must be done under otherwise constant conditions for reversible transfer of one mole of uncharged species from the gaseous state with unit fugacity (assigned purely conventionally as zero on the chemical potential scale) into the bulk of the given phase.

If one mole of charged particles is to be transferred from the given reference state, further specified by the absence of an electric field, into the bulk of an electrically charged phase, then work must be expended not only to overcome chemical bonding forces but also to overcome electric forces. The work expended is given by $\tilde{\mu}_i$, termed the *electrochemical potential* by Guggenheim and also defined by Eq. (3.1.1), where G' and U' are replaced by G and U; these are analogous to G' and U' and apply to an electrically charged phase.

In general, the electrochemical potential cannot be separated into chemical and electrical components, as the chemical interaction of a species with its environment is also electric in nature. Nonetheless, the separation of the electrochemical potential is frequently made according to the equation

$$\tilde{\mu}_i = \mu_i + zF\phi \tag{3.1.2}$$

In this equation, the second term describes purely electrostatic work, connected with an infinitely slow transfer of charge zF from infinity in a vacuum into the bulk of the second phase, i.e. to a point with electric potential ϕ. Here only electric charge is transferred, not a material species with which it might be connected. The ratio of this work to the transferred charge is equal to the *inner electrical potential* ϕ of the given phase.

Equation (3.1.2) would imply separation of the effect of short-range forces (also including dipole interactions) and of the individual ionic atmospheres, related to μ_i, from the long-range forces related to ϕ, identical with purely coulombic interaction between excess charges. It will be seen later that such splitting, although arbitrary, is very useful.

The inner electrical potential ϕ may consist of two components. Firstly, the phase may possess some excess electrical charge supplied from outside. This charge produces an *outer electrical potential* ψ. This is defined as the limit of the ratio w/q for $q \to 0$, where w is the work expended for the infinitely slow transfer of charge q from an infinite distance to a point in the vacuum adjacent to the surface of the given phase and just outside the range of image forces. A particle transferred from this point further on in the direction of the other phase must overcome the effect of electric forces in the interphase. This component is called the *surface electrical potential* χ.

The component χ is defined as the limit of the ratio w'/q for $q \to 0$, where w' is the work expended in the transfer of charge q, from a point at which the outer electrical potential ψ is defined, into the bulk of the phase.

The surface potential χ consists of the contributions of ions present in the interphase $\chi(\text{ion})$ and contributions from dipoles oriented in this region $\chi(\text{dip})$:

$$\chi = \chi(\text{ion}) + \chi(\text{dip}) \tag{3.1.3}$$

(It should be recalled that the term 'surface potential' is used quite often in membranology in rather a different sense, i.e. for the potential difference in a diffuse electric layer on the surface of a membrane, see page 443.) It holds that $\phi = \psi + \chi$ (this equation is the definition of the inner electrical potential ϕ). Equation (3.1.2) can then be written in the form

$$\tilde{\mu}_i = \mu_i + zF\psi + zF\chi \tag{3.1.4}$$

For uncharged species $z = 0$ and $\tilde{\mu}_i = \mu_i$. The first term in Eq. (3.1.2) is thus the chemical potential of a charged species in the phase considered.

If two phases α and β, with common species i, are brought into contact, then for uncharged species the tendency to pass from phase α into phase β is expressed by the difference $\mu_i(\alpha) - \mu_i(\beta)$, and for charged species by the difference $\tilde{\mu}_i(\alpha) - \tilde{\mu}_i(\beta)$. For equilibrium of uncharged species, $\mu_i(\alpha) = \mu_i(\beta)$, while the condition for equilibrium of charged species is

$$\tilde{\mu}_i(\alpha) = \tilde{\mu}_i(\beta) \tag{3.1.5}$$

The result of the establishment of phase equilibrium is the formation of differences of inner and outer potentials of both phases. These differences are defined by the equations

$$\Delta_\alpha^\beta \phi = \phi(\beta) - \phi(\alpha) = \frac{\mu_i(\alpha) - \mu_i(\beta)}{z_i F} \tag{3.1.6}$$

$$\Delta_\alpha^\beta \psi = \Delta_\alpha^\beta \phi - \Delta_\alpha^\beta \chi \tag{3.1.7}$$

It should be noted that phase equilibrium can be attained for charged species through far smaller transfer of particles from one phase into the other than for uncharged species, as the potential difference formed, $\phi(\beta) - \phi(\alpha)$, decreases the original tendency for transfer between the two phases much faster than the changes in the original chemical potentials. The difference of the inner electrical potentials of the phases, $\Delta_\alpha^\beta \phi$, is also called the *Galvani potential difference* and the analogous difference of the outer electrical potentials, $\Delta_\alpha^\beta / \psi$, is termed the *Volta potential difference*.

3.1.2 *The Galvani potential difference*

Equations (3.1.5) and (3.1.6) hold for any contact equilibrium of two phases with a common charged species. This species must be exchanged sufficiently rapidly between the two phases, so that the applicability of Eq. (3.1.5) is dependent on species i being transferred very quickly from phase α to phase β and back again.

It should be noted that the Galvani potential difference consists of the contribution of ions $g(\text{ion})$ and of the oriented dipoles $g(\text{dipole})$:

$$\Delta_\alpha^\beta \phi = \Delta_{\alpha}^{\beta} g(\text{ion}) + \Delta_{\alpha}^{\beta} g(\text{dipole}) \tag{3.1.8}$$

In solid-state physics, the electrochemical potential of the electron $\tilde{\mu}_e(\alpha)$ is mostly replaced by the equivalent energy of the Fermi level ε_F. While the electrochemical potential is usually related to one mole of particles, the Fermi energy is related to a single electron, so that

$$\tilde{\mu}_e = \varepsilon_F N_A \tag{3.1.9}$$

The lowest level of the conduction band for metals V_c and the highest level of the valence band for semiconductors is very often used as a reference point for the energies of the Fermi levels; in this book, however, the energy of a free electron at rest in a vacuum will be used as the reference point for the scale of the Fermi levels (cf. Fig. 3.2.).

The occupation of the energy levels of the conduction band in metals is described by the Fermi function

$$f(\varepsilon) = \left(1 + \exp \frac{\varepsilon - \varepsilon_F}{kT}\right)^{-1} \tag{3.1.10}$$

where ε is the energy of the given level. The Fermi energy corresponds to

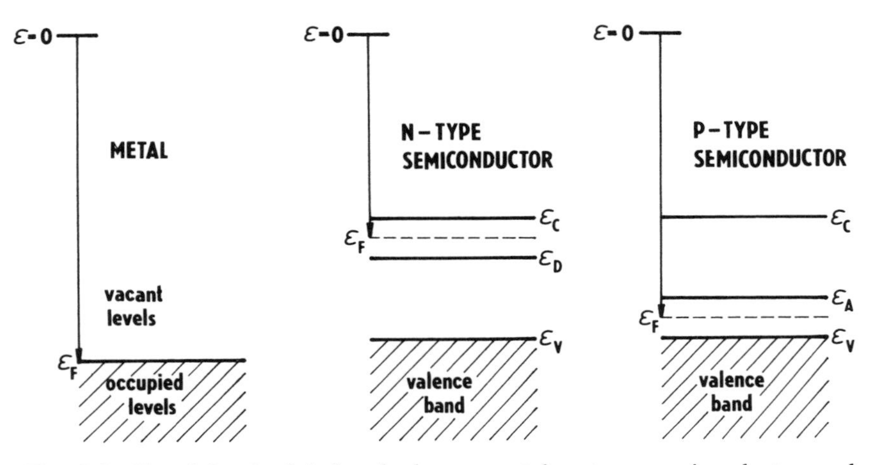

Fig. 3.2 Fermi level of isolated phases, metal, n-type semiconductor and p-type semiconductor, in a vacuum

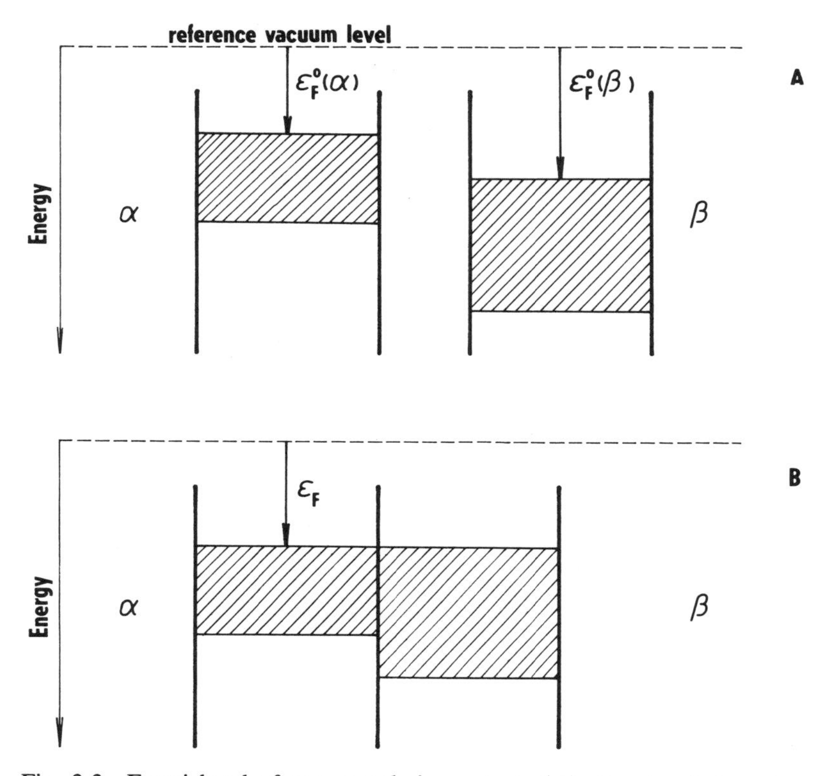

Fig. 3.3 Fermi level of two metals in contact: (A) situation before the contact when both phases have different Fermi levels, $\varepsilon_F^0(\alpha)$ and $\varepsilon_F^0(\beta)$; (B) after coming into contact both the phases assume the same Fermi level, ε_F

the just half-filled level. For semiconductors, the Fermi level lies in the forbidden band between the conductance and valence bands depending on the presence of donor and acceptor levels.

At the contact of two electronic conductors (metals or semiconductors—see Fig. 3.3), equilibrium is attained when the Fermi levels (and thus the electrochemical potentials of the electrons) are identical in both phases. The chemical potentials of electrons in metals and semiconductors are constant, as the number of electrons is practically constant (the charge of the phase is the result of a negligible excess of electrons or holes, which is incomparably smaller than the total number of electrons present in the phase). The values of chemical potentials of electrons in various substances are of course different and thus the Galvani potential differences between various metals and semiconductors in contact are non-zero, which follows from Eq. (3.1.6). According to Eq. (3.1.2) the electrochemical potential of an electron in phase α is

$$\tilde{\mu}_e(\alpha) = \mu_e^0(\alpha) - F\phi(\alpha) \tag{3.1.11}$$

Thus, for the Galvani potential difference between the metals α and β we have

$$\Delta_\beta^\alpha\phi = \phi(\alpha) - \phi(\beta) = \frac{\mu_e^0(\alpha) - \mu_e^0(\beta)}{F} \tag{3.1.12}$$

If an electronic conductor is in contact with an electrolyte solution containing the components of a simple redox system consisting of the ionic species Ox (charge numbers z_{Ox}) and Red (charge number z_{Red})

$$\mathrm{Ox} + ze = \mathrm{Red} \tag{3.1.13}$$

where $z = z_{Ox} - z_{Red}$ (Fig. 3.4) then the relationship for the energy of the Fermi level in the solution $\varepsilon_F(\beta)$ can be formulated as follows. The occupied

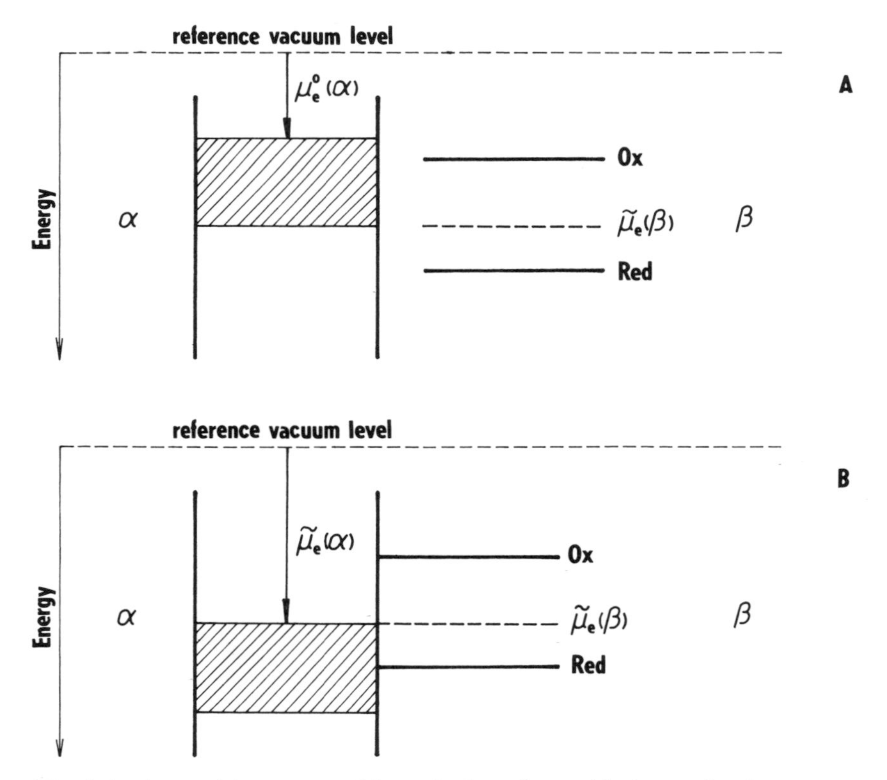

Fig. 3.4 A metal in contact with a solution of an oxidation–reduction system. (A) Situation before the contact when the electrochemical potential of electrons in the electronic conductor ($\tilde{\mu}_e^0(\alpha) \equiv \varepsilon_F(\alpha)$) has a different value from the electrochemical potential of electrons in the oxidation–reduction system. (B) When the phases are in contact the electrochemical potential of electrons becomes identical in both α and β by charge transfer between them

energy levels are considered to be electron donor levels in the reduced form, Red, whose number is proportional to the activity of the reduced form, a_{Red}, while the unoccupied energy levels can be identified with the electron acceptor levels in the oxidized form, Ox. Their number is proportional to the activity of the oxidized form, a_{Ox}. For the equilibrium of reaction (3.1.13) we have

$$\tilde{\mu}_{Ox}(\beta) + z\tilde{\mu}_e(\beta) = \tilde{\mu}_{Red}(\beta) \qquad (3.1.14)$$

Equation (3.1.2) and the definition of the chemical potential yield the equation for the electrochemical potential of species i with activity $a_i(\beta)$ and charge z_i in phase β in the form

$$\tilde{\mu}_i(\beta) = \mu_i^0(\beta) + RT \ln a_i(\beta) + z_i F\phi(\beta) \qquad (3.1.15)$$

In view of Eqs (3.1.13) and (3.1.14)

$$\tilde{\mu}_e(\beta) = \frac{\mu_{Red}^0(\beta) - \mu_{Ox}^0(\beta)}{z} + \frac{RT}{zF} \ln\left(\frac{a_{Red}}{a_{Ox}}\right) - F\phi(\beta) \qquad (3.1.16)$$

Combination of Eqs (3.1.2) and (3.1.16) yields a relationship for the Galvani potential difference between the metal and the solution,

$$\begin{aligned}\Delta_\beta^\alpha \phi &= \phi(\alpha) - \phi(\beta) \\ &= \frac{\mu_{Ox}^0(\beta) - \mu_{Red}^0(\beta) + z\mu_e^0(\alpha)}{zF} + \frac{RT}{zF} \ln\left(\frac{a_{Ox}}{a_{Red}}\right)\end{aligned} \qquad (3.1.17)$$

We shall look more closely at this equation. On one hand, the standard chemical potentials of Ox and Red depend on their standard Gibbs solvation energies, $\Delta G_{s,Ox}^0$ and $\Delta G_{s,Red}^0$, and, on the other hand, on the standard Gibbs energy of ionization of Red in the gas phase, $\Delta G_{ion,Red}^0$. This quantity is connected with the ionization potential of Red, I_{Red}, which is, however, a sort of enthalpy so that it must be supplemented by the entropy term, $-T\Delta S_{ion,Red}^0$. Thus, Eq. (3.1.17) is converted to the form

$$\begin{aligned}\Delta_\beta^\alpha \phi = &\frac{\Delta G_{s,Ox}^0(\beta) + I_{Red} - T\Delta S_{ion,Red}^0 - \Delta G_{s,Red}^0(\beta) + z\mu_e^0(\alpha)}{zF} \\ &+ \frac{RT}{zF} \ln\left(\frac{a_{Ox}}{a_{Red}}\right)\end{aligned} \qquad (3.1.18)$$

When the species common to the metal and the solution are metal ions M^{z+}, which are rapidly exchanged between the two phases, Eqs (3.1.5) and (3.1.15) readily yield the relationship for the Galvani potential difference between the metal and the solution

$$\Delta_\beta^\alpha \phi = \frac{-\mu_{M^{z+}}^0(\alpha) + \mu_{M^{z+}}^0(\beta)}{z_+ F} + \frac{RT}{z_+ F} \ln a_{M^{z+}}(\beta) \qquad (3.1.19)$$

The equilibrium

$$M \rightleftarrows M^{z+} + z_+ e$$

exists in the metal phase, so that the standard chemical potential of the metal ions in the metal phase can be expressed as

$$\mu^0_{M^{z+}}(\alpha) = \mu^0_M(\alpha) + z_+\mu^0_e(\alpha) \tag{3.1.20}$$

Substitution into Eq. (3.1.19) yields

$$\Delta^\alpha_\beta\phi = \frac{-\mu^0_M(\alpha) + z_+\mu^0_e(\alpha) + \mu^0_{M^{z+}}(\beta)}{z_+F} + \frac{RT}{z_+F}\ln a_{M^{z+}}(\beta) \tag{3.1.21}$$

If two electrolyte solutions in immiscible solvents are in contact and contain a common ion B^{z+} and the anions are not soluble in the opposite phase, then the relationship for the Galvani potential difference follows from Eq. (3.1.5):

$$\Delta^\alpha_\beta\phi = \frac{\mu^0_{B^{z+}}(\beta) - \mu^0_{B^{z+}}(\alpha)}{z_+F} + \frac{RT}{z_+F}\ln\left[\frac{a_{B^{z+}}(\beta)}{a_{B^{z+}}(\alpha)}\right] \tag{3.1.22}$$

When a solid electrolyte is in contact with an electrolyte solution, with common anion $A^{z-}(z_- < 0)$, then the Galvani potential difference between the solid electrolyte and the solution is given by the equation

$$\Delta^\alpha_\beta\phi = \frac{\mu^0_{A^{z-}}(\beta) - \mu^0_{A^{z-}}(\alpha)}{z_-F} + \frac{RT}{z_-F}\ln a_{A^{z-}}(\beta) \tag{3.1.23}$$

Finally, consider a metal (α)–solid electrolyte (β)–electrolyte solution (γ) system where the solid electrolyte and the electrolyte solution contain a common anion A^- and the metal and the solid electrolyte contain a common cation B^+ (for simplicity, only univalent ions are considered). The potential difference between phase α (metal) and phase γ (solution) is

$$\begin{aligned}
\Delta^\alpha_\gamma\phi &= \Delta^\alpha_\beta\phi + \Delta^\beta_\gamma\phi \\
&= \frac{-\mu^0_{B^+}(\alpha) + \mu^0_{B^+}(\beta) + \mu^0_{A^-}(\beta) - \mu^0_{A^-}(\gamma)}{F} - \frac{RT}{F}\ln a_{A^-}(\gamma) \\
&= \frac{-\mu^0_{B^+}(\alpha) + \mu^0_{BA} - \mu^0_{A^-}(\gamma)}{F} - \frac{RT}{F}\ln a_{A^-}(\gamma)
\end{aligned} \tag{3.1.24}$$

The chemical potential of insoluble salt BA, μ^0_{BA}, depends on its solubility product according to the relationship (cf. Section 1.5.1)

$$\mu^0_{BA} = RT\ln P_{BA} + \mu^0_{B^+}(\gamma) + \mu^0_{A^-}(\gamma) \tag{3.1.25}$$

so that Eq. (3.1.24) can be rearranged to give

$$\begin{aligned}
\Delta^\alpha_\gamma\phi &= \frac{-\mu^0_{B^+}(\alpha) + \mu^0_{B^+}(\gamma)}{F} + \frac{RT}{F}\ln P_{BA} - \frac{RT}{F}\ln a_{A^-}(\gamma) \\
&= \frac{-\mu^0_B(\alpha) + \mu^0_e(\alpha) + \mu^0_{B^+}(\gamma)}{F} + \frac{RT}{F}\ln P_{BA} - \frac{RT}{F}\ln a_{A^-}(\gamma)
\end{aligned} \tag{3.1.26}$$

Two solid electrolytes in contact with a common ion show behaviour analogous to two metals in contact.

As pointed out by J. W. Gibbs in 1875, thermodynamic procedures cannot be used to measure Galvani potential differences between chemically different phases (see page 144). Thus Eqs (3.1.12), (3.1.17), (3.1.18), (3.21) to (3.1.24) and (3.1.26) cannot be directly verified by measuring equilibrium quantities. However, their derivation was not pointless, because, as will be shown in Section 3.2, they have a form analogous to the relationships for equilibrium electrode potentials, consisting of several Galvani potential differences.

Equilibrium measurements can be used only to determine the Galvani potential differences between chemically identical phases, which is strictly possible only for two identical electronic conductors. Then

$$\Delta_\beta^\alpha \phi = \frac{\Delta \tilde{\mu}_i}{z_i F} \tag{3.1.27}$$

as the 'chemical' parts of the electrochemical potential are identical. Potentiometric measurements (see Section 3.3.1) yield the differences $\Delta\phi$ for conditions expressed by Eq. (3.1.27). It will be demonstrated later that useful relationships have been deduced for potential differences between phases of different chemical composition on the basis of certain extrathermodynamic assumptions. This is especially true for measurements in solutions with liquid junctions (see Sections 2.5.3 and 3.1.4) and direct measurement of the potential difference between two immiscible electrolyte solutions (Section 3.2.8).

3.1.3 *The Volta potential difference*

The outer electrical potential of a phase is the electrostatic potential given by the excess charge of the phase. Thus, if a unit electric charge is brought infinitely slowly from infinity to the surface of the conductor to a distance that is negligible compared with the dimensions of the conductor considered (for a conductor with dimensions of the order of centimetres, this distance equals about 10^{-4} cm), work is done that, by definition, equals the outer electric potential ψ.

In contrast to differences in the inner potentials (the Galvani potential difference), the difference in the outer potentials $\Delta_\beta^\alpha \psi$ (the Volta potential difference or the contact potential) and the *real potential* α_i can be measured. The real potential α_i is defined as

$$\alpha_i(\alpha) = \mu_i(\alpha) + z_i F \chi(\alpha) \tag{3.1.28}$$

The electrochemical potential can thus also be defined by the equation

$$\tilde{\mu}_i(\alpha) = \alpha_i(\alpha) + z_i F \psi(\alpha) \tag{3.1.29}$$

The real potential of electron times -1, $-\alpha_e(\alpha)$, is termed the *electron work function* $\Phi(\alpha)$. It should be realized that all these cases involve measurement of differences between two states. Since, however, one of these states

Table 3.1 Electron work function $\Phi_e(\alpha)$ of some metals. (*From J. Hölzl, F. K. Schultze and H. Wagner*)

Metal (with the Miller indices of the crystal face)	$\Phi_e(\alpha)$ (eV)	Method
Ag (100)	4.64 ± 0.02	Photoelectric
Ag (110)	4.52 ± 0.02	Photoelectric
Ag (111)	4.74	Photoelectric
Cs	2.00	Contact potential versus Ta (110)
Cu (100)	4.59 ± 0.03	Photoelectric
Cu (111)	4.94 ± 0.05	Photoelectric
Cu (110)	4.48 ± 0.03	Photoelectric
Cu (112)	4.53 ± 0.03	Photoelectric
K	2.26	Contact potential versus Ta (110)
Na	2.46	Contact potential versus Ta (110)
Rb	2.05	Photoelectric
Ta (110)	$4.8 + 0.6 \times 10^{-4}$ T	Thermionic emission
Ta (111)	$4.15 + 2 \times 10^{-4}$ T	Thermionic emission
Ta (111)	$4.00 + 2.6 \times 10^{-4}$ T	Thermionic emission

is assigned a zero value by convention (as mentioned above, this value is assigned to a point infinitely far away from all the conductors), then we can say that potentials α_i and ψ are measurable just as if they were absolute values. However, both the inner potential and the Galvani potential difference cannot be measured in the sense of 'absolute' values using purely thermodynamic procedures.

As a function of the surface potential χ, the electron work function for a given material depends on the state of the surface of that material (adsorption, the presence of surface compounds, etc.). For crystalline substances (see Table 3.1), various crystal faces have various electron work function values, which can be measured for single crystals. For polycrystalline substances, the final value of the electron work function depends on the contribution of the individual crystal faces to the entire area of the phase and the corresponding electron work functions; the final value of the work function, however, is strongly dependent on the experimental method used for the measurement.

The electron work function is very important for the physics of the solid phase and for its application to electronics. In electrochemistry, it is especially important for electrode processes involving adsorption of some of the species participating in the electrode reaction.

If two different, electrically uncharged phases are brought into contact, then the electrochemical potentials of their electrons become equal (i.e.

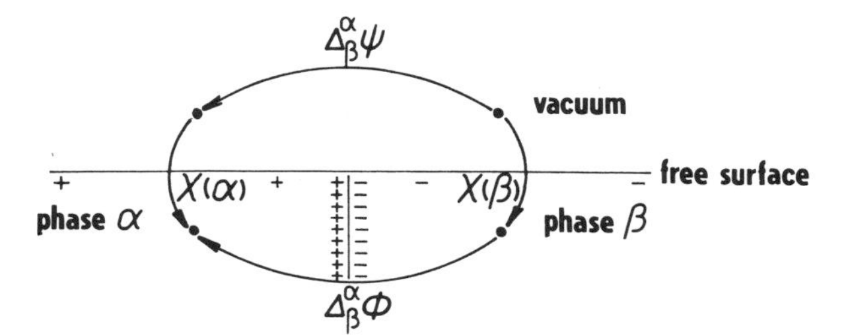

Fig. 3.5 The significance of the contact potential (Volta potential difference) $\Delta^{\alpha}_{\beta}\psi$ between two metals in contact. (According to S. Trasatti and R. Parsons)

their Fermi levels are brought to the same level—see Fig. 3.2). An electrical double layer is formed on the contact surface and the outer potentials of the free surfaces of both phases attain different values; i.e. a Volta potential difference (contact potential) is formed between them—see Fig. 3.5.

Measurement yields both the differences between the outer potentials and the work functions (real potentials). If two phases α an β with a common species (index i) come into contact, at equilibrium $\tilde{\mu}_i(\alpha) = \tilde{\mu}_i(\beta)$, that is $\alpha_i(\alpha) - \alpha_i(\beta) = z_i F \Delta^{\alpha}_{\beta}\psi$. These quantities are mostly measured using the vibrating condenser, thermoionic, calorimetric, and photoelectric methods.

The principle of the *vibrating condenser method*, originally proposed by Thomson, is depicted in Fig. 3.6. Space A between the metal phases α, α'

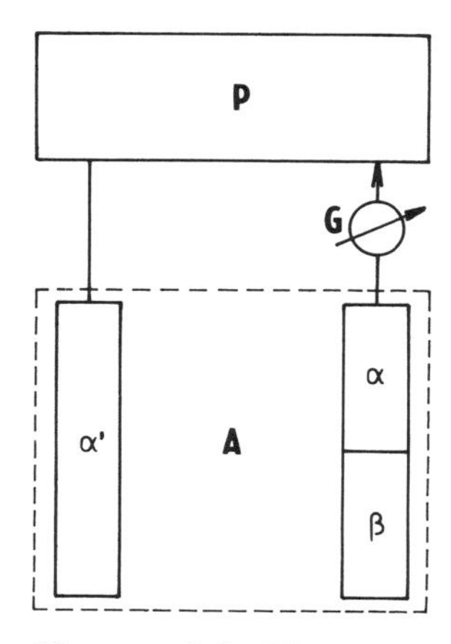

Fig. 3.6 Thomson's vibrating condenser method. A, space filled with an inert gas at low pressure; P, potentiometer; G, null instrument

and β is filled with an inert gas at very low pressure. Phases α and β are in contact. Phases α and α' are chemically identical. As they are surrounded by the same medium, $\chi(\alpha)=\chi(\alpha')$ and both the chemical potentials and the electron work functions are identical for both these phases, that is $\mu_e(\alpha)=\mu_e(\alpha')$; $\alpha_e(\alpha)=\alpha_e(\alpha')$. The outer potentials of phases α' and β are different, reflected in the fact that current flows through galvanometer G on a change in the distance between α' and β or on ionization of atoms of the gas (e.g. by irradiation with a radioactive substance) in space A. This current can be compensated by the voltage U derived from potentiometer P, for which

$$U=\phi(\alpha)-\phi(\alpha')=-F^{-1}(\tilde{\mu}_e(\alpha)-\tilde{\mu}_e(\alpha')) \tag{3.1.30}$$

The condition at equilibrium is that $\psi(\alpha')=\psi(\beta)$ and $\tilde{\mu}_e(\alpha)=\tilde{\mu}_e(\beta)$ (the latter equation describes the condition for contact equilibrium between the electrons in phase α and β). The measured compensating voltage U,

$$U=F^{-1}(\alpha_e(\alpha)-\alpha_e(\beta))=\psi(\alpha)-\psi(\beta) \tag{3.1.31}$$

thus yields both the difference of the electron work functions and the difference of the outer electrical potentials of the two phases in contact, α and β.

The surface of the phases must be clean (one of the greatest experimental difficulties of this method); otherwise, the values of χ would change along with the electron work functions.

The presence of impurities between phases α and β in contact implies the introduction of a new phase (γ). If current I passes and this new phase has resistance R, then a further potential drop RI appears in the system. However, at equilibrium the presence of phase γ has no effect on the measured results, as it still holds that $\tilde{\mu}_e(\alpha)=\tilde{\mu}_e(\beta)=\tilde{\mu}_e(\gamma)$. Similarly, the final equation is not affected by the fact that the leads and coils of the potentiometer are not made of the same metal as phase α'. It can also be readily seen that the validity of Eq. (3.1.31) is retained even in the presence of a new phase γ between phase α or phase α' and the potentiometer.

The difference in the outer potentials between the metal and the electrolyte solution is measured similarly. In Fig. 3.6, phase β will now designate an electrolyte solution in contact with the metal phase α. Also here $\mu_e(\alpha)=\mu_e(\alpha')$ and $\psi(\alpha')=\psi(\beta)$, and U is again expressed by Eq. (3.1.30); however, $\tilde{\mu}_e(\alpha)\neq\tilde{\mu}_e(\beta)$ as the communicating species between phases α and β are not electrons but metal ions, M^{z+}. Thus,

$$\tilde{\mu}_+(\alpha)=\tilde{\mu}_+(\beta) \tag{3.1.32}$$

where the subscript + refers to the metal cations. However, in phase α,

$$\mu_M(\alpha)=\tilde{\mu}_+(\alpha)+z_+\tilde{\mu}_e(\alpha) \tag{3.1.33}$$

The subscript M refers to metal ions. An analogous procedure to that employed above yields

$$U = \frac{\alpha_e(\alpha)}{F} - \frac{\mu_M(\alpha) - \alpha_+(\beta)}{z_+ F}$$
$$= \psi(\alpha) - \psi(\beta) \qquad (3.1.34)$$

Thus the compensating voltage U yields the difference between the outer electrical potentials of the metal and the solution with which it is in contact.

The *thermionic method* is based on thermoemission of electrons in a diode. At elevated temperatures, metals emit electrons that are collected at a gauze electrode placed opposite the metal surface and charged to a high positive potential. C. W. Richardson has given the following relationship for the saturation current I_s:

$$I_s = AT^2 \exp\left(\frac{-\alpha_e}{RT}\right) \qquad (3.1.35)$$

The temperature dependence of the current I_s yields the electron work function $-\alpha_e$.

The *calorimetric method* is based on measuring the energy that must be supplied to the metal to maintain stationary emission of electrons. When a metal wire is heated to a high temperature, no electron emission starts unless the anode is connected to a high voltage source. Then, sudden emission of electrons produces a sharp temperature decrease as a result of energy consumption for release of the electrons. The difference between the amount of energy required to heat the wire to a certain temperature with and without electron emission depends on the work function for the electrons of the given metal at that temperature.

The *photoelectric method* is based on the photoelectric effect. The kinetic energy of the electrons emitted during illumination of a metal with light having a frequency ν obeys the Einstein equation

$$\tfrac{1}{2}mv^2 = h\nu - h\nu_0 = h\nu - \frac{-\alpha_e}{N_A} \qquad (3.1.36)$$

where ν_0 is the limiting frequency below which no photoemission takes place. The product of this frequency with the Planck's constant h gives the work function for a single electron, which can be determined when the wavelength of the light irradiating the metal is continuously changed.

3.1.4 *The EMF of galvanic cells*

A galvanic cell is a system that can perform electrical work when its energy is consumed at the expense of chemical or concentration changes that occur inside the system. There are also systems analogous to galvanic cells based on conversion of other types of energy than chemical or osmotic into electrical work. Photovoltaic electrochemical cells will be discussed in

Section 5.9.2. Thermogalvanic cells where electrical work is gained from heat transfer will not be considered in this book and the reader is advised to inspect Agar's monograph quoted on page 169.

The electrode is considered to be a part of the galvanic cell that consists of an electronic conductor and an electrolyte solution (or fused or solid electrolyte), or of an electronic conductor in contact with a solid electrolyte which is in turn in contact with an electrolyte solution. This definition differs from Faraday's original concept (who introduced the term electrode) where the electrode was simply the boundary between a metal and an electrolyte solution.

For example, consider a system in which metallic zinc is immersed in a solution of copper(II) ions. Copper in the solution is replaced by zinc which is dissolved and metallic copper is deposited on the zinc. The entire change of enthalpy in this process is converted to heat. If, however, this reaction is carried out by immersing a zinc rod into a solution of zinc ions and a copper rod into a solution of copper ions and the solutions are brought into contact (e.g. across a porous diaphragm, to prevent mixing), then zinc will pass into the solution of zinc ions and copper will be deposited from the solution of copper ions only when both metals are connected externally by a conductor so that there is a closed circuit. The cell can then carry out work in the external part of the circuit. In the first arrangement, reversible reaction is impossible but it becomes possible in the second, provided that the other conditions for reversibility are fulfilled.

In the reversible case, the total change of the Gibbs energy during the reaction occurring in the cell will be transformed into electrical work, while in the irreversible case the electric work will be smaller and can be zero. For reversible behaviour in a galvanic cell two requirements need to be fulfilled; there must be material and energetic reversibility. Most practical galvanic cells, however, show a certain degree of irreversibility.

Material reversibility can be demonstrated by balancing the EMF of the cell with an equal and opposite electrical potential difference from a potentiometer, and then causing the reaction in the cell (the *cell reaction*) to proceed, first in one direction and then in the reverse direction, by adjusting the balancing electrical potential difference to be greater than and then less than the cell EMF.

Energetic reversibility is achieved when the same amount of electrical work is supplied from the cell reaction proceeding in one direction as is gained from the reaction proceeding to the same degree in the opposite direction.

Reversible operation of the cell requires that no other process occurs in the cell than that connected with the current flow. An electrochemical process that need not be always connected with the passage of current is the dissolution of a metal in an acid (e.g. zinc in sulphuric acid in the Volta cell) or the dissolution of a gas in an electrolyte solution (e.g. in a cell consisting of hydrogen and chlorine electrodes, hydrogen and chlorine are dissolved and hydrogen chloride can be formed in solution even without current flow; this reaction, however, is strongly inhibited kinetically). Another process that can occur in the absence of current passage is diffusion. In a concentration cell (see Section 3.2.1), the concentrations of the two solutions can be equalized, even in the absence of current passage, through diffusion alone. However, this part will consider cells for which both of the criteria of thermodynamic reversibility are satisfied.

The irreversible processes described must not occur even on open circuit. In a reversible cell, a definite equilibrium must be established and this may be defined in terms of the intensive variables in a similar way to the description of phase and chemical equilibria of electroneutral components.

Further, the passage of a finite amount of charge corresponds to a certain degree of energetic irreversibility, as the electrical work is converted into Joule heat because of the internal resistance of the cell and also in the external circuit; in addition, electrode processes involve a certain overpotential (see Chapter 5). Thus, the cell can work reversibly only on passage of negligible current, i.e. when the difference between the balancing external voltage and the EMF of the cell approaches zero.

It should be noted that the reversibility of the galvanic cell has so far been considered from a purely thermodynamic point of view. 'Reversible electrode processes' are sometimes considered in electrochemistry in a rather different sense, as will be described in Chapter 5.

A galvanic cell is usually depicted in terms of chemical symbols. The phase boundary is designated by a vertical line. For aqueous solutions, the dissolved substances and their concentrations are indicated, and for non-aqueous solutions, also the solvent. For example, the cell

$$\mathrm{Hg, Zn(12\%)\,|\,ZnSO_4(sat.), Hg_2SO_4(sat.)\,|\,Hg} \tag{3.1.37}$$

consists of a zinc amalgam electrode and a mercury electrode; the common electrolyte is a saturated solution of zinc and mercury sulphates.

As has already been mentioned, the *EMF* (*the electromotive force*) of a cell is given by the potential difference between leads of identical metallic material. In view of this, a galvanic cell is represented schematically as having identical metallic phases at either end.

The phases in the scheme are always numbered from the left to the right, but the cell can be depicted graphically in two ways. For example, there are two possibilities for the Daniell cell:

$$\begin{matrix} 1 & 2 & 3 & 4 & 1' \\ \mathrm{Zn}\,| & \mathrm{Zn^{2+}}\,| & \mathrm{Cu^{2+}}\,| & \mathrm{Cu}\,| & \mathrm{Zn} \end{matrix} \tag{a}$$

$$\begin{matrix} 1 & 2 & 3 & 4 & 1' \\ \mathrm{Cu}\,| & \mathrm{Cu^{2+}}\,| & \mathrm{Zn^{2+}}\,| & \mathrm{Zn}\,| & \mathrm{Cu} \end{matrix} \tag{b}$$

The choice between (a) and (b) is not specified, but does determine the sign of the EMF of the cell. This quantity is always defined so that the electrical potential of the last phase on the left (1) is subtracted from that of the last phase on the right (1′). Thus, for the two given schemes of the Daniell cell, $E_a = -E_b$. For practical reasons it is preferable to place the positive electrode on the right, as in scheme (a).

This section will consider only 'chemical galvanic cells' in which the chemical energy is converted to electrical energy.

The *cell reaction* is the sum of the electrochemical reactions taking place at both electrodes. The cell reaction may be written in two ways which are dependent on the sequence of phases in the graphical scheme of the cell. The representation of the cell reaction should correspond to the flow of positive charge through the cell (in a graphical scheme) from left to right:

$$Cu^{2+} + Zn \rightarrow Cu + Zn^{2+} \qquad \text{(c)}$$

$$Cu + Zn^{2+} \rightarrow Cu^{2+} + Zn \qquad \text{(d)}$$

Thus, Eq. (c) corresponds to scheme (a) and Eq. (d) to scheme (b). The flow of the charge is due to the chemical forces described by the affinity of the cell reaction, and at the same time to the electrical forces measured by the electrical potential difference. The affinity of the reaction is given by the algebraic sum of the chemical potentials of all the substances participating in the reaction multiplied by the corresponding stoichiometric coefficients (which are positive for reactants and negative for products); it is thus given by the decrease in the Gibbs free energy of the system $-\Delta G$. The sign of the affinity, however, is reversed when the reaction is written in the opposite direction. Thus, for reactions (c) and (d),

$$-\Delta G_c = +\Delta G_d \qquad (3.1.38)$$

If the reaction in the cell proceeds to unit extent, then the charge nF corresponding to integral multiples of the Faraday constant is transported through the cell from the left to the right in its graphical representation. Factor n follows from the stoichiometry of the cell reaction (for example $n = 2$ for reaction c or d). The product nFE is the work expended when the cell reaction proceeds to a unit extent and at thermodynamic equilibrium and is equal to the affinity of this reaction. Thus,

$$E_a = \frac{-\Delta G_c}{zF} \quad \text{and} \quad E_b = \frac{-\Delta G_d}{zF} \qquad (3.1.39)$$

Considering that $(-\Delta G_d) > 0$ when $(-\Delta G_c) < 0$, then it is apparent that the sign of the EMF depends on the direction in which the scheme of the cell is written and, consequently, on the direction in which the reaction is written. The EMF is positive when the reaction and the scheme are formulated so as to correspond to a spontaneous reaction (the reaction would occur spontaneously in the direction indicated if the cell were short-circuited). Thus cell (a) has a positive EMF and cell (b) a negative one.

Consider a galvanic cell consisting of hydrogen and silver chloride electrodes:

$$\mathrm{Pt}\,|\,\mathrm{H_2}\,|\,\mathrm{HCl(c)}\,|\,\mathrm{AgCl(s)}\,|\,\mathrm{Ag}\,|\,\mathrm{Pt} \tag{3.1.40}$$

In this cell, the following independent phases must be considered: platinum, silver, gaseous hydrogen, solid silver chloride electrolyte, and an aqueous solution of hydrogen chloride. In order to be able to determine the EMF of the cell, the leads must be made of the same material and thus, to simplify matters, a platinum lead must be connected to the silver electrode. It will be seen in the conclusion to this section that the electromotive force of a cell does not depend on the material from which the leads are made, so that the whole derivation could be carried out with different, e.g. copper, leads. In addition to Cl^- and H_3O^+ ions (further written as H^+), the solution also contains Ag^+ ions in a small concentration corresponding to a saturated solution of silver chloride in hydrochloric acid. Thus, the following scheme of the phases can be written (the parentheses enclose the species present in the given phase):

$$\overset{1}{\mathrm{Pt(e)}}\,|\,\overset{2}{\mathrm{H_2}}\,|\,\overset{3}{\mathrm{solution(H^+, Cl^-, Ag^+)}}\,|\,\overset{4}{\mathrm{AgCl(Ag^+, Cl^-)}}\,|\,\overset{5}{\mathrm{Ag(Ag^+, e)}}\,|\,\overset{1'}{\mathrm{Pt(e)}} \tag{3.1.41}$$

The electrodes employed will be considered in greater detail in Section 3.2. Here it is sufficient that this cell can yield electrical energy, because the cell reaction

$$\tfrac{1}{2}\mathrm{H_2(2)} + \mathrm{AgCl(4)} \rightarrow \mathrm{H^+(3)} + \mathrm{Cl^-(3)} + \mathrm{Ag(5)} \tag{3.1.42}$$

takes place spontaneously.

Depending on the processes occurring during the cell reaction at the individual electrodes, the cell reaction can be separated into two *half-cell reactions* formulated as reduction by electrons. For the cell reaction described by Eq. (3.1.42), these reactions are

$$\mathrm{AgCl(4)} + \mathrm{e} = \mathrm{Ag(5)} + \mathrm{Cl^-(3)} \tag{3.1.43}$$

$$\mathrm{H^+(3)} + \mathrm{e} = \tfrac{1}{2}\mathrm{H_2(2)} \tag{3.1.44}$$

Subtraction of reaction (3.1.44) from reaction (3.1.43) yields reaction (3.1.42).

The EMF, E, is given by the difference in the inner potentials of phases 1′ and 1:

$$E = \phi(1') - \phi(1) \tag{3.1.45}$$

If this EMF is compensated by an external electrical potential difference of the same magnitude and of opposite polarity, then no current flows and the system is at equilibrium. The phase equilibria of communicating species are described as

$$\tilde{\mu}_{\text{Cl}^-,3} = \tilde{\mu}_{\text{Cl}^-,4}; \qquad \tilde{\mu}_{\text{Ag}^+,4} = \tilde{\mu}_{\text{Ag}^+,5}; \qquad \tilde{\mu}_{\text{e},1} = \tilde{\mu}_{\text{e},5} \tag{3.1.46}$$

In the solid phases we have

$$\begin{aligned} \mu_{\text{AgCl},4} &= \tilde{\mu}_{\text{Ag}^+,4} + \tilde{\mu}_{\text{Cl}^-,4} \\ \mu_{\text{Ag},5} &= \tilde{\mu}_{\text{Ag}^+,2} + \tilde{\mu}_{\text{e},2} \end{aligned} \tag{3.1.47}$$

For the hydrogen half-cell reaction (Eq. 3.1.44),

$$2\tilde{\mu}_{\text{H}^+,3} + 2\tilde{\mu}_{\text{e},1} = \mu_{\text{H}_2,2} \tag{3.1.48}$$

All the electrochemical potentials are expanded by using Eq. (3.1.1) into the chemical potential and the inner electrical potential; the difference (3.1.45) is then found from Eqs (3.1.46) to (3.1.48), to be

$$FE = -\mu_{\text{H}^+,3} - \mu_{\text{Cl}^-,3} - \mu_{\text{Ag},5} + \mu_{\text{AgCl},4} + \tfrac{1}{2}\mu_{\text{H}_2,2} \tag{3.1.49}$$

The right-hand side of this equation expresses the affinity of reaction (3.1.42). In general, for a cell reaction proceeding according to a given stoichiometric equation with stoichiometric factors ν_i and with transfer of charge equal to nF, that

$$-\Delta G = -\sum_i \nu_i \mu_i = nFE \tag{3.1.50}$$

If the galvanic cell drives an external current, it supplies work to its surroundings. For a reversible system, the difference between the external electrical potential difference and the EMF approaches zero and, in the limiting case, these two potentials differ in sign only. Thus, the electrical work supplied to the surroundings is given by the expression $nFE = -\Delta G$, i.e. the decrease in the Gibbs free energy of the system.

It follows from Eq. (3.1.50) that the product nFE can be expressed by the Gibbs–Helmholtz equation:

$$nFE = -\Delta H + nFT\left(\frac{\partial E}{\partial T}\right)_{p,\text{composition}} \tag{3.1.51}$$

If the dependence of the EMF on the temperature is known, then the reaction enthalpy ΔH of the reaction proceeding outside the cell can be found. Cells with a positive EMF, E, and a negative temperature quotient $\partial E/\partial T$ release heat in addition to electrical energy during reversible processes

$$-T\,\Delta S = -nFT\left(\frac{\partial E}{\partial T}\right)_{p,\text{composition}} \tag{3.1.52}$$

where ΔS is the entropy change of the cell reaction.

When producing electrical current on reversible operation, the cells with both positive E and $\partial E/\partial T$ absorb heat from their surroundings, as then $-\Delta H > nFE$.

If the dependence of the EMF of a cell on temperature is known, the same value of the reaction enthalpy ΔH can be determined as if the reaction took place outside the cell.

3.1.5 *The electrode potential*

Equation (3.1.50) can be developed further. Consider once again cell (3.1.40), together with Eq. (3.1.49). We shall assume that hydrogen gas is under standard pressure;† in addition, metallic silver, solid silver chloride and gaseous hydrogen at standard pressure are selected as standards, that is

$$\mu_{H_2,2}(p=1)=\mu^0_{H_2}, \qquad \mu_{Ag,5}=\mu^0_{Ag} \qquad \text{and} \qquad \mu_{AgCl,4}=\mu^0_{AgCl}$$

Since also

$$\mu_{H^+}(w)=\mu^0_{H^+}(w)+RT\ln a_{H^+}(w),$$
$$\mu_{Cl^-}(w)=\mu^0_{Cl^-}(w)+RT\ln a_{Cl^-}(w), \qquad \mu^0_{H^+}(w)+\mu^0_{Cl^-}(w)=\mu^0_{HCl}(w)$$

and $a_{H^+}(w)a_{Cl^-}(w)=a^2_{\pm,HCl}(w)$, we can write

$$FE=-\mu^0_{H^+}(w)-\mu^0_{Cl^-}(w)-2RT\ln a_{\pm,HCl}(w) - \mu^0_{Ag}+\mu^0_{AgCl}+\tfrac{1}{2}\mu^0_{H_2} \qquad (3.1.53)$$

(where the symbol w indicates that the substances are present in aqueous solution). It is further assumed that the HCl activity in cell (3.1.40) equals unity. The quantity

$$E^0=-\frac{\mu^0_{H^+}(w)+\mu^0_{Cl^-}(w)+\mu^0_{Ag}-\mu^0_{AgCl}-\frac{1}{2}\mu^0_{H_2}}{F} \qquad (3.1.54)$$

is the *standard EMF* of the cell. In the present case where the electrode on the left in the graphical scheme is the standard hydrogen electrode, E^0 is termed the *standard electrode potential.* Standard electrode potentials are denoted as E^0 with a subscript in which the initial substances in the half-cell reaction are given first, separated from the products by a slanted line; e.g. the standard potential of the silver–silver chloride electrode is designated as $E^0_{AgCl/Ag,Cl^-}$. For typographical reasons, however, it is prefable to include the reactants in parentheses as follows:

$$E^0_{AgCl/Ag,Cl^-}\equiv E^0(AgCl+e=Ag+Cl^-)\equiv E^0_{AgCl/Ag,Cl^-} \qquad (3.1.55)$$

It follows directly from the definition of the standard potential that, for the hydrogen electrode,

$$E^0_{H^+/H_2}=E^0(H^++e=\tfrac{1}{2}H_2)\equiv E^0_{H^+/H_2}=0 \qquad (3.1.56)$$

If the equilibrium constant of the cell reaction is denoted as K, then it

† It follows from Section 1.3.1 that the expression for the chemical potential must contain the pressure referred to the standard pressure; thus $p=1$ corresponds to standard pressure, 10^2 kPa. The activities in these expressions are also dimensionless and correspond to units of moles per cubic decimetre because of the choice of the standard state.

follows from Eq. (3.1.50) and from the equation for the standard reaction Gibbs energy change $\Delta G^0 = -RT \ln K$ that

$$E^0 = (RT/F) \ln K \tag{3.1.57}$$

The relationships of the type (3.1.54) and (3.1.57) imply that the standard electrode potentials can be derived directly from the thermodynamic data (and vice versa). The values of the standard chemical potentials are identified with the values of the standard Gibbs energies of formation, tabulated, for example, by the US National Bureau of Standards. On the other hand, the experimental approach to the determination of standard electrode potentials is based on the cells of the type (3.1.41) whose EMFs are extrapolated to zero ionic strength.

For example, the tabulated values for the reactants of the cell reaction (3.1.42) are $\Delta G^0_{f(\mathrm{AgCl})} = -109.68\,\mathrm{kJ\cdot mol^{-1}}$ and $\Delta G^0_{f(\mathrm{HCl,w})} = -131.14\,\mathrm{kJ\cdot mol^{-1}}$, both at 298.15 K.† By definition, the standard Gibbs energies of formation of H_2 and Ag are equal to zero. Thus, the standard potential of the silver–silver chloride electrode is equal to

$$E^0_{\mathrm{AgCl/Ag,Cl^-}} = (-1.0968\times 10^5 + 1.3114\times 10^5) : 9.648\times 10^4 = 0.2224\ \mathrm{V}.$$

It should be noted that Eq. (3.1.52) can be obtained from the expression for the Galvani potential difference formulated in Section 3.1.2. The designation of the phases in the symbols for the individual Galvani potential differences in cell (3.1.41) will be given in brackets. The overall EMF, E, is given by the expression

$$\begin{aligned} E &= \phi(1') - \phi(1) \\ &= \phi(1') - \phi(5) + \phi(5) - \phi(4) + \phi(4) - \phi(3) + \phi(3) - \phi(1) \\ &= \Delta_5^{1'}\phi + \Delta_4^5\phi + \Delta_3^4\phi + \Delta_1^3\phi \end{aligned} \tag{3.1.58}$$

The relationship for the individual Galvani potential differences can be obtained from Eqs (3.1.12), (3.1.24) and (3.1.17) so that Eq. (3.1.58) is converted to the form ($p_{H_2} = 1$)

$$\begin{aligned} E &= \frac{\mu_e^0(\mathrm{Pt}) - \mu_e^0(\mathrm{Ag})}{F} + \frac{-\mu^0_{\mathrm{Ag^+}}(\mathrm{Ag}) + \mu^0_{\mathrm{AgCl}} - \mu^0_{\mathrm{Cl^-}}(3)}{F} \\ &\quad - \frac{RT}{F}\ln a_{\mathrm{Cl^-}}(3) - \frac{\mu^0_{\mathrm{H^+}}(3) - \frac{1}{2}\mu^0_{\mathrm{H_2}}(2) + \mu_e^0(\mathrm{Pt})}{F} - \frac{RT}{F}\ln a_{\mathrm{H^+}} \\ &= \frac{-\mu^0_{\mathrm{Ag}} + \mu^0_{\mathrm{AgCl}} - \mu^0_{\mathrm{Cl^-}} - \mu^0_{\mathrm{H^+}} + \frac{1}{2}\mu^0_{\mathrm{H_2}}}{F} - \frac{2RT}{F}\ln a_{\pm,3,\mathrm{HCl}} \end{aligned} \tag{3.1.59}$$

i.e. a relationship identical with Eq. (3.1.53). The correct formulation of the relationships for the Galvani potential differences leads to the same results

† The data have been recalculated from the previous standard pressure of 1.01325×10^5 to 10^5 Pa and those determined for unit activity on the molal scale to unit activity on molar scale (cf. Appendix A).

as the procedure following from Eq. (3.1.39) or from Eqs (3.1.45) to (3.1.49).

It should be noted that all terms concerning the electrons in the metals as well as those connected with the metals not directly participating in the cell reaction (Pt) have disappeared from the final Eq. (3.1.49). This result is of general significance, i.e. the EMFs of cell reactions involving oxidation–reduction processes *do not depend on the nature of the metals* where those reactions take place. The situation is, of course, different in the case of a metal directly participating in the cell reaction (for example, silver in the above case).

For cell reactions in general, both approaches yield the equation for the EMF:

$$E = E^0 - \frac{RT}{nF} \ln Q \tag{3.1.60}$$

where E^0 is the standard EMF (standard cell reaction potential), n is the charge number and Q is the quotient of the activities of the reactants in the cell reaction raised to the appropriate stoichiometric coefficients; this quotient has the same form as the equilibrium constant of the cell reaction.

For practical reasons it is often useful to separate the EMF of a galvanic cell into two terms and assign each of them to one of the electrodes. The half-cell reactions provide a basis for unambiguous separation. The equation for the overall EMF is converted to the difference between two expressions, in which the expression corresponding to the electrode on the left is subtracted from the expression corresponding to the electrode on the right. Thus, for example, it follows for Eq. (3.1.59), with inclusion of a term containing the relative hydrogen pressure, that

$$\begin{aligned} E &= \frac{-\mu^0_{Ag} + \mu^0_{AgCl} - \mu^0_{Cl^-}(w) - RT \ln a_{Cl^-}(w)}{F} \\ &\quad - \frac{-\frac{1}{2}\mu^0_{H_2} - \frac{1}{2}RT \ln p_{H_2} + \mu^0_{H^+}(w) + RT \ln a_{H^+}(w)}{F} \\ &= E_{AgCl/Ag,Cl^-} - E_{H^+/H_2} \end{aligned} \tag{3.1.61}$$

The terms $E_{AgCl/Ag,Cl^-}$ and E_{H^+/H_2} are designated as the *electrode potentials*. These are related to the standard electrode potentials and to the activities of the components of the system by the *Nernst equations*. By a convention for the standard Gibbs energies of formation, those related to the elements at standard conditions are equal to zero. According to a further convention, cf. Eq. (3.1.56),

$$\mu^0_{H^+}(w) \equiv \Delta G^0_f(H^+, w) = 0 \tag{3.1.62}$$

Then

$$\mu^0_{Cl^-}(w) = \mu^0_{HCl}(w) \tag{3.1.63}$$

It follows from Eqs (3.1.53) and (3.1.55) that

$$E_{\mathrm{AgCl/Ag,Cl^-}} = E^0_{\mathrm{AgCl/Ag,Cl^-}} - \frac{RT}{F} \ln a_{\mathrm{Cl^-}}(\mathrm{w}) \qquad (3.1.64)$$

$$E_{\mathrm{H^+/H_2}} = \frac{RT}{F} \ln a_{\mathrm{H^+}} - \frac{RT}{2F} \ln p_{\mathrm{H_2}} \qquad (3.1.65)$$

which are the Nernst equations. They will be discussed in detail in Section 3.2. It should be recalled that, in contrast to standard electrode potentials, which are thermodynamic quantities, the electrode potentials must be calculated by using the extrathermodynamic expressions described in Sections 1.3.1 to 1.3.4. The electrode potentials do not have the character of Galvani potential differences, but are simply operational quantities permitting simple calculation or interpretation of the EMF of a galvanic cell and are well suited to the application of electrochemistry in analytical chemistry, technology and biology.

In the subsequent text the half-cell reactions will be used to characterize the electrode potentials instead of the cell reactions of the type of Eq. (3.1.42) under the tacit assumption that such a half-cell reaction describes the cell reaction in a cell with the standard hydrogen electrode on the left-hand side.

The EMF of a cell is calculated from the electrode potentials (expressed for both electrodes with respect to the same reference electrode) as the difference of the potentials of these electrodes written on the right and left in the scheme:

$$E_{\mathrm{cell}} = E_{\mathrm{rhs}} - E_{\mathrm{lhs}} \qquad (3.1.66)$$

In practice, it is very often necessary to determine the potential of a test (indicator) electrode connected in a cell with a well defined second electrode. This *reference electrode* is usually a suitable electrode of the second kind, as described in Section 3.2.2. The potentials of these electrodes are tabulated, so that Eq. (3.1.66) can be used to determine the potential of the test electrode from the measured EMF. The *standard hydrogen electrode* is a hydrogen electrode saturated with gaseous hydrogen with a partial pressure equal to the standard pressure and immersed in a solution with unit hydrogen ion activity. Its potential is set equal to zero by convention. Because of the relative difficulty involved in preparing this electrode and various other complications (see Section 3.2.1), it is not used as a reference electrode in practice.

The term 'electrode potential' is often used in a broader sense, e.g. for the potential of an ideally polarized electrode (Chapter 4) or for potentials in non-equilibrium systems (Chapter 5).

Similar to electrode potentials, standard electrode potentials have so far been referred to the standard hydrogen electrode (SHE). These data are thus designated by 'vs. SHE' after the symbol V, that is $E^0_{\mathrm{AgCl/Ag,Cl^-}} =$

0.2224 V vs. SHE. Sometimes electrode potentials are referred to other reference electrodes, such as the saturated calomel electrode (SCE), etc.

So far, a cell containing a single electrolyte solution has been considered (*a galvanic cell without transport*). When the two electrodes of the cell are immersed into different electrolyte solutions in the same solvent, separated by a liquid junction (see Section 2.5.3), this system is termed a *galvanic cell with transport.* The relationship for the EMF of this type of a cell is based on a balance of the Galvani potential differences. This approach yields a result similar to that obtained in the calculation of the EMF of a cell without transport, plus the liquid junction potential value $\Delta\phi_L$. Thus Eq. (3.1.66) assumes the form

$$E_{cell} = E_{rhs} - E_{lhs} + \Delta\phi_L \tag{3.1.67}$$

However, in contrast to the EMF of a galvanic cell, the resultant expressions contain the activities of the individual ions, which must be calculated by using the extrathermodynamic approach described in Section 1.3.

Concentration cells are a useful example demonstrating the difference between galvanic cells with and without transfer. These cells consist of chemically identical electrodes, each in a solution with a different activity of potential-determining ions, and are discussed on page 171.

It is very often necessary to characterize the redox properties of a given system with unknown activity coefficients in a state far from standard conditions. For this purpose, *formal* (*conditional*) *potentials* are introduced, defined in terms of concentrations. Definitions are not given unambiguously in the literature; the following would seem most suitable. The formal (conditional) potential is the potential assumed by an electrode immersed in a solution with unit concentrations of all the species appearing in the Nernst equation; its value depends on the overall composition of the solution. If the solution also contains additional species that do not appear in the Nernst equation (indifferent electrolyte, buffer components, etc.), their concentrations must be precisely specified in the formal potential data. The formal potential, denoted as $E^{0\prime}$, is best characterized by an expression in parentheses, giving both the half-cell reaction and the composition of the medium, for example $E^{0\prime}(Zn^{2+} + 2e = Zn, 10^{-3} M\ H_2SO_4)$.

Coming back to equations for Galvani potential differences, (3.1.11) to (3.1.26), we find that equation (3.1.67) can be written in the form

$$E_{cell} = \Delta_{S_2}^{M_2}\phi - \Delta_{S_1}^{M_1}\phi + \Delta\phi_L \tag{3.1.68}$$

where $\Delta_{S_2}^{M_2}\phi$ is the Galvani potential difference between the right-hand side electrode and the solution S_2 in which it is immersed and $\Delta_{S_1}^{M_1}\phi$ is the analogous quantity for the left-hand side electrode. When the left-hand side electrode is the standard electrode and $\Delta\phi_L$ is kept constant (see page 114), then E_{cell} can be identified with a formal electrode potential given by the approximate relationship

$$E_{\text{formal}} \approx \Delta_{S_2}^{M_2}\phi + \text{constant} \tag{3.1.69}$$

Electrode potentials are relative values because they are defined as the EMF of cells containing a reference electrode. A number of authors have attempted to define and measure absolute electrode potentials with respect to a universal reference system that does not contain a further metal–electrolyte interface. It has been demonstrated by J. E. B. Randles, A. N. Frumkin and B. B. Damaskin, and by S. Trasatti that a suitable reference system is an electron in a vacuum or in an inert gas at a suitable distance from the surface of the electrolyte (i.e. under similar conditions as those for measuring the contact potential of the metal–electrolyte system). In this way a reference system is obtained that is identical with that employed in solid-state physics for measuring the electronic energy of the bulk of a phase.

The system

$$M_r' \mid S \mid M \mid M_r \tag{3.1.70}$$

where M is the studied electrode metal, S is the electrolyte solution and $M_r = M_r'$ is the reference electrode metal, has the EMF

$$E = [\phi(M_r) - \phi(M)] + [\phi(M) - \phi(S)] + [\phi(S) - \phi(M_r')] \tag{3.1.71}$$

Examination of Eq. (3.1.12) yields

$$\phi(M_r) - \phi(M) = \frac{\mu_e^0(M_r)}{F} - \frac{\mu_e^0(M)}{F} \tag{3.1.72}$$

so that substitution into Eq. (3.1.69) yields

$$E = \left[\Delta_S^M\phi - \frac{\mu_e^0(M)}{F}\right] - \left[\Delta_S^{M_r'}\phi - \frac{\mu_e^0(M_r)}{F}\right] \tag{3.1.73}$$

Thus the EMF has been separated into two terms, each containing a quantity related to a single electrode. If the surface potential of the electrolyte $\chi(S)$ is added to each of the two expressions in brackets in Eq. (3.1.73), then the expression for the EMF contains the difference in the absolute electrode potentials; for the absolute electrode potential of metal M we have

$$E_M(\text{abs}) = \Delta_S^M\phi - \frac{\mu_e(M)}{F} + \chi(S) \tag{3.1.74}$$

This quantity corresponds to the minimal work required to transfer an electron from the bulk of metal M through the electrolyte solution into a vacuum. Equation (3.1.74) can be readily modified to yield (cf. the definition of the electron work function on page 153)

$$E_M(\text{abs}) = \frac{\Phi_e(M)}{F} + \Delta_S^M \psi \tag{3.1.75}$$

Both quantities, the electron work function $\Phi_e(M)$ and the contact potential of the metal–electrolyte system $\Delta_S^M \psi$, are measurable quantities.

Unfortunately, as shown by Trasatti, because of water adsorption both these values cannot escape a certain ambiguity. The absolute electrode potential of the standard hydrogen electrode is most often reported as 4.44 ± 0.02 V. Recent studies showed, however, that values between 4.2 and 4.8 V can also be considered.

References

Agar, J. H., Thermogalvanic cells. *AE,* **3,** 31 (1964).

Hölzl, J., F. K. Schultze, and H. Wagner, *Solid Surface Physics,* Springer-Verlag, Berlin, 1979.

Kittel, C., *Introduction to Solid State Physics,* John Wiley & Sons, New York, 1966.

Manual of symbols and terminology for physico-chemical quantities and units. Appendix III. Electrochemical nomenclature. *Pure Appl. Chem.*, **37,** 499 (1974).

Parsons, R., Equilibrium properties of electrified interfaces, *MAE,* **1,** 103 (1954).

Reiss, H., The Fermi level and the redox potential, *J. Phys. Chem.* **89,** 3783 (1985).

Trasatti, S., The work function in electrochemistry, *AE,* **10,** 213 (1977).

Trasatti, S., The electrode potential, *CTE,* **1,** 45 (1980).

Trasatti, S., Structuring of the solvent at metal/solution interfaces and components of electrode potential, *J. Electroanal. Chem.*, **150,** 1 (1983).

Trasatti, S., and R. Parsons, Interphases in systems of conducting phases, *Pure Appl. Chem.*, **55,** 1251 (1983).

Trasatti, S., Structure of the metal/electrolyte solution interface: New data for theory, *Electrochim. Acta,* **36,** 1659 (1991).

3.2 Reversible Electrodes

Although from the thermodynamic point of view one can speak only about the reversibility of a process (cf. Section 3.1.4), in electrochemistry the term reversible electrode has come to stay. By this term we understand an electrode at which the equilibrium of a given reversible process is established with a rate satisfying the requirements of a given application. If equilibrium is established slowly between the metal and the solution, or is not established at all in the given time period, the electrode will in practice not attain a defined potential and cannot be used to measure individual thermodynamic quantities such as the reaction affinity, ion activity in solution, etc. A special case that is encountered most often is that of electrodes exhibiting a mixed potential, where the measured potential depends on the kinetics of several electrode reactions (see Section 5.8.4).

As demonstrated in Section 3.1, electrode equilibrium is a distribution equilibrium of charged species (including the electron) between the metal and the solution, so that both phases must have at least one charged species in common. The chemical potentials of this species in the two phases then determine the magnitude of the potential difference between the metal and the solution. Thus, the concept of a reversible electrode does not include the limiting case where the concentration of the species determining the electrode potential is zero in one of the phases (see Chapter 4).

We will now consider the principal electrodes that may be considered reversible, and some of their combinations in galvanic cells. Reversible electrodes may be divided into three groups:

1. *Electrodes of the first kind.* These include cationic electrodes (metal, amalgam and, of the gas electrodes, the hydrogen electrode), at which equilibrium is established between atoms or molecules of the substance and the corresponding cations in solution (see Eqs 3.1.21 and 3.1.65), and anionic electrodes, at which equilibrium is established between molecules and anions.
2. *Electrodes of the second kind.* These electrodes consist of three phases. The metal is covered by a layer of its sparingly soluble salt, usually with the character of a solid electrolyte, and is immersed in a solution containing the anions of this salt. The solution contains a soluble salt of this anion. Because of the two interfaces, equilibrium is established between the metal atoms and the anions in solution through two partial equilibria: between the metal and its cation in the sparingly soluble salt and between the anion in the solid phase of the sparingly soluble salt and the anion in solution (see Eqs (3.1.24), (3.1.26) and (3.1.64)).
3. *Oxidation–reduction electrodes.* An inert metal (usually Pt, Au, or Hg) is immersed in a solution of two soluble oxidation forms of a substance. Equilibrium is established through electrons, whose concentration in solution is only hypothetical and whose electrochemical potential in solution is expressed in terms of the appropriate combination of the electrochemical potentials of the reduced and oxidized forms, which then correspond to a given energy level of the electrons in solution (cf. page 151). This type of electrode differs from electrodes of the first kind only in that both oxidation states can be present in variable concentrations, while, in electrodes of the first kind, one of the oxidation states is the electrode material (cf. Eqs 3.1.19 and 3.1.21).

Ion-selective electrodes are based on membrane systems and will be dealt with in Section 6.3.

3.2.1 *Electrodes of the first kind*

Electrodes of the first kind can be divided into anionic and cationic. The system of a *gas electrode* includes a gas interacting with a suitable metal or semiconductor surface in the cell reaction. However, gas electrodes can also be considered as oxidation–reduction electrodes, as one of the forms can be a gas dissolved, perhaps in small concentration, in the solution. In *amalgam electrodes,* the metal is present as an amalgam. The electrode potential then also depends on the gas pressure or on the activity of the metal in the amalgam.

The potential of a *cationic electrode* of the first kind corresponding to the half-cell reaction

$$M^{z+} + z_+e = M$$

is described by the Nernst equation

$$E = E^0 + \frac{RT}{z_+F}\ln a_+ \tag{3.2.1}$$

where z_+ is the number of electrons required for reduction of one metal ion, i.e. its charge number (e.g. zinc in a solution of zinc(II) ions, copper in a solution of cupric ions, silver in a solution of silver(I) ions, etc.). It can readily be demonstrated by the procedure described in Section 3.1 (page 163) that the standard potential of a cationic electrode is given by the expression

$$E^0 = \frac{-\mu_M^0 + \mu_{M^{z+}}^0}{z_+F} \tag{3.2.2}$$

The standard potentials of some electrodes of the first kind are listed in Table 3.2.

The combination of two electrodes of the same material differing only in the solution concentration yields a *concentration cell,* for example

$$Ag\,|\,AgNO_3(c_1)\,|\,AgNO_3(c_2)\,|\,Ag \tag{3.2.3}$$

Combination of Eqs (2.5.33), (3.1.67) and (3.2.1) and rearrangement (the concentrations in Eq. (2.5.33) are replaced by the activities) yields the EMF for a *concentration cell with transport* in the form (for $z_+ = z_- = 1$):

$$E = \frac{RT}{F}\ln\frac{a_{+,1}}{a_{+,2}} - t_+\frac{RT}{F}\ln\frac{a_{+,1}}{a_{+,2}} - t_-\frac{RT}{F}\ln\frac{a_{-,1}}{a_{-,2}} \tag{3.2.4}$$

When eliminating the liquid junction potential by one of the methods described in Section 2.5.3, we obtain a *concentration cell without transport.* The value of its EMF is given simply by the difference between the two electrode potentials. More exactly than by the described elimination of the liquid junction potential, a concentration cell without transport can be obtained by using amalgam electrodes or electrodes of the second kind.

In amalgam electrodes, the metal is dissolved in mercury, so that not only the concentration of metal cations in the solution but also the concentration of metal in the amalgam is variable. The potential of an amalgam electrode

Table 3.2 Standard potentials of electrodes of the first kind at 25°C. (From the *CRC Handbook of Chemistry and Physics,* recalculated to the standard pressure of 10^2 kPa)

Electrode	E^0 (V)	Electrode	E^0 (V)
Li^+/Li	−3.0403	$OH^-/O_2(g)$	0.401
Rb^+/Rb	−2.98	$Cl^-/Cl_2(g)$	1.35793
Cs^+/Cs	−2.92	$F^-/F_2(g)$	2.866
K^+/K	−2.931		
Ba^{2+}/Ba	−2.912		
Sr^{2+}/Sr	−2.89		
Ca^{2+}/Ca	−2.868		
Na^+/Na	−2.71		
Mg^{2+}/Mg	−2.372		
Be^{2+}/Be	−1.847		
Al^{3+}/Al	−1.662		
Zn^{2+}/Zn	−0.7620		
Fe^{2+}/Fe	−0.447		
Cd^{2+}/Cd	−0.4032		
In^{3+}/In	−0.3384		
Tl^+/Tl	−0.336		
Sn^{2+}/Sn	−0.1377		
Pb^{2+}/Pb	−0.1264		
Cu^{2+}/Cu	0.3417		
Hg^{2+}/Hu	0.851		
Ag^+/Ag	0.7994		

is given as

$$E = E^0 + \frac{RT}{zF} \ln \frac{a_+}{a} \tag{3.2.5}$$

where a is the activity of the metal in the amalgam and a_+ is the activity of its ions in the solution. Combination of two amalgam electrodes with the same activity of cations in solution and different activities of the metal in the amalgam yields a different type of concentration cell, for example

$$\text{K,Hg}(a_1) \mid \text{KCl(m)} \mid \text{K,Hg}(a_2)$$

with

$$E = \frac{RF}{F} \ln \frac{a_1}{a_2} \tag{3.2.6}$$

If $a_1 > a_2$, then short circuiting of this cell results in potassium dissolution at the left-hand electrode and incorporation into the amalgam at the right-hand electrode. Amalgam electrodes can be used as reversible electrodes, even for metals as reactive as the alkali metals, especially in some non-aqueous solvents.

Theoretically, the most important cationic electrode of the first kind is the *hydrogen electrode.* This is a gas electrode at which equilibrium is established between hydrogen and oxonium ions in solution through a platinum electrode coated with platinum black. This coating is obtained by electrolytic deposition of platinum from an acidic solution of hexachloroplatinic(IV) acid and is saturated with hydrogen at a specific (usually atmospheric) pressure, p_{H_2}. The platinum black has a double function. Firstly, its high specific surface area ensures the presence of a sufficient amount of hydrogen atoms adsorbed on the electrode; secondly, it facilitates rapid establishment of the equilibrium between the oxonium ions and elemental hydrogen (cf. Section 5.7.1). The hydrogen must be carefully purified to remove traces of sulphur(II), arsenic and mercury compounds that would poison the electrode (cf. Section 5.7.2).

The potential of the standard hydrogen electrode, i.e. the potential value at $p_{H_2} = 1$ and $a_{H_3O^+} = 1$ is set by convention as zero (cf. Eqs 3.1.62 and 3.1.65). It can be seen from Eq. (3.1.65) that, when the activity of hydrogen ions increases tenfold, the potential of the hydrogen electrode increases by $2.303RT/F = 0.0591$ V at 25°C. If the hydrogen pressure increases tenfold, then the potential decreases by $2.303RT/2F = 0.0295$ V. Variation of the hydrogen pressure in the range of ±5 per cent corresponds to a potential change from +0.63 to −0.66 mV; these differences are negligible for some measurements.

The hydrogen electrode is one of the electrodes suitable for pH measurements. If a hydrogen electrode saturated at unit pressure is combined with a reference electrode, then the EMF of this cell (where the reference electrode is written on the left-hand side in the scheme of the cell) is given by the equation

$$E = E_{H^+/H_2} - E_{ref} = \frac{2.303RT}{F} \log a_{H^+} - E_{ref} \tag{3.2.7}$$

and thus

$$\text{pH} = -\frac{(E + E_{ref})F}{2.303RT} \tag{3.2.8}$$

It would appear from Eq. (3.2.8) that the pH, i.e. the activity of a single type of ion, can be measured exactly. This is not, in reality, true; even if the liquid junction potential is eliminated the value of E_{ref} must be known. This value is always determined by assuming that the activity coefficients depend only on the overall ionic strength and not on the ionic species. Thus the mean activities and mean activity coefficients of the electrolyte must be employed. The use of this assumption in the determination of the value of E_{ref} will, of course, also affect the pH value found from Eq. (3.2.8). Thus, the potentiometric determination of the pH is more difficult than would appear at first glance and will be considered in the special Section 3.3.2.

The hydrogen electrode can be used to measure pH values over the whole pH region. However, it is not applicable to reducing or oxidizing media or

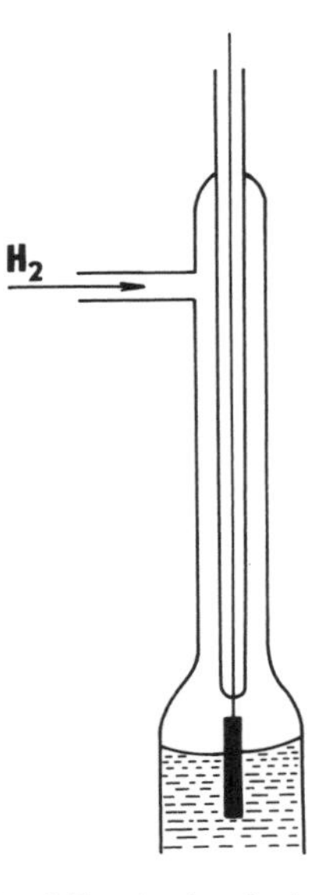

Fig. 3.7 A simple immersion hydrogen electrode

in the presence of substances that act as catalytic poisons for platinum black. Poisoning of the electrode (p. 173) is indicated by fluctuation of the electrode potential which differs from the value predicted by the Nernst equation.

Hydrogen electrodes can be constructed in variously shaped cells according to the purpose. An example is given in Fig. 3.7. The measurement is described in detail in practical manuals. However, for practical measurements, the glass electrode (Section 6.3) is employed.

The potential of an *anionic electrode of the first kind* is given by the relationship

$$E = E^0 + \frac{RT}{z_- F} \ln a_- \tag{3.2.9}$$

where z_- is the charge number of the corresponding anion ($z_- < 0$). The equation also often contains a term including the effect of the gas pressure on a gas anionic electrode of the first kind.

The chlorine electrode contains gaseous chlorine in equilibrium with atomic chlorine adsorbed on the platinum black and a solution of chloride ions. Its potential is given by the equation

$$E_{Cl_2/Cl^-} = E^0_{Cl_2/Cl^-} + \frac{RT}{F} \ln \sqrt{p_{Cl_2}} - \frac{RT}{F} \ln a_{Cl^-} \tag{3.2.10}$$

The chlorine electrode behaves reversibly, i.e. electrode equilibrium is

Table 3.3 Standard electrode potentials in eutectic melts referred to a chlorine reference electrode. (According to R. W. Laity)

	$-E^0$ (V)	
System	KCl–NaCl, 450°C	KCl–LiCl, 450°C
Mn^{2+}/Mn	2.135	2.065
Cd^{2+}/Cd	1.535	1.532
Tl^{+}/Tl		1.586
Co^{2+}/Co	1.277	1.207
Pb^{2+}/Pb	1.352	1.317
Cu^{+}/Cu	1.145	1.067
Ag^{+}/Au	0.905	0.853

established rapidly; however, side reactions also occur to a small degree: the attack on the platinum electrode by chlorine and the reaction of chlorine with water ($Cl_2 + H_2O \rightleftarrows HOCl + H^+ + Cl^-$), which can be suppressed in sufficiently acidic medium.

In chloride melts, the chlorine electrode (with graphite instead of platinum) is used as a reference electrode (see Table 3.3).

Anionic electrodes of the first kind are rarely used in practice; other, more important, sorts of electrode exhibiting a reversible response to anions are the electrodes of the second kind.

3.2.2 *Electrodes of the second kind*

The expression for the potential of electrodes of the second kind on the hydrogen scale can be derived from the affinity of the reaction occurring in a cell with a standard hydrogen electrode. For example, for the silver chloride electrode with the half-cell reaction

$$AgCl(s) + e = Ag + Cl^- \tag{3.2.11}$$

$$E_{AgCl/Ag} = E^0_{AgCl/Ag} - \frac{RT}{F} \ln a_{Cl^-} \tag{3.2.12}$$

The same expression is obtained if electrodes of the second kind are considered as electrodes of the first kind, where the activity of the metal cations depends on the solubility product of the given insoluble salt (cf. Eq. 3.1.26):

$$E^0_{AgCl/Ag} = E^0_{Ag^+/Ag} + \frac{RT}{F} \ln P_{AgCl} \tag{3.2.13}$$

It can be seen from this equation that the solubility product of silver chloride can be calculated from the known standard potentials of the silver and silver chloride electrodes. Besides the silver chloride electrode the following electrodes belong to this group:

The calomel electrode:

$$\mathrm{KCl(m)} \mid \mathrm{Hg_2Cl_2(s)} \mid \mathrm{Hg}$$

$$E_{\mathrm{Hg_2Cl_2/Hg}} = E^0_{\mathrm{Hg_2Cl_2/Hg}} - \frac{RT}{F} \ln a_{\mathrm{Cl^-}} \tag{3.2.14}$$

The mercurous sulphate electrode:

$$\mathrm{K_2SO_4(m)} \mid \mathrm{Hg_2SO_4(s)} \mid \mathrm{Hg}$$

$$E_{\mathrm{Hg_2SO_4/Hg}} = E^0_{\mathrm{Hg_2SO_4/Hg}} - \frac{RT}{2F} \ln a_{\mathrm{SO_4^{2-}}} \tag{3.2.15}$$

The mercuric oxide electrode:

$$\mathrm{KOH(m)} \mid \mathrm{HgO(s)} \mid \mathrm{Hg}$$

$$E_{\mathrm{HgO/Hg}} = E^0_{\mathrm{HgO/Hg}} - \frac{RT}{F} \ln a_{\mathrm{OH^-}} \tag{3.2.16}$$

The latter does not yield a completely reproducible potential; equilibrium is established after about two days. Nonetheless, it is often used in research on technically important galvanic cells working in alkaline media.

The described electrodes, and especially the silver chloride, calomel and mercurous sulphate electrodes are used as reference electrodes combined with a suitable indicator electrode. The calomel electrode is used most frequently, as it has a constant, well-reproducible potential. It is employed in variously shaped vessels and with various KCl concentrations. Mostly a concentration of KCl of 0.1 mol · dm^{-3}, 1 mol · dm^{-3} or a saturated solution is used (in the latter case, a salt bridge need not be employed); sometimes 3.5 mol · dm^{-3} KCl is also employed. The potentials of these calomel electrodes at 25°C are as follows (according to B. E. Conway):

0.1 M KCl | Hg_2Cl_2(s) | Hg	0.3337 V
1 M KCl | Hg_2Cl_2(s) | Hg	0.2812 V
KCl(sat.) | Hg_2Cl_2(s) | Hg	0.2422 V

Table 3.4 Standard potentials of electrodes of the second kind. (From the *CRC Handbook of Chemistry and Physics*, recalculated to the standard pressure 10^2 kPa)

Electrode	E^0 (V)
HgO, OH^-/Hg	0.0975
AgCl, Cl^-/Ag	0.22216
Hg_2Cl_2, Cl^-/Hg	0.26791
Hg_2SO_4, SO_4^{2-}/Hg	0.6123
PbO_2, $PbSO_4$, SO_4^{2-}/Pb	1.6912

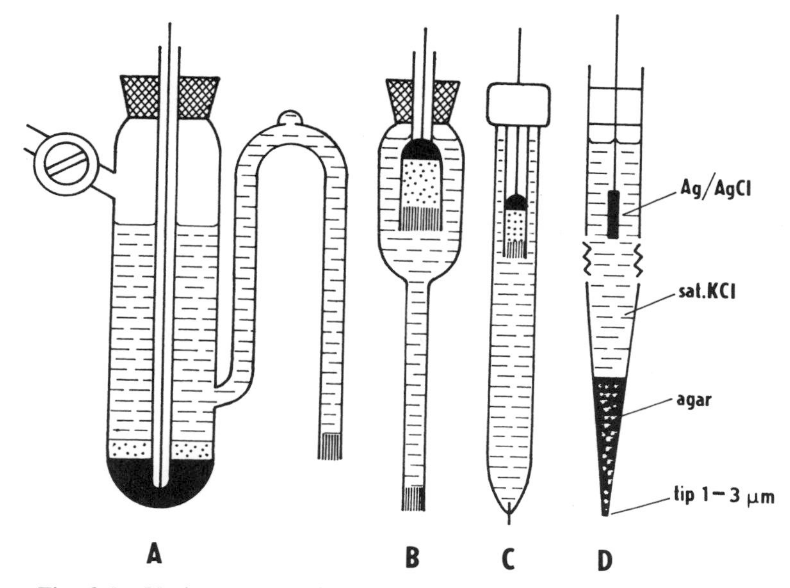

Fig. 3.8 Various types of reference electrode vessels: (A) laboratory-type calomel electrode; (B), (C) portable calomel electrodes, (D) a Ag/AgCl micropipette with a KCl electrolyte immobilized with agar in the tip of the capillary

The silver chloride electrode is preferable for precise measurements. The standard potentials of some electrodes of the second kind are listed in Table 3.4. Various electrode constructions are shown in Fig. 3.8.

In addition to their use as reference electrodes in routine potentiometric measurements, electrodes of the second kind with a saturated KCl (or, in some cases, with sodium chloride or, preferentially, formate) solution as electrolyte have important applications as *potential probes*. If an electric current passes through the electrolyte solution or the two electrolyte solutions are separated by an electrochemical membrane (see Section 6.1), then it becomes important to determine the electrical potential difference between two points in the solution (e.g. between the solution on both sides of the membrane). Two silver chloride or saturated calomel electrodes are placed in the test system so that the tips of the liquid bridges lie at the required points in the system. The value of the electrical potential difference between the two points is equal to that between the two probes. Similar potential probes on a microscale are used in electrophysiology (the tips of the salt bridges are usually several micrometres in size). They are termed *micropipettes* (Fig. 3.8D.)

3.2.3 *Oxidation–reduction electrodes*

Oxidation–reduction electrodes, abbreviated to redox electrodes, consist of an inert metal (Pt, Au, Hg) immersed in a solution containing two forms of a single substance in different oxidation states. The metal merely acts as a medium for the transfer of electrons between the two forms (cf. Section 3.1.2). The redox electrode potential can be derived in an analogous manner as for the previous types of electrodes (cf. Eq. (3.1.17)). For the cell reaction

$$\mathrm{Ox} + z\mathrm{e} = \mathrm{Red} \tag{3.2.13}$$

where z is the number of electrons necessary for reduction of one particle of Ox we obtain the Nernst equation

$$E = E^0_{\mathrm{Ox/Red}} + \frac{RT}{zF}\ln\frac{a_{\mathrm{Ox}}}{a_{\mathrm{Red}}} \tag{3.2.17}$$

This equation has been verified experimentally by Peters and is sometimes named after him.

For example, for the system Fe^{3+}–Fe^{2+} the Nernst equation is

$$E_{\mathrm{Fe^{3+}/Fe^{2+}}} = E^0_{\mathrm{Fe^{3+}/Fe^{2+}}} + \frac{RT}{F}\ln\frac{a_{\mathrm{Fe^{3+}}}}{a_{\mathrm{Fe^{2+}}}} \tag{3.2.18}$$

In the half-cell reaction

$$\mathrm{Ox} + \mathrm{mH^+} + z\mathrm{e} = \mathrm{Red} + nX \tag{3.2.19}$$

the change in the oxidation state is only displayed in the particles Ox and Red which may also function as acids or bases or as coordination compounds X denoting the ligand. Under these circumstances the Nernst equation has the form

$$E_{\mathrm{Ox/Red}} = E^0_{\mathrm{Ox,Red}} + (RT/zF)\ln\frac{a_{\mathrm{Ox}}a^{\mathrm{m}}_{\mathrm{H^+}}}{a_{\mathrm{Red}}a^n_x} \tag{3.2.20}$$

The quinhydrone electrode (Section 3.2.5) is an example for such a more complicated redox reaction.

The standard potential is a measure of the reducing or oxidizing ability of a substance. If $E^0_1 > E^0_2$, then system 1 is a stronger oxidant than system 2 (similarly for metals, metal 1 would be more noble than metal 2), i.e. in an experiment in the bulk where originally $a_{\mathrm{Red},1} = a_{\mathrm{Ox},1}$ and $a_{\mathrm{Red},2} = a_{\mathrm{Ox},2}$ equilibrium would be established for which $a_{\mathrm{Red},1} > a_{\mathrm{Ox},1}$ and $a_{\mathrm{Red},2} < a_{\mathrm{Ox},2}$.

The *formal potential of a reduction–oxidation electrode* is defined as the equilibrium potential at the unit concentration ratio of the oxidized and reduced forms of the given redox system (the actual concentrations of these two forms should not be too low). If, in addition to the concentrations of the reduced and oxidized forms, the Nernst equation also contains the concentration of some other species, then this concentration must equal unity. This is mostly the concentration of hydrogen ions. If the concentration of some species appearing in the Nernst equation is not equal to unity, then it must be precisely specified and the term *apparent formal potential* is then employed to designate the potential of this electrode.

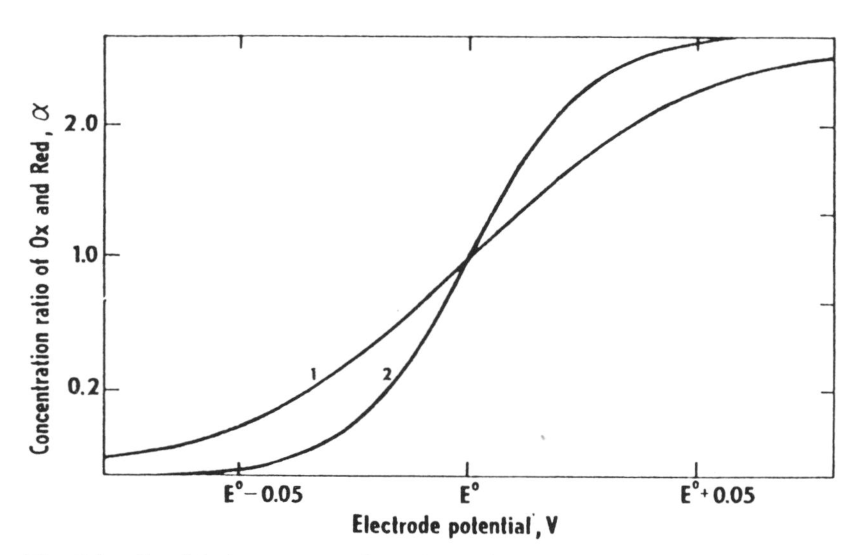

Fig. 3.9 Graphical representation of Eq. (3.2.21) with n indicated at each curve

Equation (3.2.17) can be expressed in terms of the degree of oxidation α, if the activities are set equal to the concentrations:

$$E_{\mathrm{Ox/Red}} = E^0_{\mathrm{Ox/Red}} + \frac{RT}{zF}\ln\frac{\alpha}{1-\alpha} \tag{3.2.21}$$

This equation is depicted graphically in Fig. 3.9. Differentiation of the potential E with respect to α yields the following relationship at the point $\alpha = \frac{1}{2}$:

$$\left(\frac{\mathrm{d}E}{\mathrm{d}\alpha}\right)_{\alpha=1/2} = \frac{4RT}{zF}; \qquad \left(\frac{\mathrm{d}^2E}{\mathrm{d}\alpha_2}\right)_{\alpha=1/2} = 0 \tag{3.2.22}$$

The slope of the tangent to the curve at the inflection point where $\alpha = \frac{1}{2}$ is thus inversely proportional to the number of electrons n. The E–α curves are similar to the titration curves of weak acids or bases (pH–α). For neutralization curves, the slope $\mathrm{dpH}/\mathrm{d}\alpha$ characterizes the buffering capacity of the solution; for redox potential curves, the differential $\mathrm{d}E/\mathrm{d}\alpha$ characterizes the redox capacity of the system. If $\alpha = \frac{1}{2}$ for a buffer, then changes in pH produced by changes in α are the smallest possible. If $\alpha = \frac{1}{2}$ in a redox system, then the potential changes produced by changes in α are also minimal (the system is 'well poised').

It can be seen from Eq. (3.2.21) that, if the solution contains one of the forms, oxidized or reduced, in zero concentration, then the potential would attain infinite absolute value. These extreme values change to finite values in the presence of a trace of the other form. This always occurs as a result of

Table 3.5 Standard oxidation–reduction potentials. (From the *CRC Handbook of Chemistry and Physics*)

Half-cell reaction	E^0 (V)
$Cr^{3+} + e \rightleftarrows Cr^{2+}$	−0.407
$V^{3+} + e \rightleftarrows V^{2+}$	−0.255
$O_2 + H_2O + 2e \rightleftarrows HO_2^- + OH^-$	−0.076
$TiOH^{3+} + H^+ + e \rightleftarrows Ti^{3+} + H_2O$	−0.055
$Sn^{4+} + 2e \rightleftarrows Sn^{2+}$	0.151
$Cu^{2+} + e \rightleftarrows Cu^+$	0.153
$Fe(CN)_6^{3-} + e \rightleftarrows Fe(CN)_6^{4-}$	0.358
Ferricinium $^+$ + e $\rightleftarrows$ ferrocene	0.400
$I_2 + 2e \rightleftarrows 2I^-$	0.5353
$Fe^{3+} + e \rightleftarrows Fe^{2+}$	0.771
$HO_2^- + H_2O + 2e \rightleftarrows 3OH^-$	0.878
Fe(phenanthroline)$_3^{3+}$ + e $\rightleftarrows$ Fe(phenanthroline)$_3^{2+}$	1.147
$Tl^{3+} + 2e \rightleftarrows Tl^+$	1.2152
$Mn^{3+} + e \rightleftarrows Mn^{2+}$	1.5413
$Ce^{4+} + e \rightleftarrows Ce^{3+}$	1.610

charge transfer from the electrode double layer to the component present in solution.

The standard potentials of some redox systems are listed in Table 3.5.

3.2.4 *The additivity of electrode potentials, disproportionation*

Electrode potentials are determined by the affinities of the electrode reactions. As the affinities are changes in thermodynamic functions of state, they are additive. The affinity of a given reaction can be obtained by linear combination of the affinities for a sequence of reactions proceeding from the same initial to the same final state as the direct reaction. Thus, the principle of linear combination must also be valid for electrode potentials. The electrode oxidation of metal Me to a higher oxidation state $z_{+,2}$ can be separated into oxidation to a lower oxidation state $z_{+,1}$ and subsequent oxidation to the oxidation state $z_{+,2}$. The affinities of the particular oxidation processes are equivalent to the electrode potentials $E_{2\text{-}0}$, $E_{1\text{-}0}$, and $E_{2\text{-}1}$:

$$\begin{aligned} E_{2\text{-}0} &= \frac{-\Delta G_{2\text{-}0}}{(z_+)_2 F} \\ E_{1\text{-}0} &= \frac{-\Delta G_{1\text{-}0}}{(z_+)_1 F} \qquad (3.2.23) \\ E_{2\text{-}1} &= \frac{-\Delta G_{2\text{-}1}}{(z_+)_2 - (z_+)_1 F} \end{aligned}$$

It follows from the additivity of affinities, $-\Delta G_{2-0} = -\Delta G_{2-1} - \Delta G_{1-0}$, that

$$(z_{+,2} - z_{+,1})E^0_{2-1} = z_{+,2}E^0_{2-0} - z_{+,1}E^0_{1-0} \tag{3.2.24}$$

This relationship (sometimes called Luther's law) for the transfer of several electrons permits us to calculate one redox potential if the others are known. Obviously, this is an analogy of the Hess law in thermodynamics. Equation (3.2.24) is not restricted to the case where the lowest oxidation state is a metal.

Consider the reactions

$$\mathrm{A} + \mathrm{e} \rightleftarrows \mathrm{B} \qquad \text{and} \qquad \mathrm{B} + \mathrm{e} \rightleftarrows \mathrm{C} \tag{3.2.25}$$

The sum of these reactions yields a reaction termed *disproportionation*:

$$2\mathrm{B} \rightleftarrows \mathrm{A} + \mathrm{C}, \qquad K = \frac{a_\mathrm{A} a_\mathrm{C}}{a_\mathrm{B}^2} \tag{3.2.26}$$

where K is the disproportionation constant. An inert electrode immersed in a solution containing species A, B and C will attain a potential corresponding to both of equilibria (3.2.25):

$$E = E^0_{\mathrm{A,B}} + \frac{RT}{F}\ln\frac{a_\mathrm{A}}{a_\mathrm{B}} = E^0_{\mathrm{B,C}} + \frac{RT}{F}\ln\frac{a_\mathrm{B}}{a_\mathrm{C}} \tag{3.2.27}$$

If the system initially contains only form C with concentration c_0, to which the oxidant is added, then the overall degree of oxidation is $\beta = (2[\mathrm{A}] + [\mathrm{B}])/c_0$. If activities are set equal to concentrations, then the potential E can be expressed in terms of β:

$$E = \frac{1}{2}(E^0_{\mathrm{A,B}} + E^0_{\mathrm{B,C}}) + \frac{RT}{2F}\ln\frac{\beta}{2-\beta} + \frac{RT}{2F}\ln\frac{\beta - 1 \pm [(\beta-1)^2 + 4\beta(2-\beta)K]^{1/2}}{-\beta + 1 \pm [(\beta-1)^2 + 4\beta(2-\beta)K]^{1/2}} \tag{3.2.28}$$

This dependence is shown in Fig. 3.10. The usual dependence for $n = 2$ is obtained for complete disproportionation, that is $K \gg 1$. With decreasing K, the E–β curve changes and at $K = \frac{1}{4}$ assumes the usual shape for $n = 1$. For small K values three inflection points appear on the curve. This situation often occurs for organic quinones. Form B is then termed the *semiquinone*.

A typical inorganic redox system of this type involves the equilibrium between metallic copper and copper(II) ions. In the absence of a complex-former the disproportionation constant is large and a solution of Cu^+ ions is very unstable. However, in the presence of ammonia, copper(I) ions are bonded in a complex, constant K decreases and the E–β curve exhibits three inflection points.

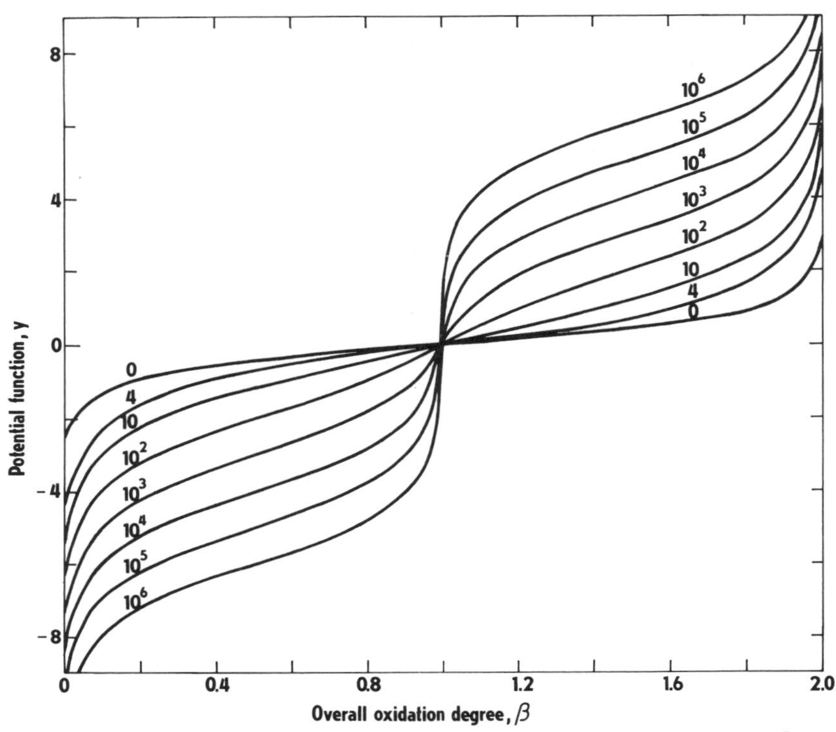

Fig. 3.10 Dependence of the potential function $y = (2F/2.3RT) \cdot [E - \frac{1}{2}(E^0_{A,B} + E^0_{B,C})]$ on the overall oxidation degree β for a system with disproportionation. The semiquinone formation constant $K' = 1/K$ is indicated at each curve

3.2.5 *Organic redox electrodes*

Quinhydrone, a solid-state associate of quinone and hydroquinone, decomposes in solution to its components. The quinhydrone electrode is an example of more complex organic redox electrodes whose potential is affected by the pH of the solution. If the quinone molecule is denoted as Ox and the hydroquinone molecule as H_2Red, then the actual half-cell reaction

$$\text{Ox} + 2\text{e} \rightleftarrows \text{Red}^{2-} \tag{3.2.29}$$

is accompanied by reactions between the hydroquinone anion Red^{2-} and the solvent. Hydroquinone is a weak dibasic acid, with the dissociation constants

$$\begin{aligned} \text{H}_2\text{Red} &\rightleftarrows \text{HRed}^- + \text{H}^+, \qquad K_1' = \frac{[\text{H}^+][\text{HRed}^-]}{[\text{H}_2\text{Red}]} \\ \text{HRed}^- &\rightleftarrows \text{Red}^{2-} + \text{H}^+, \qquad K_2' = \frac{[\text{H}^+][\text{Red}^{2-}]}{[\text{HRed}^-]} \end{aligned} \tag{3.2.30}$$

For the quinhydrone electrode, the equation for the electrode potential has the form

$$E_{qh} = E^{0\prime} + \frac{RT}{2F} \ln \frac{[Ox]}{[Red^{2-}]} \tag{3.2.31}$$

where $E^{0\prime}$ is the formal potential. The concentration of the anion of the reduced form is substituted from the equation for the dissociation constants and from the equation for the overall concentration of hydroquinone, c_{Red},

$$\begin{aligned} c_{Red} &= [H_2Red] + [HRed^-] + [Red^{2-}] \\ &= \frac{[H^+]^2[Red^{2-}]}{K_1'K_2'} + \frac{[H^+][Red^{2-}]}{K_2'} + [Red^{2-}] \end{aligned} \tag{3.2.32}$$

so that $[Red^{2-}] = c_{Red}K_1'K_2'/([H^+]^2 + K_1'[H^+] + K_1'K_2')$ and

$$\begin{aligned} E_{qh} = E^{0\prime} &+ \frac{RT}{2F} \ln \frac{[Ox]}{c_{Red}} - \frac{RT}{2F} \ln K_1'K_2' \\ &+ \frac{RT}{2F} \ln ([H^+]^2 + K_1'[H^+] + K_1'K_2') \end{aligned} \tag{3.2.33}$$

As $[Ox] = c_{Red}$ for quinhydrone, the second term on the right-hand side of this equation equals zero and the first and third are combined in the constant $E_{qh}^{0\prime}$. The fourth term is simplified in different pH ranges as follows:

(a) In an acid medium, $[H^+]^2 \gg K_1'K_2' + K_1'[H^+]$, so that

$$E_{qh} = E_{qh}^{0\prime} - \frac{2.303RT}{F} \text{pH} \tag{3.2.34}$$

(b) If $K_1'[H^+] \gg [H^+]^2 + K_1'K_2'$, then

$$E_{qh} = E_{qh}^{0\prime} + \frac{RT}{2F} \ln K_1' - \frac{2.303RT}{2F} \text{pH} \tag{3.2.35}$$

(c) In a rather alkaline region $K_1'K_2' \gg K_1'[H^+] + [H^+]^2$, dissociation is complete and the potential is independent of pH:

$$E_{qh} = E_{qh}^{0\prime} + \frac{RT}{2F} \ln K_1'K_2' \tag{3.2.36}$$

Thus a plot of the dependence of E_{qh} on the pH (Fig. 3.11) will yield a curve with three linear regions (the second of which is poorly defined), with slopes of 0.0591, 0.0295 and 0 (at 25°C). The intersections of each two neighbouring extrapolated linear portions yield the pK_1' and pK_2' values, as follows from comparison of Eqs (3.2.34) to (3.2.36).

Figure 3.12 gives examples of the dependence of the apparent formal potential on the pH for other organic systems (see also p. 465).

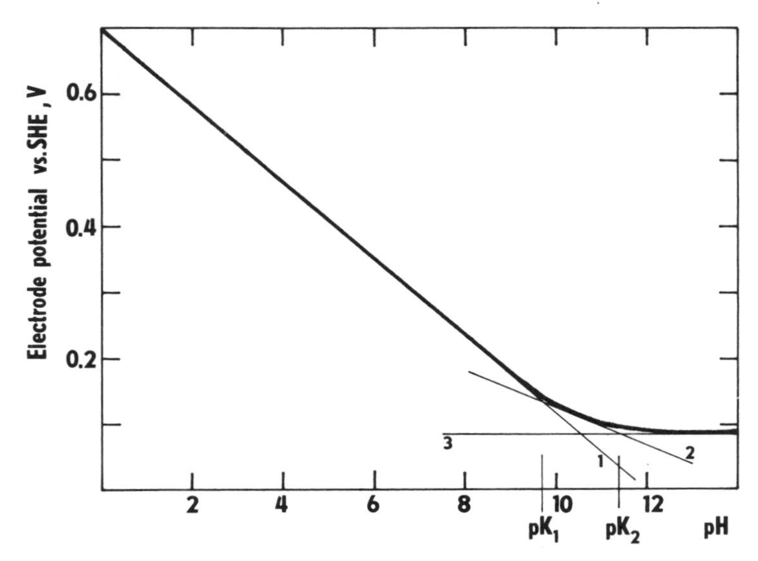

Fig. 3.11 The dependence of the potential of the quinhydrone electrode E_{qh}(V) on pH. The straight line (1) corresponds to Eq. (3.2.34), (2) to (3.2.35) and (3) to (3.2.36)

In cases where the potential of an inert electrode, in the presence of the redox system of interest, is established slowly, another redox system may then be added as a 'mediator'. The latter system should react rapidly with the former and quickly establish an equilibrium at the electrode. As a very small amount of the mediator is added, concentration changes in the measured system compared to the original state can be neglected. Suitable mediators are Ce^{4+}–Ce^{3+}, methylene blue–leucoform, etc.

Standard redox potentials can be determined approximately from the titration curves for suitably selected pairs of redox systems. However, these curves always yield only the difference between the standard potentials and a term containing the activity coefficients, i.e. the formal potential. The large values of the terms containing the activity coefficients lead to a considerable difference between the formal potential and the standard potential (of the order of tens of millivolts).

3.2.6 *Electrode potentials in non-aqueous media*

While the laws governing electrode potentials in non-aqueous media are basically the same as for potentials in aqueous solutions, the standardization in this case is not so simple. Two approaches can be adopted: either a suitable standard electrode can be selected for each medium (e.g. the hydrogen electrode for the protic medium, the bis-diphenyl chromium(II)/ bis-diphenyl chromium(I) redox electrode for a wide range of organic

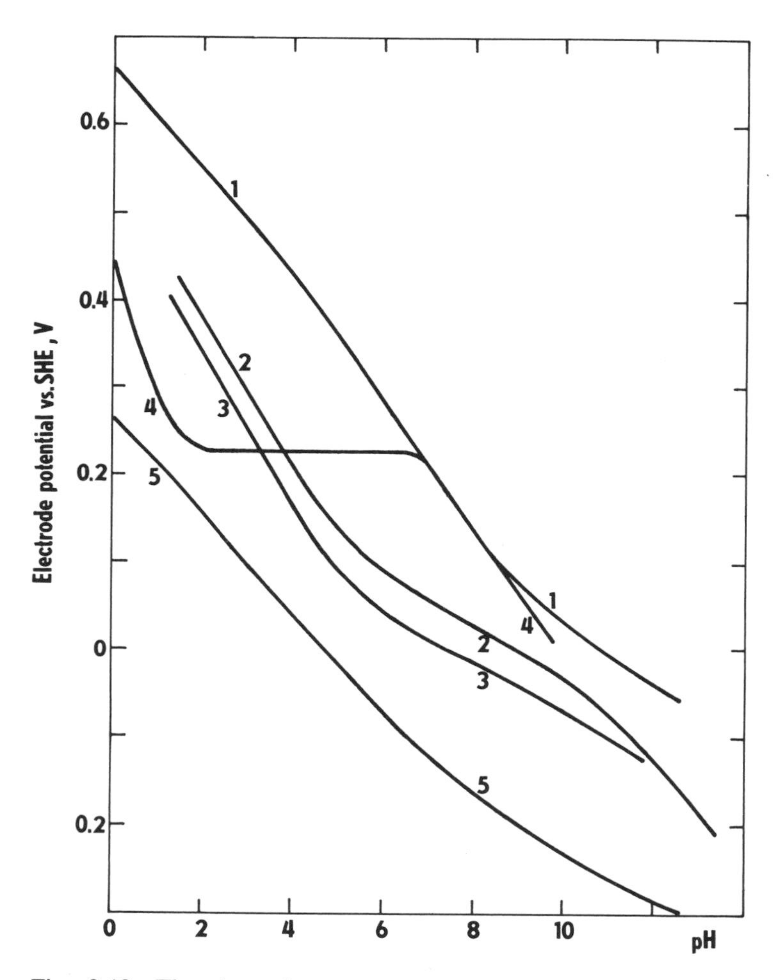

Fig. 3.12 The dependence on pH of the oxidation–reduction potential for $c_{\text{Ox}} = c_{\text{Red}}$: (1) 6-dibromphenol indophenol, (2) Lauth's violet, (3) methylene blue, (4) ferricytochrome *c*/ferrocytochrome *c*, (5) indigo–carmine

solvents or, for example, the chlorine electrode for some melts—cf. Section 3.2.1), or, on the other hand, all the potentials can be related to the aqueous standard hydrogen electrode on the basis of a suitable convention. The first approach is based on the discussion in Section 3.1.5 and requires no further explanation.

To decide whether a unified electrode potential scale is of some advantage, consider cells (3.1.40) in both aqueous medium and in protic medium s:

$$\text{Pt} \mid \text{H}_2 \mid \text{HCl}\ a_\pm, \text{w} \mid \text{AgCl} \mid \text{Ag} \mid \text{Pt}$$

$$\text{Pt} \mid \text{H}_2 \mid \text{HCl}\ a_\pm, \text{s} \mid \text{AgCl} \mid \text{Ag} \mid \text{Pt}$$

The EMFs of these cells are given by the relationships

$$E(\mathrm{w}) = \frac{-\mu^0_{\mathrm{Ag}} + \mu^0_{\mathrm{AgCl}} - \mu^0_{\mathrm{Cl^-}}(\mathrm{w}) - \mu^0_{\mathrm{H^+}}(\mathrm{w}) + \frac{1}{2}\mu^0_{\mathrm{H_2}}}{F} - \frac{2RT}{F}\ln a_{\pm,\mathrm{HCl}}(\mathrm{w}) \tag{3.2.37}$$

$$E(\mathrm{s}) = \frac{-\mu^0_{\mathrm{Ag}} + \mu^0_{\mathrm{AgCl}} - \mu^0_{\mathrm{Cl^-}}(\mathrm{s}) - \mu^0_{\mathrm{H^+}}(\mathrm{s}) + \frac{1}{2}\mu^0_{\mathrm{H_2}}}{F} - \frac{2RT}{F}\ln a_{\pm,\mathrm{HCl}}(\mathrm{s}) \tag{3.2.38}$$

It is assumed that the activities are based on the same concentration scale and that $\lim_{c_i \to 0} a_i/c_i = 1$. For $a_{\pm,\mathrm{HCl}}(\mathrm{s}) = a_{\pm,\mathrm{HCl}}(\mathrm{w})$, the difference between $E(\mathrm{s})$ and $E(\mathrm{w})$ is given by the relationship

$$\begin{aligned} E(\mathrm{w}) - E(\mathrm{s}) &= \frac{\mu^0_{\mathrm{Cl^-}}(\mathrm{s}) - \mu^0_{\mathrm{Cl^-}}(\mathrm{w}) + \mu^0_{\mathrm{H^+}}(\mathrm{s}) - \mu^0_{\mathrm{H^+}}(\mathrm{w})}{F} \\ &= \frac{1}{F}\Delta G^{0,\mathrm{w}\to\mathrm{s}}_{\mathrm{tr,HCl}} \\ &= \frac{\Delta G^{0,\mathrm{w}\to\mathrm{s}}_{\mathrm{tr,H^+}} + \Delta G^{0,\mathrm{w}\to\mathrm{s}}_{\mathrm{tr,Cl^-}}}{F} \end{aligned} \tag{3.2.39}$$

where $\Delta G^{0,\mathrm{w}\to\mathrm{s}}_{\mathrm{tr,HCl}}$ is the overall standard Gibbs energy of transfer of HCl from water to the solvent s and $\Delta G^{0,\mathrm{w}\to\mathrm{s}}_{\mathrm{tr,H^+}}$ and $\Delta G^{0,\mathrm{w}\to\mathrm{s}}_{\mathrm{tr,Cl^-}}$ are the individual standard Gibbs energies of transfer of H^+ and Cl^- from water to the solvent s (Eq. 1.4.35). It should be noted that, so far, the derivation has been concerned with the transfer of a species between two pure solvents, so that in this case, the standard Gibbs transfer energy will be somewhat different from the values measured for distribution equilibria. This type of standard Gibbs transfer energy must be determined on the basis of the solubility of the given electrolyte, present as a solid or gaseous phase, in equilibrium with each of the pair of solvents.

The derivation for equilibrium between a solution and a gaseous phase is based on the Henry law constant k_i, defined on page 5. The standard Gibbs energy of the transfer of HCl from a solution into water is

$$\Delta G^{0,\mathrm{w}\to\mathrm{s}}_{\mathrm{tr,HCl}} = RT\ln\left(\frac{k_{\mathrm{HCl}}(\mathrm{s})}{k_{\mathrm{HCl}}(\mathrm{w})}\right) \tag{3.2.40}$$

For the transfer equilibrium between a solution and a solid phase of the electrolyte BA it holds that

$$\Delta G^{0,\mathrm{w}\to\mathrm{s}}_{\mathrm{tr,BA}} = RT\ln\left(\frac{P_{\mathrm{BA}}(\mathrm{w})}{P_{\mathrm{BA}}(\mathrm{s})}\right) = 2RT\ln\frac{m_{\mathrm{BA}}(\mathrm{w})\gamma_{\pm}(\mathrm{w})}{m_{\mathrm{BA}}(\mathrm{s})\gamma_{\pm}(\mathrm{s})} \tag{3.2.41}$$

where the P_{BA}'s are the solubility products, the m_{BA}'s the solubilities and $\gamma_{\pm}$'s the mean activity coefficients in each of the two phases.

As separation of the standard Gibbs transfer energy of the electrolyte as a whole into the individual contributions for the anion and the cation can, of course, not be carried out by a thermodynamic procedure, a suitable extrathermodynamic assumption must be selected.

A suitable extrathermodynamic approach is based on structural considerations. The oldest assumption of this type was based on the properties of the rubidium(I) ion, which has a large radius but low deformability. V. A. Pleskov assumed that its solvation energy is the same in all solvents, so that the Galvani potential difference for the rubidium electrode (cf. Eq. 3.1.21) is a constant independent of the solvent. A further assumption was the independence of the standard Galvani potential of the ferricinium–ferrocene redox system (H. Strehlow) or the bis-diphenyl chromium(II)–bis-diphenyl chromium(I) redox system (A. Rusina and G. Gritzner) of the medium.

While the validity of these assumptions has been criticized, that adopted by A. J. Parker has received the widest acceptance, i.e. that tetraphenylarsonium tetraphenylborate (Ph_4AsPh_4B),

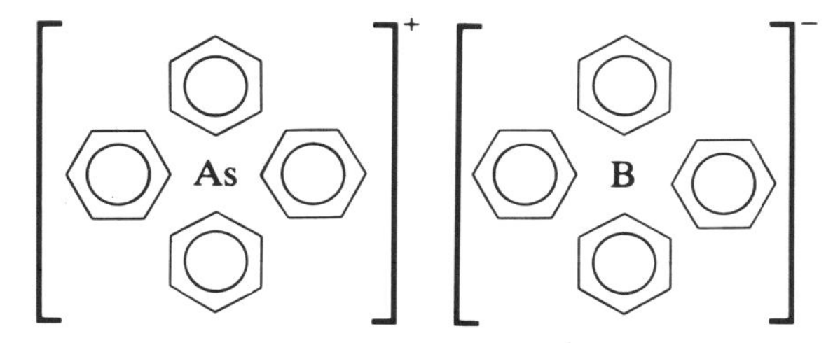

yields ions with the same standard Gibbs energy of solvation in the same medium. In other words, the standard Gibbs energy for the transfer of the tetraphenylarsonium cation and of the tetraphenylborate anion between an arbitrary pair of solvents is always the same (called the TATB assumption), that is

$$\Delta G^{0,\alpha\to\beta}_{tr,Ph_4As^+} = \Delta G^{0,\alpha\to\beta}_{tr,Ph_4B^-} = \tfrac{1}{2}\Delta G^{0,\alpha\to\beta}_{tr,Ph_4AsPh_4B} \tag{3.2.42}$$

The structure of the ions, where the bulky phenyl groups surround the central ion in a tetrahedron, lends validity to the assumption that the interaction of the shell of the ions with the environment is van der Waals in nature and identical for both ions, while the interaction of the ionic charge with the environment can be described by the Born approximation (see Section 1.2), leading to identical solvation energies for the anion and cation.

The determination of the standard Gibbs transfer energy for an arbitrary ion and arbitrary pair of solvents is based on the determination of the solubility product of Ph_4AsPh_4B in both solvents or on the determination of the distribution coefficient of Ph_4AsPh_4B between the two solvents. These experimental data then yield the standard Gibbs energies for the individual ions, Ph_4As^+ and Ph_4B^-. If, for example, the standard Gibbs energy is to be determined for the transfer of the arbitrary cation C^+ between a pair of solvents, the experimental data are determined for the salt CPh_4B and the required quantity is found from the equation

$$\Delta G^{0,\alpha\to\beta}_{\mathrm{tr,C^+}} = \Delta G^{0,\alpha\to\beta}_{\mathrm{tr,CPh_4B}} - \Delta G^{0,\alpha\to\beta}_{\mathrm{tr,Ph_4B^-}} \qquad (3.2.43)$$

These quantities can be used to rearrange the equation for the EMF of a cell in non-aqueous medium into the form of a difference between the electrode potentials on the hydrogen scale for aqueous solutions with additional terms.

Equations (3.2.37), (3.2.38) and (3.2.39) yield the potential of the silver–silver chloride electrode in non-aqueous medium

$$E_{\mathrm{AgCl/Ag,Cl^-}}(\mathrm{s}) = \frac{-\mu^0_{\mathrm{Ag}} + \mu^0_{\mathrm{AgCl}} - \mu^0_{\mathrm{Cl^-}}(\mathrm{w}) - \Delta G^{0,\mathrm{w}\to\mathrm{s}}_{\mathrm{tr,Cl^-}}}{F} - \frac{RT}{F}\ln a_{\mathrm{Cl^-}}(\mathrm{s})$$

$$= E^0_{\mathrm{AgCl/Ag,Cl^-}}(\mathrm{w}) - \frac{\Delta G^{0,\mathrm{w}\to\mathrm{s}}_{\mathrm{tr,Cl^-}}}{F} - \frac{RT}{F}\ln a_{\mathrm{Cl^-}}(\mathrm{s}) \qquad (3.2.44)$$

and, considering Eq. (3.1.56), the potential of the hydrogen electrode in the same medium ($p = 1$),

$$E_{\mathrm{H^+/H_2}}(\mathrm{s}) = \frac{-\frac{1}{2}\mu^0_{\mathrm{H_2}} + \mu^0_{\mathrm{H^+}}(\mathrm{w}) + \Delta G^{0,\mathrm{w}\to\mathrm{s}}_{\mathrm{tr,H^+}}}{F} + \frac{RT}{F}\ln a_{\mathrm{H^+}}(\mathrm{s})$$

$$= \frac{\Delta G^{0,\mathrm{w}\to\mathrm{s}}_{\mathrm{tr,H^+}}}{F} + \frac{RT}{F}\ln a_{\mathrm{H^+}}(\mathrm{s}) \qquad (3.2.45)$$

Thus, these relationships can be used to define a pH scale for non-aqueous protic media, consistent with the pH scale for aqueous solutions. For standard hydrogen pressure, the potential of the hydrogen electrode depends on the pH(s) according to the relationship

$$E_{\mathrm{H^+/H_2}}(\mathrm{s}) = -2.303RT/F\,\mathrm{pH(s)} \qquad (3.2.46)$$

Equation (3.2.45) then gives

$$\mathrm{pH(s)} = -\log a_{\mathrm{H^+}}(\mathrm{s}) - 0.4343\Delta G^{0,\mathrm{w}\to\mathrm{s}}_{\mathrm{tr,H^+}} \qquad (3.2.47)$$

The values of the standard Gibbs transfer energies for H^+ then determine the solvent affinity for protons.

3.2.7 *Potentials at the interface of two immiscible electrolyte solutions*

The procedure described in the preceding section can form a basis for unambiguous determination of the Galvani potential difference between two immiscible electrolyte solutions (the Nernst potential) considering Eqs (3.1.22) and (1.4.34), e.g. for univalent ions,

$$\Delta^\alpha_\beta\phi = \frac{\pm\Delta G^{0,\alpha\to\beta}_{\mathrm{tr},i}}{F} \pm \frac{RT}{F}\ln\frac{a_i(\beta)}{a_i(\alpha)} = \Delta^\alpha_\beta\phi^0_i + \frac{RT}{F}\ln\frac{a_i(\beta)}{a_i(\alpha)} \qquad (3.2.48)$$

where the upper sign is valid for a cation and the lower sign for an anion and $\Delta\phi^0_i$ is the standard potential difference for the transfer of ion i from

Table 3.6 Standard Gibbs transfer energies and standard electric potential differences of transfer

Ion	$\Delta G_{tr,i}^{0,w\rightarrow o}$ (kJ · mol^{-1})	$\Delta_o^w \phi_i^0$ (V)
(a) Water–nitrobenzene		
Li^+	−38.4	0.389
Na^+	−34.5	0.358
Ca^{2+}	−68.3	0.354
Sr^{2+}	−67.2	0.348
H^+	−32.5	0.337
Ba^{2+}	−63.3	0.328
K^+	−24.3	0.252
Rb^+	−19.9	0.206
Cs^+	−15.5	0.161
Me_4N^+	−3.4	0.035
Bu_4N^+	24.2	−0.251
Ph_4As^+	35.9	−0.372
Cl^-	−30.5	−0.316
Br^-	−28.5	−0.295
NO_3^-	−24.4	−0.253
I^-	−18.8	−0.195
Picrate	4.6	0.048
Tetraphenylborate	35.9	0.372
(b) Water–dichloroethane		
Me_4N^+	−17.6	0.182
Bu_4N^+	21.8	−0.225
Ph_4As^+	35.1	−0.364
Cl^-	−46.4	−0.481
Br^-	−39.3	−0.408
I^-	−26.4	−0.273
Picrate	−6.7	−0.069

phase β to phase α. The values of these quantities and the standard Gibbs energies for the transfer of ions from nitrobenzene or 1,2-dichloroethane into water are listed in Table 3.6. The values of standard Gibbs transfer energies in the water/organic solvent give a quantitative meaning to the terms, *hydrophilic* and *hydrophobic*. For hydrophilic ions $\Delta G_{tr,i}^{0,\alpha\rightarrow\beta} \gg 0$ and the opposite relation is valid for hydrophobic ions. Sometimes the ions with their standard Gibbs transfer energies around zero are called 'semihydrophobic'.

Consider a system of two solvents in contact in which a single electrolyte BA is dissolved, consisting of univalent ions. A distribution equilibrium is established between the two solutions. Because, in general, the solvation energies of the anion and cation in the two phases are different so that the ion with a certain charge has a greater tendency to pass into the second phase than the ion of opposite charge, an electrical double layer appears at the interface (see Section 4.5.3) and a Galvani potential difference is formed between the two phases, called the *distribution potential.* The value of the distribution potential follows from Eq. (3.2.48). If the activities of the cation and anion in a given phase are set equal to one another and equations of this type calculated once for cations, and then for anions, are added, it follows that

$$\Delta_{\beta}^{\alpha}\phi_{\text{distr}} = \frac{\Delta G_{\text{tr},+}^{0,\alpha\rightarrow\beta} - \Delta G_{\text{tr},-}^{0,\alpha\rightarrow\beta}}{2F} \tag{3.2.49}$$

The distribution potential $\Delta_{\beta}^{\alpha}\phi_{\text{distr}}$ is thus independent of the ion concentration.

Potential differences at the interface between two immiscible electrolyte solutions (ITIES) are typical Galvani potential differences and cannot be measured directly. However, their existence follows from the properties of the electrical double layer at the ITIES (Section 4.5.3) and from the kinetics of charge transfer across the ITIES (Section 5.3.2). By means of potential differences at the ITIES or at the aqueous electrolyte–solid electrolyte phase boundary (Eq. 3.1.23), the phenomena occurring at the membranes of ion-selective electrodes (Section 6.3) can be explained.

References

Antelman, M. S., and F. J. Harris, Jr., *The Encyclopedia of Chemical Electrode Potentials,* Plenum Press, New York, 1982.

Bard, A. J. (Ed.), *Electrochemistry of Elements,* Plenum Press, New York, individual volumes appear since 1974.

Bard, A. J., J. Jordan, and R. Parsons (Eds), *Oxidation–Reduction Potentials in Aqueous Solutions,* Blackwell, Oxford, 1986.

Butler, J. N., Reference electrodes in aprotic solvents, *AE,* **7,** 77 (1970).

Clark, W. M., *Oxidation Reduction Potentials of Organic Systems,* Williams and Wilkins, Baltimore, 1960.

Ives, D. J., and G. J. Janz (Eds), *Reference Electrodes, Theory and Practice,* Academic Press, New York, 1961.

Karpfen, F. M., and J. E. B. Randles, Ionic equilibria and phase-boundary potentials in oil–water systems, *Trans. Faraday Soc.,* **49,** 823 (1953).

Koryta, J., Electrochemical polarisation of the interface of two immiscible electrolyte solutions I, *Electrochim. Acta,* **24,** 293 (1979); II, *Electrochim. Acta,* **29,** 445 (1984); III, *Electrochim. Acta,* **33,** 189 (1988).

Laity, R. W., *J. Chem. Educ.,* **39,** 67 (1962).

Latimer, W. M., *Oxidation Potentials,* Prentice-Hall, New York, 1952.

Michaelis, L., *Oxidation Reduction Potentials,* Lippincott, Philadelphia, 1930.

Michaelis, L., Occurrence and significance of semiquinone radicals, *Ann. New York Acad. Sci.,* **40,** 39 (1940).

Parker, A. J., Solvation of ions—enthalpies, entropies and free energies of transfer, *Electrochim. Acta,* **21,** 671 (1976).

Vanýsek, P., *Electrochemistry at Liquid–Liquid Interfaces,* Springer-Verlag, Berlin, 1985.

Wawzonek, S., Potentiometry: oxidation reduction potentials, *Techniques of Chemistry,* (Eds. A. Weissberger and B. W. Rossiter), Vol. I, Part IIa, Wiley–Interscience, New York, 1971.

3.3 Potentiometry

Potentiometry is used in the determination of various physicochemical quantities and for quantitative analysis based on measurements of the EMF of galvanic cells. By means of the potentiometric method it is possible to determine activity coefficients, pH values, dissociation constants and solubility products, the standard affinities of chemical reactions, in simple cases transport numbers, etc. In analytical chemistry, potentiometry is used for titrations or for direct determination of ion activities.

3.3.1 *The principle of measurement of the EMF*

The EMF of a galvanic cell is a thermodynamic equilibrium quatity. Thus, the potential of a cell must be measured under equilibrium conditions, i.e. without current flow. The measured EMF must be compensated by a known external potential difference. The measurement of the EMF of a cell is thus based on determination of a potential difference that exactly compensates the measured potential difference so that no current passes. This is easily achieved by the Poggendorf compensation method (see Fig. 3.13).

At present, potentiometry is performed primarily using electronic instruments with solid-state elements. The EMF can be measured by compensation measurement using a transistor voltmeter, i.e. according to the scheme in Fig. 3.13, where the galvanometer is replaced by an amplifier and meter. A second possibility is to measure the tiny current passed on connecting the cell with a large external resistance. Typical instruments of this kind consist of three parts. The first part is the input circuit, acting as an impedance transducer. Mostly MOSFET elements or capacity diodes are used here. The second part contains a power amplifier, permitting the use of a meter with large energy consumption. Depending on whether the EMF measured by using an a.c. amplifier is modulated in the input circuit (by a d.c. amplifier), an electronic demodulation circuit must be connected in the second part to rectify the amplified alternating voltage. The third part is the indicator with a digital display.

If a cell is to be used as a potential standard, then it must be prepared as simply as possible from chemicals readily available in the required purity and, in the absence of current passage, it must have a known, defined, constant EMF that is practically independent of temperature. In this case the efficiency, power, etc., required for cells used as electrochemical power sources is of no importance. The electrodes of the standard cell must not be polarizable by the currents passing through them when the measuring circuit is not exactly compensated.

The mostly used Weston cell consists of mercurous sulphate and cadmium amalgam electrodes:

$$\mathrm{Hg} \mid \mathrm{Hg_2SO_4(s)} \mid \mathrm{3CdSO_4.8H_2O(sat.)} \mid \mathrm{Cd,Hg(12.5\ weight\ \%\ Cd)} \quad (3.3.1)$$

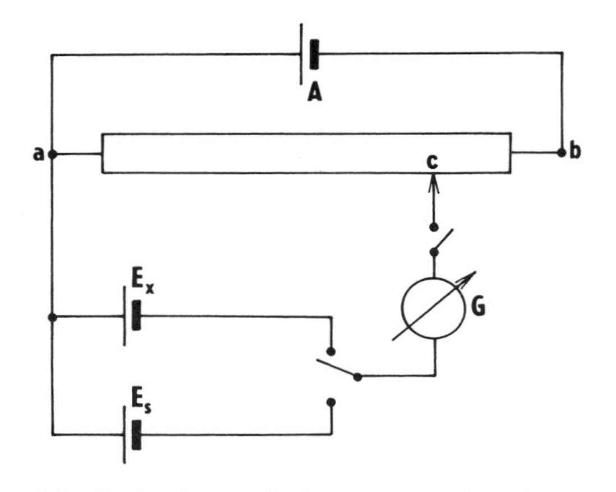

Fig. 3.13 Poggendorf's compensation circuit. A uniform resistance wire is connected at its ends (points a and b) to a stable voltage source (e.g. a storage battery). A part of this voltage between one end of the potentiometric wire and the sliding contact c is fed into the other circuit containing the measured source of voltage E_x, the null instrument G and the key. If no current flows through G the electrical potential difference compensates the EMF, E_x. The voltage fed to the potentiometric wire is calibrated by means of a standard cell, usually Weston cell, E_S by means of the same compensation procedure

3.3.2 *Measurement of pH*

The potentiometric measurement of physicochemical quantities such as dissociation constants, activity coefficients and thus also pH is accompanied by a basic problem, leading to complications that can be solved only if certain assumptions are accepted. Potentiometric measurements in cells without liquid junctions lead to mean activity or mean activity coefficient values (of an electrolyte), rather than the individual ionic values.

The EMF of the cell

$$\mathrm{Ag} \mid \mathrm{AgCl(s)} \mid \mathrm{buffer,KCl(m)} \mid \mathrm{H_2,Pt} \tag{3.3.2}$$

suitable for pH measurements is given by an expression containing the mean activity

$$\log a_{\mathrm{H_3O^+}} = 0.4343F/RT(E + E^0_{\mathrm{Ag/AgCl}}) - \log (m_{\mathrm{Cl^-}}\gamma_{\mathrm{Cl^-}}) \tag{3.3.3}$$

The pH value obtained in this way is accompanied by an error resulting from the approximation used for the calculation of $\gamma_{\mathrm{Cl^-}}$. Although this error may be small in dilute solutions, the pH values obtained in this manner are not exactly equal to the values corresponding to the absolute definition of the pH in Eq. (1.4.39). The pH values given by this absolute definition cannot, in principle, be exactly measured.

For current practice, the described method of pH measurement is too tedious. Moreover, not hydrogen but glass electrodes are used for routine pH measurements (see Section 6.3). Then the expression for the EMF of the cell consisting of the glass and reference electrodes contains a constant term from Eq. (6.3.10), in addition to the terms present in Eq. (3.3.3); this term must be obtained by calibration. Further, a term describing the liquid junction potential between the reference electrode and the measured solution must also be included.

In view of these difficulties an *operational definition of* pH has been introduced (cf. Section 1.4.6). This definition is based on the concept that the pH is measured with a hydrogen electrode, combined to the measuring galvanic cell with a suitable reference electrode. The reference electrode solution is connected with the solution of the hydrogen electrode through a salt bridge with KCl of at least $3.5\ \text{mol} \cdot \text{kg}^{-1}$ concentration. The hydrogen electrode is immersed first into the test solution X with pH(X) and the EMF E(X) is measured, and then into a standard solution S of pH(S), corresponding to potential E(S). It is assumed that the dependence of E on the pH is linear with the Nernst slope, $2.303RT/F$. Under these conditions,

$$\text{pH(X)} = \text{pH(S)} + \frac{E(\text{S}) - E(\text{X})}{2303RT/F} \tag{3.3.4}$$

Now the origin of the scale must be defined, i.e. a pH value must be selected for a standard (as close as possible to the value expected on the basis of definition 1.4.46). A solution of potassium hydrogen phthalate with a molality of $0.05\ \text{mol} \cdot \text{kg}^{-1}$ has been selected as the reference value pH standard (RVS).

The pH of the saturated solution of potassium hydrogen phthalate containing chloride molarity m_{Cl^-} is given by the equation

$$\text{pH} = -0.4343F(E + E^0_{\text{Ag/AgCl}}) + \log\left[m_{\text{Cl}^-}\gamma_{\text{Cl}^-}\right] \tag{3.3.5}$$

where E is the EMF of the cell (3.3.2). The first two terms on the right-hand side can be measured. If the pH of the cell (3.3.2) is determined for various values of m_{Cl^-} and if the sum of these two terms if plotted vs m_{Cl^-}, a nearly linear dependence is obtained which may easily be extrapolated to zero concentration of chloride ions. For the pH(RVS), the pH of the standard phthalate solution alone, we have then

$$\text{pH(RVS)} = -\lim_{m_{\text{Cl}^-} \to 0}\left[0.4343F(E + E_{\text{Ag/AgCl}})/RT - \log m_{\text{Cl}^-}\right] + \lim_{m_{\text{Cl}^-} \to 0} \log \gamma_{\text{Cl}^-}. \tag{3.3.6}$$

The value of $\lim_{m_{\text{Cl}^-} \to 0} \log \gamma_{\text{Cl}^-}$ may not be put equal to zero, as the overall ionic strength of the solution is not equal to zero, but it may be calculated using the Bates–Guggenheim equation (1.3.35). The values of pH(RVS) obtained in this way are listed in Table 3.7.

Table 3.7 Values of pH(RVS) for the reference value standard of $0.05\ \text{mol} \cdot \text{kg}^{-1}$ potassium hydrogen phthalate at various temperatures

°C	pH(RVS)	°C	pH(RVS)	°C	pH(RVS)
0	4.000	35	4.018	65	4.097
5	3.998	37	4.022	70	4.116
10	3.997	40	4.027	75	4.137
15	3.998	45	4.038	80	4.159
20	4.001	50	4.050	85	4.183
25	4.005	55	4.064	90	4.210
30	4.011	60	4.080	95	4.240

For practical measurements, six further solutions were measured as primary standards and fifteen additional solutions as operational standards (the difference between these two types of standards lies in the presence or absence of a liquid junction; they need not be distinguished for routine measurements).

In practice, the pH is mostly measured with a glass electrode (see Section 6.3), connected with a calomel electrode (see Section 3.2.2). The measuring system is calibrated by using a single standard S, with a pH(S) value lying as close as possible to the pH(X) value. The pH(X) value is then calculated from $E(\mathrm{S})$, $E(\mathrm{X})$ and pH(S) by Eq. (3.3.4). It is preferable to use two standards S_1 and S_2, selected so that $\mathrm{pH}(S_1)$ is smaller and $\mathrm{pH}(S_2)$ larger than pH(X) (both the pH(S) values should be as close to pH(X) as possible). The value of pH(X) is then calculated from the usual formula for linear interpolation:

$$\frac{\mathrm{pH(X)} - \mathrm{pH}(S_1)}{\mathrm{pH}(S_2) - \mathrm{pH}(S_1)} = \frac{E(\mathrm{X}) - E(S_1)}{E(S_2) - E(S_1)} \tag{3.3.7}$$

Analogously to water, standards are measured for a mixture of methanol and water (50 per cent by weight) as well as for heavy water, $\mathrm{pD} = -\log a(D_3O^+)$.

An operational approach to the determination of the acidity of solutions in deuterium oxide (heavy water) was suggested by Glasoe and Long. This quantity, pD, is determined in a cell consisting of an aqueous (H_2O) glass electrode and a saturated aqueous calomel reference electrode on the basis of the equation

$$\mathrm{pD} = \mathrm{pH}_{\mathrm{pHmeter\ reading}} + 0.4 \tag{3.3.8}$$

where the subscript pHmeter reading denotes the pH value indicated on the conventional pHmeter.

Determination of the pH in non-aqueous solvents is discussed in Section 3.2.7.

3.3.3 *Measurement of activity coefficients*

Mean activity coefficients can be measured potentiometrically, mostly in a concentration cell with or without transfer. Consider, for example, the cell (with a non-aqueous electrolyte solution)

$$Ag \mid AgCl(s) \mid KCl(m_1) \mid K,Hg \mid KCl(m_2) \mid AgCl(s) \mid Ag$$

On passing a positive charge from the left to the right in the graphical scheme of this cell, silver is oxidized to form silver chloride; potassium passes through the amalgam into the other solution. Here, silver chloride is reduced to metallic silver and chloride ions. The overall reaction is the transfer of KCl from a region of higher concentration to a region of lower concentration, so that the EMF of the cell is given by the equation

$$E = \frac{2RT}{F} \ln \frac{a_{\pm,1}}{a_{\pm,2}} \tag{3.3.9}$$

where $a_{\pm}^2 = a_{K^+}a_{Cl^-}$ is the mean activity of KCl. On rearrangement we obtain

$$\begin{aligned} \frac{2RT}{F} \ln a_{\pm,1} &= \frac{2RT}{F} \ln a_{\pm,2} + E \\ &= \frac{2RT}{F} \ln m_2 + \frac{2RT}{F} \ln \gamma_{\pm,2} + E \end{aligned} \tag{3.3.10}$$

The concentration of solution 1 is kept constant while E is measured for different concentrations of solution 2. The expression $(2RT/F)\ln m_2 + E$ is plotted against m_2. The value of the ordinate at point $m_2 = 0$ yields the term $(2RT/F)\ln a_{\pm,1}$ as $\ln \gamma_{\pm,2} = 0$ at this point. Once the value of $a_{\pm,1}$ is known, then Eq. (3.3.10) and the measured E values can be used to calculate the actual mean activity of the electrolyte at an arbitrary concentration.

3.3.4 *Measurement of dissociation constants*

The dissociation constants of acids and bases are determined either exactly, by means of a suitable cell without liquid junction and without measuring the pH directly, or approximately on the basis of a pH measurement in a cell with liquid junction, the potential of which is reduced to a minimum with the help of a salt bridge. In the former case we shall use, for example, the cell

$$Pt,H_2(p_{H_2} = 1) \mid HA(m_1),NaA(m_2),NaCl(m_3) \mid AgCl(s) \mid Ag$$

whose EMF is given by the expression (where H_3O^+ is replaced by H^+ for simplicity)

$$E = E^0_{\text{AgCl/Ag}} - \frac{RT}{F} \ln (a_{H^+} a_{Cl^-}) \tag{3.3.11}$$

The dissociation constant, K_A, of the acid HA is given by the equation

$$K_A = \frac{a_{H^+} a_{A^-}}{a_{HA}} = \frac{m_{H^+} m_{A^-}}{m_{HA}} \frac{\gamma_{H^+} \gamma_{A^-}}{\gamma_{HA}} = K'_A \frac{\gamma_{H^+} \gamma_{A^-}}{\gamma_{HA}} \tag{3.3.12}$$

where the apparent dissociation constant K'_A can be found, for example, conductometrically. It holds for the individual concentrations that

$$m_{Cl^-} = m_3, \quad m_{HA} = m_1 - (m_{H^+} - m_{OH^-}) = m_1 - m_{H^+} + \frac{K_w}{m_{H^+}},$$

$$m_{A^-} = m_2 + m_{H^+} - m_{OH^-} = m_2 + m_{H^+} - \frac{K_w}{m_{H^+}},$$

so that

$$K'_A = m_{H^+} \frac{m_2 + m_{H^+} - K_w/m_{H^+}}{m_1 - m_{H^+} + K_w/m_{H^+}} \tag{3.3.13}$$

If the activity a_{H^+} is substituted from Eq. (3.3.11) into Eq. (3.3.10) then rearrangement yields

$$X = \frac{F(E - E^0_{\text{AgCl/Ag}})}{2.303RT} + \log \frac{m_{HA} m_{Cl^-}}{m_{A^-}} = -\log \frac{\gamma_{HA} \gamma_{Cl^-}}{\gamma_{A^-}} - \log K_A \tag{3.3.14}$$

The concentration in the second term on the left-hand side of this equation is expressed in terms of the known analytical concentrations, m_1, m_2 and m_3, and of the concentration m_{H^+}, calculated from the apparent dissociation

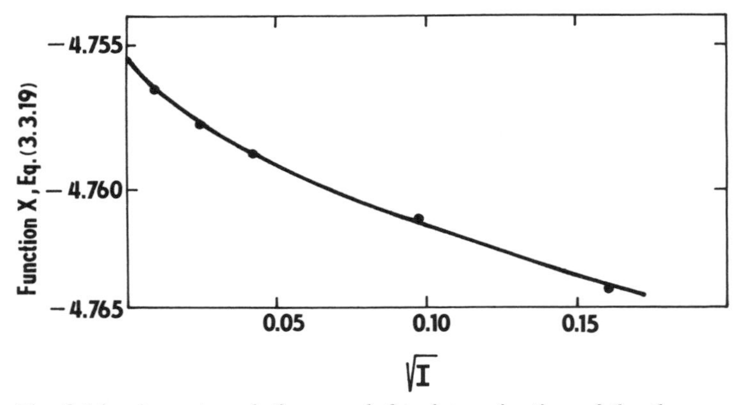

Fig. 3.14 An extrapolation graph for determination of the thermodynamic dissociation constant of acetic acid using Eq. (3.3.14)

constant using Eq. (3.3.13). A series of measurements of E is carried out for a single m_1/m_2 ratio and varying sodium chloride concentrations m_3. Then the expression on the left-hand side of Eq. (3.3.14) is plotted against $\sqrt{I}$ and the dependence is extrapolated to $I \rightarrow 0$. The intercept on the ordinate axis yields the value of $-\log K_A$ (see Fig. 3.14). However, the value of $E^0_{AgCl/Ag}$ must be the same as that employed for standard pH measurements.

References

Bates, R. G., *Determination of pH. Theory, Practice.* John Wiley & Sons, New York, 1973.

Covington, A. K., R. G. Bates and R. A. Durst, Definition of pH scales, standard reference values, measurement of pH and related terminology, *Pure Appl. Chem.*, **57,** 531 (1985).

Glasoe, P. K., and F. A. Long, Use of glass electrode to measure acidities in deuterium oxide, *J. Phys. Chem.*, **64,** 188 (1960).

Harned, H. S., and B. B. Owen, *The Physical Chemistry of Electrolytic Solutions,* Reinhold, New York, 1950.

Robinson, R. A., and R. H. Stokes, *Electrolyte Solutions,* Butterworths, London, 1959.

Serjeant, E. P., *Potentiometry and Potentiometric Titrations,* John Wiley & Sons, New York, 1984.

Wawzonek, S., see page 190.

Chapter 4
The Electrical Double Layer

Metal–solution interfaces lend themselves to the exact study of the double layer better than other types because of the possibility of varying the potential difference between the phases without varying the composition of the solution. This is done through the use of a reference electrode and a potentiometer which fixes the potential difference in question. In favourable cases there is a range of potentials for which a current does not flow across the interface in a system of this kind, the interface being similar to a condenser of large specific capacity. The capacity of this condenser gives a fairly direct measure of the electronic charge on the metallic surface, and this, in turn, leads to other information about the double layer. No such convenient and informative procedure is possible with other types of interfaces.

D. C. Grahame, 1947

4.1 General Properties

In the interphase the cohesion forces binding the individual particles together in the bulk of each condensed phase are significantly reduced. Particles that had a certain number of nearest neighbours in the bulk of the phase have a smaller number of such neighbours at the interface. However, particles from the other phase can also become new neighbours. This change in the equilibrium of forces affecting particles at the interface can lead to a new lateral force, termed the *interfacial tension*. In addition, the interphase usually has different electrical properties than the bulk phase. The situation becomes relatively simple when the phase is electrically charged. The free charge is then centred in the interphase. Orientation of dipoles in the interphase can also lead to a change in the electrical properties. Further charges can enter the interphase through adsorption of ions and/or dipoles. Excess charge in the interphase resulting from the presence of ions, electrons and dipoles produces an electric field. The region in which these charges are present is termed the *electrical double layer*.

The presence of electrical charge affects the interfacial tension in the interphase. If one of the phases considered is a metal and the other is an electrolyte solution, then the phenomena accompanying a change in the interfacial tension are included under the term of *electrocapillarity*.

While the formation of an electrical double layer at interfaces is a general phenomenon, the electrode–electrolyte solution interface will be considered first. If the electrode has a charge of $Q(\mathrm{m})$, then this charge is distributed uniformly over the interface with the solution. If part of the electrode is not immersed in the solution and is in contact with the air, then the portion of the charge corresponding to the surface in contact with the air is negligible, as this interface has minute capacitance. The great majority of the charge is located at the metal–solution interface.

Excess charge in the metal $Q(\mathrm{m})$ must be compensated in the solution by a charge of the same magnitude but of opposite sign. This charge is attracted from the solution by electrostatic forces. The general relationship

$$Q(\mathrm{m}) + Q(\mathrm{s}) = 0 \tag{4.1.1}$$

is then valid, where $Q(\mathrm{m})$ is the charge corresponding to the surface area A of the interface on the electrode side and $Q(\mathrm{s})$ is the charge in that part of the double layer in the solution.

According to the oldest hypothesis, put forward by H. L. F. Helmholtz, the electrical double layer has the character of a plate condenser, whose plates are represented by homogeneously distributed charge in the metal and ions of the opposite charge lying in a parallel plane in the solution at a minimal distance from the surface of the metal. According to modern conceptions, the electron cloud in the metal extends to a certain degree into the region of the layer of solvent molecules in the immediate vicinity of the metal surface. In this layer, the dipoles of the solvent molecules are oriented to various degrees towards the metal surface. Ions can gather in the solution portion of the interphase as a result of the electrostatic field of the electrode ('electrostatic adsorption'). They can also, however, be adsorbed specifically on the electrode through van der Waals, hydrophobic and chemical forces.

If they are less polar than the solvent molecules, uncharged molecules from the solution can also be adsorbed and replace solvent molecules that were originally in direct contact with the electrode.

In the electrode–solution interphase, the adsorption of these substances is also affected by the influence of the electric field in the double layer on their dipoles. Substances that collect in the interphase as a result of forces other than electrostatic are termed surface-active substances or surfactants.

In the simple case of electrostatic attraction alone, electrolyte ions can approach to a distance given by their primary solvation sheaths, where a monomolecular solvent layer remains between the electrode and the solvated ions. The plane through the centres of the ions at maximum approach under the influence of electrostatic forces is called the *outer Helmholtz plane* and the solution region between the outer Helmholtz plane and the electrode surface is called the *Helmholtz* or *compact layer*. Quantities related to the outer Helmholtz plane are mostly denoted by symbols with the subscript 2.

However, electrostatic forces cannot retain ions at a minimal distance from the electrode, as thermal motion continually disperses ions to a greater distance from the phase boundary. In this way the *Gouy* or *diffuse layer* is formed, i.e. the region between the outer Helmholtz plane and the bulk of the solution. When only electrostatic forces act on the ions, the entire charge is concentrated in this diffuse layer (see Fig. 4.1).

The charge accumulated by specific adsorption is partly compensated by a

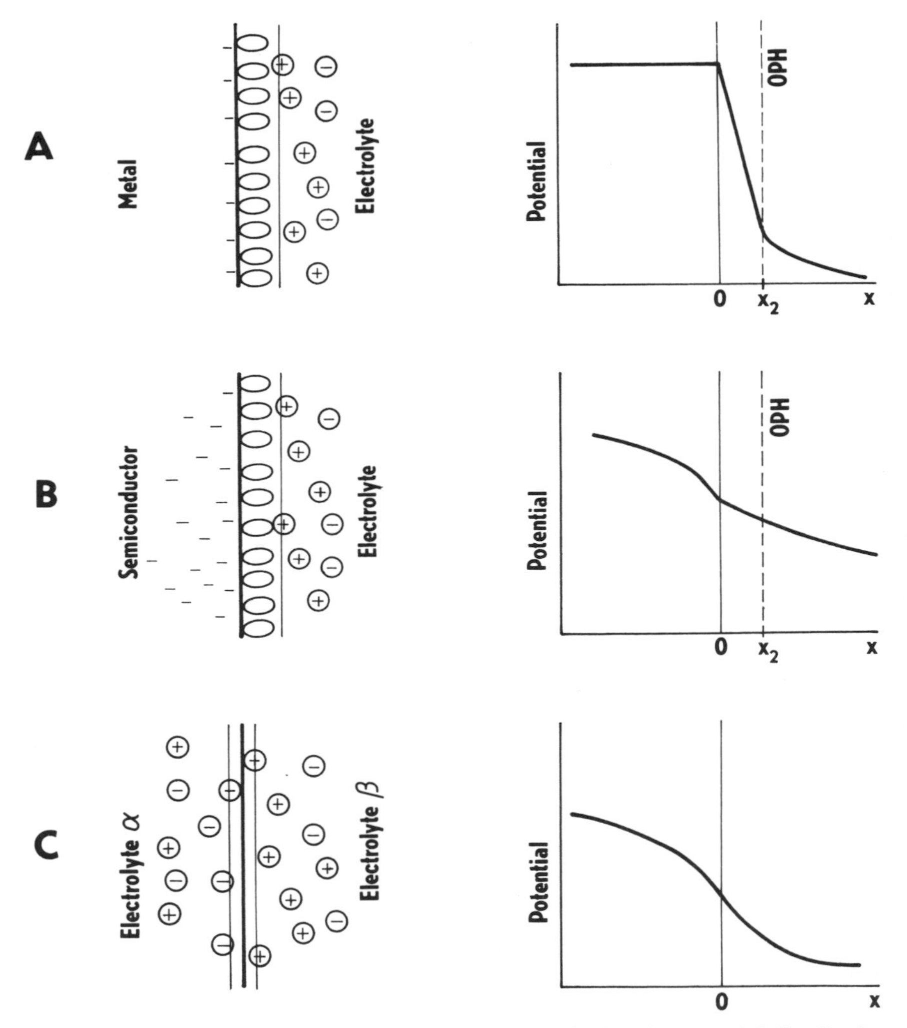

Fig. 4.1 Structure of the electric double layer and electric potential distribution at (A) a metal–electrolyte solution interface, (B) a semiconductor–electrolyte solution interface and (C) an interface of two immiscible electrolyte solutions (ITIES) in the absence of specific adsorption. The region between the electrode and the outer Helmholtz plane (OHP, at the distance x_2 from the electrode) contains a layer of oriented solvent molecules while in the Verwey and Niessen model of ITIES (C) this layer is absent

change in the charge in the diffuse layer. A plane through the centres of the adsorbed ions is termed the *inner Helmholtz plane* (and quantities connected with this plane are denoted by the subscript 1).

In order to be able to derive the basic relationships between quantities characterizing the electrical double layer, the concept of an *ideal polarized electrode* must be introduced. Reversible electrodes, considered in Chapter 3, are characterized by a potential that, at equilibrium at a given temperature and pressure, is unambiguously determined by the composition of the solution and electrodes, i.e. by the appropriate activities. If a charge is passed through an ideal reversible electrode, processes occur that immediately restore the original equilibrium; in this sense, such an electrode is an *ideal non-polarizable electrode.* In contrast, an ideal polarized electrode can assume an arbitrary potential difference versus a reference electrode by the application of an external voltage source and maintain this potential difference even after the voltage source is disconnected. Similarly to the EMF of a reversible galvanic cell, this potential is an equilibrium quantity, as will be demonstrated in Section 4.2. In contrast to a galvanic cell, the potential of an ideal polarized electrode can be arbitrarily changed through a change in its charge without disturbing the electrode from the equilibrium state. Thus this electrode has one degree of freedom more than a non-polarizable electrode whose potential is determined by the composition of the solution. An ideal polarized electrode is equivalent to a perfect condenser without leakage.

The difference between an ideal polarized and ideal non-polarizable electrode is best demonstrated in terms of the original Lippmann definition. The *surface charge density* of an ideal polarized electrode $\sigma(\mathrm{m})$ is equal to the charge, Q, that must be supplied to the electrode from a voltage source so that its original potential against a reference electrode is maintained during a reversible increase in its surface area by one area unit. Charge $\sigma(\mathrm{m})$ is, of course, a function of the potential of the ideal polarized electrode and can even equal zero. For an ideal non-polarizable electrode, no charge need be supplied from an external source during a reversible increase in its surface area. Clearly, an electrode whose potential is determined by the given ions or oxidation–reduction system in solution has the properties of a condenser (with leakage), as an electrical double layer with a certain capacity is formed at its surface. Nonetheless, for a very slow increase in the surface of an ideal non-polarizable electrode, the half-cell reaction† is a sufficient source of charge for loading the electrical double layer. The supply of charge from an external source is then unnecessary for maintenance of a constant electrode potential.

Real electrodes, at which at least one electrode reaction occurs or can occur with finite velocity (see Chapter 5), exhibit behaviour between that of

† This term is used deliberately, rather than 'electrode reaction' (Chapter 5), as a thermodynamically reversible process is involved.

ideal polarized and ideal non-polarizable electrodes, especially during current passage.

Another definition of an ideal polarized electrode is based on the practical form of this electrode. At an ideal polarized electrode either no exchange of charged particles takes place between the electrode and the solution or—if thermodynamically feasible—exchange occurs very slowly as a result of the large activation energy.

Grahame gives the mercury electrode in a 1 M KCl solution as a suitable example of a nearly ideal polarized electrode. At a potential of −0.556 V (versus the normal calomel electrode) the following reactions can occur: $2Hg \rightleftarrows Hg_2{}^{2+} + 2e$, $K^+ + Hg + e \rightleftarrows K,Hg$ and $2Cl^- \rightleftarrows Cl_2 + 2e$. However, the concentration of mercury(I) ions in solution is equal to 10^{-36} mol · dm^{-3}, that of potassium in the amalgam equals to 10^{-45} mol · dm^{-3} and the partial pressure of chlorine is 10^{-57} Pa. Clearly, such minute quantities will not measurably influence the electrode potential. A further possible reaction, $2H_2O + 2e \rightleftarrows H_2 + 2OH^-$, corresponding to a relatively large partial pressure of hydrogen of 1.6 Pa, practically does not occur at the given potential because of the high overpotential of hydrogen. Thus, this electrode fulfils almost perfectly the condition that no transfer of charged particles occurs between the electrode and the solution.

Under real conditions it is very difficult to fulfil the condition of constant electrode charge after disconnecting the external source. For example, a negatively charged electrode is discharged by the reduction of traces of impurities such as metal ions or oxygen.

Electrical double layers are also characteristic of the semiconductor–electrolyte solution, solid electrolyte or insulator–electrolyte solution interface and for the interface between two immiscible electrolyte solutions (ITIES) (Section 4.5).

References

Adamson, A. W., *Physical Chemistry of Surfaces,* 2nd ed. Interscience, New York, 1967.

Barlow, C. A., The electrical double layer, *PCAT,* **IX A,** 167 (1970).

Blank, M. (Ed.), *Electrical Double Layer in Biology,* Plenum Press, New York, 1986.

Butler, J. A. V., *Electrical Phenomena at Interfaces,* Butterworths, London, 1951.

Davies, J. T., and E. K. Rideal, *Interfacial Phenomena,* 2nd ed. Academic Press, New York, 1963.

Delahay, P., *Double Layer and Electrode Kinetics,* John Wiley & Sons, New York, 1965.

Frumkin, A. N., V. S. Bagotzky, Z. A. Iofa, and B. N. Kabanov, *Kinetics of Electrode Processs* (in Russian), Publishing House of the Moscow State University, Moscow, 1952.

Grahame, D. C., Electric double layer, *Chem. Rev.,* **41,** 441 (1947).

Marynov, G. A., and R. R. Salem, *Electrical Double Layer at Metal–Dilute Solution Interface, Lecture Notes in Chemistry,* Vol. 33, Springer-Verlag, Berlin, 1983.

Mohilner, D. M., The electric double layer, in *Electroanalytical Chemistry* (Ed. A. J. Bard), Vol. 1, p. 241, M. Dekker, New York, 1966.
Parsons, R., Equilibrium properties of electrified interfaces, *MAE,* **1,** 103 (1954).
Parsons, R., The structure of electrical double layer and its influence on the rates of electrode reactions, *AE,* **1,** 1 (1961).
Parsons, R., Faradaic and non-faradaic processes, *AE,* **7,** 177 (1970).
Payne, R., The electrical double layer in nonaqueous solutions, *AE,* **7,** 1 (1970).
Perkins, R. S., and T. N. Andersen, Potentials of zero charge of electrodes, *MAE,* **5,** 203 (1969).
Spaarnay, M. J., *The Electrical Double Layer, The International Encyclopedia of Physical Chemistry and Chemical Physics,* Topic 14, Vol. 4, Pergamon Press, Oxford, 1972.

4.2 Electrocapillarity

The interfacial tension at an interface is a force acting on a unit length of the interface against an increase in the interface area. The region around the interface in which interfacial tension is produced is very narrow. Tolman, Kirkwood and Buff state that this distance is only about 0.1–0.3 nm for the liquid–vapour interface, corresponding to a monolayer at the surface of the liquid, still affecting the next nearest layer. In order to derive a relationship that would qualitatively characterize the formation of interfacial tension at the interface between two homogeneous single-component phases, the following cycle will be considered. First a free surface area of A is formed in each of phases α and β. These surfaces are then brought into contact, forming the interphase s. The work required to separate these two phases and to return the system to the original state is $-\gamma A$, where γ is the interfacial tension. Thus,

$$\gamma A = \Delta G(\alpha) + \Delta G(\beta) - \Delta G(\mathrm{s}) \tag{4.2.1}$$

where $\Delta G(\alpha)$ and $\Delta G(\beta)$ are the Gibbs energies required for reversible separation (i.e. the formation of two free surfaces A) of the phases α and β both divided by two. $\Delta G(\mathrm{s})$ is the Gibbs energy required for interrupting the contact between phases α and β. The ΔGs are given by the equations

$$\Delta G(\alpha) = \left\{-[n_0(\alpha) - n_\mathrm{s}(\alpha)]\frac{\varepsilon_\mathrm{c}(\alpha)}{2} + \varepsilon_\mathrm{r}(\alpha) + \varepsilon_\mathrm{v}(\alpha)\right\}N(\alpha)A$$

$$\Delta G(\beta) = \left\{-[n_0(\beta) - n_\mathrm{s}(\beta)]\frac{\varepsilon_\mathrm{c}(\beta)}{2} + \varepsilon_\mathrm{r}(\beta) + \varepsilon_\mathrm{v}(\beta)\right\}N(\beta)A$$

$$\Delta G(\mathrm{s}) = [\varepsilon_\mathrm{c}(\mathrm{s}) + \varepsilon_\mathrm{v}(\mathrm{s})]N(\mathrm{s})A \tag{4.2.2}$$

where $n_0(\alpha)$ and $n_0(\beta)$ are the numbers of nearest neighbours that the particle has in the bulk of phases α and β, respectively, $n_\mathrm{s}(\alpha)$ and $n_\mathrm{s}(\beta)$ are the numbers of nearest neighbours from phase α or β that the particle has at the interface, $\varepsilon_\mathrm{c}(\alpha)$ and $\varepsilon_\mathrm{c}(\beta)$ are the bond energies between this particle and its nearest neighbour, ε_r's are the energies required for rearrangement of the particle in the interphase in order to be able to form a bond to a particle in the neighbouring phase, $\varepsilon_v(\alpha)$, $\varepsilon_v(\beta)$, and $\varepsilon_v(s)$ are the changes in the thermal vibrational energies of the particle at the interface as a result of formation of the free surface and contact between the phases, $N(\alpha)$ and $N(\beta)$ are the numbers of particles per unit surface area for the free surface of phases α and β, $\varepsilon_c(s)$ is the bond energy between the particles of phases α and β in the interphase and $N(s)$ is the number of such bonds per unit surface area at the interphase. If we assume that $N(\alpha) = N(\beta) = N(s)$, that species at the surface have only one free bond ($n_0 - n_s = 1$) and that $\varepsilon_c \gg \varepsilon_r$, ε_v, then

$$\gamma = \left[\varepsilon_c(s) - \frac{\varepsilon_c(\alpha) - \varepsilon_c(\beta)}{2} \right] N(s) \tag{4.2.3}$$

This expression can be roughly interpreted as the difference between the Gibbs energy of adhesion of the two phases and the sum of the Gibbs energies of cohesion for the two phases.

These considerations can also be used to derive the Dupré equation, where $\Delta G(s)$ is the Gibbs energy of adsorption of the solvent per unit area of the metal surface:

$$\Delta G(s) = \gamma_{m/a} + \gamma_{l/a} - \gamma \tag{4.2.4}$$

where $\gamma_{m/a}$ is the interfacial tension of the metal–air interface, $\gamma_{l/a}$ is the interfacial tension of the solution–air interface and γ is the interfacial tension of the metal–solution interface.

Now the relationship between the interfacial tension and the composition of the two phases in contact will be analysed thermodynamically by using the approach of J. W. Gibbs.

The interphase can be considered as a particular phase s of thickness h. This phase differs from the homogeneous phases only in that the effect of pressure is accompanied by the effect of the interfacial tension γ. Consider a rectangle with sides h (perpendicular to the interphase) and l (parallel with the interphase) located perpendicular to the interphase. The force acting on the rectangle is not equal to the product phl (as for an area in the bulk of the solution) but $phl - \gamma l$. If the volume of the interphase $V(s)$ is increased by $dV(s)$ by increasing the thickness of the interphase by dh, then area $A = V(s)/h$ increases by dA. The overall work, W, connected with this process consists of volume work accompanying the increase in the thickness of the interphase and volume and surface work connected with an increase of the surface area:

$$W(s) = -pA\, dh + (-ph + \gamma)\, dA = -p\, dV(s) + \gamma\, dA \tag{4.2.5}$$

Because of this formation, a different definition of the enthalpy must be introduced for the interphase, differing from the usual expression for a homogeneous phase:

$$H(s) = U(s) + pV(s) - \gamma A \tag{4.2.6}$$

where $U(\mathrm{s})$ is the internal energy of the interphase. For the differential Gibbs energy of the interphase we have

$$\begin{aligned} \mathrm{d}G(\mathrm{s}) &= \mathrm{d}[H(\mathrm{s}) - TS(\mathrm{s})] \\ &= \mathrm{d}U(\mathrm{s}) + p\,\mathrm{d}V(\mathrm{s}) - V(\mathrm{s})\,\mathrm{d}p - \gamma\,\mathrm{d}A - A\,\mathrm{d}\gamma \\ &\quad - T\,\mathrm{d}S(\mathrm{s}) - S(\mathrm{s})\,\mathrm{d}T + \sum_{i=0}^{n} \mu_i\,\mathrm{d}n_i(\mathrm{s}) \end{aligned} \tag{4.2.7}$$

where $n_i(\mathrm{s})$ is the amount of the ith component of the system in phase s and n is the total number of components. As

$$\mathrm{d}U(\mathrm{s}) + p\,\mathrm{d}V(\mathrm{s}) - \gamma\,\mathrm{d}A - T\,\mathrm{d}S(\mathrm{s}) = 0 \tag{4.2.8}$$

it follows that

$$\mathrm{d}G(\mathrm{s}) = -S(\mathrm{s})\,\mathrm{d}T + V(\mathrm{s})\,\mathrm{d}p - A\,\mathrm{d}\gamma + \sum_{i=0}^{n} \mu_i\,\mathrm{d}n_i(\mathrm{s}) \tag{4.2.9}$$

The appropriate Gibbs–Duhem equation has the form

$$-S(\mathrm{s})\,\mathrm{d}T + V(\mathrm{s})\,\mathrm{d}p - A\,\mathrm{d}\gamma - \sum_{i=0}^{n} n_i(\mathrm{s})\,\mathrm{d}\mu_i = 0 \tag{4.2.10}$$

Introduction of the surface concentrations of the components,

$$\Gamma_i^* = \frac{n_i(\mathrm{s})}{A} \tag{4.2.11}$$

yields together with the relation $V(\mathrm{s})/A = h$

$$\mathrm{d}\gamma = -\frac{S(\mathrm{s})}{A}\,\mathrm{d}T + h\,\mathrm{d}p - \sum_{i=0}^{n} \Gamma_i^*\,\mathrm{d}\mu_i \tag{4.2.12}$$

This relationship is termed the Gibbs adsorption equation.

It is often useful (e.g. for dilute solutions) to express the adsorption of components with respect to a predominant component, e.g. the solvent. The component that prevails over m components is designated by the subscript 0 and the case of constant temperature and pressure is considered. In the bulk of the solution, the Gibbs–Duhem equation, $\sum_i n_i\,\mathrm{d}\mu_i = 0$, is valid, so that

$$-\mathrm{d}\mu_0 = \sum_{i=1}^{n} \frac{n_i}{n_0}\,\mathrm{d}\mu_i \tag{4.2.13}$$

The sum in Eq. (4.2.12) can be modified using the relationship

$$\begin{aligned} \sum_{i=0}^{n} \Gamma_i^*\,\mathrm{d}\mu_i &= \sum_{i=1}^{n} \Gamma_i^*\,\mathrm{d}\mu_i + \Gamma_0^*\,\mathrm{d}\mu_0 \\ &= \sum_{i=1}^{n} \left(\Gamma_i^* - \frac{n_i}{n_0}\Gamma_0^*\right)\mathrm{d}\mu_i \end{aligned} \tag{4.2.14}$$

The surface excess of component i over component 0 is designated as Γ_i:

$$\Gamma_i = \Gamma_i^* - \frac{n_i}{n_0}\Gamma_0^* \tag{4.2.15}$$

and the Gibbs adsorption equation assumes the form

$$\mathrm{d}\gamma = -\sum_{i=1}^{n} \Gamma_i \, \mathrm{d}\mu_i \tag{4.2.16}$$

In very dilute solutions, where $n_0 \gg n_i$, $\Gamma_i \approx \Gamma_i^*$.

When the adsorbed components are electrically charged, then the partial molar Gibbs energy of the charged component depends on the charge of the given phase, and thus the chemical potentials in the above relationships must be replaced by the electrochemical potentials. The Gibbs adsorption isotherm then has the form

$$\mathrm{d}\gamma = -\sum_{i=1}^{n} \Gamma_i \, \mathrm{d}\tilde{\mu}_i \tag{4.2.17}$$

The interfacial tension always depends on the potential of the ideal polarized electrode. In order to derive this dependence, consider a cell consisting of an ideal polarized electrode of metal M and a reference non-polarizable electrode of the second kind of the same metal covered with a sparingly soluble salt MA. Anion A^- is a component of the electrolyte in the cell. The quantities related to the first electrode will be denoted as m, the quantities related to the reference electrode as m′ and to the solution as l. For equilibrium between the electrons and ions M^+ in the metal phase, Eq. (4.2.17) can be written in the form ($s = n - 2$)

$$\begin{aligned} \mathrm{d}\gamma = {} & -\Gamma_\mathrm{e}\, \mathrm{d}\tilde{\mu}_\mathrm{e}(\mathrm{m}) - \Gamma_{\mathrm{M}^+}\, \mathrm{d}\tilde{\mu}_{\mathrm{M}^+}(\mathrm{m}) \\ & -\sum_{i=1}^{s} \Gamma_i \, \mathrm{d}\mu_i(\mathrm{l}) - F\sum_{i=1}^{s} \Gamma_i z_i \, \mathrm{d}\phi(\mathrm{l}) \end{aligned} \tag{4.2.18}$$

The overall potential difference E and the potential of the reference electrode $E(\mathrm{m}')$ are given by the equations

$$\begin{aligned} E &= -\frac{1}{F}[\tilde{\mu}_\mathrm{e}(\mathrm{m}) - \tilde{\mu}_\mathrm{e}(\mathrm{m}')] = E_\mathrm{p} - E(\mathrm{m}') \\ E(\mathrm{m}') &= E^0(\mathrm{m}') - \frac{RT}{F}\ln a_{\mathrm{A}^-}(\mathrm{l}) = E^0(\mathrm{m}') - \frac{1}{F}[\mu_{\mathrm{A}^-}(\mathrm{l}) - \mu^0_{\mathrm{A}^-}(\mathrm{l})] \end{aligned} \tag{4.2.19}$$

where E_p is the electrode potential of the ideal polarized electrode on the hydrogen scale (this is a generalization of the electrode potential; see Section 3.1.5).

The surface charge densities $\sigma(\mathrm{m})$ for the metal phase and $\sigma(\mathrm{l})$ for the solution phase are expressed by the relationships

$$\sigma(\mathrm{m}) = (\Gamma_{\mathrm{M}^+} - \Gamma_{\mathrm{e}})F$$
$$\sigma(\mathrm{l}) = -\sigma(\mathrm{m}) = F\sum_{i=1}^{s} z_i\Gamma_i \tag{4.2.20}$$

It holds in the metal phase that $\mu_{\mathrm{M}}(\mathrm{m}) = \tilde{\mu}_{\mathrm{M}^+}(\mathrm{m}) + \tilde{\mu}_{\mathrm{e}}(\mathrm{m})$ and in the salt phase MA (denoted r) that $\mu_{\mathrm{MA}}(\mathrm{r}) = \tilde{\mu}_{\mathrm{M}^+}(\mathrm{r}) + \tilde{\mu}_{\mathrm{A}^-}(\mathrm{r})$. As $\mu_{\mathrm{M}}(\mathrm{m})$ and $\mu_{\mathrm{MA}}(\mathrm{r})$ are constants and because at equilibrium $\tilde{\mu}_{\mathrm{A}^-}(\mathrm{r}) = \tilde{\mu}_{\mathrm{A}^-}(\mathrm{l})$, $\tilde{\mu}_{\mathrm{M}^+}(\mathrm{r}) = \tilde{\mu}_{\mathrm{M}^+}(\mathrm{m}')$, it follows that

$$\mathrm{d}\tilde{\mu}_{\mathrm{M}^+}(\mathrm{m}) + \mathrm{d}\tilde{\mu}_{\mathrm{e}}(\mathrm{m}) = 0$$
$$\mathrm{d}\tilde{\mu}_{\mathrm{M}^+}(\mathrm{m}') + \mathrm{d}\tilde{\mu}_{\mathrm{A}^-}(\mathrm{l}) = \mathrm{d}\mu_{\mathrm{A}^-}(\mathrm{l}) - F\,\mathrm{d}\phi(\mathrm{l}) - \mathrm{d}\tilde{\mu}_{\mathrm{e}}(\mathrm{m}') = 0 \tag{4.2.21}$$

Combination of Eqs (4.2.20) and (4.2.21) with Eq. (4.2.18) yields

$$\mathrm{d}\gamma = \frac{\sigma(\mathrm{m})}{F}[\mathrm{d}\tilde{\mu}_{\mathrm{e}}(\mathrm{m}) - \mathrm{d}\tilde{\mu}_{\mathrm{e}}(\mathrm{m}') + \mathrm{d}\mu_{\mathrm{A}^-}(\mathrm{l})] - \sum_{i=1}^{s}\Gamma_i\,\mathrm{d}\mu_i(\mathrm{l}) \tag{4.2.22}$$

It can further be seen from Eq. (4.2.19) that $[\mathrm{d}\tilde{\mu}_{\mathrm{e}}(\mathrm{m}) - \mathrm{d}\tilde{\mu}_{\mathrm{e}}(\mathrm{m}')]/F = -\mathrm{d}E_{\mathrm{p}} + \mathrm{d}E(\mathrm{m}')$ and also that $\mathrm{d}E(\mathrm{m}') = -\mathrm{d}\mu_{\mathrm{A}^-}(\mathrm{l})/F$. The dependence of the interfacial tension on the electrode potential is then given by the *Gibbs–Lippmann equation,*

$$\mathrm{d}\gamma = -\sigma(\mathrm{m})\,\mathrm{d}E_{\mathrm{p}} - \sum_{i=1}^{s}\Gamma_i\,\mathrm{d}\mu_i(\mathrm{l}) \tag{4.2.23}$$

The dependence of the interfacial tension on the potential is termed the *electrocapillary curve.* It is convex to the axis of potential and often reminiscent of a parabola (see Fig. 4.2).

The first differential of this curve yields the *surface charge density* $\sigma(\mathrm{m}) = -\sigma(\mathrm{l})$:

$$-\left(\frac{\partial\gamma}{\partial E}\right)_{p,T,\mu_i} = \sigma(\mathrm{m}) \tag{4.2.24}$$

The second differential of the electrocapillary curve yields the *differential capacity* of the electrode, C:

$$-\left(\frac{\partial^2\gamma}{\partial E_{\mathrm{p}}^2}\right)_{p,T,\mu_i} = \left(\frac{\partial\sigma(\mathrm{m})}{\partial E_{\mathrm{p}}}\right)_{p,T,\mu_i} = C \tag{4.2.25}$$

This quantity is, in general, a function of the electrode potential.

An important point on the electrocapillary curve is its maximum. It follows from Eq. (4.2.24) that $\sigma(\mathrm{m}) = \sigma(\mathrm{l}) = 0$ at the potential of the electrocapillary maximum.

This potential is termed the *zero-charge potential* and is denoted as E_{pzc}. In earlier usage, this potential was also called the potential of the electrocapillary zero; this designation is not suitable, as E_{pzc} is connected with the zero charge $\sigma(\mathrm{m})$ rather than the zero potential.

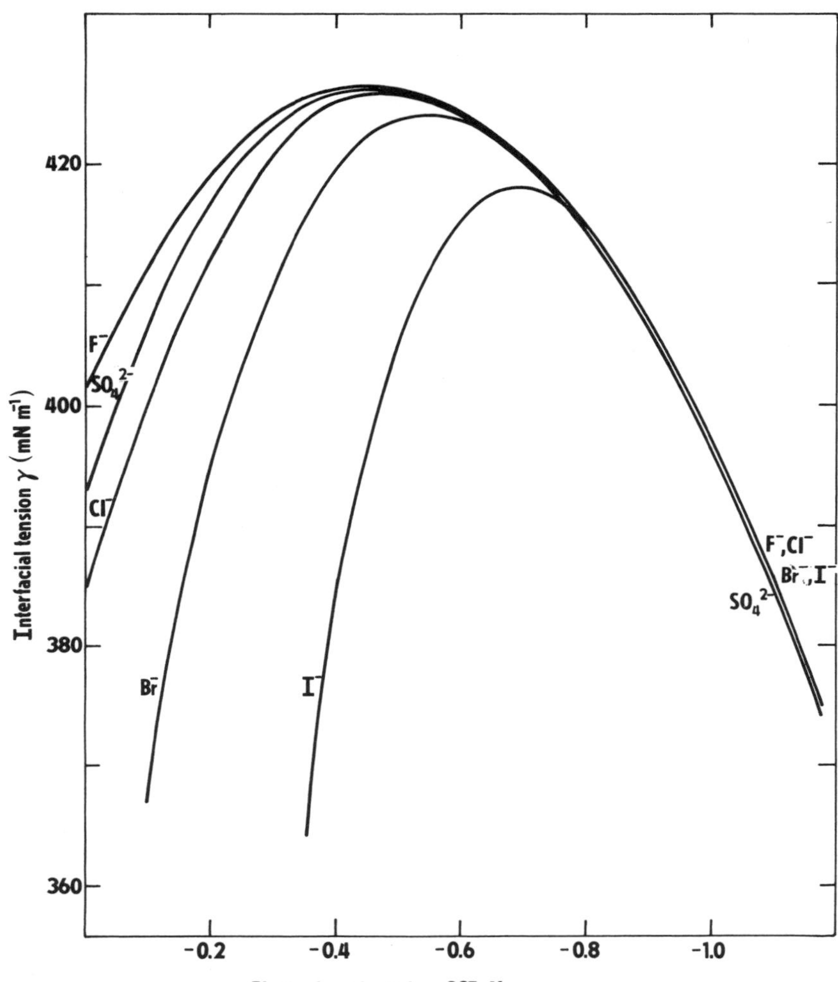

Fig. 4.2 Electrocapillary curves of 0.1 M aqueous solutions of KF, KCl, KBr, KI and K_2SO_4 obtained by means of the drop-time method (page 233). The slight deviation of the right-hand branch of the SO_4^{2-} curve is caused by a higher charge number of sulphate. (By courtesy of L. Novotný)

Further, the concept of the integral capacity K will be introduced:

$$K = \frac{\sigma(\mathrm{m})}{E_\mathrm{p} - E_\mathrm{pzc}} = \frac{1}{E_\mathrm{p} - E_\mathrm{pzc}} \int_{E_\mathrm{pzc}}^{E_\mathrm{p}} C\,\mathrm{d}E \tag{4.2.26}$$

The differential capacity is given by the slope of the tangent to the curve of the dependence of the electrode charge on the potential, while the integral capacity at a certain point on this dependence is given by the slope of the radius vector of this point drawn from the point $E_\mathrm{p} = E_\mathrm{pzc}$.

The zero-charge potential is determined by a number of methods (see Section 4.4). A general procedure is the determination of the differential capacity minimum which, at low electrolyte concentration, coincides with E_{pzc} (Section 4.3.1). With liquid metals (Hg, Ga, amalgams, metals in melts) E_{pzc} is directly found from the electrocapillary curve.

As a material function, the zero-charge potential of various metals (and even of different crystallographic faces of the same metal) extends over a wide range of values (Table 4.1). How it is related to other material functions like the electron work function Φ_e has not yet been elucidated. However, S. Trasatti found two empirical equations, one for *sp* metals with the exception of Ga and Zn (i.e. for Sb, Hg, Sn, Bi, In, Pb, Cd, Ti), $E_{pzc}(H_2O) = \Phi_e/F - 4.69$ V versus SHE, and the other one for the transition metals (Ti, Ta, Nb, Co, Ni, Fe, Pd), $E_{pzc} = \Phi_e/F - 5.01$ V versus SHE.

For solutions of simple electrolytes, the surface excess of ions can be determined by measuring the interfacial tension. Consider the valence-symmetrical electrolyte BA ($z_+ = -z_- = z$). The Gibbs–Lippmann equation then has the form

$$-\mathrm{d}\gamma = \sigma(\mathrm{m})\,\mathrm{d}E + \Gamma_{\mathrm{A}^-}\,\mathrm{d}\mu_{\mathrm{A}^-}(\mathrm{l}) + \Gamma_{\mathrm{B}^+}\,\mathrm{d}\mu_{\mathrm{B}^+}(\mathrm{l}) \tag{4.2.27}$$

while

$$\sigma(\mathrm{l}) = zF(\Gamma_{\mathrm{B}^+} - \Gamma_{\mathrm{A}^-}) = \sigma(\mathrm{m}) \tag{4.2.28}$$

Substitution of this equation into Eq. (4.2.27) and incorporation of the relationship $\mathrm{d}\mu_{\mathrm{B}^+}(\mathrm{l}) + \mathrm{d}\mu_{\mathrm{A}^-}(\mathrm{l}) = RT\,\mathrm{d}\ln(a_{\mathrm{B}^+}a_{\mathrm{A}^-}) = RT\,\mathrm{d}\ln a_\pm^2$ yields

$$\begin{aligned}-\mathrm{d}\gamma &= \sigma(\mathrm{m})\left[\mathrm{d}E_\mathrm{p} + \frac{1}{F}\mathrm{d}\mu_{\mathrm{A}^-}(\mathrm{l})\right] + \Gamma_{\mathrm{B}^+}[\mathrm{d}\mu_{\mathrm{B}^+}(\mathrm{l}) + \mathrm{d}\mu_{\mathrm{A}^-}(\mathrm{l})]\\ &= \sigma(\mathrm{m})\,\mathrm{d}E_\mathrm{A} + RT\Gamma_{\mathrm{B}^+}\,\mathrm{d}\ln a_\pm^2\end{aligned} \tag{4.2.29}$$

where E_A is the potential of an ideal polarized electrode related to the potential of a reference electrode reversible to anions A^-, $E_\mathrm{A} = E_\mathrm{p} - E^0_{\mathrm{MA/A}} + (RT/F)\ln a_{\mathrm{A}^-}$. For the surface excess Γ_{B^+} we have

$$\Gamma_{\mathrm{B}^+} = -\frac{1}{RT}\left(\frac{\partial\gamma}{\ln a_\pm^2}\right)_{E_\mathrm{A}} \tag{4.2.30}$$

In this manner, the surface excess of ions can be found from the experimental values of the interfacial tension determined for a number of electrolyte concentrations. These measurements require high precision and are often experimentally difficult. Thus, it is preferable to determine the surface excess from the dependence of the differential capacity on the concentration. By differentiating Eq. (4.2.30) with respect to E_A and using Eqs (4.2.24) and (4.2.25) in turn we obtain the Gibbs–Lippmann equation

$$\begin{aligned}\frac{1}{RT}\frac{\partial\sigma(\mathrm{m})}{\partial\ln a_\pm^2} &= \frac{\partial\Gamma_{\mathrm{B}^+}}{\partial E_\mathrm{A}},\\ \frac{1}{RT}\frac{\partial^2\sigma(\mathrm{m})}{\partial E_\mathrm{A}\,\partial\ln a_\pm^2} = \frac{1}{RT}\frac{\partial C}{\partial\ln a_\pm^2} &= \frac{\partial^2\Gamma_{\mathrm{B}^+}}{\partial E_\mathrm{A}^2}\end{aligned} \tag{4.2.31}$$

Table 4.1 Zero-charge potentials (vs. SCE). (According to A. N. Frumkin)

Electrode	E_{pzc} (V)	Electrolyte	Method of measurement
PbO_2	1.60	5 mM H_2SO_4	Differential capacity minimum
Au (110)	−0.05	5 mM NaF	Differential capacity minimum
Pt (H)	−0.12	0.025 mM H_2SO_4	Electrokinetic potential
C (act.)	−0.17	0.5 M Na_2SO_4 + 5 mM H_2SO_4	Ion adsorption
Cu	−0.19	0.1 M NaOH	Contact (wetting) angle
Sb	−0.39	2 mM $KClO_4$	Differential capacity minimum
Hg	−0.435	All NaF	Electrocapillarity Differential capacity minimum
Fe	−0.61	5 mM H_2SO_4	Differential capacity minimum
Sn	−0.62	1 mM K_2SO_4	Differential capacity minimum
Bi (111)	−0.66	10 mM KF	Differential capacity minimum
Ag (111)	−0.695	5 mM KPF_6	Differential capacity minimum
Ag (100)	−0.870	5 mM KPF_6	Differential capacity minimum
Ag (110)	−0.975	5 mM KPF_6	Differential capacity minimum
Pb	−0.80	1 mM NaF	Differential capacity minimum
Ga	−0.85	1 M $NaClO_4$ + 0.1 M $HClO_4$	Electrocapillarity
In	−0.89	3 mM NaF	Differential capacity minimum
Tl	−0.95	1 mM NaF	Differential capacity minimum
Cd	−0.99	1 mM NaF	Differential capacity minimum

Data for different Ag faces were taken from A. Hamelin, L. Stoicoviciu, L. Doubova, and S. Trasatti, *Surface Science Letters*, **201,** L498 (1988).

Double integration with respect to E_A yields the surface excess Γ_{B^+}; however, the calculation requires that the value of this excess be known, along with the value of the first differential $\partial\Gamma_{B^+}/\partial E_A$ for a definite potential. This value can be found, for example, by measuring the interfacial tension, especially at the potential of the electrocapillary maximum. The surface excess is often found for solutions of the alkali metals on the basis of the assumption that, at potentials sufficiently more negative than the zero-charge potential, the electrode double layer has a diffuse character without specific adsorption of any component of the electrolyte. The theory of diffuse electrical double layer is then used to determine Γ_{B^+} and $\partial\Gamma_{B^+}/\partial E_A$ (see Section 4.3.1).

In practical measurements, the differential capacity values are determined with respect to a reference electrode connected to the studied electrolyte through a salt bridge. The measured data are then recalculated for an anionic reference electrode by adding the value $RT/F \ln a_\pm^2$ to the E_p value. Figure 4.3 gives an example of the measured values of the surface excess for an electrolyte that is not adsorbed (KF) and that is adsorbed (KCl). It can be seen that, at potentials $E_p < -0.8$ V (vs. NCE), the curves for the surface excesses for both electrolytes merge, indicating that the two salts behave identically in this region of electrode potentials (because chloride ions are no longer adsorbed, the double layer is diffuse).

The quantity $\partial\gamma/\partial \ln a_\pm^2$ at the potential of the electrocapillary maximum is of basic importance. As the surface charge of the electrode is here equal to zero, the electrostatic effect of the electrode on the ions ceases. Thus, if no specific ion adsorption occurs, this differential quotient is equal to zero and no surface excess of ions is formed at the electrode. This is especially true for ions of the alkali metals and alkaline earths and, of the anions, fluoride at low concentrations and hydroxide. Sulphate, nitrate and perchlorate ions are very weakly surface active. The remaining ions decrease the surface tension at the maximum on the electrocapillary curve to a greater or lesser degree.

In the case of a specific ('superequivalent') adsorption we have

$$-\frac{1}{RT}\left(\frac{\partial\gamma}{\partial \ln a_\pm^2}\right)_{E_p=E_{pzc}} = \Gamma_{A^-} + \Gamma_{B^+} = \Gamma_{salt} \qquad (4.2.32)$$

where Γ_{salt} is the surface excess for both components of the electrolyte. One of them may be specifically adsorbed and the second compensates the corresponding excess charge by its excess charge of opposite sign in the diffuse layer.

References

Frumkin, A. N., *Potentials of Zero Charge* (in Russian), Nauka, Moscow, 1979.

Frumkin, A. N., O. A. Petrii, and B. B. Damaskin, Potential of zero charge, *CTE*, **1**, 221 (1980).

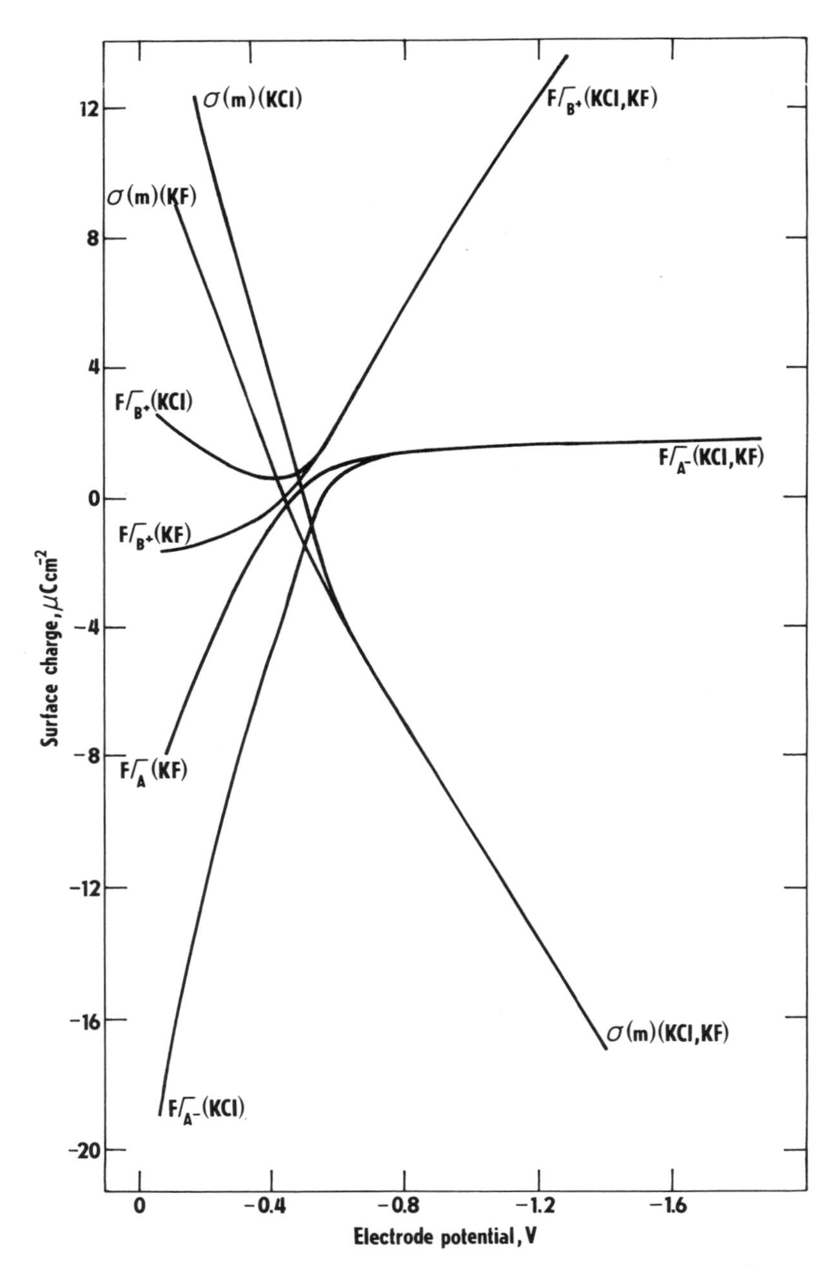

Fig. 4.3 Surface charge of the mercury electrode $\sigma(\mathrm{m})$ ($\mu\mathrm{C}\cdot\mathrm{cm}^{-2}$) and surface charges due to cations ($\mathbf{F}\Gamma_{\mathrm{B}^+}$) and anions ($F\Gamma_{\mathrm{A}^-}$) in 0.1 M KCl and 0.1 M KF at 25°C. (According to D. C. Grahame)

Parsons, R., Thermodynamic methods for the study of interfacial regions in electrochemical systems, *CTE*, **1,** 1 (1980).

4.3 Structure of the Electrical Double Layer

The thermodynamic theory of electrocapillarity considered above is simultaneously the thermodynamic theory of the electrical double layer and yields, in its framework, quantitative data on the double layer. However, further clarification of the properties of the double layer must be based on a consideration of its structure.

At present it is impossible to formulate an exact theory of the structure of the electrical double layer, even in the simple case where no specific adsorption occurs. This is partly because of the lack of experimental data (e.g. on the permittivity in electric fields of up to $10^9\,\mathrm{V \cdot m^{-1}}$) and partly because even the largest computers are incapable of carrying out such a task. The analysis of a system where an electrically charged metal in which the positions of the ions in the lattice are known (the situation is more complicated with liquid metals) is in contact with an electrolyte solution should include the effect of the electrical field on the permittivity of the solvent, its structure and electrolyte ion concentrations in the vicinity of the interface, and, at the same time, the effect of varying ion concentrations on the structure and the permittivity of the solvent. Because of the unsolved difficulties in the solution of this problem, simplifying models must be employed: the electrical double layer is divided into three regions that interact only electrostatically, i.e. the electrode itself, the compact layer and the diffuse layer.

On the basis of this model, the overall differential capacity C for a system without specific adsorption, i.e. if the compact layer does not contain ions, is divided into two capacities in series, one corresponding to the compact layer C_c and the other to the diffuse layer C_d:

$$\frac{1}{C} = \frac{1}{C_c} + \frac{1}{C_d} \tag{4.3.1}$$

It is assumed that the quantity C_c is not a function of the electrolyte concentration c, and changes only with the charge σ, while C_d depends both on σ and on c, according to the diffuse layer theory (see below). The validity of this relationship is a necessary condition for the case where the adsorption of ions in the double layer is purely electrostatic in nature. Experiments have demonstrated that the concept of the electrical double layer without specific adsorption is applicable to a very limited number of systems. Specific adsorption apparently does not occur in LiF, NaF and KF solutions (except at high concentrations, where anomalous phenomena occur). At potentials that are appropriately more negative than E_{pzc}, where adsorption of anions is absent, no specific adsorption occurs for the salts of

Li^+, Na^+ and K^+ and the ions of some alkaline earths (cf. Fig. 4.3).

4.3.1 *Diffuse electrical layer*

The *diffuse layer* is formed, as mentioned above, through the interaction of the electrostatic field produced by the charge of the electrode, or, for specific adsorption, by the charge of the ions in the compact layer. In rigorous formulation of the problem, the theory of the diffuse layer should consider:

(a) The individual ion dimensions
(b) The effect of the electric field in the diffuse layer on the dielectric properties of the solvent
(c) The effect of the ions on the dielectric properties of the solvent

A rigorous solution of this problem was attempted, for example, in the 'hard sphere approximation' by D. Henderson, L. Blum, and others. Here the discussion will be limited to the classical Gouy–Chapman theory, describing conditions between the bulk of the solution and the outer Helmholtz plane and considering the ions as point charges and the solvent as a structureless dielectric of permittivity ε. The inner electrical potential $\phi(1)$ of the bulk of the solution will be taken as zero and the potential in the outer Helmholtz plane will be denoted as ϕ_2. The space charge in the diffuse layer is given by the Poisson equation

$$\operatorname{div} \mathbf{D} = -\rho \tag{4.3.2}$$

where $\mathbf{D}$ is the electrical displacement given by the product of the permittivity of the solution and of the electric field grad ϕ. If ε is not a function of the coordinates and a linear problem is involved (i.e. if in the diffuse layer the potential ϕ, charge density ρ and ion concentrations c_i are functions only of the coordinate x perpendicular to the outer Helmholtz plane), then Eq. (4.3.2) becomes

$$\operatorname{div} \operatorname{grad} \phi = \frac{\mathrm{d}^2\phi}{\mathrm{d}x^2} = -\frac{\rho}{\varepsilon} \tag{4.3.3}$$

The space charge density is given by the sum of the charges of all the ions present:

$$\rho = \sum_{i=1}^{s} (c_i)_\phi z_i F \tag{4.3.4}$$

where $(c_i)_\phi$ denotes concentration at potential ϕ. At equilibrium, the electrochemical potential of any ion in any point of the solution is constant, $(\tilde{\mu}_i)_\phi = (\tilde{\mu}_i)_{\phi=0}$. If, to a first approximation, the activities are replaced by concentrations, then

$$\mu_i^0 + RT \ln (c_i)_{\phi=0} = \mu_i^0 + RT \ln (c_i)_\phi + z_i F\phi$$

$$(c_i)_\phi = (c_i)_{\phi=0} \exp\left(-\frac{z_i F\phi}{RT}\right) \tag{4.3.5}$$

Substitution for $(c_i)_\phi$ from Eq. (4.3.5) into Eq. (4.3.4) and then for ρ in Eq. (4.3.3) yields the Poisson–Boltzmann differential equation

$$\frac{d^2\phi}{dx^2} = -\frac{1}{\varepsilon}\sum_{i=1}^{s}(c_i)_{\phi=0} z_i F \exp\left(-\frac{z_i F\phi}{RT}\right) \tag{4.3.6}$$

with the boundary conditions $d\phi/dx = 0$, $\phi = 0$ for $x \rightarrow \infty$. This differential equation can be readily solved after multiplying both sides by $d\phi/dx$. Integration from x to infinity then gives (the concentration in the bulk of the solution being denoted as c_i rather than $(c_i)_{\phi=0}$)

$$\frac{d\phi}{dx} = \sqrt{\frac{2RT}{\varepsilon}\sum_{i=1}^{s} c_i\left[\exp\left(-\frac{z_i F\phi}{RT}\right) - 1\right]}$$

while from the outer Helmholtz plane $x = x_2$ to infinity

$$\left(\frac{d\phi}{dx}\right)_{x=x_2} = \sqrt{\frac{2RT}{\varepsilon}\sum_{i=1}^{s} c_i\left[\exp\left(-\frac{z_i F\phi_2}{RT}\right) - 1\right]} \tag{4.3.7}$$

The Gauss theorem

$$\left(\frac{d\phi}{dx}\right)_{x=x_2} = \frac{\sigma_d}{\varepsilon} \tag{4.3.8}$$

allows calculation of the charge σ_d in a column with unit cross-section in the diffuse layer:

$$\sigma_d = \int_{x=x_2}^{\infty} \rho\, dx = \sqrt{2\varepsilon RT\sum_{i=1}^{s} c_i\left[\exp\left(-\frac{z_i F\phi_2}{RT}\right) - 1\right]} \tag{4.3.9}$$

If this equation is employed for a solution of a single valence-symmetrical electrolyte ($z_+ = z_- = z$), then

$$\begin{aligned}\frac{d\phi}{dx} &= \sqrt{\frac{2RTc}{\varepsilon}\left[\exp\left(-\frac{zF\phi}{RT}\right) - 2 + \exp\left(\frac{zF\phi}{RT}\right)\right]}\\ &= -\sqrt{\frac{2RTc}{\varepsilon}}\left[\exp\left(\frac{zF\phi}{2RT}\right) - \exp\left(-\frac{zF\phi}{2RT}\right)\right]\\ &= -\sqrt{\frac{8RTc}{\varepsilon}}\sinh\left(\frac{zF\phi}{2RT}\right)\end{aligned} \tag{4.3.10}$$

$$\sigma_d = -\sqrt{8\varepsilon RTc}\sinh\left(\frac{zF\phi_2}{2RT}\right) \tag{4.3.11}$$

Thus, for water and a temperature of 25°C, after substitution for the constants, when concentrations are expressed in moles per cubic decimetre, it follows that $\sigma_d = 11.72c^{1/2}\sinh(19.46z\phi_2)$, $\mu C \cdot cm^{-2}$. The quantity σ_d expresses the total charge per square centimetre in the diffuse layer. If no specific adsorption occurs, i.e. if no ions are present within the compact part of the double layer, then $\sigma_d = \sigma(l) = -\sigma(m)$.

For small $\phi_2 \ll RT/F$, in Eq. (4.3.11) all the terms in the infinite series for the hyperbolic sine except the first can be neglected, so that

$$-\sigma_d = zF\sqrt{\frac{2\varepsilon c}{RT}}\,\phi_2 = \frac{\varepsilon}{L_D}\,\phi_2 \tag{4.3.12}$$

where $L_D = (\varepsilon RT/2c)^{1/2}/2F$ is the Debye length (see Eq. 1.3.17). Under these conditions, the diffuse layer acts as a plate capacitor with charge σ_d and thickness L_D.

The differential capacity of the diffuse layer is defined by the relationship $C_d = -d\sigma_d/d\phi_2$. According to this definition we obtain, from Eq. (4.3.11),

$$C_d = -\frac{d\sigma_d}{d\phi_2} = zF\sqrt{\frac{2\varepsilon c}{RT}}\cosh\left(\frac{zF\phi_2}{2RT}\right) \tag{4.3.13}$$

The total differential capacity can be either measured directly (see Section 4.4) or calculated from Eq. (4.2.25). Because $d[\phi(m) - \phi(l)] = dE_p$, the differential capacity of the diffuse layer C_d can be calculated from Eq. (4.3.13), with substitution of the potential ϕ_2 from Eq. (4.3.11). As for zero specific adsorption $q_d = -q(m)$, the potential ϕ_2 is given by the relationship

$$\phi_2 = \frac{2RT}{zF}\sinh^{-1}\left[\frac{\sigma(m)}{(8\varepsilon RTc)^{1/2}}\right]$$

$$= \frac{RT}{zF}\ln\left[\frac{\sigma(m)}{(8\varepsilon RTc)^{1/2}} + \sqrt{\frac{\sigma(m)^2}{(8\varepsilon RTc)} + 1}\right] \tag{4.3.14}$$

Figure 4.4 depicts the dependence of ϕ_2 on $E - E_{pzc}$ (the 'rational potential') for various electrolyte concentrations.

The charge density on the electrode $\sigma(m)$ is mostly found from Eq. (4.2.24) or (4.2.26) or measured directly (see Section 4.4). The differential capacity of the compact layer C_c can be calculated from Eq. (4.3.1) for known values of C and C_d. It follows from experiments that the quantity C_c for surface inactive electrolytes is a function of the potential applied to the electrode, but is not a function of the concentration of the electrolyte. Thus, if the value of C_c is known for a single concentration, it can be used to calculate the total differential capacity C at an arbitrary concentration of the surface-inactive electrolyte and the calculated values can be compared with experiment. This comparison is a test of the validity of the diffuse layer theory. Figure 4.5 provides examples of theoretical and experimental capacity curves for the non-adsorbing electrolyte NaF. Even at a concentration of $0.916\,\text{mol}\cdot\text{dm}^{-3}$, the C_d value is not sufficient to permit us to set $C \approx C_c$.

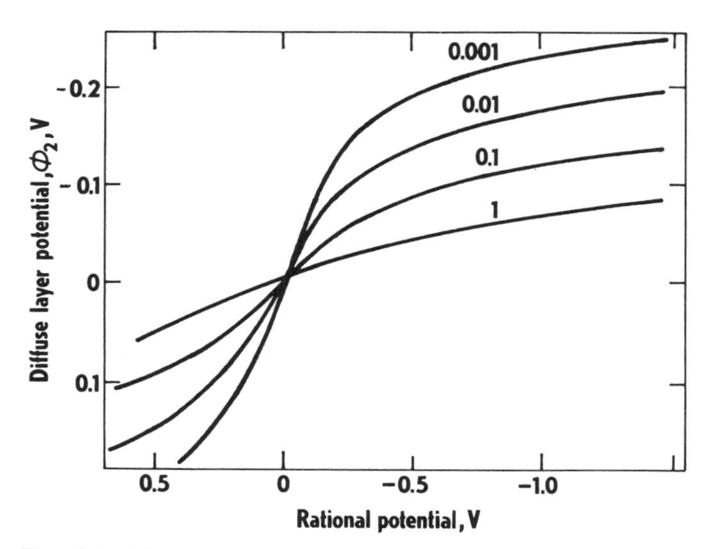

Fig. 4.4 The dependence of the potential difference in the diffuse layer on the difference $E - E_{pzc}$ (the rational potential) for various concentrations of the surface inactive electrolyte KF. (According to R. Parsons)

According to Eq. (4.3.13) the differential capacity of the diffuse layer C_d has a minimum at $\phi_2 = 0$, i.e. at $E = E_{pzc}$. It follows from Eq. (4.3.1) and Fig. 4.5 that the differential capacity of the diffuse layer C_d has a significant effect on the value of the total differential capacity C at low electrolyte concentrations. Under these conditions, a capacity minimum appears on the experimentally measured C–E curve at $E = E_{pzc}$. The value of E_{pzc} can thus be determined from the minimum of C at low electrolyte concentrations (millimolar or lower).

This theory of the diffuse layer is satisfactory up to a symmetrical electrolyte concentration of $0.1\,\text{mol}\cdot\text{dm}^{-3}$, as the Poisson–Boltzmann equation is valid only for dilute solutions. Similarly to the theory of strong electrolytes, the Gouy–Chapman theory of the diffuse layer is more readily applicable to symmetrical rather than unsymmetrical electrolytes.

4.3.2 *Compact electrical layer*

The structure of the *compact layer* depends on whether specific adsorption occurs (ions are present in the compact layer) or not (ions are absent from the compact layer). In the absence of specific adsorption, the surface of the electrode is covered by a monomolecular solvent layer. The solvent molecules are oriented and their dipoles are distorted at higher field strengths. The permittivity of the solvent in this region is only an operational quantity, with a value of about 12 at the E_{pzc} in water,

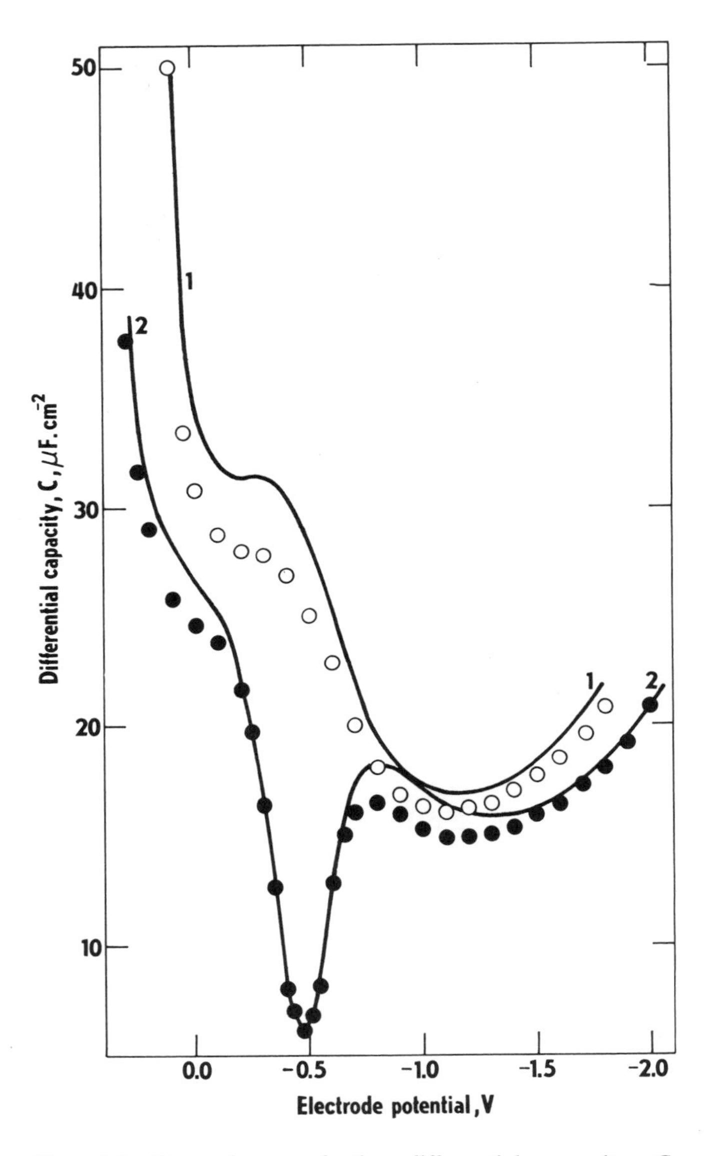

Fig. 4.5 Dependence of the differential capacity C ($\mu F \cdot cm^{-2}$) of a mercury electrode in 0.916 M NaF (○) and in 1 mM NaF (●) on the electrode potential (vs. NCE). The curves correspond to calculations according to Eqs (4.3.1) and (4.3.13); curve 1 as C_c, curve 2 as $C = (1/C_c + 1/C_d)^{-1}$. (According to D. C. Grahame)

decreasing to as little as 6 at potentials more distant from E_{pzc} (for the mercury–water interface; these values vary with the electrode metal).

It can be seen from Fig. 4.5 that a maximum (or 'hump') appears at potentials somewhat more positive than the potential of the electrocapillary maximum on the differential capacity–potential curve. This capacity maximum can be explained by postulating that, in the corresponding potential range, the orientation of the water molecules is minimal, i.e. the permittivity is maximal; at more positive potentials, the water dipoles are oriented so that their negative ends (i.e. the oxygen atoms) are directed towards the electrode surface while at more negative potentials this orientation is reversed. The disagreement between the potential of the 'hump' and the potential of the electrocapillary maximum is apparently a result of weak chemical interaction between the water molecules and the surface of the electrode. It should be noted that the dependence of the differential capacity for some mercury–solvent interfaces is simpler than that for water (e.g. for methanol or formic acid) and for others is more complicated (propylene carbonate, dimethylsulphoxide, sulpholane, etc.). In water, the hump disappears at higher temperatures.

The picture of the compact double layer is further complicated by the fact that the assumption that the electrons in the metal are present in a constant concentration which discontinuously decreases to zero at the interface in the direction towards the solution is too gross a simplification. Indeed, Kornyshev, Schmickler, and Vorotyntsev have pointed out that it is necessary to assume that the electron distribution in the metal and its surroundings can be represented by what is called a 'jellium': the positive metal ions represent a fixed layer of positive charges, while the electron plasma spills over the interface into the compact layer, giving rise to a surface dipole. This surface dipole, together with the dipoles of the solvent molecules, produces the total capacity value of the compact double layer.

Specific adsorption occurs, i.e. ions enter the compact layer, in a considerable majority of cases. The most obvious result of specific adsorption is a decrease and shift in the maximum of the electrocapillary curve to negative values because of adsorption of anions (see Fig. 4.2) and to positive values for the adsorption of cations. A layer of ions is formed at the interface only when specific adsorption occurs.

By means of the thermodynamic theory of the double layer and the theory of the diffuse layer it is possible to determine the charge density σ_1 corresponding to the adsorbed ions, i.e. ions in the inner Helmholtz plane, and the potential of the outer Helmholtz plane ϕ_2 in the presence of specific adsorption.

The charge density of specifically adsorbed ions (assuming that anions are adsorbed specifically while cations are present only in the diffuse layer) for a valence-symmetrical electrolyte ($z_+ = -z_- = z$) is

$$\sigma_1 = -zF(\Gamma_{A^-} - \Gamma_{A^-,d}) = -zF\Gamma_{A^-,1} \tag{4.3.15}$$

where $\Gamma_{A^-,d}$ is the surface anion concentration in the diffuse layer. Depending on its sign, the charge of the specifically adsorbed anions either increases or decreases the charge of the metal part of the double layer, $\sigma(m)$. The resultant charge, given by the sum $\sigma_1 + \sigma(m)$, affects the double layer electrostatically. As

$$\sigma_1 + \sigma(m) = -\sigma_d \tag{4.3.16}$$

Eq. (4.3.11) assumes the form

$$\sigma_1 + \sigma(m) = \sqrt{8\varepsilon RTc}\,\sinh\left(\frac{zF\phi_2}{2RT}\right) \tag{4.3.17}$$

and Eq. (4.3.14) becomes

$$\phi_2 = \frac{2RT}{zF}\sinh^{-1}\left[\frac{\sigma_1 + \sigma(m)}{(8\varepsilon RTc)^{1/2}}\right] \tag{4.3.18}$$

The adsorbed ions (anions are considered) are located at the inner Helmholtz plane. In describing the distribution of the electrical potential in the compact layer, the discrete structure of the charge in this layer must be considered, rather than continuous charge distribution, as was pointed out by O. A. Essin, B. V. Ershler, and D. C. Grahame. The inner Helmholtz plane is no longer an equipotential surface. As demonstrated by A. N. Frumkin, the potential at the centre of the adsorbed ion plays the greatest role in the determination of the adsorption energy of ions. Ershler called this the *micropotential.* More exactly, it is the potential at the point where the given ion is adsorbed after subtracting the value of the coulombic potential formed by this ion.

If the solution is considered to be a good electric conductor, then it is useful to introduce the concept of electric image induced in an arbitrary conductor when a point charge is located close to its surface. The electric image has the opposite charge with the same absolute value as the point charge and is situated inside the conductor at the same distance from the surface as the original charge on the other side of the interface. The discrete negative charges induce a positive image charge in the solution, which is reflected on the outer Helmholtz plane (Fig. 4.6). The actual electric cloud of ions behind the outer Helmholtz plane is compressed so that it can be replaced by this electric image.

The introduction of the concept of the micropotential permits derivation of various expressions for the potential difference produced by the adsorbed anions, i.e. for the potential difference between the electrode and the solution during specific adsorption of ions. It has been found that, with small coverage of the surface by adsorbed species, the micropotential depends almost linearly on the distance from the surface. The distance between the inner and outer Helmholtz planes is denoted as x_{1-2} and the distance between the surface of the metal and the outer Helmholtz plane as x_2. The micropotential, i.e. the potential difference between the inner and

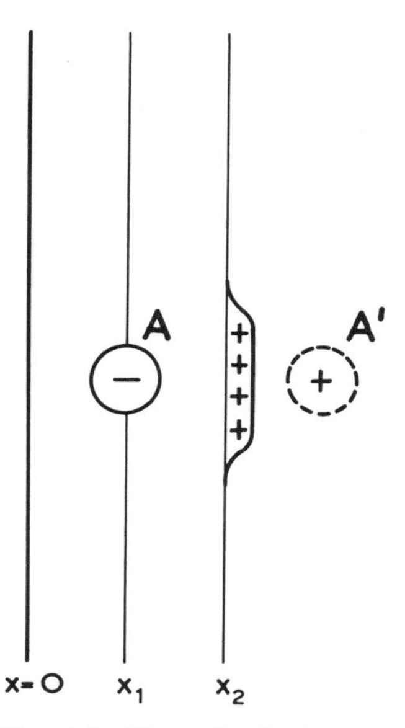

Fig. 4.6 The adsorbed anion A induces a positive charge excess at the outer Helmholtz plane, which can be represented as its image A′. The plane of imaging is identical with the outer Helmholtz plane. (According to B. B. Damaskin, O. A. Petrii and V. V. Batrakov)

outer Helmholtz planes, $\phi_1 - \phi_2$, for the simple case of the electrocapillary maximum, where there is no effect of the electrode charge, is given approximately by the equation

$$\phi_1 - \phi_2 = \frac{x_{1-2}}{x_2}[\phi(\mathrm{m}) - \phi_2] \tag{4.3.19}$$

In view of the assumed linear dependence of the electrical potential in the whole compact layer, the above authors derived the following expression for $\phi(\mathrm{m}) - \phi_2$ when $\sigma(\mathrm{m}) = 0$:

$$\phi(\mathrm{m}) - \phi_2 = \frac{\sigma_1 x_{1-2}}{\varepsilon_\mathrm{c}} \tag{4.3.20}$$

where ε_c is the permittivity in the compact layer, differing from the permittivity in the bulk of the solution. For $\phi_1 - \phi_2$ and for ϕ_1 alone the substitution from Eq. (4.3.20) into Eq. (4.3.19) gives

$$\begin{aligned}\phi_1 - \phi_2 &= \frac{\sigma_1}{\varepsilon_c}\frac{x_{1-2}^2}{x_2} \\ \phi_1 &= \frac{\sigma_1}{\varepsilon_c}\frac{x_{1-2}^2}{x_2} + \phi_2\end{aligned} \tag{4.3.21}$$

The adsorption of ions is determined by the potential of the inner Helmholtz plane ϕ_1, while the shift of E_{pzc} to more negative values with increasing concentration of adsorbed anions is identical with the shift in $\phi(m)$. Thus, the electrocapillary maximum is shifted to more negative values on an increase in the anion concentration more rapidly than would follow from earlier theories based on concepts of a continuously distributed charge of adsorbed anions over the electrode surface (Stern, 1925). Under Stern's assumption, it would hold that $\phi(m) = \phi_1$ (where, of course, ϕ_1 no longer has the significance of the potential at the inner Helmholtz plane).

According to Ershler, ϕ_1 must appear in the expression for the electrochemical potential of ions adsorbed on the inner Helmholtz plane. If their electrochemical potential is expressed by the equation

$$\tilde{\mu}_{i,1} = \mu_i^0 + RT \ln \Gamma_{i,1} + \Delta G_{ads}^0 + z_i F \phi_1 \tag{4.3.22}$$

which is equal to the electrochemical potential for the same kind of ions in the bulk of the solution (the activity of the ions in solution is a_i and the potential $\phi(l)$ is again set equal to zero), then the equation for the adsorption isotherm is obtained:

$$\Gamma_{i,1} = a_i \exp\left(\frac{-\Delta G_{ads}^0}{RT}\right) \exp\left(-\frac{z_i F \phi_1}{RT}\right) \tag{4.3.23}$$

Here ΔG_{ads}^0 is in the standard Gibbs energy change due to specific adsorption. To a first approximation, this quantity can be considered independent of the activity of the adsorbed ions; in general, ΔG_{ads}^0 is a function of the electrode charge and, at a given electrode charge ($\sigma(m) = 0$ in the present case), is constant. The surface charge density σ_1 for adsorbed anions with charge number $z_- = -1$ is given by the expression

$$\sigma_1 = -F\Gamma_{-,1} = Ka_- \exp\left(-\frac{F\phi_1}{RT}\right) \tag{4.3.24}$$

Taking logarithms of Eq. (4.3.24) and differentiating with respect to $\ln a_-$ yields

$$\left(\frac{\partial \ln |\sigma_1|}{\partial \ln a_-}\right)_{\sigma(m)} = 1 + \frac{F}{RT}\left(\frac{\partial \phi_1}{\partial \ln a_-}\right)_{\sigma(m)} \tag{4.3.25}$$

$$\left(\frac{\partial \phi_1}{\partial \ln a_-}\right)_{\sigma(m)} = -\frac{RT}{F}\left[1 - \left(\frac{\partial \ln |\sigma_1|}{\partial \ln a_-}\right)_{\sigma(m)}\right] \tag{4.3.26}$$

From Eq. (4.3.26) the potential ϕ_1 at the electrocapillary maximum can be calculated:

$$\phi_1 = \frac{RT}{F}\int\left[\left(\frac{\partial \ln|\sigma_1|}{\partial \ln a_-}\right)_{\sigma(\mathrm{m})} - 1\right] d\ln a_- \qquad (4.3.27)$$

The integration constant is found from the condition that $\phi_1 = \phi_2$ for the limiting value of the expression $\partial \ln|\sigma_1|/\partial \ln a_- \to 0$ and $\sigma_1 \to 0$ (see Eq. 4.3.21).

If, for simplicity, ϕ_2 is neglected in Eqs (4.3.19) to (4.3.21) (valid for higher electrolyte concentrations), then

$$\phi_1 \approx \frac{x_{1-2}}{x_2}\phi(\mathrm{m}) \quad \text{and} \quad \sigma_1 = \frac{\varepsilon_c}{x_{1-2}}\phi(\mathrm{m}) \qquad (4.3.28)$$

Substitution into Eq. (4.3.24) yields

$$\phi(\mathrm{m}) = -K'a_- \exp\left(\frac{Fx_{1-2}}{RTx_2}\phi(\mathrm{m})\right) \qquad (4.3.29)$$

where $K' = x_{1-2}K/\varepsilon_c$. Hence

$$\frac{\partial\phi(\mathrm{m})}{\partial \log a_-} = -\frac{(2.303RT/F)x_2/x_{1-2}}{1-(RT/F)(x_2/x_{1-2})1/\phi(\mathrm{m})} \qquad (4.3.30)$$

As mentioned above, the quantity $\phi(\mathrm{m})$ is identified with the shift in the potential of the electrocapillary maximum during adsorption of surface-active anions. For large values of this shift ($\phi(\mathrm{m}) \gg RT/F$)

$$\frac{\partial\phi(\mathrm{m})}{\partial \log a_-} \approx -\frac{2.303RT}{F}\frac{x_2}{x_{1-2}} < -\frac{2.303RT}{zF} \qquad (4.3.31)$$

Indeed, for adsorption of iodide, Essin and Markov found a shift in the electrocapillary maximum $\partial\phi/\partial \log a_-$ of approximately -100 mV, in agreement with the theory.

At potentials far removed from the potential of zero charge, the electrical properties of the compact layer are determined by both the charge of the adsorbed ions and the actual electrode charge. The simplest model for this system is one which assumes independent action of these two types of charge. The quantity $\phi(\mathrm{m}) - \phi_2$ can then be separated into two parts, $[\phi(\mathrm{m}) - \phi_2]_{\sigma(\mathrm{m})}$ and $[\phi(\mathrm{m}) - \phi_2]_{\sigma_1}$, each of which is a function of the corresponding charge alone:

$$\phi(\mathrm{m}) - \phi_2 = [\phi(\mathrm{m}) - \phi_2]_{\sigma(\mathrm{m})} + [\phi(\mathrm{m}) - \phi_2]_{\sigma_1} \qquad (4.3.32)$$

These two potential differences can both be expressed in terms of the corresponding charges and two integral capacities:

$$\phi(\mathrm{m}) - \phi_2 = \frac{\sigma(\mathrm{m})}{(K_c)_{\sigma(\mathrm{m})}} + \frac{\sigma_1}{(K_c)_{\sigma_1}} \qquad (4.3.33)$$

It follows from the measurement that the quantity $(K_c)_{\sigma(m)}$ is a function of the charge $\phi(m)$, but not of the concentration and the sort of the electrolyte. The quantity $(K_c)_{\sigma_1}$ changes with charges $\sigma(m)$ and σ_1 and has different values for different electrolytes ($70\ \mu F \cdot cm^{-2}$ in iodide solutions, $120\ \mu F \cdot cm^{-2}$ in chloride solutions).

Opinions differ on the nature of the metal-adsorbed anion bond for specific adsorption. In all probability, a covalent bond similar to that formed in salts of the given ion with the cation of the electrode metal is not formed. The behaviour of sulphide ions on an ideal polarized mercury electrode provides evidence for this conclusion. Sulphide ions are adsorbed far more strongly than halide ions. The electrocapillary quantities (interfacial tension, differential capacity) change discontinuously at the potential at which HgS is formed. Thus, the bond of specifically adsorbed sulphide to mercury is different in nature from that in the HgS salt. Some authors have suggested that specific adsorption is a result of partial charge transfer between the adsorbed ions and the electrode.

4.3.3 *Adsorption of electroneutral molecules*

Electroneutral substances that are less polar than the solvent and also those that exhibit a tendency to interact chemically with the electrode surface, e.g. substances containing sulphur (thiourea, etc.), are adsorbed on the electrode. During adsorption, solvent molecules in the compact layer are replaced by molecules of the adsorbed substance, called surface-active substance (surfactant).† The effect of adsorption on the individual electrocapillary terms can best be expressed in terms of the difference of these quantities for the original (base) electrolyte and for the same electrolyte in the presence of surfactants. Figure 4.7 schematically depicts this dependence for the interfacial tension, surface electrode charge and differential capacity and also the dependence of the surface excess on the potential. It can be seen that, at sufficiently positive or negative potentials, the surfactant is completely desorbed from the electrode. The strong electric field leads to replacement of the less polar particles of the surface-active substance by polar solvent molecules. The desorption potentials are characterized by sharp peaks on the differential capacity curves.

For molecules with small dipoles, the adsorption region is distributed symmetrically around the potential of the electrocapillary maximum. However, if chemisorption interaction occurs between one end of the dipole (e.g. sulphur in thiourea) and the electrode, the adsorption region is shifted to the negative or positive side of the electrocapillary maximum.

The basic quantity in the study of adsorption is the surface excess of the surface-active substance. In the formation of a monomolecular film of the

† Surfactants also include a number of ions with hydrophobic groups, such as tetraalkylammonium ions with long-chain alkyl groups or dodecylsulphate.

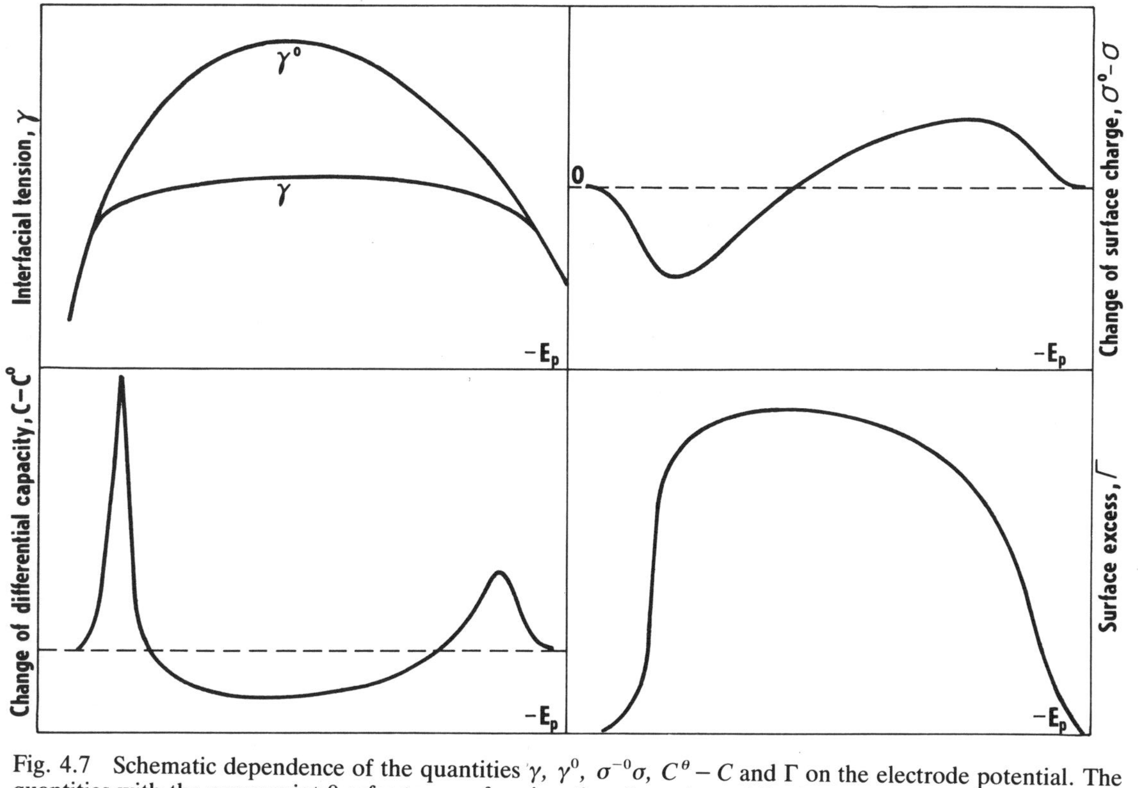

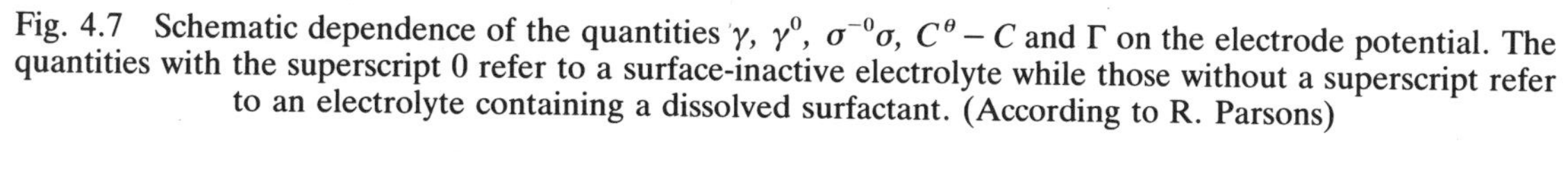
Fig. 4.7 Schematic dependence of the quantities γ, γ^0, $\sigma^{-0}\sigma$, $C^\theta - C$ and Γ on the electrode potential. The quantities with the superscript 0 refer to a surface-inactive electrolyte while those without a superscript refer to an electrolyte containing a dissolved surfactant. (According to R. Parsons)

adsorbed substance, the maximum value of the surface excess Γ_{max} is attained at complete coverage of the interface. The relative coverage Θ, an important term, is defined by the relationship

$$\Theta = \frac{\Gamma}{\Gamma_{max}} \tag{4.3.34}$$

The fact that the electrocapillary quantities are measured relative to their value in the base electrolyte can also be expressed in the formulation of the Gibbs–Lippmann equation. If quantities referred to the base electrolyte are primed and quantities referred to the studied surface-active substance are denoted by the subscript 1, then

$$\begin{aligned} \mathrm{d}(\gamma' - \gamma) = \mathrm{d}\pi = -[\sigma'(\mathrm{m}) - \sigma(\mathrm{m})]\,\mathrm{d}E_\mathrm{p} \\ - \sum_{i=2}^{s} (\Gamma_i' - \Gamma_i)\,\mathrm{d}\mu_i(\mathrm{l}) + \Gamma_1\,\mathrm{d}\mu_1(\mathrm{l}) \end{aligned} \tag{4.3.35}$$

where $\pi = \gamma' - \gamma$ is the surface pressure of the adsorbed surface-active substance. This is the force acting on a unit length in the direction of an increase of the interface area, i.e. against the surface tension. The Gibbs adsorption isotherm can be derived from Eq. (4.3.65):

$$\Gamma_1 = \left(\frac{\partial \pi}{\partial \mu_1}\right)_{T,p,E_\mathrm{p},\mu_i \neq \mu_1} \tag{4.3.36}$$

which acquires the following form for dilute solutions:

$$\Gamma_1 = \frac{c_1}{RT}\left(\frac{\partial \pi}{\partial c_1}\right)_{T,p,E_\mathrm{p},c_i \neq c_1} \tag{4.3.37}$$

Equation (4.3.37) can be used to determine the function $\Gamma_1 = \Gamma_1(c_1)$, which is the adsorption isotherm for the given surface-active substance. Substitution for c_1 in the Gibbs adsorption isotherm and integration of the differential equation obtained yields the equation of state for a monomolecular film $\Gamma_1 = \Gamma_1(\pi)$.

The simplest adsorption isotherm is that of Henry's law (linear adsorption isotherm),

$$\Gamma_1 = \beta c_1 \tag{4.3.38}$$

where β is the adsorption coefficient. Substitution into Eq. (4.3.37) and integration yields

$$\pi = RT\Gamma_1 \tag{4.3.39}$$

a two-dimensional analogy of the 'ideal gas law'.

The limited number of free sites for the adsorbed substance on the surface is considered by the Langmuir isotherm,

$$\Gamma_1 = \frac{\beta' c_1 \Gamma_{max}}{1 + \beta' c_1} \quad \text{or} \quad \beta' c_1 = \frac{\Theta}{1 - \Theta} \tag{4.3.40}$$

The basic assumption of the Langmuir adsorption isotherm is that the adsorbed molecules do not interact. This condition is not always fulfilled for adsorption, particularly on electrodes. The Frumkin adsorption isotherm includes interaction between molecules in the adsorption film,

$$\beta' c_1 = \frac{\theta}{1-\theta} \exp(-a\theta) \tag{4.3.41}$$

where the interaction coefficient a has a positive value for attraction between molecules and consequent favouring of adsorption, and a negative value for repulsion between molecules. It can be seen that the Langmuir adsorption isotherm is a special case of the Frumkin adsorption isotherm for $a = 0$, and for low surface coverage ($\theta \to 0$) both are simplified into the linear adsorption isotherm (see Fig. 4.8).

The adsorption coefficient β or β' is a function of the standard Gibbs adsorption energy alone when the molecules do not interact and the adsorption occurs without limitation, i.e. at very low electrode coverage. This dependence is expressed by the relationship

$$\beta = \exp\left(-\frac{\Delta G^0_{\text{ads}}}{RT}\right) \tag{4.3.42}$$

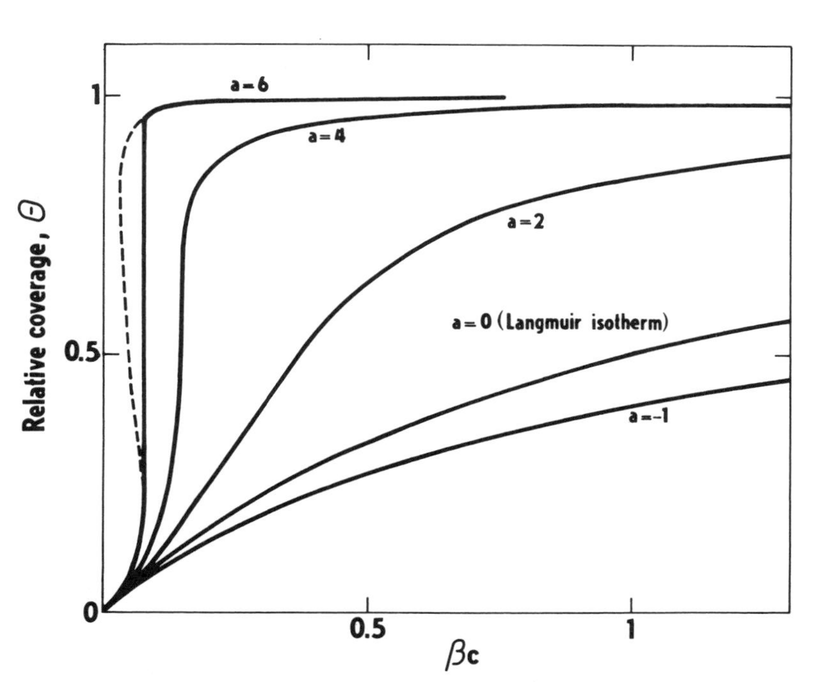

Fig. 4.8 The Frumkin adsorption isotherm. The value of the interaction coefficient a is indicated at each curve. The transition in the metastable region is indicated by a dashed line

The surface of a *solid electrode* is not homogeneous; even an apparently smooth surface as observed in an optical microscope contains corners and edges of the crystal structure of the metal and dislocations, i.e. sites where the regular crystal structure is disordered (see Section 5.5.5).

Adsorption on a physically heterogeneous surface differs from that on a physically homogeneous surface. The heat of adsorption and Gibbs energy of adsorption differ from one site to another, and the sites with the lowest Gibbs energy of adsorption are occupied first. M. Temkin derived an adsorption isotherm considering a physically heterogeneous surface. The derivation was based on the assumption of a continuously heterogeneous surface, where the free energy of adsorption per unit fraction dS of a unit surface area increases linearly with increasing S (the unit surface is divided into elements dS, each of which is considered to be homogeneous):

$$\Delta G_{\text{ads}} = \Delta G^0_{\text{ads}} + \alpha S$$
$$\mathrm{d}\Delta G_{\text{ads}} = \alpha \, \mathrm{d}S \tag{4.3.43}$$

where ΔG^0_{ads} is the value of the Gibbs adsorption energy for $S = 0$, $\Delta G^0_{\text{ads}} + \alpha$ for $S = 1$. If it is assumed that adsorption proceeds according to the Langmuir isotherm on each homogeneous, planar element, then the covered fraction of planar element dS is

$$\Theta(S) = \frac{\beta c}{1 + \beta c} \tag{4.3.44}$$

where

$$\beta = \exp\left(-\frac{\Delta G_{\text{ads}}}{RT}\right) = \exp\left(-\frac{\Delta G^0_{\text{ads}} + \alpha S}{RT}\right)$$
$$= \beta_0 \exp\left(-\frac{\alpha S}{RT}\right) \tag{4.3.45}$$

For the overall coverage θ we have

$$\Theta = \int_0^1 \theta(S)\, \mathrm{d}S = \frac{RT}{\alpha} \ln \frac{1 + \beta_0 c}{1 + \beta_0 c \exp(-\alpha/RT)} \tag{4.3.46}$$

which is the basic form of the Temkin isotherm. In the central region of the isotherm, where $\beta_0 c \gg 1 \gg \beta_0 c \exp(-\alpha/RT)$, a simpler form can be used, called the logarithmic isotherm,

$$\theta = \frac{RT}{\alpha} \ln \beta_0 c \tag{4.3.47}$$

The standard Gibbs energy of adsorption ΔG^0_{ads} is mostly a function of the electrode potential. In the simplest model, the *adsorption of a neutral substance* can be conceived as the replacement of a dielectric with a larger dielectric constant (the solvent) by a dielectric with a smaller dielectric constant (the surfactant) in a plate condenser, corresponding to the electrode double layer. On the basis of this model, A. N. Frumkin and J. A. V. Butler derived a theoretical relationship for the dependence of ΔG^0_{ads} on the electrode potential:

$$\Delta G^0_{ads} = (\Delta G^0_{ads})_{max} + a(E_p - E_{max})^2 \tag{4.3.48}$$

where E_{max} is the potential at which the quantity $-\Delta G^0_{ads}$ has the maximum value ($a > 0$). Substitution for $-\Delta G^0_{ads}$ into Eq. (4.3.42) yields

$$\beta = \beta_0 \exp\left(-\frac{a(E_p - E_{max})^2}{RT}\right) \tag{4.3.49}$$

The appearance of peaks on the differential capacity curves can be derived from this potential dependence in the following manner. The Gibbs–Lippmann equation (see Eq. 4.2.23) gives

$$\left(\frac{\partial \sigma(\mathrm{m})}{\partial \mu_1}\right)_{E_p, \mu_i \neq \mu_1} = \left(\frac{\partial \Gamma_1}{\partial E_p}\right)_{\mu_i} \tag{4.3.50}$$

Differentiation with respect to E_p yields

$$\left(\frac{\partial C}{\partial \mu_1}\right)_{E_p, \mu_i \neq \mu_1} = \left(\frac{\partial^2 \Gamma_1}{\partial E_p^2}\right)_{\mu_i} \tag{4.3.51}$$

Hence for the simplest case of a linear adsorption isotherm

$$\frac{1}{RT}\frac{\partial C}{\partial c_1} = \frac{\partial^2 \beta}{\partial E_p^2} \tag{4.3.52}$$

Integration from $c_1 = 0$ to $c_1 = c$ gives

$$\frac{1}{RT}(C - C') = \frac{\partial^2 \beta}{\partial E_p^2} c \tag{4.3.53}$$

where C' is the differential capacity of the electrode for $c_1 = 0$. The dependence of capacity C on the concentration and on the electrode potential is determined by the quantity $\partial^2\beta/\partial E_p^2$, which, considering Eq. (4.3.49), is given by the relationship

$$\partial^2\beta/\partial E_p^2 = 2a\beta_0/(RT)[2a(E_p - E_m)^2/(RT) - 1] \exp[-a(E_p - E_{max})^2/(RT)] \tag{4.3.54}$$

Thus, the dependence of the differential capacity on the potential in this case has a minimum and two maxima. The potential minimum has a value of $E_p = E_{max}$. As $\partial^2\beta/\partial E_p^2 < 0$ in the vicinity of the minimum, it follows from Eq. (4.3.53) that the differential capacity decreases with increasing concentration c in this region. The potentials of the maxima, given by the relationship $E_p = E_{max} \pm \sqrt{3RT/2a}$, lie symmetrically on either side of E_{max}. A qualitatively similar picture would be obtained for a more complex isotherm (see Fig. 4.9), but the peak potentials would depend on the surfactant concentration.

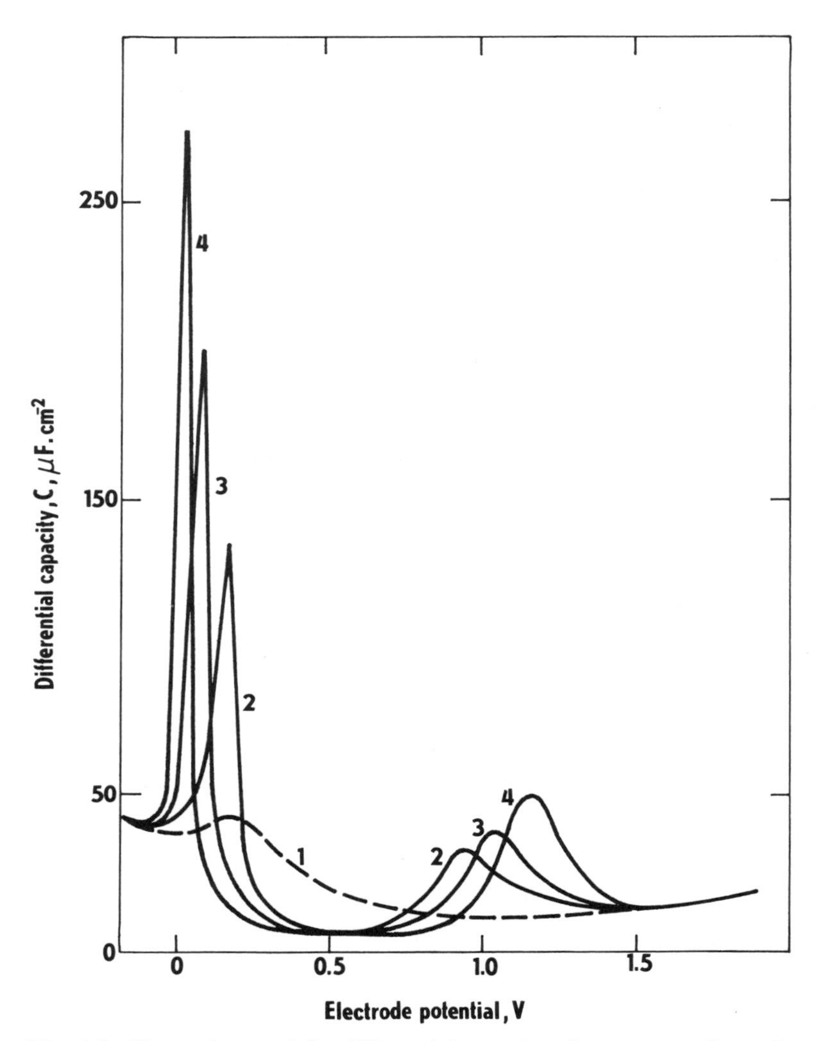

Fig. 4.9 Dependence of the differential capacity of a mercury electrode on the electrode potential E_p (vs. NCE) in a solution of 0.5 M NaF. Concentrations of t-$C_5H_{11}OH$: (1) 0, (2) 0.01, (3) 0.02, (4) 0.04 mol · dm^{-3}. (According to B. B. Damaskin)

The above relationships were derived for low electrode coverages by the adsorbed substance, where a linear adsorption isotherm could be used. Higher electrode coverages are connected with a marked change in the surface charge. The two-parallel capacitor model proposed by Frumkin and described by the equation

$$\sigma = \sigma_0(1 - \Theta) + \sigma_1\Theta \tag{4.3.55}$$

describes this situation. Here σ_0 is the surface charge in the absence of the surfactant and σ_1 is the surface charge for complete coverage of the electrode by the surfactant.

The described adsorption phenomena are characteristic for electrodes of *sp* metals. Transition metal electrodes are usually connected with irreversible chemisorption phenomena, discussed in Section 5.7.

References

Barlow, C. A., and J. R. MacDonald, Theory of discreteness of charge effects in the electrolyte compact double layer, *AE,* **6,** 1 (1967).

Damaskin, B. B., and V. E. Kazarinov, The adsorption of organic molecules, *CTE,* **1,** 353 (1980).

Damaskin, B. B., O. A. Petrii, and V. V. Batrakov, *Adsorption of Organic Compounds on Electrodes,* Plenum Press, New York, 1971.

Frumkin, A. N., and B. B. Damaskin, Adsorption of organic compounds at electrodes, *MAE,* **3,** 149 (1964).

Grahame, D. C., see page 203.

Henderson, D., L. Blum, and M. Lozada-Cassou, The statistical mechanics of the electrical double layer, *J. Electroanal. Chem.,* **150,** 291 (1983).

Kornyshev, A. A., W. Schmickler, and M. A. Vorotyntsev, *Phys. Rev. B,* **25,** 5244 (1982).

Marynov, A., and R. R. Salem, *Electrical Double Layer at a Metal–Dilute Electrolyte Interface,* Vol. 33, Lecture Notes in Chemistry, Springer-Verlag, Berlin, 1983.

Mittal, K. L. (Ed.), *Surfactants in Solution,* Vols 1–3, Plenum Press, New York, 1984.

Parsons, R., Structural effect on adsorption at the solid metal/electrolyte interface, *J. Electroanal. Chem.,* **150,** 51 (1983).

Parsons, R., Inner layer structure and the adsorption of organic compounds at metal electrodes, *J. Electroanal. Chem.,* **29,** 1563 (1984).

Reeves, R., The double layer in the absence of specific adsorption, *CTE,* **1,** 83 (1980).

Schmickler, W., A jellium-dipole model for the double layer, *J. Electroanal. Chem.,* **150,** 19 (1983).

Vetter, K. J., and J. W. Schultze, Potential dependence of electrosorption equilibria and the electrosorption valence (in German), *Ber. Bunsenges.,* **76,** 920 (1972).

Void, R. D., and M. J. Void, *Colloid and Interface Chemistry,* Addison-Wesley Publishing Company, Reading, Mass., 1983.

Vorotyntsev, M. A., and A. A. Kornyshev, Models for description of collective properties of the metal/electrolyte contact in the electrical double-layer theory, *Elektrokhimiya,* **20,** 3 (1984).

4.4 Methods of the Electrical Double-layer Study

Of the quantities connected with the electrical double layer, the interfacial tension γ, the potential of the electrocapillary maximum E_{pzc}, the differential capacity C of the double layer and the surface charge density $q(\mathrm{m})$ can be measured directly. The latter quantity can be measured only in extremely pure solutions. The great majority of measurements has been carried out at mercury electrodes.

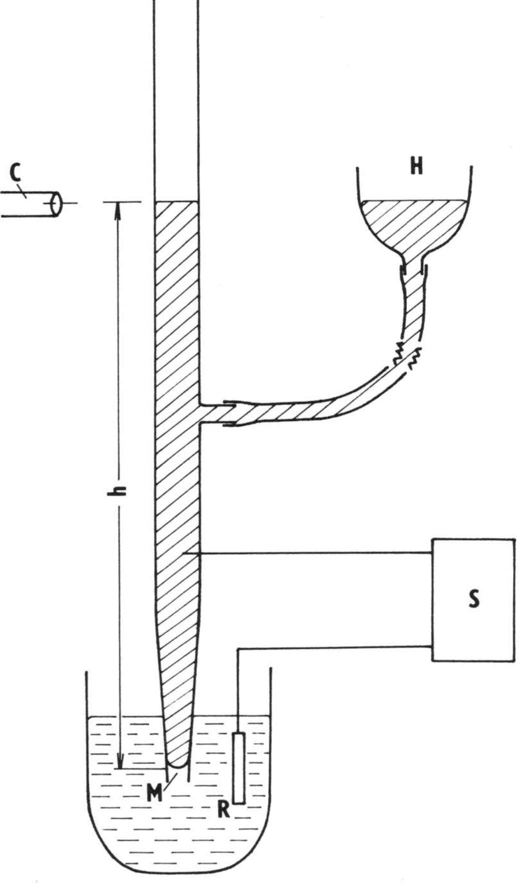

Fig. 4.10 Capillary electrometer. The basic component is the cell consisting of an ideally polarized electrode (formed by the mercury meniscus M in a conical capillary) and the reference electrode R. This system is connected to a voltage source S. The change of interfacial tension is compensated by shifting the mercury reservoir H so that the meniscus always has a constant position. The distance between the upper level in the tube and the meniscus h is measured by means of a cathetometer C. (By courtesy of L. Novotný)

G. Lippmann introduced the *capillary electrometer* to measure the surface tension of mercury (Fig. 4.10). A slightly conical capillary filled with mercury under pressure from a mercury column (or from a pressurized gas) is immersed in a vessel containing the test solution. The weight of the mercury column of height h is compensated by the surface tension according to the Laplace equation

$$p = \frac{2\gamma}{r} \approx h\rho_{Hg}g \tag{4.4.1}$$

where p is the hydrostatic pressure at the mercury meniscus, r is the radius of the capillary, ρ_{Hg} is the density of mercury and g is the standard acceleration of gravity. Both the polarized electrode and the reference electrodes are connected to a suitable external voltage source. The precise position of the meniscus in the capillary is adjusted by using a microscope. If the potential is changed, the interfacial tension of the electrode also changes and the original position of the meniscus is restored by the change in height h.

The *dropping electrode method* was suggested by B. Kučera. The arrangement is similar to the Lippmann method, but the height of the column h is sufficiently great for the mercury to drop from the end of the capillary. The mercury drops are then collected and weighed or, as the flow rate of mercury through the capillary may easily be determined, the drop-time is measured. Prior to dropping off, the mercury drop at the tip of the capillary with radius r is held by the force $2\pi r\gamma$. It thus falls off when its weight w in solution is such that it equals the force $2\pi r\gamma$:

$$w\left(1 - \frac{\rho_{sol}}{\rho_{Hg}}\right)g = 2\pi r\gamma \tag{4.4.2}$$

where ρ_{sol} is the density of the solution. This equation is not exactly obeyed by the dropping electrode, but, nonetheless, Kučera's method can be used as a relative method. The measuring apparatus must be calibrated by means of a standard solution with a known γ–E dependence.

The *wetting* (*contact*) *angle* method is used for solid surfaces. If a gas bubble sticks to a metal surface, then the individual interfacial tensions are distributed as shown in Fig. 4.11. It holds at equilibrium that

$$\gamma_{sg} = \gamma_{sl} + \gamma_{lg} \cos\theta \tag{4.4.3}$$

where θ is the wetting angle.

The quantity γ_{sl} is a function of the electrode potential, the quantity γ_{lg} is independent of the potential and γ_{sg} should not depend on the potential, provided that the surface is dry below the bubble. As there is always a trace of moisture below the bubble in this arrangement, the value of γ_{sg} changes slightly with potential, but far less than γ_{sl}. The further the electrode potential is from the potential of the electrocapillary maximum, the more γ_{sl} decreases, i.e. $\cos\theta$ increases and the wetting angle θ decreases. Obviously,

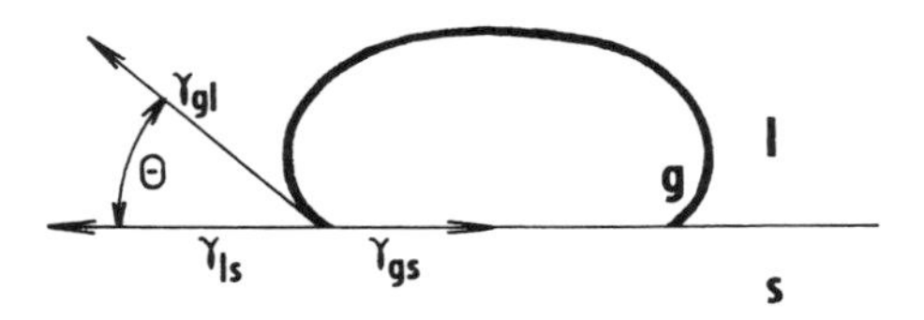

Fig. 4.11 Interfacial tensions between a solid, liquid and gaseous phase, γ_{gs}, γ_{ls} and γ_{gl}. θ denotes the wetting angle

this method cannot be used to find absolute γ_{sl} values, but is at least useful for determining E_{pzc}.

The differential capacity can be measured primarily with a capacity bridge, as originally proposed by W. Wien (see Section 5.5.3). The first precise experiments with this method were carried out by M. Proskurnin and A. N. Frumkin. D. C. Grahame perfected the apparatus, which employed a dropping mercury electrode located inside a spherical screen of platinized platinum. This platinum electrode has a high capacitance compared to a mercury drop and thus does not affect the meaurement, as the two capacitances are in series. The capacity component is measured for this system. As the flow rate of mercury is known, then the surface of the electrode A(square centimetres) is known at each instant:

$$A = 0.85m^{2/3}t^{2/3} \tag{4.4.4}$$

where m is the flow rate of mercury ($g \cdot s^{-1}$) and t is the time elapsed since the beginning of formation of the drop. The capacitance of the system is proportional to this surface area. The values of the components of the bridge have to reach a balance during the droptime, and the time from the beginning of formation of the drop is measured, i.e. from detachment of the previous drop up to bridge balancing.

The surface charge of the electrode can be determined directly, e.g. from the polarographic charging current. During the growth of the drop electrode, its capacitance increases, as does its charge. The charge that must be supplied to the electrode at constant potential produces the charging (capacity) current. The instantaneous value of this current is given, with reference to Eq. (4.4.4), as

$$I_c = \sigma(\mathrm{m})\frac{\mathrm{d}A}{\mathrm{d}t} + A\frac{\mathrm{d}\sigma(\mathrm{m})}{\mathrm{d}t} \tag{4.4.5}$$

In most inorganic electrolytes at constant potential $\sigma(\mathrm{m}) = \text{constant}$, giving for the instantaneous electrode charge

$$Q(\mathrm{t}) = \int_0^t I_c\,\mathrm{d}t = 0.85m^{2/3}t^{2/3}\sigma(\mathrm{m}) \tag{4.4.6}$$

The potential of the electrocapillary maximum can be found from the electrocapillary curve while the Paschen method is an alternative. In this case a mercury jet is injected into the solution, which has been thoroughly purified and deoxygenated. In the absence of a sufficiently fast electrode reaction the potential of this mercury electrode indicates E_{pzc}. The Paschen method is especially suitable for dilute solutions.

The surface excesses of solution components can be determined from electrocapillary measurements using the Gibbs–Lippmann equation (see Eq. 4.2.23) or the Gibbs adsorption isotherm (Eq. 4.3.36). Non-thermodynamic methods have been proposed for study of surfactant adsorption, especially of high-molecular-weight organic substances, based on Frumkin's two-parallel capacitor model. Differentiating Eq. (4.3.55) with respect to E_p gives

$$C = C_0(1 - \Theta) + C_1\Theta + (\sigma_1 - \sigma_0)\frac{d\Theta}{dE_p} \tag{4.4.7}$$

where C_0 is the differential capacity in the base electrolyte alone and C_1 is the differential capacity for complete electrode coverage by the surfactant. As we can set $d\Theta/dE_p \approx 0$ at potentials sufficiently distant from the desorption potential, the relative electrode coverage is then

$$\Theta = \frac{C - C_0}{C_1 - C_0} \tag{4.4.8}$$

Other methods, based on kinetic methods of measuring surface excesses, will be discussed in Section 5.7.

References

Bard, A. J., and L. R. Faulkner, *Electrochemical Methods,* John Wiley & Sons, New York, 1980.
Grahame, D. C., see page 202.
Sluyters-Rehbach, M., and J. H. Sluyters, A.C. techniques, *CTE,* **9,** 177 (1984).
Southampton Electrochemistry Group, *Instrumental Methods in Electrochemistry,* Ellis Horwood, Chichester, 1985.

4.5 The Electrical Double Layer at the Electrolyte–Non-metallic Phase Interface

4.5.1 *Semiconductor–electrolyte interfaces*

The interfaces between a semiconductor and another semiconductor (e.g. the very important *p*/*n* junction, the interface between *p*- and *n*-type semiconductors), between a semiconductor and a metal (the Schottky barrier) and between a semiconductor and an electrolyte are the subject of solid-state physics, using a nomenclature different from electrochemical terminology.

The basic difference between metal–electrolyte and semiconductor–electrolyte interfaces lies primarily in the fact that the concentration of charge carriers is very low in semiconductors (see Section 2.4.1). For this reason and also because the permittivity of a semiconductor is limited, the semiconductor part of the electrical double layer at the semiconductor–electrolyte interface has a marked diffuse character with Debye lengths of the order of 10^{-4}–10^{-6} cm. This layer is termed the *space charge region* in solid-state physics.

The description of the properties of this region is based on the solution of the Poisson equation (Eqs 4.3.2 and 4.3.3). For an intrinsic semiconductor where the only charge carriers are electrons and holes present in the conductivity or valence band, respectively, the result is given directly by Eq. (4.3.11) with the electrolyte concentration c replaced by the ratio n_e^0/N_A, where n_e^0 is the concentration of electrons in 1 cm^3 of the semiconductor in a region without an electric field (in solid-state physics, concentrations are expressed in terms of the number of particles per unit volume).

For doped semiconductors, it is assumed that the charge density in the semiconductor, e.g. of type n, is

$$\rho(x) = e[n_e(x) - n_e^0] \tag{4.5.1}$$

where $n_e(x)$ is the concentration of electrons and n_e^0 is the electron concentration in the absence of an electric field, practically equivalent to the concentration of electron donors in the semiconductor. This is identical to the concentration of positive ions, independent of the field in the semiconductor. It is assumed that the concentration of electron acceptors is much larger than the concentration of electron donors and of electrons and holes for the original undoped intrinsic semiconductor.

It follows from Eq. (4.3.5) that

$$n_e(x) = n_e^0 \exp\left\{\frac{e[\phi(x) - \phi_{sc}]}{kT}\right\} \tag{4.5.2}$$

where ϕ_{sc} is the inner potential of the bulk of the semiconductor. The procedure on page 214 can be used to yield

$$\frac{\partial\phi}{\partial x} = -\sqrt{\frac{2kTn_e^0}{\varepsilon}}\left\{\exp\left[-\frac{e(\phi - \phi_{sc})}{kT}\right] + \frac{e}{kT}(\phi - \phi_{sc}) - 1\right\}^{1/2} \tag{4.5.3}$$

Equation (4.5.3) gives, for $\phi \ll kT/e$ with respect to Eqs (4.3.8) and (4.3.12)

$$\partial\phi/\partial x = [\varepsilon kT/(2e^2 n_e^0)]^{1/2}(\phi - \phi_{sc}) = (\phi - \phi_{sc})/L_{sc} \tag{4.5.4}$$

where L_{sc} is the Debye length in the semiconductor.

In view of Eqs (4.3.8) and (4.5.3), where $\phi - \phi_{sc}$ is replaced by the overall potential difference between the bulk semiconductor and the interface, $\phi_{sc} - \phi_s = \Delta_s^{sc}\phi$, we find for the surface charge of the semiconductor

$$\sigma_{sc} = \mp\sqrt{2kT\varepsilon n_e^0}\left[\exp\left(-\frac{e\Delta_s^{sc}\phi}{kT}\right) + \frac{e\Delta_s^{sc}\phi}{kT} - 1\right]^{1/2} \tag{4.5.5}$$

The overall Galvani potential difference between the bulk of the semiconductor and the bulk of the electrolyte solution can be separated into three parts:

$$\Delta_l^{sc}\phi = \Delta_s^{sc}\phi + \Delta_2^{s}\phi + \Delta_2^{l}\phi \tag{4.5.6}$$

where $\Delta_s^{sc}\phi$ is the potential difference between the bulk of the semiconductor and the semiconductor–electrolyte interface, $\Delta_2^{s}\phi$ is the potential difference in the compact layer (i.e. between the interface and the outer Helmholtz plane) and $\Delta_l^{2}\phi$ is the potential difference between the outer Helmholtz plane and the bulk of the electrolyte solution. Obviously,

$$\begin{aligned}\Delta_l^{sc}\phi &= E_{sc}L_{sc} + E_1 d_1 + E_l L_l \\ &= E_{sc}\left(L_{sc} + \frac{d_1\varepsilon_H}{\varepsilon_{sc}} + \frac{L_l\varepsilon_l}{\varepsilon_{sc}}\right)\end{aligned} \tag{4.5.7}$$

where E_{sc}, E_l and E_1 are the electric field strengths in the semiconductor, in the compact layer and in the diffuse layer in the electrolyte, respectively, L_l is the Debye length in the solution, d_1 is the thickness of the compact layer and ε_{sc}, ε_H and ε_l are the permittivities of the semiconductor, compact layer and electrolyte, respectively. At sufficiently large electrolyte concentrations, the second and third terms in parentheses in Eq. (4.5.7) can be neglected. Under these conditions, the Galvani potential difference between the semiconductor and the electrolyte solution is given by the potential difference in the space charge region (diffuse layer) in the semiconductor, which also holds for larger potential differences than RT/F.

The distribution of charge carriers in the space charge region of an n-type semiconductor is shown in Fig. 4.1. It is assumed that the semiconductor contains not only electrons (majority charge carriers) but also much lower concentrations of holes (minority charge carriers). If the negative potential of an n-type semiconductor electrode increases, then the negative charge of the electrode increases through an increase in the concentration of electrons in the interphase (accumulation layer). As the electron energy in the space charge region $e\phi(x)$ depends on the distance from the interface, the energy bands bend downwards in the direction towards the interface in the negatively charged space charge region (Fig. 4.12A).

When the surface charge decreases to zero, the energy bands become horizontal. The corresponding flat-band potential $\Delta_l^{sc}\phi_{fb}$ is an analogy of the zero-charge potential E_{pzc} (Fig. 4.12B).

When the difference $\Delta_l^{sc}\phi - \Delta_l^{sc}\phi_{fb}$ becomes positive, the interphase is depleted of majority charge carriers (electrons in this case) forming a

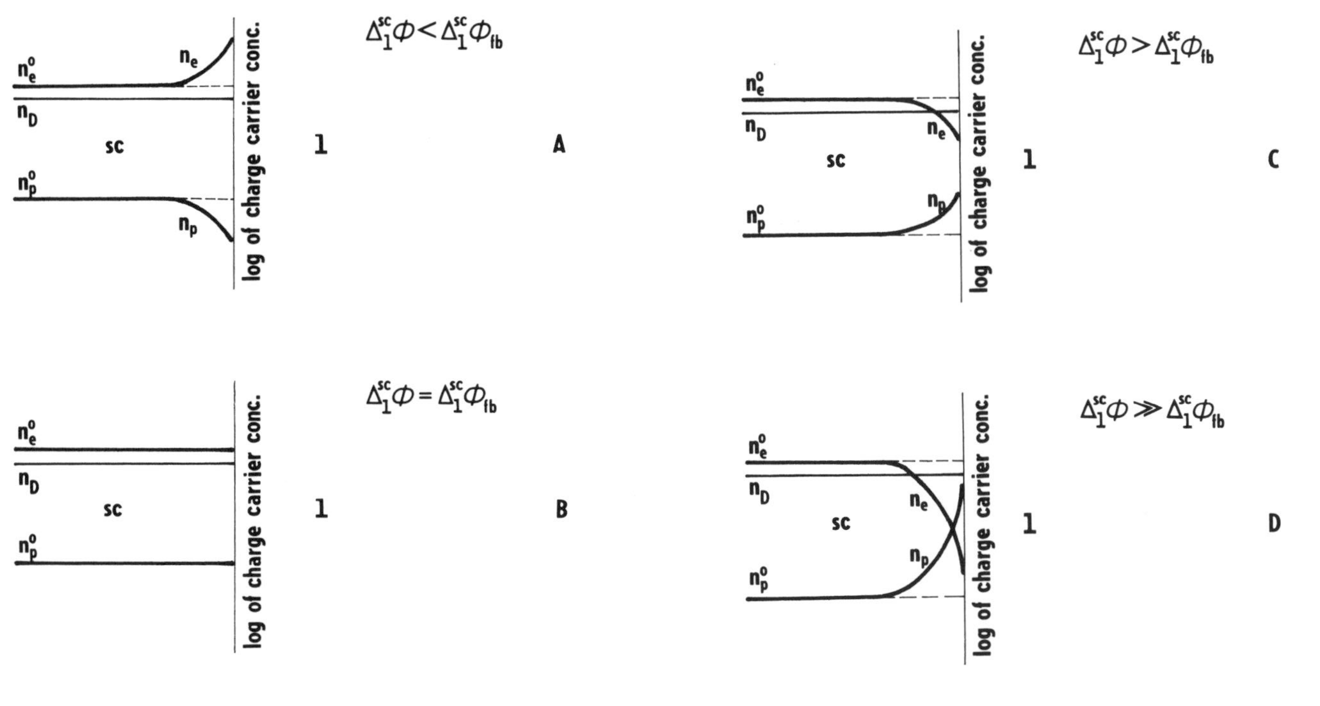

Fig. 4.12 Dependence of concentrations of negative charge carriers (n_e) and positive charge carriers (n_p) on distance from the interface between the semiconductor (sc) and the electrolyte solution (l) in an n-type semiconductor. These concentration distributions markedly differ if the semiconductor/electrolyte potential difference $\Delta_l^{sc}\phi$ is (A) smaller than the flat-band potential $\Delta_l^{sc}\phi_{fb}$, (B) equal to the flat-band potential, (C) larger and (D) much larger than the flat-band potential. n_D denotes the donor concentration. (According to Yu. V. Pleskov)

depletion or Mott–Schottky layer (Fig. 4.12C). However, Eq. (4.5.5) is a simplification, as it neglects the presence of minority charge carriers, which was originally very low. The following relationship holds between the concentration of electrons and the concentration of holes n_p:

$$n_p n_e = n_0^2 \tag{4.5.8}$$

where n_0 is the concentration of electrons in the same semiconductor in the absence of impurities and of an electric field inside the semiconductor, i.e. in the corresponding intrinsic semiconductor. According to the quantum theory of semiconductors, the quantity n_0 is given by the relationship

$$n_0 = \frac{2(2\pi m_e kT)^{3/2}}{h^3} \exp\left(-\frac{\varepsilon_g}{2kT}\right) \tag{4.5.9}$$

where m_e is the electron mass, h is the Planck constant and ε_g is the width of the forbidden band.

For a sufficiently large potential increase, the charge in the interphase finally corresponds to the minority charge carriers (Fig. 4.12D). The greater the width of the forbidden band ε_g, the broader is the potential range $\Delta_1^{sc}\phi$ in which the space charge region has the character of a depletion layer, i.e. is formed by ionized impurity atoms.

It should be noted that the study of the properties of an electrical double layer at semiconductor electrodes corresponds to study of the cell (in the simplest formulation)

$$\text{M} \mid \text{S} \mid \text{electrolyte solution} \mid \text{M} \tag{4.5.10}$$

where M denotes the metal phase and S is the semiconductor. The electrolyte solution has a composition such that the interface with the metal is unpolarizable. An electrical double layer is formed at the interface between the metal and the semiconductor (called the Schottky barrier), but this interface is unpolarized, so that

$$\Delta_m^{sc}\phi = [\mu_e^0(\text{s}) - \mu_e^0(\text{m})]/e \tag{4.5.11}$$

The discussion of Eq. (4.5.7) has shown that the potential of a polarized semiconductor electrode at sufficiently high electrolyte concentrations is

$$E_p \approx \Delta_s^{sc}\phi + \text{constant} \tag{4.5.12}$$

In Eq. (4.5.5), describing an n-type semiconductor strongly doped with electron donors, the first and third terms in brackets can be neglected for the depletion layer ($\Delta_s^{sc}\phi \gg kT/e$). Thus, the Mott–Schottky equation is obtained for the depletion layer,

$$\sigma_{sc} = \sqrt{2\varepsilon n_e^0 e \Delta_s^{sc}\phi} \tag{4.5.13}$$

The differential capacity of an n-type semiconductor electrode in the

Mott–Schottky approximation is

$$C = \frac{d\sigma_{sc}}{d\Delta_s^{sc}\phi} = \sqrt{\frac{\varepsilon n_e^0 e}{2\Delta_s^{sc}\phi}} \tag{4.5.14}$$

The Mott–Schottky plot following from Eqs (4.5.12) and (4.5.14) is the relationship

$$C^{-2} = \frac{2}{\varepsilon n_e^0 e} E_p + \text{constant} \tag{4.5.15}$$

which can be used to determine the concentration of electron donors $N_D \approx n_e^0$.

It has so far been assumed that the semiconductor–electrolyte interphase does not contain either ions adsorbed specifically from the electrolyte or electrons corresponding to an additional system of electron levels. These *surface states* of electrons are formed either through adsorption (the Shockley levels) or through defects in the crystal lattice of the semiconductor (the Tamm levels). In this case—analogously as for specific adsorption on metal electrodes—three capacitors in series cannot be used to characterize the semiconductor–electrolyte interphase system and Eq. (4.5.6) must include a term describing the potential difference for surface states.

The presence of surface states results in a certain compression of the space charge region and leads to a decrease in the band bending.

The situation of the electric double layer at a semiconductor/electrolyte solution interface affected by light radiation will be dealt with in Section 5.10.

4.5.2 *Interfaces between two electrolytes*

The electrical double layer has also been investigated at the interface between two immiscible electrolyte solutions and at the solid electrolyte–electrolyte solution interface. Under certain conditions, the interface between two immiscible electrolyte solutions (ITIES) has the properties of an ideally polarized interphase. The dissolved electrolyte must have the following properties:

(a) The electrolyte in the aqueous phase, B_1A_1 (assumed to dissociate completely to form cation B_1^+ and anion A_1^-), is strongly hydrophilic, while the electrolyte in the organic phase, B_2A_2 (completely dissociated to cation B_2^+ and anion A_2^-), is strongly hydrophobic. These properties can be expressed quantitatively by the conditions for the distribution coefficients between water and the organic phase $k_{B_1A_1}^{w/o}$ and $k_{B_2A_2}^{w/o}$:

$$k_{B_1A_1}^{w/o} \ll 1, \qquad k_{B_2A_2}^{w/o} \gg 1 \tag{4.5.16}$$

(b) The equilibria of the exchange reactions

$$\begin{aligned} B_1^+(w) + B_2^+(o) &\rightleftarrows B_1^+(o) + B_2^+(w) \\ A_1^-(w) + A_2^-(o) &\rightleftarrows A_1^-(o) + A_2^-(w) \end{aligned} \tag{4.5.17}$$

are shifted to the left. In other words, the standard potential difference between water and the organic phase for the transfer of the cation of the aqueous phase and the anion of the organic phase must be as positive as possible and the corresponding potential difference for the cation of the organic phase and the anion of the aqueous phase must be as negative as possible. In addition, there must be a sufficiently large difference between the less positive value of $\Delta_o^w \phi_i^o$ for the first pair and the less negative value for the second pair (cf. Section 3.2.8).

Under these conditions, the potential difference between the two phases $\Delta_o^w \phi$ can be changed by charge injection from an external electric current source. The appropriate experimental arrangement is shown in Fig. 5.18.

E. J. W. Verwey and K. F. Niessen described the electric double layer at ITIES using a simplifying assumption that consists of only two diffuse electric layers, each in one of the phases (see Fig. 4.1C). The overall potential difference between the two phases, $\Delta_o^w \phi$, is thus given by the relationship

$$\Delta_o^w \phi = \phi_2(\mathrm{o}) - \phi_2(\mathrm{w}) \tag{4.5.18}$$

where the potential differences in the diffuse layers, $\phi_2(\mathrm{o})$ and $\phi_2(\mathrm{w})$, are defined such that the potential in the bulk of the phase is subtracted from the potential at the interface. The terms $\phi_2(\mathrm{o})$ and $\phi_2(\mathrm{w})$ are given by Eq. (4.3.14), where the appropriate permittivity, $\varepsilon(\mathrm{w})$ or $\varepsilon(\mathrm{o})$, is substituted for each phase, along with the electrolyte concentrations $c(\mathrm{w})$ and $c(\mathrm{o})$. As the surface charges $\sigma(\mathrm{w})$ and $\sigma(\mathrm{o})$ are equal except for the sign (cf. Eq. 4.1.1), Eqs (4.5.18) and (4.3.14) can be used to calculate the dependence of $\phi_2(\mathrm{o})$ and $\phi_2(\mathrm{w})$ on $\Delta_o^w \phi$. The electrical double layer has, of course, a similar structure when a single electrolyte is present in the distribution equilibrium in the system.

It is interesting that the experimentally measured zero-charge potential is practically identical with the value of $\Delta_o^w \phi = 0$, calculated using the TATB assumption (3.2.64). This fact helps to justify the use of this assumption.

The electrical double layer at the *solid electrolyte–electrolyte solution interface* has been studied primarily in colloid suspensions, especially for silver halides. The potential difference between the solid and liquid phases can be changed by changing the concentration of the 'potential-determining ions', i.e. either the silver or halide ions. In solid oxide suspensions, hydrogen ions act as potential-determining ions. The zero-charge potential of this system can be found from the dependence of the electrophoretic mobility (see Section 4.5.4) on the concentration of potential-determining ions, i.e. it corresponds to zero electrophoretic mobility. With this type of interface the structure of the electrical double layer depends to a marked degree on the preparation and thus also on the final structure of the solid phase. Two cases are most often observed:

(a) Ions are adsorbed from solution on the surface of the solid phase and counterions form a diffuse layer.

(b) The electrical double layer has a structure similar to ITIES, but the diffuse layer in the solid becomes a simple Helmholtz layer because of the high concentration of ions.

It should, however, be noted that the electrical double layer at the *metal–fused electrolyte interface* does not have this character, in spite of the ion concentration being high. In this system, the space charge includes several ion layers at the interface.

4.5.3 *Electrokinetic phenomena*

The adsorption of ions at insulator surfaces or ionization of surface groups can lead to the formation of an electrical double layer with the diffuse layer present in solution. The ions contained in the diffuse layer are mobile while the layer of adsorbed ions is immobile. The presence of this mobile space charge is the source of the *electrokinetic phenomena*.† Electrokinetic phenomena are typical for insulator systems or for a poorly conductive electrolyte containing a suspension or an emulsion, but they can also occur at metal–electrolyte solution interfaces.

Consider a solid surface in contact with a dilute electrolyte solution. The plane where motion of the liquid can commence is parallel to the outer Helmholtz plane but shifted in the direction into the bulk of the solution. The electric potential in this plane with respect to the solution is termed the *electrokinetic potential* $\zeta(|\zeta| \leqq |\phi_2|)$.

Of the four electrokinetic phenomena, two (electroosmotic flow and the streaming potential) fall into the region of membrane phenomena and will thus be considered in Chapter 6. This section will deal with the electrophoresis and sedimentation potentials.

If the electric field $\mathbf{E}$ is applied to a system of colloidal particles in a closed cuvette where no streaming of the liquid can occur, the particles will move with velocity $\mathbf{v}$. This phenomenon is termed *electrophoresis*. The force acting on a spherical colloidal particle with radius r in the electric field $\mathbf{E}$ is $4\pi\varepsilon r\mathbf{E}\phi_2$ (for simplicity, the potential in the diffuse electric layer is identified with the electrokinetic potential). The resistance of the medium is given by the Stokes equation (2.6.2) and equals $6\pi\eta r\mathbf{v}$. At a steady state of motion these two forces are equal and, to a first approximation, the electrophoretic mobility $\mathbf{v}/\mathbf{E}$ is

$$\frac{\mathbf{v}}{\mathbf{E}} = \frac{\varepsilon\phi_2}{3\eta} \tag{4.5.19}$$

In closer approximations, correction must be made for conductivity effects (relaxation and electrophoretic) and for the real shape of the particles. Thus, the velocity of electrophoretic motion depends on the composition of the

† This term must be distinguished from the concept of electrochemical kinetics, discussed in Chapter 5.

solution, on the properties of the surface of the particles and also on the charge of the particles themselves. If ampholytic particles are involved, then it also depends markedly on the pH, as the particles obtain a charge through dissociation that is dependent on the pH. This fact was utilized by A. Tiselius to develop an electrophoretic method that is especially useful for proteins. At a given pH various proteins are ionized to a different degree and also have different mobilities. The original single sharp boundary between a solution of a protein mixture in a suitable buffer and the pure buffer separates into several boundaries in an electric field, corresponding to differently mobile components. Thus electrophoresis permits analysis of a mixture of proteins without destructive chemical reactions. The experimental methods for determining the boundary position in either classical or free electrophoresis are the same as in the study of diffusion (see Section 2.5.5). In addition, electrophoresis can be carried out by saturating a suitable porous carrier, e.g. filter paper, with a pure buffer and applying the studied solution as spots or bands. The evaluation methods are analogous to those used in paper chromatography. Electrophoresis can be used preparatively to separate the components of a mixture, to concentrate fine suspensions in solution, etc.

A further electrokinetic phenomenon is the inverse of the former according to the Le Chatelier–Brown principle: if motion occurs under the influence of an electric field, then an electric field must be formed by motion (in the presence of an electrokinetic potential). During the motion of particles bearing an electrical double layer in an electrolyte solution (e.g. as a result of a gravitational or centrifugal field), a potential difference is formed between the top and the bottom of the solution, called the *sedimentation potential.*

References

Andrews, A. T., *Electrophoresis, Theory, Techniques and Biochemical and Clinical Applications,* Oxford University Press, Oxford, 1982.

Boguslavsky, L. I., Insulator/electrolyte interface, *CTE,* **1,** 329 (1980).

Butler, J. A. V., see page 202.

Gaal, O., G. Medgyesi, and L. Vereczkey, *Electrophoresis in the Separation of Biological Macromolecules,* John Wiley & Sons, New York, 1980.

Gerischer, H., Semiconductor electrode reactions, *AE,* **1,** 31 (1961).

Gerischer, H., Electrochemical photo and solar cells. Principles and some experiments, *J. Electroanal. Chem.,* **58,** 263 (1975).

Green, M., Electrochemistry of the semiconductor–electrolyte interface, *MAE,* **2,** 343 (1959).

Hunter, R. J., The double layer in colloidal systems, *CTE,* **1,** 397 (1980).

Lyklema, J., The electrical double layer on oxides, *Croatica Chem. Acta,* **43,** 249 (1971).

Mareček, V., Z. Samec, and J. Koryta, Electrochemical phenomena at the interface of two immiscible electrolyte solutions, *Advances in Interfacial and Colloid Science,* **29,** 1 (1988).

Morrison, S. R., *Electrochemistry of the Semiconductor and Oxidized Metal Electrodes,* Plenum Press, New York, 1980.
Myamlin, V. A., and Yu. V. Pleskov, *Electrochemistry of Semiconductors,* Plenum Press, New York, 1967.
Pleskov, Yu. V., Electric double layer on semiconductor electrode, *CTE,* **1,** 291 (1980).
Samec, Z., The electrical double layer at the interface of two immiscible electrolyte solutions, *Chem. Revs,* **88,** 617 (1988).
Spaarnay, M. J., see page 203.
Vanýsek, P., see page 191.
Verwey, E. J. W., and J. Th. G. Overbeek, *Theory of the Stability of Lyophobic Colloids,* Elsevier, Amsterdam, 1948.

Chapter 5

Processes in Heterogeneous Electrochemical Systems

In the case of electrodes with low overpotential the process of molecular hydrogen evolution is particularly more complicated than with electrodes showing high overpotential where a single assumption of slow discharge step could successfully elucidate all experimental results. It can be taken for sure, however, that in the case of electrodes with low overpotential it is necessary to consider, as a slow step, the process of removal of molecular hydrogen from the electrode together with the discharge process.

A. N. Frumkin, V. S. Bagotzky, and Z. A. Iofa, 1952

5.1 Basic Concepts and Definitions

This chapter will be concerned with the kinetics of charge transfer across an electrically charged interface and the transport and chemical processes accompanying this phenomenon. Processes at membranes that often have analogous features will be considered in Chapter 6. The interface that is most often studied is that between an electronically conductive phase (mostly a metal electrode) and an electrolyte, and thus these systems will be dealt with first.

A system consisting of two electrodes in an electrolyte medium is called an *electrolytic cell.* It should be realized here that there is no basic difference between the concepts of a 'galvanic cell' and an 'electrolytic cell'. In common usage, the term 'galvanic cell' is understood to refer to a system in the absence of current flow (it can be at equilibrium, as mentioned in the previous chapter, or only in a steady state, as will be demonstrated in Section 5.8.4 on mixed potentials) or to a system that yields electric work to its surroundings; an 'electrolytic cell' is then a system receiving energy from its surroundings to carry out the required chemical conversions. In fact, however, a given system sometimes can have both functions depending on the electrical potential difference between the electrodes.

If an electrode reaction at a given electrode results in the transfer of a positive electric charge from the electrolyte to the electrode material or a negative charge from the electrode material to the electrolyte, then the corresponding current is defined as *cathodic* (I_c) and the process is termed a cathodic reaction or reduction. In the opposite case, the current is termed *anodic* (I_a) and the process is an anodic reaction or oxidation. The terms *cathode* and *anode* have already been defined in Section 1.1.1. For example, in a Daniel cell (see Section 3.1.4), the cathode is the copper electrode when the cell carries out work; if the applied voltage is greater than the equilibrium value, current flows in the opposite direction and the cathode becomes the zinc electrode at which zinc ions are reduced. According to the IUPAC convention, anodic current is considered to be positive and cathodic current to be negative.

Generally, an electrode that is macroscopically characterized by a smooth surface actually contains many steps and other microscopic irregularities. The real (physical) electrode surface A_r is thus mostly larger than the geometric (macroscopic) surface A_g. The current density is usually defined as the current divided by the geometric surface area. The ratio of the real and geometric surface areas is termed the roughness factor, $f_R = A_r/A_g$. The physical and geometric surface areas are identical at mercury and other liquid electrodes.

The flow of electric current through the electrolytic cell is connected with chemical, electrochemical and physical processes which, as a whole, are termed the *electrode process*. The main electrochemical step in the electrode process is the actual exchange of charged species between the electrode and the electrolyte, which will be termed the *electrode reaction* (charge transfer reaction). Substances participating directly in the charge transfer reaction are termed *electroactive*. These substances can be either soluble or insoluble in the electrolyte or electrode material. Common basic types of electrode reactions are as follows:

1. Reduction processes where the electrode is the cathode:
   (a) Reduction of ions or complexes to a lower oxidation state, the reduction of inorganic or organic molecules; the reduced form remains in solution.
   (b) Deposition of ions or complexes on the electrode with formation of a metallic or gaseous phase or of an amalgam.
   (c) Reduction of insoluble compounds or surface films with formation of a metal phase or an amalgam or an insoluble phase with different composition.
2. Oxidation processes in which the electrode is an anode:
   (a) Oxidation of ions, complexes or molecules to a soluble higher oxidation state.
   (b) Oxidation of the electrode material with formation of soluble ions or complexes.
   (c) Oxidation of the electrode material with formation of insoluble anodic films or oxidation of insoluble films or other insoluble substance to form insoluble substances with a higher oxidation state.

Of these electrode reactions, two—one oxidation and one reduction—

form a pair in a reversible charge transfer reaction (e.g. the deposition of metal ions with subsequent amalgam formation and ionization of the amalgam with formation of metal ions in solution).

A more fundamental classification considers the character of the charge transfer between the electrode and the electroactive substance:

1. The transfer of electrons or holes between the electrode and the electroactive substance present in solution.
2. Transfer of metal ions from the electrolyte into the electrode and vice versa.
3. Emission of electrons from the electrode into the solution with formation of solvated electrons and the subsequent reaction between the solvated electrons and the electron 'scavenger' in solution.

These processes have various characteristic properties when they occur at metallic or semiconductor electrodes and if they occur between partners (electroactive substances or electrons or holes in the electrode) that are in the ground or the excited state.

The basic condition for electron transfer in cathodic processes (reduction) to an electroactive substance is that this substance (Ox) be an electron acceptor. It must thus have an unoccupied energy level that can accept an electron from the electrode. The corresponding donor energy level in the electrode must have approximately the same energy as the unoccupied level in the substance Ox.

On the other hand, in oxidation processes, the electroactive substance Red must have the character of an electron donor. It must contain an occupied level with energy corresponding to that of some unoccupied level in the electrode. Oxidation occurs through transfer of electrons from the electroactive substance to the electrode or through the transfer of holes from the electrode to the electroactive substance.

Thus, an electrode redox reaction occurs according to the scheme

$$\text{Red} + \text{unoccupied level} \rightleftarrows \text{Ox} + \text{occupied level} \qquad (5.1.1)$$

This situation is depicted in Fig. 5.1. The occupied level in the substance Red_1 has an energy corresponding to the unoccupied level in the electrode. Thus, oxidation can occur (either through the transfer of an electron e^- to the electrode or of a hole h^+ from the electrode). On the other hand, the unoccupied level in the substance Ox_1 has too high an energy, so that it does not correspond to any of the occupied levels in the electrode as all these levels lie below the Fermi level ε_F, while the energy of the unoccupied level of the substance Ox_1 is far above this level. Reduction can thus not occur. The situation is the opposite for the substances Red_2 and Ox_2.

As was demonstrated in Section 3.1.2, the energy of the Fermi level is identical with the electrochemical potential of an electron in the metal. A change in the inner potential of the electrode phase by $\Delta\phi$ (attained by changing the potential difference of an external voltage source by $\Delta E = \Delta\phi$,

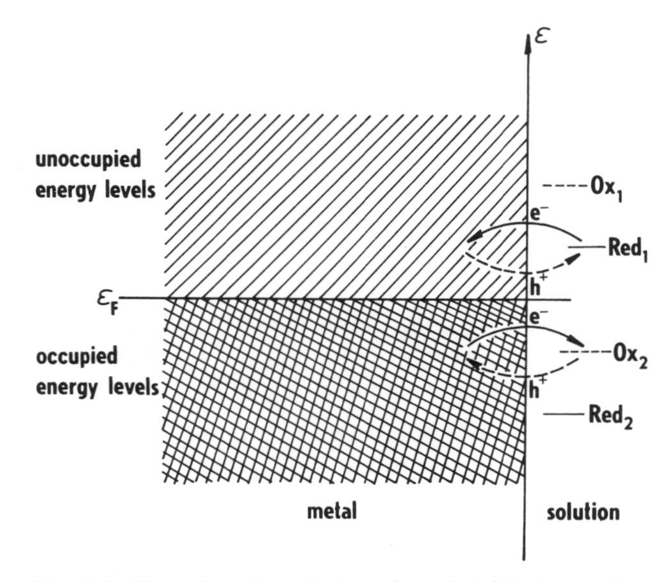

Fig. 5.1 Transfer of an electron (or a hole) between the electrode and the electroactive substance in solution. (According to H. Gerischer)

i.e. by polarizing the electrode to the potential $E + \Delta E$) can shift ε_F to higher or lower energies. If ε_F is shifted towards the occupied level in Ox_1 (that is $\Delta\phi$ is negative), then reduction of Ox_1 is made easier and oxidation of Red_1 becomes more difficult (or even completely impossible if ε_F is shifted sufficiently above the energy of Red_1). A decrease in ε_F (through a positive $\Delta\phi$) towards the occupied level in Red_2 has the opposite effect. Obviously, oxidation and reduction occur simultaneously in the vicinity of the equilibrium potential (where the Fermi levels in the metal and solution are close to one another).

Conditions are somewhat more complicated at *semiconductor electrodes*. In contrast to metal electrodes, where the donor and acceptor levels are concentrated in a single energy band, the Fermi level for semiconductors lies in the forbidden band (see Figs 2.3 and 3.2), where there are no electron energy levels. Consequently, in the vicinity of the equilibrium potential where the electrochemical potentials of the electrons in solution and in the metal (i.e. the Fermi level) are identical, no charge transfer occurs (provided that there are no electronic surface states at the interface; see Section 4.5.1). Charge transfer requires a large change in $\Delta\phi$, where the occupied electronic level in Red attains the level of the conductivity band in the semiconductor or the unoccupied level in Ox attains the level of the valence band. It should be recalled that the electrons and holes participate in charge transfer in the vicinity of the semiconductor–electrolyte interface, where their energy is different from that in the bulk of the semiconductor as a result of band 'bending' (see Sections 4.5.1 and 5.10.4).

Conditions at the *interface between two immiscible electrolyte solutions* (*ITIES*) are analogous to those at the electronic conductor–electrolyte interface. The charge carriers passing across the interface are primarily ions and they are transferred to a region where they have a lower electrochemical potential. The transfer of electrons from the species Red_1 in one phase requires that the second phase contain the species Ox_2 with an unoccupied acceptor level with similar or lower energy compared with the donor level in Red_1. A change in the electrochemical potential of the ions or electrons can be achieved through a change in the potential difference between the two phases.

The presence of a concentration gradient of electroactive substances at the phase boundary between the electrode and the solution or between two electrolytes, the presence of an electric field and, finally, the effect of mechanical forces all lead to transport processes, described in Chapter 2. It should be pointed out that investigations in electrochemical kinetics are usually carried out with solutions containing an excess of indifferent electrolyte compared to the electroactive substance, to exclude the effect of migration. The rule for the equivalence between the material flux of the charged species and the electric current (Eq. 2.3.14) is a general formulation of the Faraday law. In its classical form, this rule is connected with a special case, i.e. material flux through the electrode–solution interface. In this special case the Faraday law deals with equivalence between the material yield in an electrode reaction and the charge passed:

1. The amount of substance converted in the electrolysis is proportional to the charge passed.
2. The amounts of various substances converted by a given charge correspond to the ratio of the chemical equivalents of these substances.

The electric current connected with chemical conversion is termed the *faradaic current* in contrast to the *non-faradaic current* required to charge the electrical double layer. The equation for the electrode reaction is formulated similarly to that for the half-cell reaction (Section 3.1.4); for the cathodic reaction, the electrons are placed on the left-hand side of the equation and, for the anodic reaction, on the right-hand side.

Consider the electrode reaction of metal deposition:

$$M^{n+} + ne \rightarrow M \tag{5.1.2}$$

If current I flows through the solution for time t, then the charge It is passed, corresponding to a transfer of $-It/F$ moles of electrons. If the relative atomic mass of the metal is M, then a total amount of

$$w = -\frac{MIt}{nF} \tag{5.1.3}$$

is discharged. This equation follows directly from Eq. (2.3.14) for a single substance where $z_i = n$ and $I = jA$, where A is the area of the electrode surface.

If no side reactions occur at the electrode that would participate in the overall current flow, then the Faraday law can be used not only to measure the charge passed (i.e. in coulometres; see Section 5.5.4) but also to define the units of electric current and even to determine Avogadro's constant.

If the electrolyte components can react chemically, it often occurs that, in the absence of current flow, they are in chemical equilibrium, while their formation or consumption during the electrode process results in a *chemical reaction* leading to renewal of equilibrium. Electroactive substances mostly enter the charge transfer reaction when they approach the electrode to a distance roughly equal to that of the outer Helmholtz plane (Section 5.3.1). It is, however, sometimes necessary that they first be *adsorbed*. Similarly, adsorption of the products of the electrode reaction affects the electrode reaction and often retards it. Sometimes, the electroinactive components of the solution are also adsorbed, leading to a change in the structure of the electrical double layer which makes the approach of the electroactive substances to the electrode easier or more difficult. Electroactive substances can also be formed through *surface reactions* of the adsorbed substances. *Crystallization processes* can also play a role in processes connected with the formation of the solid phase, e.g. in the cathodic deposition of metals.

In electrode processes, the *overall* ('brutto') *reaction* must be distinguished from the actual *mechanism of the electrode process*. For example, by a cathodic reaction at a number of metal electrodes, molecular hydrogen is formed, leading to the overall reaction

$$2H_3O^+ + 2e \rightarrow H_2 + 2H_2O \qquad (5.1.4)$$

The electroreduction of aldehydes and ketones often involves the formation of dimeric hydroxy derivatives—pinacols:

$$2C_6H_5CHO + 2e + 2H^+ \rightarrow C_6H_5CHOH.CHOHC_6H_5 \qquad (5.1.5)$$

The deposition of cadmium in the form of an amalgam from a cadmium cyanide solution involves the overall reaction

$$Cd(CN)_4^{2-} + 2e \rightarrow Cd + 4CN^- \qquad (5.1.6)$$

However, this formulation yields very little information on the actual course of the reaction, i.e. of which partial processes it consists. This set of partial processes is termed the mechanism of the electrode reaction. In the first case, the electrode reaction at some electrodes involves the formation of an adsorbed hydrogen atom, followed by the recombination reaction:

$$\begin{aligned} H_3O^+ + e &\rightarrow H(ads) + H_2O \\ 2H(ads) &\rightarrow H_2 \end{aligned} \qquad (5.1.7)$$

(this process can occur through other mechanisms; see Section 5.7.1). In the second case, the reduction

$$\begin{aligned} &C_6H_5CHO + H^+ \rightarrow C_6H_5C^+HOH \\ &C_6H_5C^+HOH + e \rightarrow C_6H_5\dot{C}HOH \\ &2C_6H_5\dot{C}HOH \rightarrow C_6H_5CHOH.CHOHC_6H_5 \end{aligned} \tag{5.1.8}$$

occurs at the mercury electrode under suitable conditions (pH, potential). The deposition of cadmium from its cyanide complex at a mercury electrode is analogous. A solution containing cyanide ions in a concentration of $1\ \text{mol} \cdot \text{dm}^{-3}$ contains practically only the $Cd(CN)_4{}^{2-}$ complex. This complex cannot participate directly in the electrode reaction and thus the $Cd(CN)_2$ complex enters the electrode reaction in a given potential range; the equilibrium concentration of this complex in solution is small. The reaction scheme can then be written as

$$\begin{aligned} &Cd(CN)_4{}^{2-} \rightleftarrows Cd(CN)_3{}^- + CN^- \rightleftarrows Cd(CN)_2 + 2CN^- \\ &Cd(CN)_2 + 2e \rightarrow Cd(Hg) + 2CN^- \end{aligned} \tag{5.1.9}$$

It is obvious that the electrode process in general is more complicated than would appear from simple overall schemes of electrode processes. It often happens that the substances that are the initial components in the overall reaction are not those that participate in the actual electrode reaction; the latter are rather different substances in chemical equilibrium with the initial components.

The case where the overall reaction involves several electrode reactions is even more general. An example is the reaction of quinone (Q) at a platinum electrode. This process occurs at $pH < 5$ according to the scheme

$$\begin{aligned} &H^+ + Q \rightleftarrows HQ^+ \\ &HQ^+ + e \rightleftarrows H\dot{Q} \\ &H^+ + H\dot{Q} \rightleftarrows H_2Q^+ \\ &H_2Q^+ + e \rightleftarrows H_2Q \end{aligned} \tag{5.1.10}$$

Two chemical equilibria participate in this process and, in addition, stepwise charge transfer reactions occur. The product of the first electrode reaction undergoes a further electrode reaction.

Electrochemical kinetics is part of general chemical kinetics and has a similar purpose: to determine the mechanism of the electrode process and quantitatively describe its time dependence. Mostly the study involves several stages. Firstly it is necessary to determine the reaction path, i.e. to determine the mechanism of the actual electrode reaction (for more detail, see Section 5.3.1), and the partial steps forming the overall electrode process. The next task involves finding the step that is the slowest and, therefore, controls the overall rate of the process, i.e. the current. The other, faster steps are then not apparent as the appropriate equilibrium is established (e.g. in the chemical reaction or the electrode reaction) or because they are negligible (e.g. convective diffusion in connection with a very slow charge transfer reaction, where the concentration gradient formed is removed by stirring the solution).

For more complex electrode mechanisms, it is useful to designate the individual steps by symbols: E for an electrode reaction and C for a chemical reaction. Thus, the mechanisms of the process (5.1.7) are denoted as EC, of (5.1.9) as CE, of (5.1.8) as CEC and of (5.1.10) as CECE.

In electrochemical kinetics, the concept of the electrode potential is employed in a more general sense, and designates the electrical potential difference between two identical metal leads, the first of which is connected to the electrode under study (*test, working* or *indicator electrode*) and the second to the *reference electrode* which is in a currentless state. Electric current flows, of course, between the test electrode and the third, *auxiliary,* electrode. The electric potential difference between these two electrodes includes the *ohmic potential difference* as discussed in Section 5.5.2.

If current passes through an electrolytic cell, then the potential of each of the electrodes attains a value different from the equilibrium value that the electrode should have in the same system in the absence of current flow. This phenomenon is termed *electrode polarization.* When a single electrode reaction occurs at a given current density at the electrode, then the degree of polarization can be defined in terms of the *overpotential.* The overpotential η is equal to the electrode potential E under the given conditions minus the equilibrium electrode potential corresponding to the considered electrode reaction E_e:

$$\eta = E - E_e \tag{5.1.11}$$

However, the value of the equilibrium electrode potential is often not well defined (e.g. when the electrode reaction produces an intermediate that undergoes a subsequent chemical reaction yielding one or more final products). Often, an equilibrium potential is not established at all, so that the calculated equilibrium values must often be used.

It should also be recalled that, for many electrodes in various solutions, the open circuit electrode potential, i.e. the electrode potential in the absence of current flow, is not the thermodynamic equilibrium but rather a mixed potential, discussed in Section 5.8.4.

Polarization is produced by the slow rate of at least one of the partial processes in the overall electrode process. If this rate-controlling step is a transport process, then *concentration polarization* is involved; if it is the charge transfer reaction, then it is termed *charge transfer polarization,* etc. Electrode processes are often classified on this basis.

References

Albery, W. J., *Electrode Kinetics,* Clarendon Press, Oxford, 1975.
Bamford, C. H., and R. G. Compton (Eds), *Electrode Kinetics—Principles and Methodology,* Elsevier, Amsterdam, 1986.
Compton, R. G. (Ed.), *Electrode Kinetics,* Elsevier, Amsterdam, 1988.
Damaskin, B. B., and O. A. Petrii, *Introduction into Electrochemical Kinetics* (in Russian), Vysshaya shkola, Moscow, 1975.
Gerischer, H., Semiconductor electrode reactions, *AE,* **1,** 31 (1961).
Gerischer, H., Elektrodenreaktionen mit angeregten elektronischen Zuständen, *Ber. Bunsenges.,* **77,** 771 (1973).
Hush, N. S. (Ed.), *Reactions of Molecules at Electrodes,* Wiley–Interscience, New York, 1970.
Kinetics and Mechanisms of Electrode Processes, CTE, **7** (1983).
Krishtalik, L. I., *Charge transfer in Chemical and Electrochemical Processes,* Plenum Press, New York, 1986.
Morrison, S. R., *Electrochemistry at Semiconductor and Oxidized Metal Electrodes,* Plenum Press, New York, 1980.
Myamlin, V. A., and Yu. V. Pleskov, *Electrochemistry of Semiconductors,* Plenum Press, New York, 1967.
Southampton Electrochemistry Group (R. Greef, R. Peat, L. M. Peter, D. Pletcher, and J. Robinson), *Instrumental Methods in Electrochemistry,* Ellis Horwood, Chichester, 1985.
Vetter, K. J., *Electrochemical Kinetics,* Academic Press, New York, 1967.
Vielstich, W., and W. Schmickler, *Elektrochemie II. Kinetik elektrochemischer Systeme,* Steinkopf, Darmstadt, 1976.

5.2 Elementary Outline of Simple Electrode Reactions

5.2.1 *Formal approach*

The basic relationships of electrochemical kinetics are identical with those of chemical kinetics. Electrochemical kinetics involves an additional parameter, the electrode potential, on which the rate of the electrode reaction depends. The rate of the electrode process is proportional to the current density at the studied electrode. As it is assumed that electrode reactions are, in general, reversible, i.e. that both the anodic and the opposite cathodic processes occur simultaneously at a given electrode, the current density depends on the rate of the oxidation (anodic) process, v_a, and of the reduction (cathodic) process, v_c, according to the relationship

$$j = nF(v_a - v_c) \tag{5.2.1}$$

where n is the charge number of the electrode reaction appearing in the general equation

$$\sum_i \nu_{i,r}\,\mathrm{Red}_i \rightleftarrows \sum_i \nu_{i,o}\mathrm{Ox}_i + n\mathrm{e} \tag{5.2.2}$$

where Red_i and Ox_i are the reactants and $\nu_{i,r}$ and $\nu_{i,o}$ are the *stoichiometric coefficients.*

The rates of the partial anodic and cathodic processes at a given electrode potential can often be expressed by the relationships

$$v_a = k_a \prod_i c_i^{\nu_{i,a}}$$
$$v_c = k_c \prod_i c_i^{\nu_{i,c}} \tag{5.2.3}$$

where k_a is the potential-dependent rate constant of the anodic process, k_c is the potential-dependent rate constant of the cathodic process, c_i is the concentration of reactant i, and the *order of the electrode reaction* with respect to the ith reactant, $\nu_{i,a}$ or $\nu_{i,c}$, is given by the relationships

$$\nu_{i,a} = \frac{d \ln v_a}{d \ln c_i}$$
$$\nu_{i,c} = \frac{d \ln v_c}{d \ln c_i} \tag{5.2.4}$$

In more complicated reactions, the reaction orders $\nu_{i,a}$ and $\nu_{i,c}$ need not and often do not correspond to the stoichiometric coefficients $\nu_{i,r}$ and $\nu_{i,o}$. In contrast to the latter, the reaction orders can often be fractional or even negative. The concentration of a given reactant can sometimes appear in the expressions for both the anodic and the cathodic reaction rates.

If it is known which of the reactions determine the rate of the overall complex electrode process, then the concept of the *stoichiometric number of the electrode process* ν is often introduced. This number is equal to the number of identical partial reactions required to realize the overall electrode process, as written in an equation of type (5.2.2).† If the rate constant of this partial rate-determining reaction is k'_a, then $k_a = k'_a/\nu$. Thus, for example, if the first of reactions (5.1.7) is the rate-determining step in the overall electrode process (5.1.4) then the stoichiometric number has the value $\nu = 2$.

5.2.2 *The phenomenological theory of the electrode reaction*

As the generality of equations of type (5.2.3) should not be exaggerated (e.g. in the presence of strong adsorption, such as in electrocatalytic processes, they are no longer valid), the basic features of electrochemical kinetics will be explained by using the simple electrode reaction

$$\text{Red} \rightleftarrows \text{Ox} + ne \tag{5.2.5}$$

† A different formulation would be: the stoichiometric number gives the number of identical activated complexes formed and destroyed in the completion of the overall reaction as formulated with charge number n.

The rate of the oxidation reaction (i.e. the amount of substance oxidized per unit surface area per unit time) is given by the relationship (cf. Eq. 5.2.3)

$$v_a = k_a c_{Red} \tag{5.2.6}$$

and the rate of the reduction reaction by the relationship

$$v_c = k_c c_{Ox} \tag{5.2.7}$$

The oxidized and reduced forms can be dissolved in the electrolyte but can also be present as a solid phase (metal, insoluble compound). In the latter case, the concentration of these electroactive substances is constant and is set by convention equal to unity. The reduced form can also be dissolved in the form of an amalgam in mercury representing the electrode material.

The rate of the electrode process—similar to other chemical reactions—depends on the rate constant characterizing the proportionality of the rate to the concentrations of the reacting substances. As the charge transfer reaction is a heterogeneous process, these constants for first-order processes are mostly expressed in units of centimetres per second.

Similarly as for chemical reactions, the rate constants of the electrode reactions can be written in terms of the Arrhenius equation

$$k_a = P_a \exp\left(-\frac{\Delta\tilde{H}_a}{RT}\right) \tag{5.2.8}$$

$$k_c = P_c \exp\left(-\frac{\Delta\tilde{H}_c}{RT}\right) \tag{5.2.9}$$

where P_a and P_c are the *pre-exponential factors* which are independent of the electrode potential, and $\Delta\tilde{H}_a$ and $\Delta\tilde{H}_c$ are the *activation enthalpies* of the oxidation (anodic) and reduction (cathodic) electrode reactions, respectively. The wavy line indicates that these quantities depend on the electric potential difference between the electrode and the solution, $\Delta\phi = \phi(\mathrm{m}) - \phi(\mathrm{l})$. Since, according to Eq. (3.1.69), this quantity is identical, except for a constant, with the electrode potential, $\Delta\phi = E + \text{constant}$, the activation enthalpy depends on the electrode potential. As demonstrated in a number of experiments, the rate of the electrode reaction is approximately an exponential function of the electrode potential; the rate of the cathodic (reduction) electrode reaction increases towards negative electrode potential values, while the rate of the anodic (oxidation) electrode reaction increases with increasing (positive) electrode potential. This dependence of the rate of the electrode reaction on the electrode potential appears in a simple functional dependence of the activation enthalpy. Thus, in an approximation, the activation enthalpy of the cathodic reaction is expressed as

$$\Delta\tilde{H}_c = \Delta H_c^0 + \alpha n F E \tag{5.2.10}$$

where α is a constant termed the *charge transfer coefficient*. In general, however, this quantity is also a function of the electrode potential (see Section 5.3).

The functional dependence of the activation energy of the anodic electrode reaction can be derived as follows. According to the definition of the rate of the electrode reaction, the partial current density

$$j_a = nFk_a c_{Red} \tag{5.2.11}$$

corresponds to the anodic electrode reaction and the current density

$$-j_c = nFk_c c_{Ox} \tag{5.2.12}$$

to the cathodic electrode reaction. The overall current density is given by the relationship

$$j = j_a + j_c = nF(k_a c_{Red} - k_c c_{Ox}) \tag{5.2.13}$$

In the absence of current ($j = 0$), the system is in equilibrium but the electrode reactions nonetheless proceed. Thus, similar to chemical equilibria, the electrode equilibria have a dynamic character. Under equilibrium conditions

$$k_a c_{Red} = k_c c_{Ox} > 0 \tag{5.2.14}$$

Substitution of Eqs (5.2.8), (5.2.9) and (5.2.10) into this equation yields

$$P_a \exp\left(-\frac{\Delta\tilde{H}_a}{RT}\right) c_{Red} = P_c \exp\left(-\frac{\Delta H_c^0 + \alpha nFE}{RT}\right) c_{Ox} \tag{5.2.15}$$

and, after rearrangement,

$$\alpha E - \frac{\Delta\tilde{H}_a}{nF} = -\frac{\Delta H_c^0}{nF} + \frac{RT}{nF}\ln\frac{P_c}{P_a} + \frac{RT}{nF}\ln\frac{c_{Ox}}{c_{Red}} \tag{5.2.16}$$

The equilibrium electrode potential is given by the Nernst equation (cf. 3.2.17),

$$E = E^0 + \frac{RT}{nF}\ln\frac{a_{Ox}}{a_{Red}} = E^{0\prime} + \frac{RT}{nF}\ln\frac{c_{Ox}}{c_{Red}} \tag{5.2.17}$$

where $E^{0\prime}$ is the formal potential of the half-cell reaction (5.2.5) (cf. Section 3.1.5). These equations satisfy Eqs (5.2.8) and (5.2.9) when

$$\Delta\tilde{H}_a = \Delta H_a^0 - (1 - \alpha)nFE \tag{5.2.18}$$

Obviously $0 < \alpha < 1$. For the conditional (formal) electrode potential we obtain

$$E^{0\prime} = \frac{\Delta H_{Ox}^0 - \Delta H_{Red}^0}{nF} + \frac{RT}{nF}\ln\frac{P_c^0}{P_a^0} \tag{5.2.19}$$

If the system is at equilibrium at the formal potential, $c_{Ox} = c_{Red}$ (see Eq. 5.2.14) and thus

$$k_a = k_c = k^{\ominus} \tag{5.2.20}$$

The quantity $k^{\ominus}$ is termed the conditional (formal) rate constant of the electrode reaction. It follows from Eqs (5.2.8), (5.2.9), (5.2.10) and (5.2.18) that

$$\begin{aligned} k^{\ominus} &= P_a \exp\left(-\frac{\Delta H_a^0 - (1-\alpha)nFE^{0\prime}}{RT}\right) \\ &= P_c \exp\left(-\frac{\Delta H_c^0 + \alpha nFE^{0\prime}}{RT}\right) \end{aligned} \tag{5.2.21}$$

Substitution for $P_a \exp(-\Delta H_a^0/RT)$ and $P_c \exp(-\Delta H_c^0/RT)$ from these equations into Eqs (5.2.8) and (5.2.9), considering Eqs (5.2.10) and (5.2.18), yields the equations for the dependence of the rate constant on the electrode potential:

$$k_a = k^{\ominus} \exp\left[\frac{(1-\alpha)nF(E - E^{0\prime})}{RT}\right] \tag{5.2.22}$$

$$k_c = k^{\ominus} \exp\left[-\frac{\alpha nF(E - E^{0\prime})}{RT}\right] \tag{5.2.23}$$

A further substitution from these equations into Eq. (5.2.13) yields the *fundamental equation of electrochemical kinetics*:

$$\begin{aligned} j = nFk^{\ominus}\bigg\{ &\exp\left[\frac{(1-\alpha)nF(E - E^{0\prime})}{RT}\right] c_{Red} \\ &- \exp\left[-\frac{\alpha nF(E - E^{0\prime})}{RT}\right] c_{Ox}\bigg\} \end{aligned} \tag{5.2.24}$$

In addition to the thermodynamic quantity $E^{0\prime}$, the electrode reaction is characterized by two kinetic quantities: the charge transfer coefficient α and the conditional rate constant $k^{\ominus}$. These quantities are often sufficient for a complete description of an electrode reaction, assuming that they are constant over the given potential range. Table 5.1 lists some examples of the constant $k^{\ominus}$. If the constant $k^{\ominus}$ is small, then the electrode reaction occurs only at potentials considerably removed from the standard potential. At these potential values practically only one of the pair of electrode reactions proceeds which is the case of an *irreversible* or one-way electrode reaction.

As mentioned above, at equilibrium two opposing currents pass through the electrode, with absolute values termed the *exchange current*. The exchange current density j_0 can be expressed at an arbitrary value of the equilibrium potential as a function of the concentrations of the oxidized and

Table 5.1 Conditional electrode reaction rate constants $k^{\ominus}$ and charge transfer coefficients α. (From R. Tamamushi)

System	Electrolyte solution	Electrode	°C	$k^{\ominus}$ $(cm \cdot s^{-1})$	α	Method
$Ag^{+} + e \rightleftarrows Ag$	1 M $HClO_4$	Ag	25	0.025 ± 0.005		Chronopotentiometry
$Cd^{2+} + e \rightleftarrows Cd(Hg)$	1 M $HClO_4$	Hg	25	0.35 ± 0.03	0.14 ± 0.02	Faradaic impedance
$Ce^{4+} + e \rightleftarrows Ce^{3+}$	1 M H_2SO_4	C paste	25	3.8×10^{-4}	0.28	Current–potential curve
$Cu(NH_3)_2^{2+} + 2e \rightleftarrows Cu(Hg)$	1 M NH_3, 1 M NH_4Cl	Hg	20	1.1×10^{-2}	—	Faradaic impedance
$Eu^{III} + e \rightleftarrows Eu^{II}$	1 M KCl	Hg	20	2.1×10^{-4}	—	Faradaic impedance
$Fe^{3+} + e \rightleftarrows Fe^{2+}$	0.5 M $HClO_4$	Pt	25	9×10^{-6}	0.50	Current–potential curve
$Fe^{3+} + e \rightleftarrows Fe^{2+}$	1 M HCl	C	21	1.2×10^{-4}	0.59	Current–potential curve
$Fe(CN)_6^{3-} + e \rightleftarrows Fe(CN)_6^{4-}$	1 M KNO_3	Pt	35	6.6×10^{-2}	0.49	Faradaic rectification
$Hg^{+} + e \rightleftarrows Hg$	1.1 M $HClO_4$	Hg	24 ± 2	1.3	0.28	Faradaic rectification
$MnO_4^{-} + e \rightleftarrows MnO_4^{2-}$	1 M KOH	Pt	20	1.2×10^{-2}	—	Voltammetry at rotating disk electrode
$Pb^{2+} + 2e \rightleftarrows Pb(Hg)$	1 M $HClO_4$	Hg	22	2.0	0.50	Faradaic impedance
$Ti^{IV} + e \rightleftarrows Ti^{III}$	1 M tartaric acid	Hg	20	9×10^{-3}	—	Faradaic impedance
$Tl^{+} + e \rightleftarrows Tl(Hg)$	1 M $HClO_4$	Hg	22	1.8	—	Faradaic impedance
$V^{III} + e \rightleftarrows V^{II}$	1 M $HClO_4$	Hg	20	3.2×10^{-3}	0.52	Faradaic impedance
$Zn^{2+} + 2e \rightleftarrows Zn(Hg)$	0.014 M $NaNO_3$	Hg	22	5×10^{-2}	0.25–0.30	Polarography
	1.05 M $NaNO_3$	Hg	22	3×10^{-3}	0.25–0.30	Polarography

reduced forms of the electroactive substance. Equation (5.2.24) gives

$$\frac{j_0}{nF} = c_{Ox} k^{\ominus} \exp\left[-\frac{\alpha nF(E - E^{0\prime})}{RT}\right]$$
$$= c_{Red} k^{\ominus} \exp\left[\frac{(1-\alpha)nF(E - E^{0\prime})}{RT}\right] \tag{5.2.25}$$

Substitution for $E - E^{0\prime}$ from the Nernst equation (3.2.14) or (5.2.17) yields

$$j_0 = nFk^{\ominus} c_{Ox}^{1-\alpha} c_{Red}^{\alpha} \tag{5.2.26}$$

The charge transfer coefficient α can be found from the dependence of j_0 on c_{Ox} or c_{Red}.

It is often useful to express the current density as a function of the overpotential $\eta = E - E_e$ (cf. Eq. 5.1.11) and of the exchange current density. On substituting for $k^{\ominus}$ from Eq. (5.2.26) and for $E - E^{0\prime}$,

$$E - E^{0\prime} = \eta + \frac{RT}{nF} \ln(c_{Ox}/c_{Red}) \tag{5.2.27}$$

into Eq. (5.2.24) we obtain the equation

$$j = j_0\left\{\exp\left[\frac{(1-\alpha)nF\eta}{RT}\right] - \exp\left(-\frac{\alpha nF\eta}{RT}\right)\right\} \tag{5.2.28}$$

(see Fig. 5.2). The $j - \eta$ dependence is termed the *polarization curve* (voltammogram). If the overpotential is small ($\eta < RT/nF$), then, on expanding the exponential function and neglecting all the terms in the series except the first two, we have

$$j = \frac{j_0 nF\eta}{RT} \tag{5.2.29}$$

The value of the exchange current determines the deviation of the electrode potential from the equilibrium value during the current flow. The greater the deviation, the slower the electrode reaction. It follows from Eq. (5.2.29) that

$$\left(\frac{\partial j}{\partial \eta}\right)_{\eta \to 0} = \frac{j_0 nF}{RT} \tag{5.2.30}$$

The reciprocal value of this differential quotient,

$$R_p = \frac{RT}{nFj_0} \tag{5.2.31}$$

has the dimensions of resistance; it is termed the *polarization resistance* at the equilibrium potential. This resistance is referred to the unit electrode area.

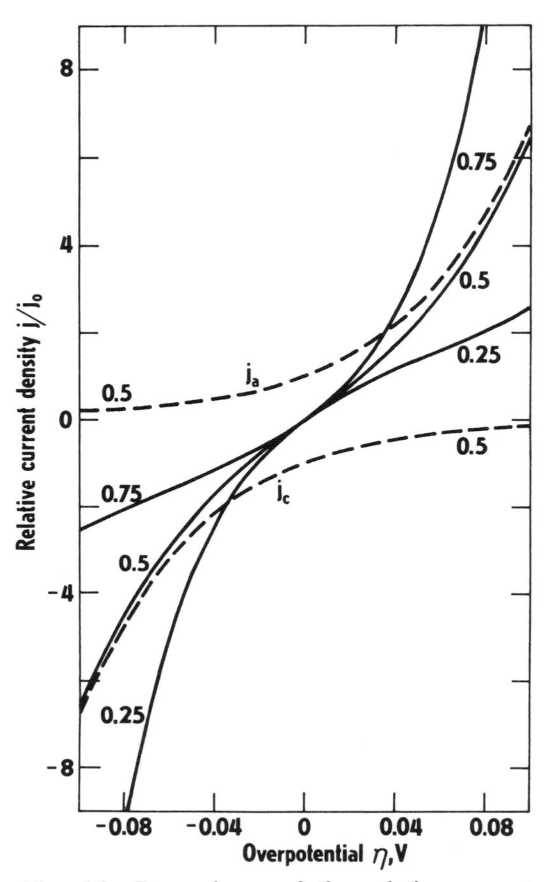

Fig. 5.2 Dependence of the relative current density j/j_0 on the overpotential η according to Eq. (5.2.28). Various values of the charge transfer coefficient α are indicated at each curve. Dashed curves indicate the partial current densities (Eqs 5.2.11 and 5.2.12 for $\alpha = 0.5$). (According to K. Vetter)

Plotting the overpotential against the decadic logarithm of the absolute value of the current density yields the 'Tafel plot' (see Fig. 5.3). Both branches of the resultant curve approach the asymptotes for $|\eta| \gg RT/F$. When this condition is fulfilled, either the first or second exponential term on the right-hand side of Eq. (5.2.28) can be neglected. The electrode reaction then becomes irreversible (cf. page 257) and the polarization curve is given by the Tafel equation

$$\eta = a + b \log |j| \tag{5.3.32}$$

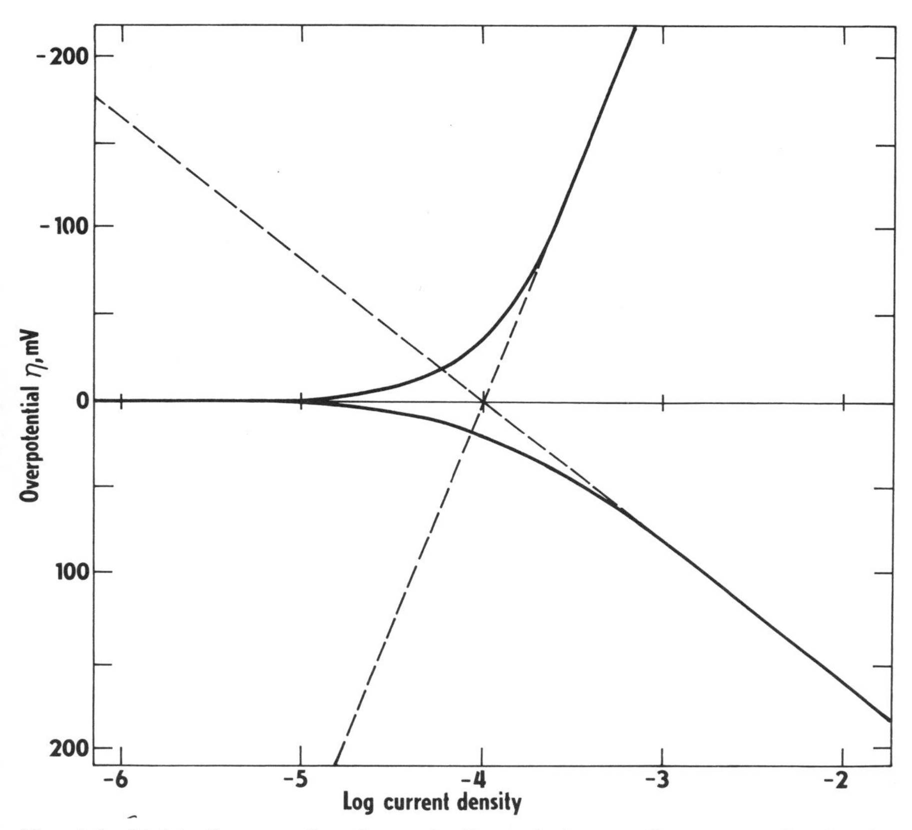

Fig. 5.3 Tafel diagrams for the cathodic and the anodic current density for $\alpha = 0.25$ and $j_0 = 10^{-4}\ \mathrm{A \cdot cm^{-2}}$

For an anodic process, the quantities a and b are given by the relationships

$$a = -\frac{2.3RT}{(1-\alpha)nF}\log j_0 \tag{5.2.33}$$

$$b = \frac{2.3RT}{(1-\alpha)nF} \tag{5.2.34}$$

and for cathodic processes,

$$\begin{aligned} a &= \frac{2.3RT}{\alpha nF}\log j_0 \\ b &= -\frac{2.3RT}{\alpha nF} \end{aligned} \tag{5.2.35}$$

Very often, the value of the formal electrode potential E_0' is not known for an irreversible electrode reaction. The overpotential η cannot, therefore, be determined and the Tafel equation is then used as an operational formula in the form

$$E = a + b \log |j| \tag{5.2.36}$$

It is often useful to replace this equation by Eq. (5.2.24) modified for an irreversible process (occurring at large overpotentials), which has the following form for cathodic processes:

$$\begin{aligned} j &= -nFk^{\ominus} \exp\left[-\frac{\alpha nF(E - E^{0\prime})}{RT}\right] c_{\mathrm{Ox}} \\ &= -nFk_{\mathrm{conv}} \exp\left(-\frac{\alpha nFE}{RT}\right) c_{\mathrm{Ox}} \end{aligned} \tag{5.2.37}$$

The quantity $k_{\mathrm{conv}} = k^{\ominus} \exp(\alpha nFE^{0\prime}/RT)$ is the value of the rate constant of the electrode reaction at the potential of the standard reference electrode and will be termed the *conventional rate constant* of the electrode reaction. It can be found, for example, by extrapolation of the dependence (5.2.36) to $E = 0$, as

$$k_{\mathrm{conv}} = \frac{-(j)_{E=0}}{nFc_{\mathrm{Ox}}} \tag{5.2.38}$$

The existence of the exchange current was clearly demonstrated by Pleskov and Miller on experiments with a solution containing metal ions labelled with a radioactive isotope. The solution was in contact with a dilute unlabelled amalgam of the same metal. The electrode reaction (proceeding in absence of an external voltage imposed on the metal) involved exchange of the labelled ions for unlabelled atoms in the amalgam. The rate of enrichment of the amalgam by the labelled isotope and its dependence on the concentrations of the solution and amalgam were completely in agreement with the relationship derived for the exchange current (5.2.26). The experiments were carried out under strong stirring, to suppress the effect of diffusion on the rate of the electrode process.

An electrode reaction in which the oxidized form accepts more than one electron usually proceeds as a series of one-electron reaction steps. As will be demonstrated below, if the formal potentials of these partial electrode reactions satisfy certain conditions, then the electrode reaction simulates the transfer of several electrons in one step (Eq. 5.2.5) and obeys Eq. (5.2.24). An example is the two-electron reaction of substance A_1, converted to substance A_3 by the transfer of two electrons, where the reaction occurs through the unstable intermediate A_2:

$$A_1 + e \underset{k_{a,1}}{\overset{k_{c,1}}{\rightleftharpoons}} A_2 \tag{5.2.39}$$

$$A_2 + e \underset{k_{a,2}}{\overset{k_{c,2}}{\rightleftharpoons}} A_3 \tag{5.2.40}$$

The charge transfer reaction (5.2.39) is characterized by the formal electrode potential $E_1^{0\prime}$, the conditional rate constant of the electrode reaction $k_1^{\ominus}$ and the charge transfer coefficient α_1, while the reaction (5.2.40) is characterized by the analogous quantities $E_2^{0\prime}$, $k_2^{\ominus}$ and α_2. If the rate constants of the electrode reactions, which are functions of the potential, are denoted as in Eqs (5.2.39) and (5.2.40) and the concentrations of substances A_1, A_2 and A_3 are c_1, c_2 and c_3, respectively, then

$$-\frac{j}{F} = k_{c,1}c_1 + k_{c,2}c_2 - k_{a,1}c_2 - k_{a,2}c_3 \tag{5.2.41}$$

The concentration of the unstable intermediate A_2 is obtained from the steady-state condition,

$$\frac{dc_2}{dt} = 0 = k_{c,1}c_1 - k_{c,2}c_2 - k_{a,1}c_2 + k_{a,2}c_3 \tag{5.2.42}$$

Substitution for c_2 in Eq. (5.2.41) yields

$$-\frac{j}{2F} = \frac{k_{c,1}k_{c,2}}{k_{c,2} + k_{a,1}} c_1 - \frac{k_{a,1}k_{a,2}}{k_{c,2} + k_{a,1}} c_3 \tag{5.2.43}$$

It will be assumed that in a given potential range

$$k_{a,1} \ll k_{c,2} \tag{5.2.44}$$

This condition expresses the fact that reaction (5.2.39) determines the rate of the overall process. On neglecting $k_{a,1}$ in the denominator of Eq. (5.2.43), substitution of

$$\begin{aligned}
k_{c,1} &= k_1^{\ominus} \exp\left[-\frac{\alpha_1 F(E - E_1^{0\prime})}{RT}\right] \\
k_{c,2} &= k_2^{\ominus} \exp\left[-\frac{\alpha_2 F(E - E_2^{0\prime})}{RT}\right] \\
k_{a,1} &= k_1^{\ominus} \exp\left[\frac{(1-\alpha_1) F(E - E_1^{0\prime})}{RT}\right] \\
k_{a,2} &= k_2^{\ominus} \exp\left[\frac{(1-\alpha_2) F(E - E_2^{0\prime})}{RT}\right]
\end{aligned} \tag{5.2.45}$$

yields

$$\frac{j}{2F} = -k_1^{\ominus}\left\{\exp\left[-\frac{\alpha_1 F(E - E_1^{0\prime})}{RT}\right]c_1 - \exp\left[\frac{(1-\alpha_1) F(E - E_1^{0\prime})}{RT}\right]\exp\left[\frac{F(E - E_2^{0\prime})}{RT}\right]c_3\right\} \tag{5.2.46}$$

Introduction of new constants

$$E^{0\prime} = \frac{E_1^0 + E_2^0}{2}$$

$$k^{\ominus} = k_1^{\ominus} \exp\left[\frac{\alpha F(E_1^{0\prime} - E_2^{0\prime})}{RT}\right]$$

$$\alpha = \frac{\alpha_1}{2} \tag{5.2.47}$$

converts Eq. (5.2.46) to Eq. (5.2.24) for $n = 2$. A similar result would be obtained if the rate of the overall reaction were controlled by the reaction (5.2.40). Thus, assuming the validity of Eqs. (5.2.42) and (5.2.44), the stepwise transfer of electrons can formally yield the same dependence of the current on the potential as for direct transfer of several electrons in one step.

This is a very common case in irreversible, multielectron electrode reactions. The rate-controlling process is often the first step in which, for example, a single electron is accepted, while the other reactions connected with the acceptance of further electrons are very fast and the rate of the reverse (here oxidation) reaction is negligible. Equation (5.2.46) gives

$$j = -nFk_1^{\ominus} \exp\left[-\frac{\alpha_1 F(E - E_1^{0\prime})}{RT}\right]c_1 \tag{5.2.48}$$

The overall charge number of the electrode reaction is then n, while the exponential term in the rate constant of the electrode reaction has a form corresponding to a one-electron reaction. If the value of $E^{0\prime}$ is not known, then the conventional rate constant of the electrode reaction is introduced, $k_{\text{conv}} = k_1^{\ominus} \exp(\alpha_1 F E_1^{0\prime}/RT)$, so that Eq. (5.2.48) can be expressed in the form

$$j = -nFk_{\text{conv}} \exp\left(-\frac{\alpha_1 FE}{RT}\right)c_1 \tag{5.2.49}$$

If the single-electron mechanism has not been demonstrated to be the rate-controlling process by an independent method, then, in the publication of the experimental results, it is preferable to replace the assumed quantity α_1 by the conventional value αn, provided that the charge number of the overall reaction is known (e.g. in an overall two-electron reaction it is preferable to replace $\alpha_1 = 0.5$ by $\alpha = 0.25$). If the independence of the charge transfer coefficient on the potential has not been demonstrated for the given potential range, then it is useful to determine it for the given potential from the relation for a cathodic electrode reaction (cf. Eq. 5.2.37):

$$\alpha_c = \frac{2.3RT}{nF}\frac{\mathrm{d}\log|j|}{\mathrm{d}E} \tag{5.2.50}$$

A similar relationship is valid for the charge transfer coefficient of the anodic electrode reaction, α_a.

The determination of the activation energies of electrode reactions is especially important for the theory of electrode reactions and for study of the relationship between the structure of the reacting substances and the electrode reaction rates.

Several descriptions of electrode reaction rates discussed on the preceding pages and the difficulty to standardize electrode potential scales with respect to different temperatures imply several definitions of *activation energies of electrode reactions*. The easiest way to determine this quantity, for example, for an irreversible cathodic process, employs Eqs (5.2.9), (5.2.10) and (5.2.12) at a constant electrode potential,

$$\left(\frac{\partial \ln |j|}{\partial T}\right)_E = \frac{\Delta H_c^0 + \alpha nFE}{RT^2} = \frac{\Delta H_1^\ddagger}{RT^2} \tag{5.2.51}$$

This apparently simple expression is actually only conventional; it must be realized that the electrode potential at various temperatures does not differ by only an additive constant when it is referred to different standard reference electrodes, but also by the product of the change in the temperature and the temperature coefficient of this additive constant (see Section 3.1.4). Often, however, the actual value of the *activation energy at constant potential* $\Delta H_1^\ddagger$ is not of interest, but rather its increase $\Delta\Delta H_1^\ddagger$, giving the change in the activation energy between two members, e.g. in a homologous series of electroactive substances (see Section 5.9.1). This quantity does not depend on the electrode potential if the charge transfer coefficient α is constant for the electrode process in the studied group of substances. In this case

$$\Delta\left(\frac{\partial \ln |j|}{\partial T}\right)_E = \frac{\Delta\Delta H_c^0}{RT^2} = \frac{\Delta\Delta H_1^\ddagger}{RT^2} \tag{5.2.52}$$

The *standard activation energy* of the electrode reaction $H_0^\ddagger$ is defined as

$$\frac{\partial \ln k^\ominus}{\partial T} = \frac{\Delta H_0^\ddagger}{RT^2} \tag{5.2.53}$$

Equation (5.2.21) gives

$$\frac{\partial \ln k^\ominus}{\partial T} = \frac{\Delta H_a^0 - (1-\alpha)nFE^{0\prime}}{RT^2} = \frac{\Delta H_c^0 + \alpha nFE^{0\prime}}{RT^2} \tag{5.2.54}$$

Elimination of $E^{0\prime}$ from this equation yields

$$H_0^\ddagger = \alpha\Delta H_a^0 + (1-\alpha)\Delta H_c^0 \tag{5.2.55}$$

The third type of activation energy is the *activation energy at constant overpotential* ($\eta \gg RT/F$), defined by the equation

$$\left(\frac{\partial \ln |j|}{\partial T}\right)_\eta = \frac{\Delta H_2^\ddagger}{RT^2} \tag{5.2.56}$$

Table 5.2 Activation energies of electrode reactions (From R. Tamamushi and from A. A. Vlček)

System	Electrolyte solution	Electrode	$\Delta H_0^{\ddagger}$ (kJ · mol^{-1})
$Cu(NH_3)_2^{2+} + 2e \rightleftarrows Cu(Hg)$	1 M NH_3, 1 M NH_4Cl	Hg	20.9
$Eu^{III} + e \rightleftarrows Eu^{II}$	1 M KCl	Hg	37.6
$MnO_4^- + e \rightleftarrows MnO_4^{2-}$	1 M KOH	Pt	19.6
$Tl^+ + e \rightleftarrows Tl(Hg)$	0.33 M K_2SO_4 5 mM H_2SO_4	Hg	8.0
$V^{III} + e \rightleftarrows V^{II}$	1 M $HClO_4$	Hg	32.6
$V^{IV} + e \rightleftarrows V^{III}$	0.1 M H_2SO_4	Hg	59.0
$Co(NH_3)_6^{3+} + e \rightleftarrows Co(NH_3)_6^{2+}$	0.14 M $HClO_4$ 1.26 M $NaClO_4$	Hg	53.0[a]

[a] $\Delta H_1^{\ddagger}$; $E = 0.333$ V vs. SHE.

For an anodic process, this quantity is given by a relationship following from Eqs (5.2.28), (5.2.26), (5.2.53) and (5.2.55):

$$\Delta H_2^{\ddagger} = \Delta H_0^{\ddagger} - (1 - \alpha)nF\eta \tag{5.2.57}$$

and for a cathodic process by the relationship

$$\Delta H_2^{\ddagger} = \Delta H_0^{\ddagger} + \alpha nF\eta \tag{5.2.58}$$

Table 5.2 lists examples of the activation energies of electrode reactions.

References

See references on page 253

Delahay, P., *Double Layer and Electrode Kinetics,* John Wiley & Sons, New York, 1965.

Parsons, R., Electrode reaction orders, charge transfer coefficients and rate constants. Extension of definitions and recommendations for publication of parameters. *Pure Appl. Chem.,* **52,** 233 (1979).

Tamamushi, R., *Kinetic Parameters of Electrode Reactions of Metallic Compounds,* Butterworths, London, 1975.

Vlček, A. A., Relation between electronic structure and polarographic behaviour of inorganic depolarizers. VII. Determination of activation energy of electrode processes. *Coll. Czech. Chem. Comm.,* **24,** 3538 (1959).

5.3 The Theory of Electron Transfer

5.3.1 *The elementary step in electron transfer*

This part is concerned with discussion of the reaction mechanism in a narrower sense, i.e. the 'microscopic' elementary step in the electrode reaction, connected with the transfer of a single electron between the electrode and the oxidized or reduced form of the electroactive substance. The possibility of adsorption interaction between the electrode and the electroactive substance is excluded.

The contemporary theory of chemical reactions attempts to describe similar microscopic steps on the basis of construction of the potential energy hypersurface of the reacting system if the applicability of the Born–Oppenheimer approximation can be assumed. In this approximation, also called the *adiabatic approximation,* the nuclear motion of the atoms (forming the molecules of the reactants and products) is supposedly independent of the motion of the electrons forming the electron shells of these atoms. The path of the reacting system across this hypersurface is termed the *system trajectory.* An advantageous system trajectory passing through a minimum in the valley of the energies of the reactants and products and through a saddle point on the hypersurface is the *reaction path of the system.* Obviously, construction of the potential energy hypersurface of the system encounters numerous difficulties and is possible only in the simplest case (reactions in the gaseous state). Thus, the discussion of elementary processes is often limited to discussion of *energy profiles,* based on a cross-section through the hypersurface along the estimated reaction pathway, where the resultant diagram, the *reaction profile,* describes only the dependence of the potential energy on the *reaction coordinate.* The reaction coordinate indicates the change of a single space coordinate of the system during the reaction.

Attempts were made to quantitatively treat the elementary process in electrode reactions since the 1920s by J. A. V. Butler (the transfer of a metal ion from the solution into a metal lattice) and by J. Horiuti and M. Polányi (the reduction of the oxonium ion with formation of a hydrogen atom adsorbed on the electrode). In its initial form, the theory of the elementary process of electron transfer was presented by R. Gurney, J. B. E. Randles, and H. Gerischer. Fundamental work on electron transfer in polar media, namely, in a homogeneous redox reaction as well as in the elementary step in the electrode reaction was made by R. A. Marcus (Nobel Prize for Chemistry, 1992), R. R. Dogonadze, and V. G. Levich.

In polar media, the strong solvation interaction between the ions and solvent dipoles, with an energy of the order of $10–10^2\,\mathrm{kJ\cdot mol^{-1}}$, must be considered for electron transfer between the electrode and the electroactive ion, or between two components of the redox system in solution. Thus, charge transfer is connected with a marked change in the solvation shell, and the activation energy of this transfer depends on the momentary configuration of the solvation shell given by its thermal fluctuations.

A comparison of the transfer between two species in a vacuum and in the condensed phase reveals marked differences. Figure 5.4A depicts the energy levels of an electron in two species, where level ε_A in species A is occupied and level ε_B in species B is empty. If these levels had different energies, as indicated in the figure, then the non-radiant energy transfer would be

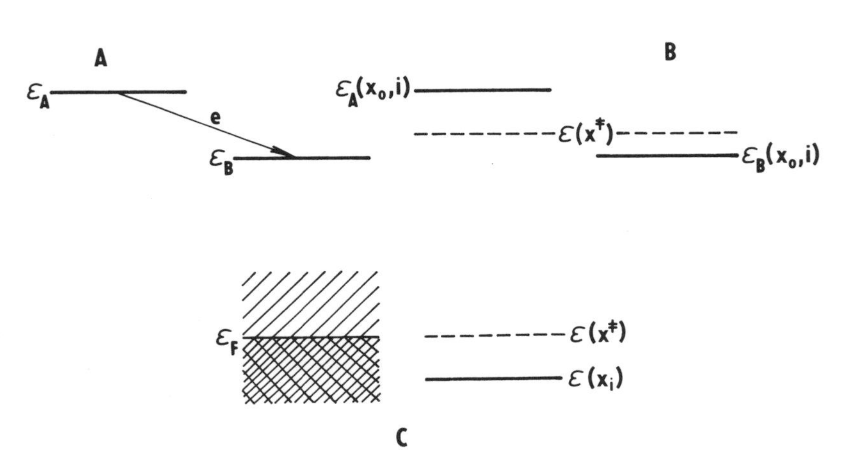

Fig. 5.4 Electron transfer *in vacuo* and in solution. (A) When the electron donor and the proton acceptor have very different corresponding energy levels the electron transfer is impossible. (B) When the reaction partners are present in a solution, then a change in their configuration can bring their corresponding energy levels close together (dashed line) so that electron transfer is possible. (C) An analogous situation is in the system of an electrode and an electroactive particle in solution

connected with an energy change that would have to appear instantaneously (during 10^{-15}–10^{-16} s) in the kinetic energy of the reacting species. However, such a rapid change of velocity of the particles is impeded by the inertia of the particles, so that electron transfer between levels of different energies is highly improbable. It becomes probable only when both the levels have approximately the same energy. This condition is termed the Franck–Condon principle (isoenergetic electron transfer). It is also valid for electron transfer in solution (where it was applied first by W. F. Libby), but in this case the time-dependent interaction of the electron donor or acceptor with the solvent dipoles plays a decisive role (Fig. 5.4B). Thus, by the momentary configuration of the solvation shell, which determines both ε_A and ε_B, the system can approach the situation $\varepsilon_A \approx \varepsilon_B$ (although in the ground state $\varepsilon_A \neq \varepsilon_B$). This is also true of the system consisting of an electron in the electrode and an electron in the electroactive particle in solution (Fig. 5.4C). Here, the energy level of the electron in the electrode is given by the Fermi level (the participation of excited states will be considered later).

The development of the theory of the rate of electrode reactions (i.e. formulation of a dependence between the rate constants k_a and k_c and the physical parameters of the system) for the general case is a difficult quantum-mechanical problem, even when adsorption does not occur. It would be necessary to consider the vibrational spectrum of the solvation shell and its vicinity and quantum-mechanical interactions between the reacting particles and the electron at various energy levels in the electrode.

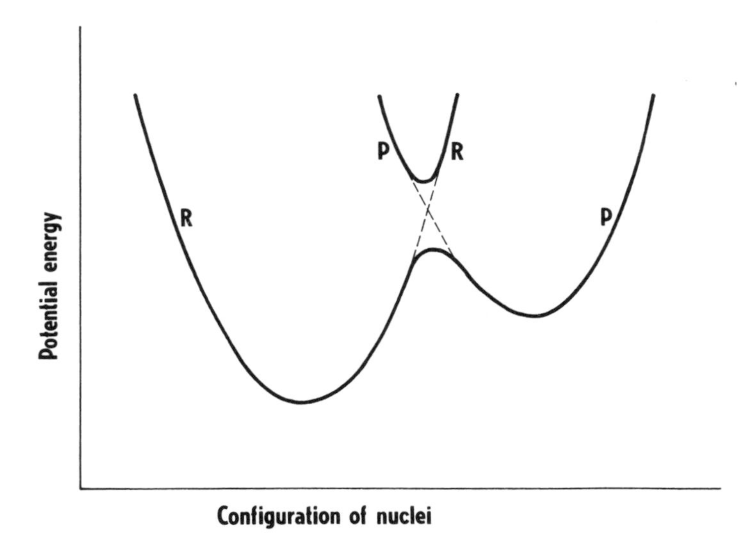

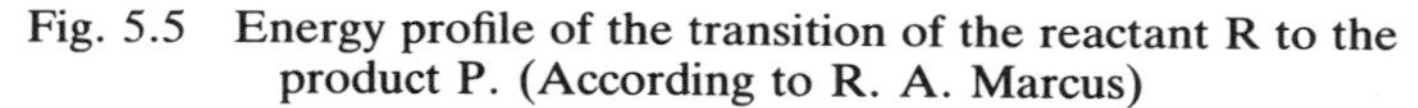

Fig. 5.5 Energy profile of the transition of the reactant R to the product P. (According to R. A. Marcus)

The limitation of the problem to a single energy profile is depicted in Fig. 5.5 as a symbolic dependence of the potential energy of the reacting system on the reaction coordinate.

If, according to the Marcus theory, the system passes rapidly through the region of intersection of the potential energy curves for the initial state (R) and for the final state (P), then there is insufficient time for the electron transfer, so that the system mostly passes through the intercept into the upper part of the R curve; the probability of electron transfer is low, characterizing *non-adiabatic electron transfer.* When the system passes through the intercept region slowly (i.e. with a slow change in the configuration of the atomic nuclei), then electron transfer almost always occurs. This is characteristic of *adiabatic electron transfer.*

The slow process is termed *reorganization of the system.* According to Marcus, the following contributions are involved:

(a) Changes in the energy and entropy of the internal structure of the species as a result of changes in the bond lengths and angles
(b) Changes in the size of the reacting particle
(c) Changes in the vibrational and orientational polarization of the vicinity of the particle

Further considerations will deal only with the simple case where the slow process involves a change in the solvation shell of the reacting species, e.g. for the $Fe^{3+}(aq)$ and $Fe^{2+}(aq)$ ions.

The applicability of the assumption of an adiabatic process depends on the vibration frequency of the slow system (i.e. of the solvation shell). The criterion is the quantity $kT/\hbar\omega$, where $\hbar$ is Planck's constant divided by 2π and ω is the angular frequency of vibration. At laboratory temperature $kT/\hbar\omega \simeq 10^{-13}$. For the vibration of the solvent shell $\omega \ll kT/\hbar$ while for the vibration of the electron shell $\omega \gg kT/\hbar$, which means that the electron transfer can be considered as adiabatic. Under these conditions, a further simplification is possible as the whole problem can be treated classically, the vibrations of the solvation shell being discussed in terms of a set of classical harmonic oscillations.

In order to describe properly the electron transfer between the electroactive ion B^{z+} and the electrode

$$B^{z+}(l) + e(m) \rightarrow B^{(z-1)+}(l) \tag{5.3.1}$$

the following assumptions are necessary:

(a) The solvation shells of the ions B^{z+} and $B^{(z-1)+}$ vibrate harmonically with identical frequency ω, so that the potential energies of both solvation shells can be expressed by identical mutually shifted parabolas.
(b) Quantum-mechanical effects need not be considered in the region around the activated complex (in the vicinity of the intersection of the two parabolas).
(c) The kinetic energy of the system is identical in the initial and final states.
(d) Only electrons from the Fermi level in the electrode and from the redox system in the solution in the ground state participate in the electrode reaction.
(e) The concentrations of the ions B^{z+} and $B^{(z-1)+}$ are equal to unity.

The profile of the potential energy E_p of the reacting system in dependence on the reaction coordinate x is shown in Fig. 5.6. The curve denoted as R corresponds to the initial state of the system. The coordinates of its minimum, i.e. the ground state of the system, are $x_0(i)$, 0. The curve for the final state P is shifted by a value corresponding to the difference between the final and initial energies of the electronic subsystem, ΔE_e. The coordinate $x_0(f)$ corresponds to the minimum on the potential curve for the final state. The potential energy of the initial state of the system is given by the equation

$$E_{p,i} = \tfrac{1}{2}m\omega^2[x - x_0(i)]^2 \tag{5.3.2}$$

(where m is the effective mass of the vibrating system) and the energy of the final state by the equation

$$E_{p,f} = \tfrac{1}{2}m\omega^2[x - x_0(f)]^2 + \Delta E_e \tag{5.3.3}$$

At the intercept, the quantities $E_{p,i}$ and $E_{p,f}$ are equal to the activation enthalpy of the cathodic electrode reaction $\Delta\tilde{H}_c$. The corresponding

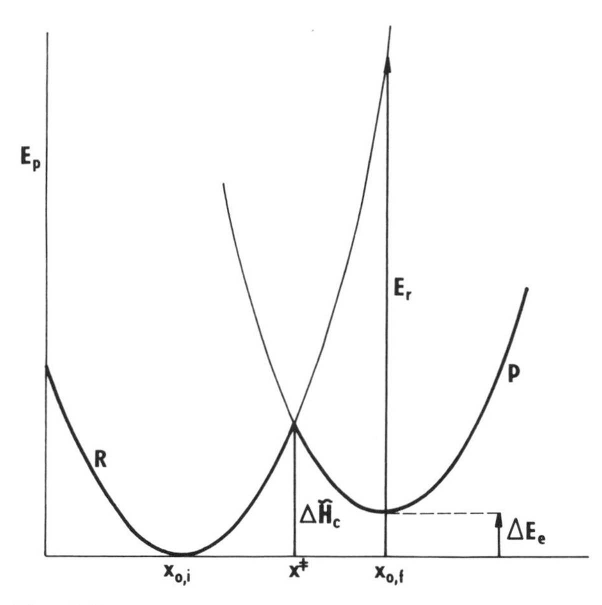

Fig. 5.6 Dependence of the potential energy E_p of the system on the reaction coordinate x during electron transfer. $\Delta\tilde{H}_c$ is the cathodic activation energy, $x^\ddagger$ the reaction coordinate corresponding to the intersection of R and P, ΔE_e the change in the electronic energy from the ground states of the system and E_r the reorganization energy

reaction coordinate is given by the equation

$$\Delta\tilde{H}_c = \tfrac{1}{2}m\omega^2[x^\ddagger - x_0(\mathrm{i})]^2 = \tfrac{1}{2}m\omega^2[x^\ddagger - x_0(\mathrm{f})]^2 + \Delta E_e \qquad (5.3.4)$$

Thus,

$$x^\ddagger = \tfrac{1}{2}[x_0(\mathrm{f}) - x_0(\mathrm{i})] + \Delta E_e[x_0(\mathrm{f}) - x_0(\mathrm{i})] \qquad (5.3.5)$$

Substitution into Eq. (5.3.4) yields

$$\Delta\tilde{H}_c = \frac{(E_r + \Delta E_e)^2}{4E_r} \qquad (5.3.6)$$

where E_r is the *reorganization energy* of the system,

$$E_r = \tfrac{1}{2}m\omega^2[x_0(\mathrm{f}) - x_0(\mathrm{i})]^2 \qquad (5.3.7)$$

The difference between the electronic energies of the final and initial states must include the energy of ionization of the ion $B^{(z-1)+}$ *in vacuo* (where its ionization potential is complemented by the entropy term $T\Delta S_i$), the interaction energy of the ions B^{z+} and $B^{(z-1)+}$ with the surroundings, i.e. the solvation Gibbs energies, and finally the energy of an electron at the Fermi level in the electrode. These quantities can be expressed most simply in terms of the electrochemical potentials of the appropriate species (cf. Eq. 3.1.17 and 3.1.18):

$$\Delta E_e = \mu^{0\prime}_{B^{(z-1)+}}(l) - \mu^{0\prime}_{B^{z+}}(l) - \mu^0_e(m) + F[\phi(m) - \phi(l)] \qquad (5.3.8)$$

where $\mu^{0\prime}$ denotes the 'formal' chemical potentials (for unit concentrations).

It is assumed for simplicity that metal m is also the material of the standard hydrogen electrode, for which (cf. Eq. 3.1.61):

$$\phi(m') - \phi(l) = \frac{\mu^0_{H^+}(l) - \frac{1}{2}\mu^0_{H_2}(g) + \mu^0_e(m)}{F} \qquad (5.3.9)$$

Substitution of $\phi(l)$ into Eq. (5.3.8) together with the relationship $\phi(m) - \phi(m') = E$ yields

$$\Delta E_e = F(E - E^{0\prime}) \qquad (5.3.10)$$

where $E^{0\prime}$ is the formal electrode potential of the redox electrode reaction defined by Eq. (5.3.1) (formulated as a reversible reaction),

$$E^{0\prime}_{Ox/Red} = -\frac{\mu^0_{B^{(z-1)+}}(l) - \mu^0_{B^{z+}}(l) + \mu^0_{H^+}(l) - \frac{1}{2}\mu^0_{H_2}(g)}{F} \qquad (5.3.11)$$

Equation (5.3.6) then assumes the form

$$\Delta\tilde{H}_c = \frac{[E_r + F(E - E^{0\prime}_{Ox/Red})]^2}{4E_r} \qquad (5.3.12)$$

The general relationship for the activation energy including both the electrode reaction and the chemical redox reaction as derived by Marcus in the form

$$\Delta H^{\ddagger} = w_r + \frac{(E_r + \Delta G^0 + w_p - w_r)^2}{4E_r} \qquad (5.3.13)$$

where w_r and w_p are the works required to bring the reactants together and to separate the products and ΔG^0 is the standard Gibbs energy change for the reaction. For electrode reactions, the quantities w_r and w_p appear in the effect of the electrical double layer on the rate of the electrode reaction (see Section 5.3.2).

If the effect of the electrical double layer is neglected (e.g. at higher indifferent electrolyte concentrations), the rate constant of the cathodic reaction is approximately given by the equation

$$k_c = r_s\rho\omega\kappa \exp\left(-\frac{\Delta\tilde{H}_c}{RT}\right)$$
$$= r_s\rho\omega\kappa \exp\left\{-\frac{[E_r + F(E - E^{0\prime}_{Ox/Red})]^2}{4RTE_r}\right\} \qquad (5.3.14)$$

where r_s is the geometric mean of the radii of the oxidized and of the reduced form of the electroactive species, $r = (r_{s,1}r_{s,2})^{1/2}$, ρ is the number of electronic states per unit area in the conductivity band of the electrode, ω is the vibrational frequency of the molecule in the solvation sheath and κ is the transmission coefficient, with a value of $\kappa \approx 1$ for an adiabatic process and $\kappa \ll 1$ for a non-adiabatic process. As it is assumed that the electron is transferred to a species whose centre lies in the outer Helmholtz plane, the volume concentration of the reactants is converted by the factor $2r_s$ to the surface concentration. The density of the electronic states available for the transfer reaction is estimated as $\rho/2$. This is a considerable simplification; in a more exact approach, the quantity ρ should be multiplied by the Fermi function (Eq. 3.1.10), $\mu_e^0(\mathrm{m}) - F\phi(\mathrm{m}) = \tilde{\mu}_e(\mathrm{m})$ in Eq. (5.3.8) should be replaced by the energy of an electron level ε and the whole expression should be integrated with respect to ε. However, the more precise solution would not lead to a great improvement because of the great many other simplifications involved in the present treatment.

In the same way the equation for the rate constant of the anodic reaction is obtained:

$$k_a = r_s\rho\omega\kappa \exp\left\{-\frac{[E_r - F(E - E^{0\prime}_{\mathrm{Ox/Red}})]^2}{4RTE_r}\right\} \tag{5.3.15}$$

The charge transfer coefficient defined by Eq. (5.2.50) is given with respect to Eq. (5.2.12) as

$$\alpha_{c,\mathrm{true}} = \frac{1}{2} + \frac{F(E - E^{0\prime}_{\mathrm{Ox,Red}})}{2E_r} \tag{5.3.16}$$

and is in general not a constant quantity. This is apparent from Fig. 5.7,

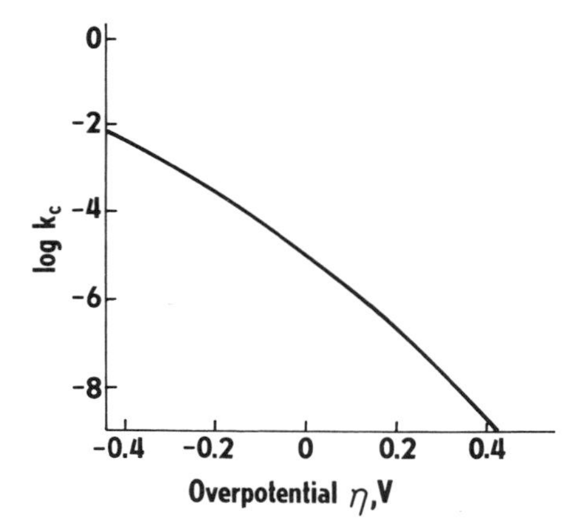

Fig. 5.7 Dependence of the rate constant of the cathodic reaction $Fe^{3+} + e \rightarrow Fe^{2+}$ in 1 M $HClO_4$ on the overpotential. (According to J. Weber, Z. Samec and V. Mareček)

depicting the dependence of the rate constant of the cathodic reduction of $Fe^{3+}(aq)$ ions on the potential.

Very often the reorganization energy is rather large (the barrier for the process is high), $E_r \gg F\,|E - E^{0'}_{Ox/Red}|$, and Eq. (5.3.14) attains a simplified form,

$$k_c = r_s \rho \omega \kappa \exp\left[-\frac{E_r/4 + \frac{1}{2}F(E - E^{0'}_{Ox/Red})}{RT}\right] \qquad (5.3.17)$$

which is identical with Eq. (5.2.23) for $k^{\ominus} = r_s \rho \omega \kappa \exp(-E_r/4RT)$, $n = 1$ and $\alpha = \frac{1}{2}$. It should be recalled that the value $\alpha = \frac{1}{2}$ is connected with the assumption of a single harmonic oscillator with the same frequency in the initial and final states of the electrode reaction. In the general case when this assumption is not fulfilled, the charge transfer coefficient can assume various values between 0 and 1, as given in the description of the phenomenological theory (Section 5.2). At a sufficiently large value of the quantity ΔE_e, the intercept between the curves R and P approaches the minimum on the curve of the final state P and, according to Eq. (5.3.16), the charge transfer coefficient approaches unity. In this situation, however, as shown by L. I. Krishtalik, the electrons from the region above the Fermi level will participate in the process so that a further increase in ΔE_e will not result in a shift of the intercept in Fig. 5.6 along curve P to the right. On the contrary, the activation energy will increase in the same way as ΔE_e increases, i.e. the charge transfer coefficient retains the value of unity. This barrierless state has been observed for a number of electrode processes, e.g. for the hydrogen evolution reaction at mercury (see page 355).

The above-described theory, which has been extended for the transfer of protons from an oxonium ion to the electrode (see page 353) and some more complicated reactions was applied in only a limited number of cases to interpretation of the experimental data; nonetheless, it still represents a basic contribution to the understanding of electrode reactions. More frequently, the empirical values n, $k^{\ominus}$ and α (Eq. 5.2.24) are the final result of the investigation, and still more often only k_{conv} and αn (cf. Eq. 5.2.49) or the corresponding constant of the Tafel equation (5.2.32) and the reaction order of the electrode reaction with respect to the electroactive substance (Eq. 5.2.4) are determined.

5.3.2 *The effect of the electrical double-layer structure on the rate of the electrode reaction*

The electrode reaction occurs in the region where by the influence of the electrode charge an electric field is formed, characterized by the distribution of the electric potential as a function of the distance from the electrode surface (see Section 4.3). This electric field affects the concentrations of the reacting substances and also the activation energy of the electrode reaction, expressed by the quantities w_r and w_p in Eq. (5.3.13). This effect can be explained in terms of the phenomenological theory (Section 5.2.2) by the procedure proposed by A. N. Frumkin. In the electrode reaction the reacting particles approach the electrode to the smallest possible distance they can reach without adsorption, i.e. to the outer Helmholtz plane. The concentrations of reacting particles at this distance from the electrode, c'_{Ox} and c'_{Red}, are given by Eq. (4.3.5) as

$$\begin{aligned} c'_{Ox} &= c_{Ox} \exp\left(-\frac{z_{Ox}F\phi_2}{RT}\right) \\ c'_{Red} &= c_{Red} \exp\left(-\frac{z_{Red}F\phi_2}{RT}\right) \\ &= c_{Red} \exp\left[-\frac{(z_{Ox}-n)F\phi_2}{RT}\right] \end{aligned} \tag{5.3.18}$$

The value of the electric potential affecting the activation enthalpy of the electrode reaction is decreased by the difference in the electrical potential between the outer Helmholtz plane and the bulk of the solution, ϕ_2, so that the activation energies of the electrode reactions are not given by Eqs (5.2.10) and (5.2.18), but rather by the equations

$$\begin{aligned} \Delta\tilde{H}_c &= \Delta H^0_c + \alpha nF(E-\phi_2) \\ \Delta\tilde{H}_a &= \Delta H^0_a - (1-\alpha)nF(E-\phi_2) \end{aligned} \tag{5.3.19}$$

Under these conditions, Eq. (5.2.24) becomes

$$\begin{aligned} j = zFk^{\ominus}\bigg\{&\exp\left[\frac{(1-\alpha)nF(E-\phi_2-E^{0\prime})}{RT}\right]\exp\left[-\frac{(z_{Ox}-n)F\phi_2}{RT}\right]c_{Red} \\ &-\exp\left[-\frac{\alpha nF(E-\phi_2-E^{0\prime})}{RT}\right]\exp\left(-\frac{z_{Ox}F\phi_2}{RT}\right)c_{Ox}\bigg\} \\ = j(\phi_2 = 0)&\exp\left[\frac{(\alpha n - z_{Ox})F\phi_2}{RT}\right] \end{aligned} \tag{5.3.20}$$

where $j(\phi_2 = 0)$ is the current density given by Eq. (5.2.24).

Figure 5.8 depicts the experimental Tafel diagram in $\log|j| - E$ coordinates for the electrode reaction of the reduction of oxygen in 0.1 M NH_4F and 0.1 M NH_4Cl at a dropping mercury electrode (after correction for the concentration overpotential, described in Section 5.4.3). As the reduction of oxygen to hydrogen peroxide is a two-electron process (see Eq. 5.7.7) with the rate controlled by the transfer of the first electron, where $k_a = 0$, it can be described by a relationship formed by combination of Eqs (5.2.48) and (5.3.20):

$$j = 2Fk_1^{\ominus}\exp\left[-\frac{\alpha_1 F(E-\phi_2-E^{0\prime})}{RT}\right]c_{Ox} \tag{5.3.21}$$

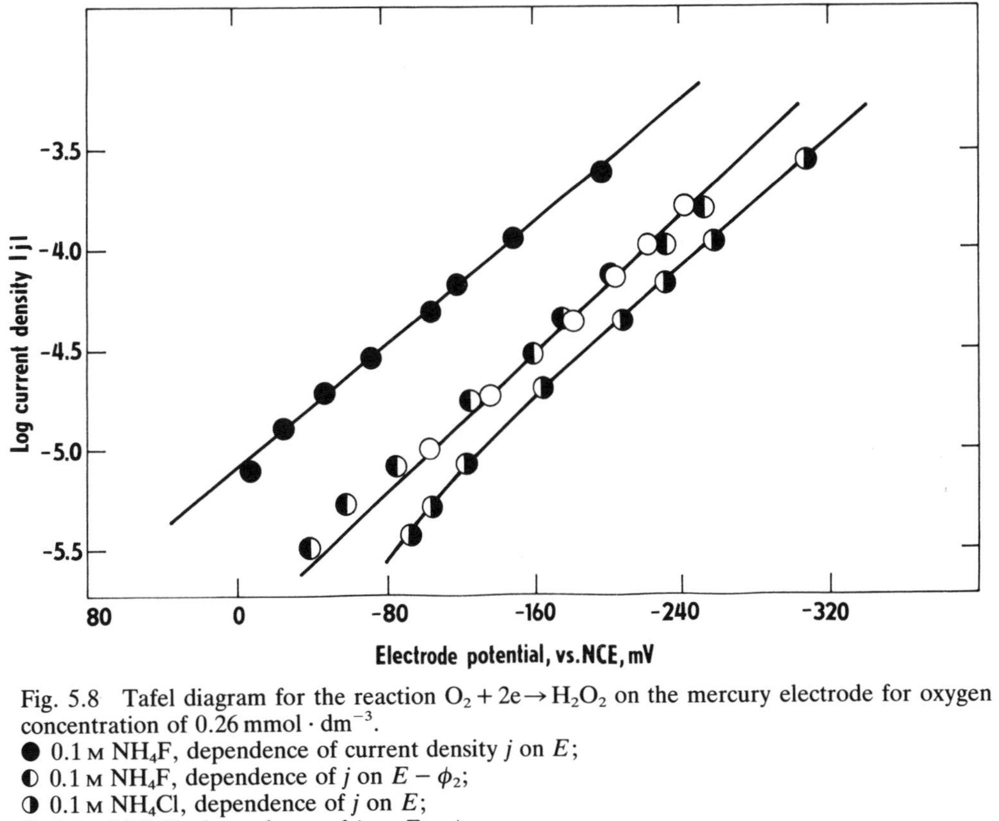

Fig. 5.8 Tafel diagram for the reaction $O_2 + 2e \rightarrow H_2O_2$ on the mercury electrode for oxygen concentration of 0.26 mmol · dm^{-3}.
● 0.1 M NH_4F, dependence of current density j on E;
◐ 0.1 M NH_4F, dependence of j on $E - \phi_2$;
◑ 0.1 M NH_4Cl, dependence of j on E;
○ 0.1 M NH_4Cl, dependence of j on $E - \phi_2$.
By means of the Frumkin correction, identical Tafel diagrams for different media are obtained. (According to J. Kůta and J. Koryta)

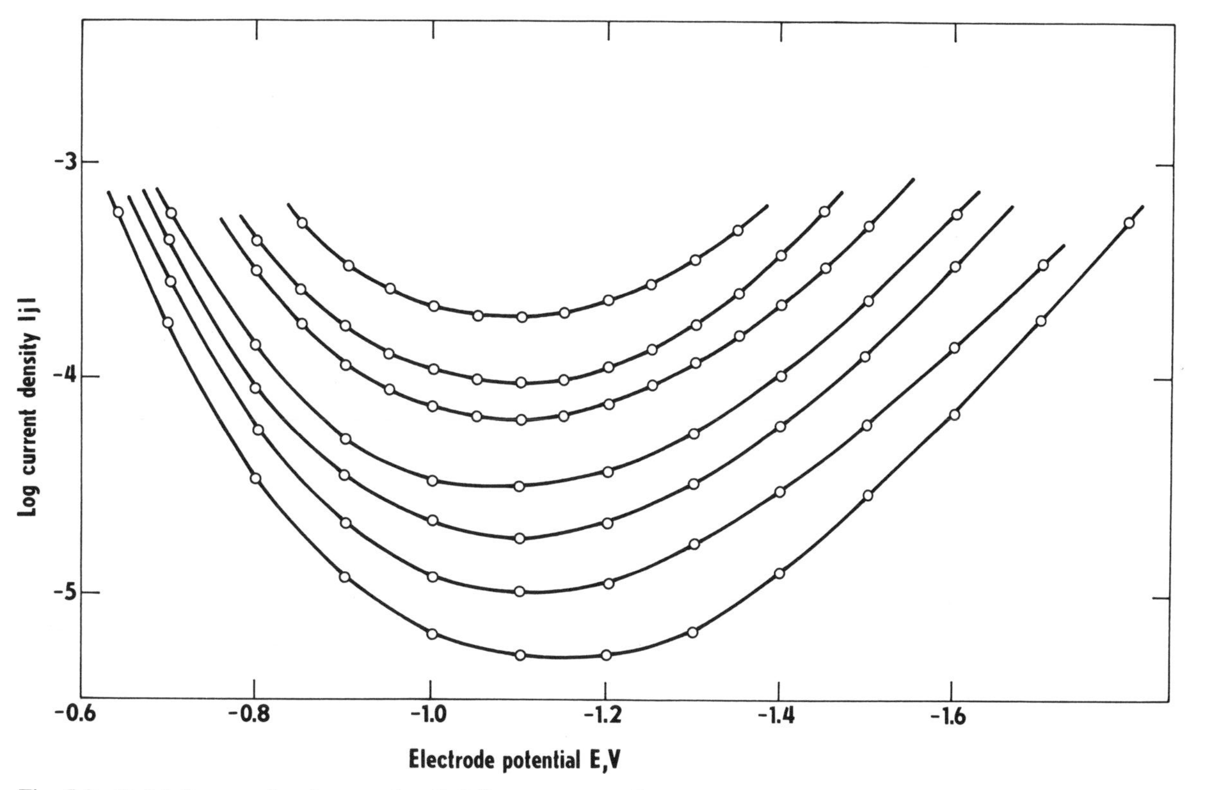

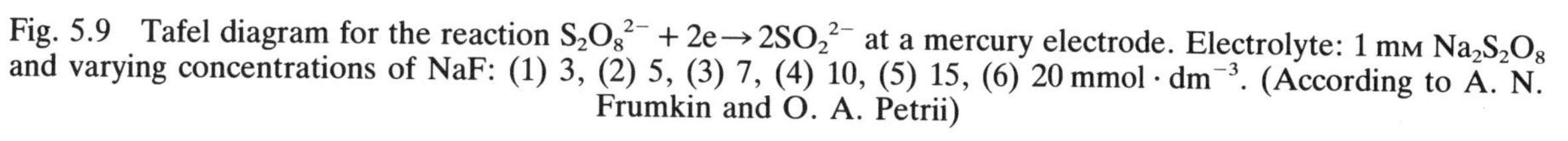
Fig. 5.9 Tafel diagram for the reaction $S_2O_8^{2-} + 2e \rightarrow 2SO_2^{2-}$ at a mercury electrode. Electrolyte: 1 mM $Na_2S_2O_8$ and varying concentrations of NaF: (1) 3, (2) 5, (3) 7, (4) 10, (5) 15, (6) 20 mmol · dm^{-3}. (According to A. N. Frumkin and O. A. Petrii)

Correction for the effect of the potential difference in the diffuse layer is carried out by plotting $\log |j|$ against $E - \phi_2$ rather than against E. It can be seen from Fig. 5.8 that the corrected curves are identical for both electrolytes.

The effect of the electrical double layer on the reduction of polyvalent anions such as $S_2O_8^{2-}$ is especially conspicuous. Here it holds that $\alpha z - z_{Ox} < 1$. At potentials corresponding to a negatively charged electrode ($E < E_{pzc}$), the potential of the outer Helmholtz plane is $\phi_2 < 0$ and its absolute value in dilute solutions increases with increasing negative potentials. This leads to a paradoxical situation: the rate of the cathodic reaction decreases with increasing negative potential (see Fig. 5.9).

The Frumkin theory of the effect of the electrical double layer on the rate of the electrode reaction is a gross simplification. For example, the electrode reaction does not occur only at the outer Helmholtz plane but also at a somewhat greater distance from the electrode surface. More detailed considerations indicate, however, that Eq. (5.3.20) can still be used to describe the effect of the electrical double layer as a good approximation.

Various approaches to the theory of electrode reactions with participation of adsorbed reactants are beyond the scope of the present book. The effect of of the electrical double layer is negligible for chemisorption (see Section 4.3.3 and 5.7.1).

For semiconductor electrodes and also for the interface between two immiscible electrolyte solutions (ITIES), the greatest part of the potential difference between the two phases is represented by the potentials of the diffuse electric layers in the two phases (see Eq. 4.5.18). The rate of the charge transfer across the compact part of the double layer then depends very little on the overall potential difference. The potential dependence of the charge transfer rate is connected with the change in concentration of the transferred species at the boundary resulting from the potentials in the diffuse layers (Eq. 4.3.5), which, of course, depend on the overall potential difference between the two phases. In the case of simple ion transfer across ITIES, the process is very rapid being, in fact, a sort of diffusion accompanied with a resolvation in the recipient phase.

References

All references from page 253

Armstrong, F. A., H. A. O. Hill, and N. J. Walton, Reactions of electron-transfer proteins at electrodes, *Quart. Rev. Biophys.*, **18**, 261 (1986).

Delahay, P., see page 266.

Dogonadze, R. R., Theory of molecular electrode kinetics, in *Reactions of Molecules at Electrodes,* p. 135, see page 253.

Dogonadze, R. R., L. I. Krishtalik, and V. G. Levich, Present state of the theory of the elementary step of electrode reactions (in Russian), *Zh. Vsesoyuz. Khim. Obsh.*, **16**, 613 (1971).

Friedrich, B., Z. Havlas, Z. Herman, and R. Zahradník, Theoretical studies of reaction mechanisms in chemistry, *Advances in Quantum Chemistry*, Vol. 19 (Ed. P. O. Lövdin), Academic Press, Orlando, 1987.
Koryta, J., see page 191.
Koryta, J., Ion transfer across water/organic solvent phase boundaries and analytical applications, Part 1, 2, *Selective El. Revs.*, **5**, 131 (1983), **13**, 133 (1991).
Levich, V. G., Kinetics of reactions with charge transport, *PChAT*, **IX B**, 985 (1970).
Marcus, R. A., Chemical and electrochemical electron-transfer theory, *Ann. Rev. Phys. Chem.*, **15**, 155 (1964).
Marcus, R. A., Electron, proton and related transfers, *Faraday Discuss. Chem. Soc.*, **74**, 7 (1982).
Marcus, R. A., and N. Sutin, Electron transfers in chemistry and biology, *Biochim. Biophys. Acta*, **811**, 265 (1985).
Mareček, V., Z. Samec, and J. Koryta, see page 243.
Newton, M. A., and N. Sutin, Electron transfer reactions in condensed phases, *Ann. Rev. Phys. Chem.*, **35**, 437 (1984).
Tsonskii, V. S., L. B. Korsunov, and L. I. Krishtalik, *Electrochim. Acta*, **36**, 411 (1991).
Vanýsek, P., see page 191.

5.4 Transport in Electrode Processes

5.4.1 *Material flux and the rate of electrode processes*

Material changes occurring during electrolysis at electrodes lead to mass transport throughout the whole system. The mathematical treatment of electrode processes involves relationships between the concentrations, their gradients and the current densities, corresponding to the rates of processes occurring directly at the electrodes (primarily the rates of electrode reactions), as boundary conditions for the differential equation describing transport processes in the electrolyte, formulated in Chapter 2. Sometimes (small currents, strong mixing), the concentration changes are so small that the effect of transport phenomena can be neglected.

According to Faraday's law, the current passing through the electrode is equivalent to the material flux of electroactive substances. The disappearance of electroactive substances in the electrode reaction is considered as their transport through the electrode surface. Consequently, only diffusion and migration but not convection flux need be considered at the electrode surface, as the electrode is impenetrable to the solution components.

Assume that both the initial substances and the products of the electrode reaction are soluble either in the solution or in the electrode. The system will be restricted to two substances whose electrode reaction is described by Eq. (5.2.1). The solution will contain a sufficient concentration of indifferent electrolyte so that migration can be neglected. The surface of the electrode is identified with the reference plane, defined in Section 2.5.1. In this plane a definite amount of the oxidized component, corresponding to the material flux J_{Ox} and equivalent to the current density j, is formed or disappears in unit time. The reduced form participates in the reverse process, connected with the material flux J_{Red}. When both forms are present in solution

$$j = nFJ_{Ox} = -nFJ_{Red} \tag{5.4.1}$$

$$j = -nFD_{Ox}\left(\frac{\partial c_{Ox}}{\partial x}\right)_{x=0} = nFD_{Red}\left(\frac{\partial c_{Red}}{\partial x}\right)_{x=0} \tag{5.4.2}$$

In general, for a larger number of electroactive substances, the partial current densities corresponding to the individual substances are additive.

It is necessary to clarify the meaning of the distance, $x = 0$, from the electrode. We must remember that the concentration changes resulting from transport processes appear up to distances considerably larger than the dimensions of the electrical double layer (the space charge region, i.e. the region of the diffuse electrical layer, is mostly spread out to a distance of at most several tens of nanometres). Thus 'zero' distance from the electrode corresponds to points lying just outside the diffuse electrical layer. At this distance, the concentrations of the electroactive substances are not yet affected by the space charge. The influence of the surface charge is shown, however, in the rate constants of the electrode reactions which appear in the boundary conditions (see Section 5.3.2).

The current density corresponding to the electrode reaction (5.2.1) is described by Eq. (5.2.13). Combination of Eqs (5.2.13) and (5.4.2) yields

$$D_{Ox}\frac{\partial c_{Ox}}{\partial x} = k_c c_{Ox} - k_a c_{Red} \tag{5.4.3}$$

$$D_{Red}\frac{\partial c_{Red}}{\partial x} = -k_c c_{Ox} + k_a c_{Red} \tag{5.4.4}$$

If the rates of the electrode reactions are large and the system is fairly close to equilibrium (the electrode potential is quite close to the reversible electrode potential), then the right-hand sides of Eqs (5.4.3) and (5.4.4) correspond to the difference between two large numbers whose absolute values are much larger than those of the left-hand sides. The left-hand side can then be set approximately equal to zero, $k_c c_{Ox} - k_a c_{Red} \approx 0$, and in view of Eqs (5.2.14) and (5.2.17),

$$\frac{c_{Ox}}{c_{Red}} = \exp\left[\frac{(E - E^{0\prime})nF}{RT}\right] = \lambda \tag{5.4.5}$$

which is a form of the Nernst equation. The electrode reaction thus proceeds approximately at equilibrium (this is often termed a 'reversible' electrode process).

When $\lambda \to \infty$ or $\lambda \to 0$, then $c_{Red} \to 0$ or $c_{Ox} \to 0$. A limiting current is then formed on the polarization curve that is independent of the electrode potential (see page 286). If the initial concentration, e.g. of the oxidized component, equals c^0, then the equations of the limiting diffusion current density j_d are obtained for various types of diffusion and convective diffusion; these equations follow from Eqs (2.5.8), (2.7.12) and (2.7.17) after multiplication of the right-hand sides by $-nF$.

In the general case, the initial concentration of the oxidized component equals c^0_{Ox} and that of the reduced component c^0_{Red}. If the appropriate differential equations are used for transport of the two electroactive forms (see Eqs 2.5.3 and 2.7.16) with the corresponding diffusion coefficients, then the relationship between the concentrations of the oxidized and reduced forms at the surface of the electrode (for linear diffusion and simplified convective diffusion to a growing sphere) is given in the form

$$D^{1/2}_{Ox}(c_{Ox})_{x=0} + D^{1/2}_{Red}(c_{Red})_{x=0} = D^{1/2}_{Ox}c^0_{Ox} + D^{1/2}_{Red}c^0_{Red} \tag{5.4.6}$$

For the sake of simplicity, it will further be assumed that $c^0_{Red} = 0$.

For the boundary conditions (5.4.5), i.e. the prescribed ratio of the concentrations of the two forms, we may use the transformation

$$c = \frac{c_{Ox} - \lambda(D_{Ox}/D_{Red})^{1/2}c^0_{Ox}}{1 + \lambda(D_{Ox}/D_{Red})^{1/2}} \tag{5.4.7}$$

Together with the boundary condition (5.4.5) and relationship (5.4.6), this yields the partial differential equation (2.5.3) for linear diffusion and Eq. (2.7.16) for convective diffusion to a growing sphere, where $D = D_{Ox}$ and $c^0 = c^0_{Ox}/[1 + \lambda(D_{Ox}/D_{Red})^{1/2}]$. As for linear diffusion, the limiting diffusion current density is given by the *Cottrell equation*

$$j_d = -nFc^0_{Ox}\left(\frac{D_{Ox}}{\pi t}\right)^{1/2} \tag{5.4.8}$$

and, for convective diffusion to a growing sphere (cf. Eq. (2.7.17)), by the *Ilkovič equation* of instantaneous current,

$$j_d = -nFc^0_{Ox}(7D_{Ox}/3\pi t)^{1/2} \tag{5.4.9}$$

the current density at an arbitrary potential for a 'reversible' electrode process is

$$j = \frac{j_d}{1 + \lambda(D_{Ox}/D_{Red})^{1/2}} \tag{5.4.10}$$

whence

$$\begin{aligned} E &= E^{0\prime} + \frac{RT}{nF}\ln\left[\left(\frac{D_{Red}}{D_{Ox}}\right)^{1/2}\frac{j_d - j}{j}\right] \\ &= E^{0\prime} + \frac{RT}{nF}\ln\left[\left(\frac{D_{Red}}{D_{Ox}}\right)^{1/2}\frac{I_d - I}{I}\right] \end{aligned} \tag{5.4.11}$$

where I and I_d are the currents corresponding to the current densities j and j_d.

For $j = \frac{1}{2}j_d$, the *half-wave potential* $E_{1/2}$ is obtained:

$$E_{1/2} = E^{0\prime} + \frac{RT}{nF} \ln \left(\frac{D_{\mathrm{Red}}}{D_{\mathrm{Ox}}} \right)^{1/2} \tag{5.4.12}$$

As the ratio of the diffusion coefficients in Eq. (5.4.12) is mostly very close to unity, $E_{1/2} \approx E^{0\prime}$.

For a 'slow' electrode reaction (boundary conditions 5.4.3 and 5.4.4) the substitutions, Eq. (5.4.6), $k_a/k_c = \lambda$, $c = [1 + \lambda(D_{\mathrm{Ox}}/D_{\mathrm{Red}})^{1/2}]c_{\mathrm{Ox}} - \lambda(D_{\mathrm{Ox}}/D_{\mathrm{Red}})^{1/2}c^0_{\mathrm{Ox}}$ and $D = D_{\mathrm{Ox}}/[1 + \lambda(D_{\mathrm{Ox}}/D_{\mathrm{Red}})^{1/2}]$ convert the boundary conditions (5.4.3) and (5.4.4) to the form

$$x = 0, \qquad t > 0, \qquad D\frac{\partial c}{\partial x} = k_c c \tag{5.4.13}$$

while the partial differential equations (2.5.3) and (2.7.16) preserve their original form. The solution of this case is

$$j = -nFc^0_{\mathrm{Ox}}k^{\ominus} \exp\left[-\frac{\alpha nF(E - E^{0\prime})}{RT}\right] \exp(Q^2 t)\, \mathrm{erfc}\,(Qt^{1/2}) \tag{5.4.14}$$

where

$$Q = \left(\frac{k^{\ominus}}{D^{1/2}_{\mathrm{Ox}}}\right)\left\{\exp\left[-\frac{\alpha nF(E - E^{0\prime})}{RT}\right] + \left(\frac{D_{\mathrm{Ox}}}{D_{\mathrm{Red}}}\right)^{1/2} \exp\left[\frac{(1-\alpha)nF(E - E^{0\prime})}{RT}\right]\right\}$$

(see Fig. 5.10)

For large values of $Qt^{1/2}$, this solution goes over to the diffusion-

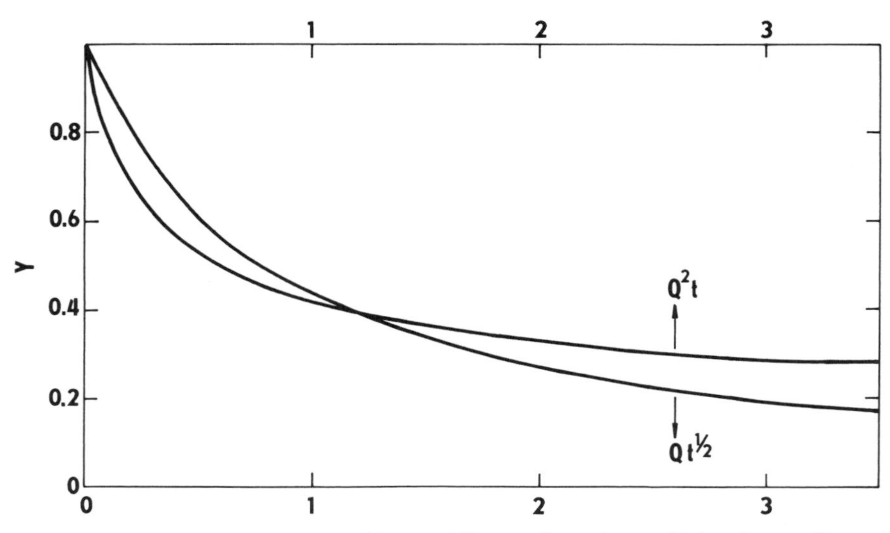

Fig. 5.10 Dependence of $Y = j/(nFc_{\mathrm{Ox}}k^{\ominus}) \exp[\alpha nF(E - E^{0\prime})/RT]$ on $Q^2 t$ and $Qt^{1/2}$. (According to W. Vielstich and H. Gerischer)

controlled case (Eq. 5.4.10). On the other hand, for small values of $Qt^{1/2}$, the rate of the process is controlled by the rate of the electrode reaction alone,

$$j = -nFk_c c^0_{Ox} \tag{5.4.15}$$

The case of the prescribed material flux at the phase boundary, described in Section 2.5.1, corresponds to the constant current density at the electrode. The concentration of the oxidized form is given directly by Eq. (2.5.11), where $K = -j/nF$. The concentration of the reduced form at the electrode surface can be calculated from Eq. (5.4.6). The expressions for the concentration are then substituted into Eq. (5.2.24) or (5.4.5), yielding the equation for the dependence of the electrode potential on time (a chronopotentiometric curve). For a reversible electrode process, it follows from the definition of the transition time τ (Eq. 2.5.13) for identical diffusion coefficients of the oxidized and reduced forms that

$$E = E^{0\prime} + \frac{RT}{nF} \ln \frac{\tau^{1/2} - t^{1/2}}{t^{1/2}} \tag{5.4.16}$$

(see Fig. 5.11). For an irreversible electrode reaction ($k_a = 0$),

$$E = E^{0\prime} + \frac{RT}{\alpha nF} \ln \frac{2k^{\ominus}}{\pi^{1/2} D_{Ox}^{1/2}} + \frac{RT}{\alpha nF} \ln (\tau^{1/2} - t^{1/2}) \tag{5.4.17}$$

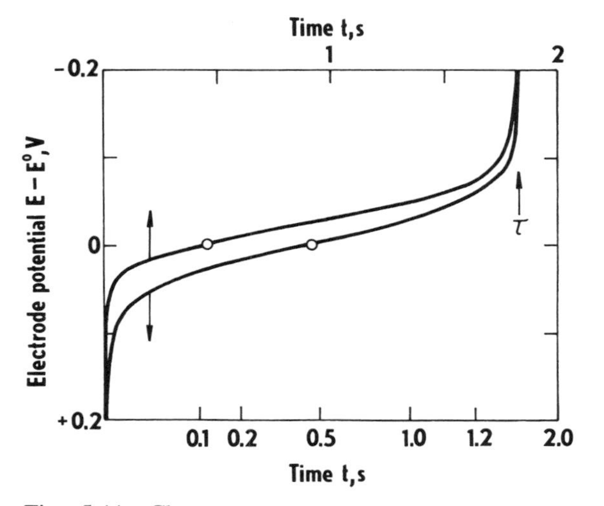

Fig. 5.11 Chronopotentiometric curves for a reversible electrode reaction (Eq. 5.4.16) in E–t (upper curve) and E–$t^{1/2}$ (lower curve) coordinates. τ denotes the transition time

5.4.2 *Analysis of polarization curves (voltammograms)*

Electrode processes are often studied under steady-state conditions, for example at a rotating disk electrode or at a ultramicroelectrode. Polarography with dropping electrode where average currents during the droptime are often measured shows similar features as steady-state methods. The distribution of the concentrations of the oxidized and reduced forms at the surface of the electrode under steady-state conditions is shown in Fig. 5.12. For the current density we have (cf. Eq. (2.7.13))

$$\begin{aligned} j &= -nFD_{\mathrm{Ox}}\left(\frac{\partial c_{\mathrm{Ox}}}{\partial x}\right)_{x=0} = nFD_{\mathrm{Ox}}\delta_{\mathrm{Ox}}^{-1}[(c_{\mathrm{Ox}})_{x=0} - c_{\mathrm{Ox}}^0] \\ &= nF\kappa_{\mathrm{Ox}}[(c_{\mathrm{Ox}})_{x=0} - c_{\mathrm{Ox}}^0] \end{aligned} \tag{5.4.18}$$

$$\begin{aligned} j &= nFD_{\mathrm{Red}}\left(\frac{\partial c_{\mathrm{Red}}}{\partial x}\right)_{x=0} = -nFD_{\mathrm{Red}}\delta_{\mathrm{Red}}^{-1}[(c_{\mathrm{Red}})_{x=0} - c_{\mathrm{Red}}^0] \\ &= -nF\kappa_{\mathrm{Red}}[(c_{\mathrm{Red}})_{x=0} - c_{\mathrm{Red}}^0] \end{aligned} \tag{5.4.19}$$

where δ_{Ox} and δ_{Red} are the thicknesses of the Nernst layer for the reduced and oxidized forms and κ_{Ox} and κ_{Red} are the mass transfer coefficients. Again, the case $c_{\mathrm{Red}}^0 = 0$ will be considered. The limiting cathodic current

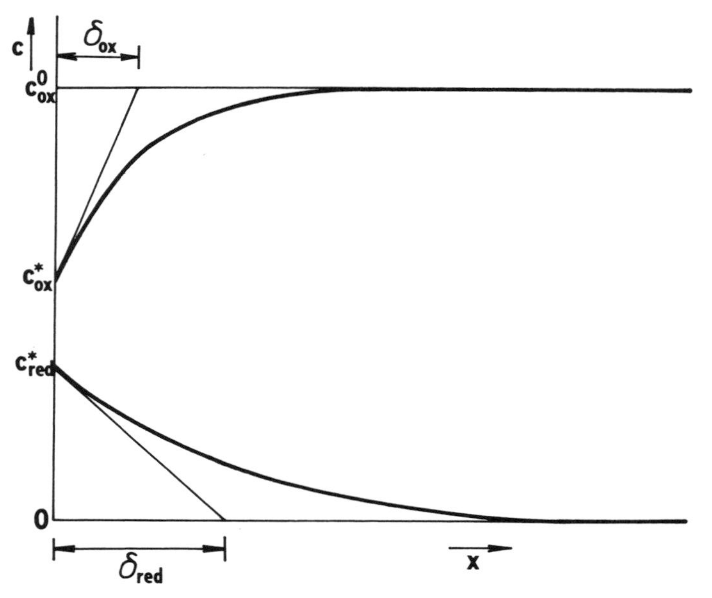

Fig. 5.12 Steady-state concentration distribution of the oxidized and of the reduced form of an oxidation–reduction system in the neighbourhood of the electrode

density is

$$j_d = -nF\kappa_{Ox}c^0_{Ox} \tag{5.4.20}$$

For the concentration at the surface of the electrode, Eq. (5.2.24) assumes the form

$$j = nFk^{\ominus}\left\{\exp\left[\frac{(1-\alpha)nF(E-E^{0\prime})}{RT}\right](c_{Red})_{x=0} - \exp\left[-\frac{\alpha nF(E-E^{0\prime})}{RT}\right](c_{Ox})_{x=0}\right\} \tag{5.4.21}$$

giving, after substitution for $(c_{Ox})_{x=0}$ and $(c_{Red})_{x=0}$ from Eqs (5.4.18) and (5.4.19) and rearrangement,

$$\frac{j_d - j}{j}\frac{\kappa_{Red}}{\kappa_{Ox}} = \exp\left[\frac{nF(E-E^{0\prime})}{RT}\right] + \frac{\kappa_{Red}}{k^{\ominus}}\exp\left[\frac{\alpha nF(E-E^{0\prime})}{RT}\right] \tag{5.4.22}$$

This equation describes the cathodic current–potential curve (polarization curve or voltammogram) at steady state when the rate of the process is simultaneously controlled by the rate of the transport and of the electrode reaction. This equation leads to the following conclusions:

1. *Limiting current.* As the negative electrode potential increases, the right-hand side decreases to zero and the current density j approaches the limiting value j_d. The limiting diffusion current does not depend either on the potential of the electrode or on the rate of the electrode reaction as it is controlled by convective diffusion alone. According to Eq. (5.4.20), it is directly proportional to the concentration of electroactive substance; this fact forms the basis of the application of limiting currents to analytical chemistry. The coefficient κ_{Ox} is a function of stirring; e.g. in the case of the rotating disk electrode (Eq. 2.7.13)

$$\kappa_{Ox} = 0.62D^{2/3}_{Ox}\nu^{-1/6}\omega^{1/2} \tag{5.4.23}$$

   where the concentration is expressed in $\text{mol}\cdot\text{cm}^{-3}$, D_{Ox} in $\text{cm}^2\cdot\text{s}^{-1}$, $\nu = \eta/\rho$ in $\text{cm}^2\cdot\text{s}^{-1}$ with ω in $\text{rad}\cdot\text{s}^{-1}$. The limiting current density apparently increases with increasing stirring rate (for the rotating disk electrode, it is proportional to the square root of the number of revolutions per second).
2. *Reversible polarization curve (voltammogram).* If the value of $k^{\ominus}$ is so large that the first term on the right-hand side of Eq. (5.4.12) is much larger than the second term, even when j approaches j_d, then

$$E = E^{0\prime} + \frac{RT}{nF}\ln\frac{\kappa_{Red}}{\kappa_{Ox}} + \frac{RT}{nF}\ln\frac{j_d - j}{j} \tag{5.4.24}$$

   This is the equation of a reversible polarization curve. The anodic polarization curve of the reduced form obeys an identical equation

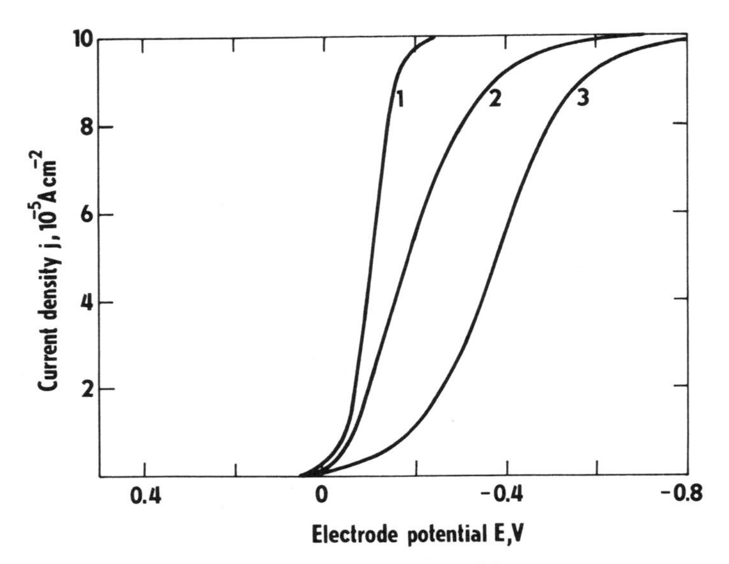

Fig. 5.13 Steady-state voltammograms: (1) reversible voltammogram (Eq. 5.4.24); (2) quasireversible voltammogram (Eq. 5.4.22) for $k^{\ominus} = 10^{-4}\,\text{cm}\cdot\text{s}^{-1}$; (3) irreversible voltammogram (Eq. 5.4.26) for $k^{\ominus} = 10^{-5}\,\text{cm}\cdot\text{s}^{-1}$. $E^{0\prime} = -0.1\,\text{V}$, $\alpha = 0.3$, $\kappa_{\text{Ox}} = \kappa_{\text{Red}} = 2\times10^{-4}\,\text{cm}\cdot\text{s}^{-1}$, $j_{\text{d}} = 10^{-4}\,\text{A}\cdot\text{cm}^{-1}$

containing the anodic limiting current. This curve corresponds to the case that equilibrium is established at the surface of the electrode between the reduced and oxidized forms according to the Nernst equation. The dependence of the current density on the potential is not a function of $k^{\ominus}$. The reversible polarization curve has the form of a wave (see Fig. 5.13, curve 1). The half-wave potential is given as

$$E_{1/2} = E^{0\prime} + \frac{RT}{nF}\ln\frac{\kappa_{\text{Red}}}{\kappa_{\text{Ox}}} \tag{5.4.25}$$

This value is close to the formal electrode potential and independent of the convection velocity. The plot of $\log(j_{\text{d}} - j)/j$ versus E is linear with the slope $nF/2.303RT$.

3. *Irreversible polarization curve (voltammogram).* If the value of $k^{\ominus}$ is so small that the first term on the right-hand side of Eq. (5.4.22) is much smaller than the second term, even when j approaches j_{d}, then the equation assumes the form

$$E = E^{0\prime} + \frac{RT}{\alpha nF}\ln\left(\frac{k^{\ominus}}{\kappa_{\text{Ox}}}\right) + \frac{RT}{\alpha nF}\ln\frac{j_{\text{d}} - j}{j} \tag{5.4.26}$$

This irreversible polarization curve also has the shape of a wave (see Fig. 5.13, curve 3) with the limiting current density j_{d}. The half-wave potential is

$$E_{1/2} = E^{0\prime} + \frac{RT}{\alpha nF} \ln \frac{k^{\ominus}}{\kappa_{Ox}} \tag{5.4.27}$$

or, considering the definition of k_{conv} in Eq. (5.2.37),

$$E_{1/2} = \frac{RT}{\alpha nF} \ln \frac{k_{conv}}{\kappa_{Ox}} \tag{5.4.28}$$

When increasing the rate of stirring the half-wave potential of a reduction reaction is shifted to more negative values.

Obviously, in the case of a cathodic reaction, the process at more positive potentials (at the 'foot' of the wave) is reversible while at more negative potentials irreversible. Increasing rate of stirring makes the irreversibility more pronounced.

The constants characterizing the electrode reaction can be found from this type of polarization curve in the following manner. The quantity $k^{\ominus}$ is determined directly from the half-wave potential value (Eq. 5.4.27) if $E^{0\prime}$ is known and the mass transfer coefficient κ_{Ox} is determined from the limiting current density (Eq. 5.4.20). The charge transfer coefficient α is determined from the slope of the dependence of $\ln [(j_d - j)/j]$ on E.

At current densities well below the limiting value, j can be neglected against j_d in the numerator of the expression after the logarithm in Eq. (5.4.26); rearrangement then yields

$$E = E^{0\prime} + \frac{RT}{\alpha nF} \ln (nFk^{\ominus}c^0_{Ox}) - \frac{RT}{\alpha nF} \ln |j| \tag{5.4.29}$$

This is the Tafel equation (5.2.32) or (5.2.36) for the rate of an irreversible electrode reaction in the absence of transport processes. Clearly, transport to and from the electrode has no effect on the rate of the overall process and on the current density. Under these conditions, the current density is termed the kinetic current density as it is controlled by the kinetics of the electrode process alone.

4. *The quasireversible case.* When Eq. (5.4.22) cannot be simplified as described under alternatives 2 and 3, then it should be considered that the first term on its right-hand side increases compared to the second term as the potential becomes more positive and decreases at more negative potentials. As the velocity of convection increases, the second term increases compared to the first term.

For the sake of a practical analysis of the polarization curve the exponential dependence of the electrode reaction rate constant need not be assumed. Then the more general form of Eq. (5.4.22) can be written as

$$\frac{1}{nFk_c c^0_{Ox}} = \frac{1}{j} - \frac{1}{j_d}\{1 - \exp [nF(E - E_{1/2})/RT]\} \tag{5.4.30}$$

This expression can be used to determine k_c directly by extrapolation for $j_d \to \infty$. For example in the case of the rotating disk this is achieved by means of the plot of j versus $\omega^{-1/2}$.

In the case of an irreversible reaction the exponential term in Eq. (5.4.30) is negligible against unity.

5.4.3 *Potential-sweep voltammetry*

The transient method characterized by linearly changing potential with time is called *potential-sweep* (*potential-scan*) *voltammetry* (cf. Section 5.5.2). In this case the transport process is described by equations of linear diffusion with the potential function

$$E = E_i \pm vt \tag{5.4.31}$$

entering the potential-dependent parameters of boundary conditions (5.4.3)–(5.4.5). Here E_i is the initial potential ($t = 0$) and v is the polarization rate dE/dt. The resultant I–E curve usually has a characteristic shape of a 'peak voltammogram' (see Fig. 5.14). At higher potentials than that of the peak the curve approaches the $j \sim t^{-1/2}$ dependence (Eq. (5.4.8)), while the part prior to the peak corresponds to a process dependent on electrode potential (electrode reaction or Nernst's equilibrium) and

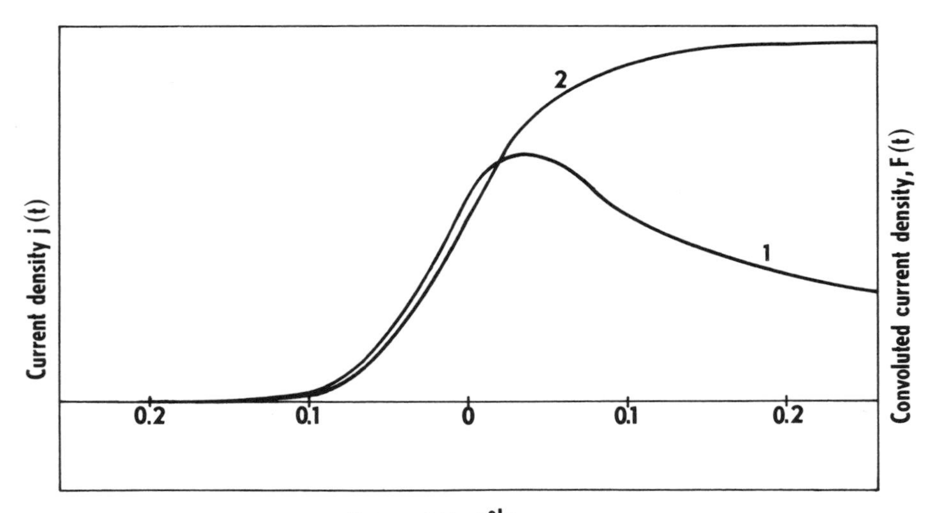

Fig. 5.14 Peak voltammogram (curve 1, $j(t) - E$ dependence) and convoluted voltammogram (curve 2, $F(t) - E$ dependence) for the reversible case

$$F(t) = \frac{1}{\pi^{1/2}} \int_0^t \frac{j(u)}{(t-u)^{1/2}} \, du$$

diffusion. The peak current density follows the Randles–Ševčík equation

$$j_p = 0.452(nF)^{3/2}(RT)^{-1/2}(Dv)^{1/2}c^0_{Ox} \quad (5.4.32)$$

The peak potential (corresponding to the peak current) for a cathodic reversible process is shifted by

$$\Delta E = 1.1RT/nF \quad (5.4.33)$$

to more negative potentials in comparison with the reversible half-wave potential (Eq. (5.4.12)).

A transformation of the peak voltammogram to the sigmoidal shape shown in the preceding section, Fig. 5.13, is achieved by the convolution analysis method proposed by K. Oldham. The experimental function $j(t) = j[\mp(E - E_i)/v]$ is transformed by convolution integration

$$F(t) = \pi^{-1/2}\int_0^t [j(u)/(t-u)^{1/2}]\,du \quad (5.4.34)$$

to a current function $F(t)$ whose dependence on potential (Fig. 5.14, curve 2) can be treated in the same way as the polarization curve in the preceding section.

5.4.4 *The concentration overpotential*

The effect of transport phenomena on the overall electrode process can also be expressed in terms of the *concentration* or *transport overpotential.* The original concentration of the oxidized form c_{Ox} decreases during the cathodic process through depletion in the vicinity of the electrode to the value $(c_{Ox})_{x=0}$ and, similarly, the concentration of the reduced form c^0_{Red} increases to the value $(c_{Red})_{x=0}$. It follows from Eqs (5.4.18), (5.4.19) and (5.4.20) that

$$(c_{Ox})_{x=0} = c^0_{Ox}\left(1 - \frac{j}{j_{d,c}}\right)$$
$$(c_{Red})_{x=0} = c^0_{Red}\left(1 - \frac{j}{j_{d,a}}\right) \quad (5.4.35)$$

where $j_{d,c}$ is the cathodic limiting current density, given by Eq. (5.4.20), and $j_{d,a}$ is the anodic limiting diffusion current density, given by the relationship

$$j_{d,a} = nF\kappa_{Red}c^0_{Red} \quad (5.4.36)$$

The overpotential of an electrode process (Eq. (5.1.11)) is given in the case of a simple first-order electrode reaction as

$$\eta = E - E_r = E - E^{0\prime} - \frac{RT}{nF}\ln\frac{c^0_{Ox}}{c^0_{Red}} \quad (5.4.37)$$

This equation can be expanded to give

$$\eta = E - \left[E^{0\prime} + \frac{RT}{nF} \ln \frac{(c_{\mathrm{Ox}})_{x=0}}{(c_{\mathrm{Red}})_{x=0}} \right] + \frac{RT}{nF} \ln \frac{(c_{\mathrm{Ox}})_{x=0}}{c^0_{\mathrm{Red}}} \frac{c^0_{\mathrm{Ox}}}{(c_{\mathrm{Red}})_{x=0}} \tag{5.4.38}$$

$$\eta = \eta_{\mathrm{r}} + \eta_{\mathrm{c}} \tag{5.4.39}$$

The first two terms on the right-hand side of this equation express the proper overpotential of the electrode reaction η_{r} (also called the activation overpotential) while the last term, η_{c}, is the EMF of the concentration cell without transport, if the components of the redox system in one cell compartment have concentrations $(c_{\mathrm{Ox}})_{x=0}$ and $(c_{\mathrm{Red}})_{x=0}$ and, in the other compartment, c^0_{Ox} and c^0_{Red}. The overpotential given by this expression includes the excess work carried out as a result of concentration changes at the electrode. This type of overpotential was called the *concentration overpotential* by Nernst. The expression for a concentration cell without transport can be used here under the assumption that a sufficiently high concentration of the indifferent electrolyte suppresses migration.

By substitution of Eq. (5.4.37) into Eq. (5.4.21) and rearrangement we obtain

$$j = nFk^{\ominus}(c^0_{\mathrm{Ox}})^{1-\alpha}(c^0_{\mathrm{Red}})^{\alpha}\left\{\exp\left[\frac{(1-\alpha)nF\eta}{RT}\right]\frac{(c_{\mathrm{Red}})_{x=0}}{c^0_{\mathrm{Red}}} - \exp\left(-\frac{\alpha nF\eta}{RT}\right)\frac{(c_{\mathrm{Ox}})_{x=0}}{c^0_{\mathrm{Ox}}}\right\} \tag{5.4.40}$$

Substitution from Eqs (5.2.26) and (5.4.35) into Eq. (5.4.40) finally yields the equation

$$j = j_0\left\{\exp\left[\frac{(1-\alpha)nF\eta}{RT}\right]\left(1 - \frac{j}{j_{\mathrm{d,c}}}\right) - \exp\left(-\frac{\alpha nF\eta}{RT}\right)\left(1 - \frac{j}{j_{\mathrm{d,a}}}\right)\right\} \tag{5.4.41}$$

This is the basic relationship of electrode kinetics including the concentration overpotential. Equations (5.4.40) and (5.4.41) are valid for both steady-state and time-dependent currents.

References

Albery, W. J., see page 253.

Bard, A. J., and L. R. Faulkner, *Electrochemical Methods, Fundamentals and Applications,* John Wiley & Sons, Chichester, 1980.

Delahay, P., *New Instrumental Methods in Electrochemistry,* Interscience, New York, 1954.

Levich, V. G., see page 81.

Southampton Electrochemistry Group, see page 253.

5.5 Methods and Materials

The study of electrochemical kinetics includes measurement of various electrical data but, at the same time, it needs information on reaction products and on the situation in the interphase. Thus, electrochemical methodology includes genuine electrochemical methods, analysis of the products of these processes, methods based on interaction of electromagnetic radiation, electrons or other charged species with the electrolysed system or with the electrode itself, etc.

Electrochemical methods have been described in a number of excellent monographs (e.g. Bard and Faulkner, Albery, the Southampton Electrochemistry Group, etc., see page 253). Thus, only the general characteristics of these methods will be given here. They mostly provide information on the functional relationships between the current density, electrode potential, time passed since the beginning of the electrolysis and sometimes also the total charge passed through the system. The use of these methods is connected with two main problems:

1. Elimination of the ohmic electric potential difference in the electrolytic cell
2. Selection of a fixed or programmed electric variable

The electrode material and its pretreatment considerably influence the course of the electrode process. The importance of *non-electrochemical methods* is constantly increasing as a result of the development of experimental techniques.

5.5.1 *The ohmic electrical potential difference*

The origin of the ohmic potential difference was described in Section 2.5.2. The ohmic potential gradient is given by the ratio of the local current density and the conductivity (see Eq. 2.5.28). If an external electrical potential difference ΔV is imposed on the system, so that the current I flows through it, then the electrical potential difference between the electrodes will be

$$\Delta V = E_1 - E_2 + \Delta V_{\mathrm{ohm}} = E_1 - E_2 + IR_i \tag{5.5.1}$$

where E_1 and E_2 are the electrode potentials, $\Delta V_{\mathrm{ohm}} = IR_i$ is the ohmic electrical potential difference and R_i is the resistance of the electrolytic cell, measured between the surfaces of the two electrodes. This approach involves the average values of these quantities; the electrode potential and ohmic electrical potential difference values can exhibit local variations in actual systems.

In the case of a cylindrical electrolytic cell with electrodes forming bases of the cylinder the ohmic potential difference is

$$\Delta V_{\mathrm{ohm}} = I\rho l/S \tag{5.5.2}$$

where I is the current, ρ the resistivity of the electrolyte, l the height and S the cross-section of the cylinder. Electrolytic cells with more diversified forms require more complicated approaches. For a centrosymmetrical system, we have

$$dR = \rho\, dr/4\pi r^2 \tag{5.5.3}$$

and for the increment of the ohmic potential difference

$$d\Delta V_{ohm} = I\rho\, dr/4\pi r2 \tag{5.5.4}$$

The ohmic potential difference in an electrolytic cell consisting of a spherical test electrode, termed, for a small radius r_0, *ultramicroelectrode*, in the centre and another very distant concentrical counter-electrode is given by the equation

$$\Delta V_{ohm} = I/4\pi \int_{r_0}^{\infty} r^{-2}\, dr = I\rho/4\pi r_0 \tag{5.5.5}$$

Since the current is

$$I = 4\pi r_0^2 j \tag{5.5.6}$$

where j is the current density, the resulting expression for the ohmic potential difference is

$$\Delta V_{ohm} = r_0 \rho j \tag{5.5.7}$$

Obviously, the ohmic potential difference does not depend on the distance of the counterelectrode (if, of course, this is sufficiently apart) being 'situated' mainly in the neighbourhood of the ultramicroelectrode. At constant current density it is proportional to its radius. Thus, with decreasing the radius of the electrode the ohmic potential decreases which is one of the main advantages of the ultramicroelectrode, as it makes possible its use in media of rather low conductivity, as, for example, in low permittivity solvents and at very low temperatures. This property is not restricted to spherical electrodes but also other electrodes with a small characteristic dimension like microdisk electrodes behave in the same way.

The kinetic investigation requires, as already stated in Section 5.1, page 252, a three-electrode system in order to programme the magnitude of the potential of the *working electrode,* which is of interest, or to record its changes caused by flow of controlled current (the ultramicroelectrode is an exception where a two-electrode system is sufficient).

Figure 5.15 shows a simple arrangement of the three-electrode system. The *reference electrode R* is usually connected to the electrolytic cell through a liquid bridge, terminated by a *Luggin capillary* which is generally filled with the solution being investigated. The tip of the Luggin capillary is placed as close as possible to the surface of the electrode, A, whose potential is to be measured (the *test, working* or *indicator electrode*). In the simple case of measurement of stationary voltammograms, the electrode potential is measured with a voltmeter with a high resistance P. The current passing through the system is controlled by the potential source *C* and the resistance P. If this resistance is much higher than the resistance of

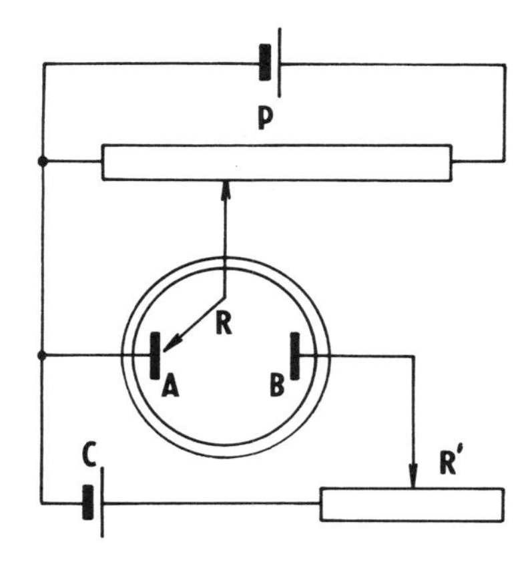

Fig. 5.15 Basic circuit for the electrode potential measurement during current flow: A is working (indicator, test), B, auxiliary and R, reference electrode connected by means of the Luggin capillary (arrow) and P, potentiometer. The constant current source consists of a battery C and a variable high-resistance resistor R′

the electrolytic cell, then the current is given by the ratio of the voltage of source C and resistance R'.

5.5.2 *Transition and steady-state methods*

The classification of methods for studying electrode kinetics is based on the criterion of whether the electrical potential or the current density is controlled. The other variable, which is then a function of time, is determined by the electrode process. Obviously, for a steady-state process, these two quantities are interdependent and further classification is unnecessary. Techniques employing a small periodic perturbation of the system by current or potential oscillations with a small amplitude will be classified separately.

The *controlled-potential methods* require special experimental instrumentation for elimination of the ohmic potential difference, called the *potentiostat* (see Fig. 5.16). The programmed voltage between the indicator electrode and the tip of the Luggin capillary is taken from a suitable source and fed to the input of the potentiostat. The current from the potentiostat with feedback control then flows between the working and auxiliary electrodes. There are two main types of potentiostat differing in their applications. High-power potentiostats control the potential of an electrode through which a strong current passes that changes only slowly with time. It

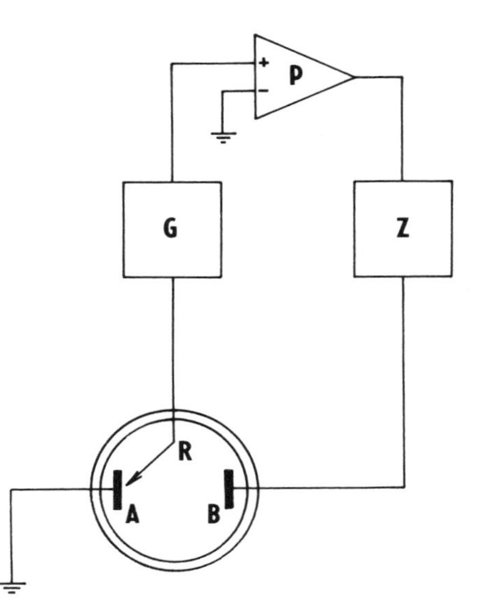

Fig. 5.16 Block diagram of a potentiostatic network: for A, B and R see Fig. 5.15; P is a potentiostat based on an operational amplifier, G, generator of programmed voltage and Z, recorder

is used in preparative electrolysis, in research on galvanic cells, etc. Fast potentiostats are used in the kinetic study of electrode processes where there are great variations in the current density during short time intervals reaching fractions of a millisecond. The main factor limiting the usefulness of this type of potentiostat is charging of the electrical double layer through the ohmic resistance of the electrolytic cell at the beginning of the pulse.

The study of processes at ITIES and in membrane electrochemistry requires elimination of two ohmic potential differences, achieved with a four-electrode potentiostat, 'voltage-clamp' (Fig. 5.17).

Various types of controlled-potential pulsing are shown in Fig. 5.18. The simplest case is the *single-pulse potentiostatic method* (Fig. 5.18A). The current–time (I–t) curves obtained by this method have already been described in Section 5.4.1, Eq. (5.4.10) and (5.4.14).

The double-pulse potentiostatic method (Fig. 5.18C) is suitable for studying the products or intermediates in electrode reactions, formed in the A pulse by means of the B pulse. For example, if an electroactive substance is reduced in pulse A and if pulse B is sufficiently more positive than pulse A, then the product can be reoxidized. The shape of the I–t curve in pulse B can indicate, for example, the degree to which the unstable product of the electrode reaction is changed in a subsequent chemical reaction.

Methods employing individual linear or triangular pulses (*potential-sweep, triangular pulse* and *cyclic voltammetry,* sometimes also called

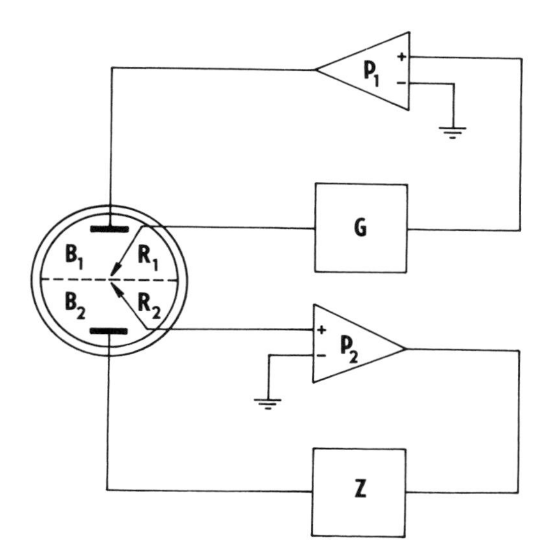

Fig. 5.17 A four-electrode potentiostatic circuit (voltage clamp). R_1 and R_2 are reference electrodes with Luggin capillaries (arrows) attached as close as possible to the membrane or ITIES (dashed line), B_1 and B_2 are auxiliary electrodes, P_1 and P_2 potentiostats, G programmed voltage generator and Z recorder

potentiodynamic methods; Fig. 5.18D, E and G) yield information on the I–E dependence. Cyclic voltammetry involves the study of the curve obtained after a greater number of pulses, when a steady voltammogram is obtained. A certain disadvantage of these popular methods is the fact that the processes occurring at a given potential can depend on those that occur at preceding potentials (i.e. on the history of the electrode). Various types of voltammetry with triangular pulses are applied to study the reaction products because each peak observed on the I–t curve usually corresponds to a definite electrode process. This is especially useful in the study of processes at solid electrodes, which are almost always complicated by adsorption (see Section 5.7) and where quantitative analysis of the given kinetic problem is often very difficult.

The term, *polarography,* is most often used for voltammetry with dropping mercury electrode (DME—Fig. 5.19A), which also belongs to potentiostatic methods, as practically constant potential is applied to the electrode during the drop growth. The corresponding average current is automatically recorded as a function of the slowly increasing potential. This method was discovered by J. Heyrovský (Nobel Prize for Chemistry, 1959).

Because of low current densities, the ohmic potential drop can often be neglected and a three-electrode system is not necessary. The same electrode can act as the auxiliary electrode and the reference electrode (sometimes a

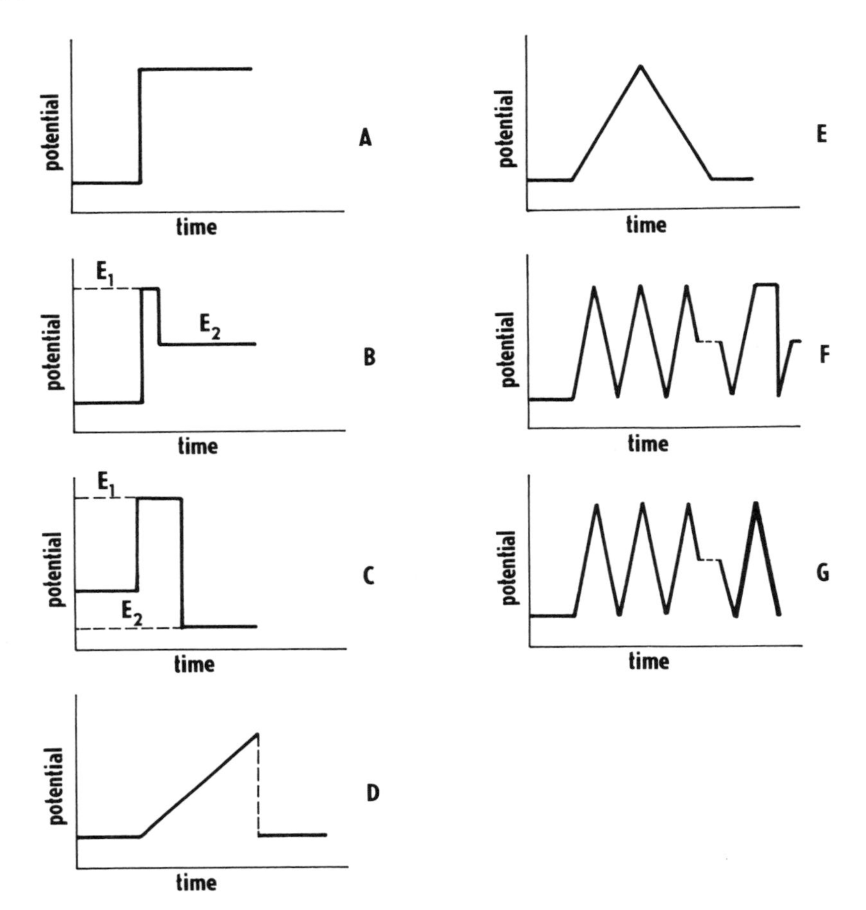

Fig. 5.18 Potentiostatic methods: (A) single-pulse method, (B), (C) double-pulse methods (B for an electrocrystallization study and C for the study of products of electrolysis during the first pulse), (D) potential-sweep voltammetry, (E) triangular pulse voltammetry, (F) a series of pulses for electrode preparation, (G) cyclic voltammetry (the last pulse is recorded), (H) d.c. polarography (the electrode potential during the drop-time is considered constant; this fact is expressed by the step function of time—actually the potential increases continuously), (I) a.c. polarography and (J) pulse polarography

large-surface area calomel electrode or mercury sulphate electrode is used). The average current at the dropping mercury electrode $\langle I \rangle$ is given by the relationship

$$\langle I \rangle = t_1^{-1} \int_0^{t_1} jA \, \mathrm{d}t \tag{5.5.8}$$

where j is the instantaneous current density, t_1 is the drop-time and A is the instantaneous surface area of the electrode (see Eq. 4.4.4). If the approximate equation (5.4.9) is used, then the *Ilkovič equation* is obtained

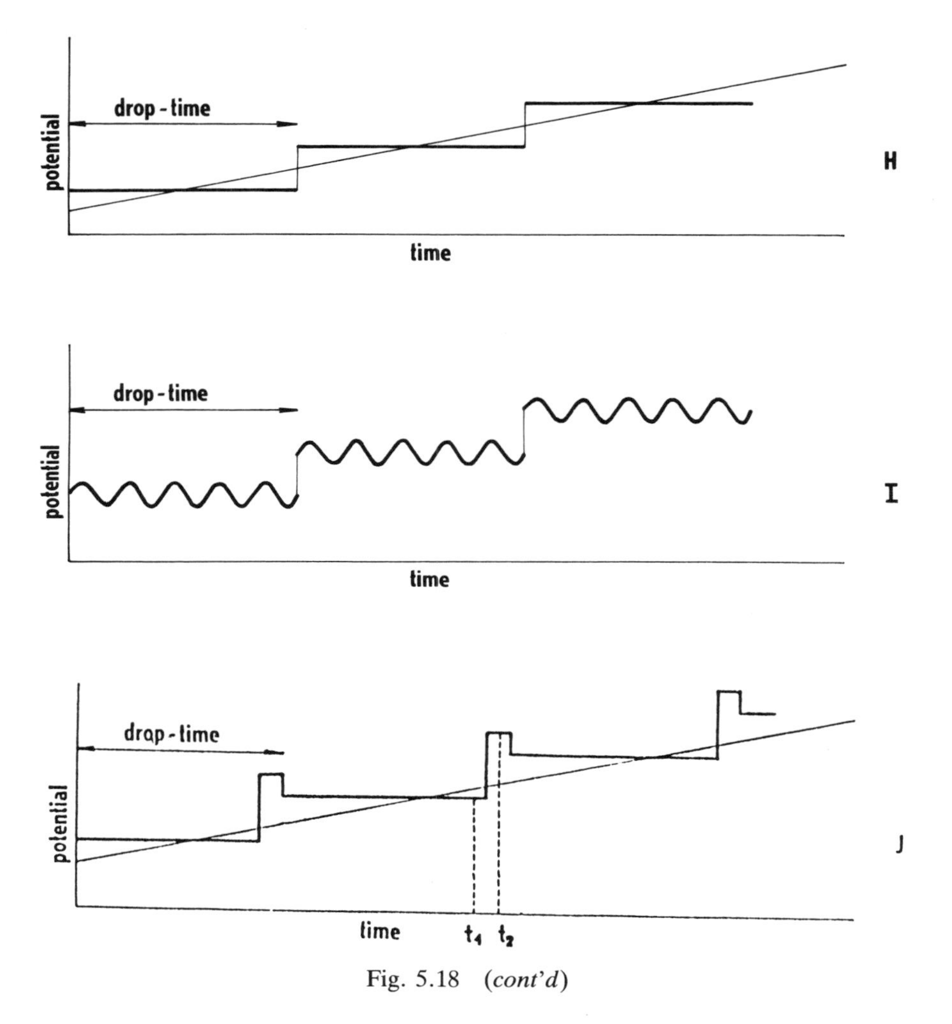

Fig. 5.18 (*cont'd*)

for the cathodic mean limiting diffusion current (the concentration is expressed in mol · dm^{-3}, m in g · s^{-1}, t_1 in s and D_{Ox} in cm^2 · s^{-1}):

$$\langle I_{\mathrm{d}} \rangle = -0.627 \times 10^{-3} nFm_1^{2/3} t_1^{1/6} D_{\mathrm{Ox}}^{1/2} c_{\mathrm{Ox}}^0 = -nF\bar{\kappa}_{\mathrm{Ox}} c_{\mathrm{Ox}}^0 \tag{5.5.9}$$

or, in general,

$$\langle I \rangle = -nF\bar{\kappa}[c_{\mathrm{Ox}}^0 - (c_{\mathrm{Ox}})_{x=0}] \tag{5.5.10}$$

This equation is analogous to Eq. (5.4.18) or (5.4.19) for the steady-state current density, although the instantaneous current depends on time. Thus, the results for a stationary polarization curve (Eqs (5.4.18) to (5.4.32)) can also be used as a satisfactory approximation even for electrolysis with the dropping mercury electrode, where the mean current must be considered instead of the current density. The quantities j and j_d must be replaced by $\langle I \rangle$ and $\langle I_d \rangle$, κ_{Ox} and κ_{Red} by $\bar{\kappa}_{Ox}$ and $\bar{\kappa}_{Red}$ ($\bar{\kappa}_{Ox}/\bar{\kappa}_{Red} = (D_{Ox}/D_{Red})^{1/2}$) and $k^{\ominus}$ by the product $k^{\ominus}\langle A \rangle$, where $\langle A \rangle$ is the mean area of the electrode (in cm^2):

$$\langle A \rangle = 0.51 m^{2/3} t^{2/3} \tag{5.5.11}$$

with m in $g \cdot s^{-1}$ and t_1 in s.

The theory of *voltammetry with rotating disk electrode* (*RDE*) was described in Sections 2.7.2 and 5.4.1.

The properties of the *voltammetric ultramicroelectrode* (*UME*) were discussed in Sections 2.5.1 and 5.5.1 (Fig. 5.19). The steady-state limiting diffusion current to a spherical UME is

$$I_d = 4\pi nFDr_0c_0 \tag{5.5.12}$$

while to a hemispherical UME placed on an insulating support

$$I_d = 2\pi nFDr_0c_0 \tag{5.5.13}$$

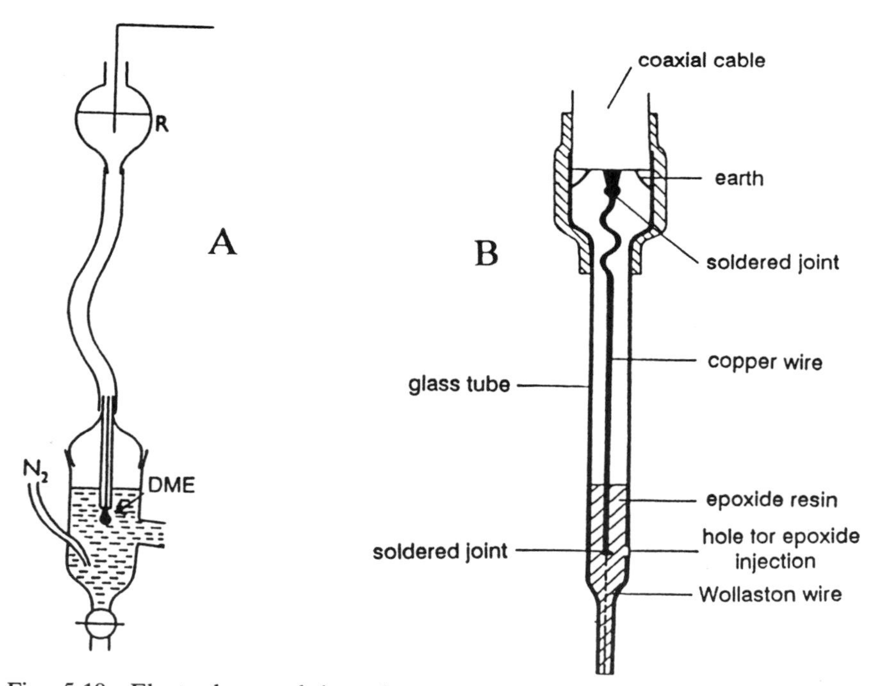

Fig. 5.19 Electrodes used in voltammetry. A—dropping mercury electrode (DME). R denotes the reservoir filled with mercury and connected by a plastic tube to the glass capillary at the tip of which the mercury drop is formed. B—ultramicroelectrode (UME). The actual electrode is the microdisk at the tip of a Wollaston wire (a material often used for UME) sealed in the glass tube

For a disk UME situated in the plane of the insulating support

$$I_d = 4nFDr_0c_0 \tag{5.5.14}$$

The non-steady state situation at UME represents a comparatively simple problem at a spherical electrode while it is very difficult to solve in the case of a disk electrode.

When comparing these three voltammetric electrodes they show different performance with respect to the purpose of their use.

1. With a DME, clean surface is quite easy to achieve whereas with the other electrodes a special treatment of the surface is required in order to obtain reproducible results (page 307). Even in solutions containing surfactants, the amount adsorbed at DME can be controlled (cf. Section 364).
2. The mass transfer coefficient for both RDE and UME can be varied in a large interval while in the case of DME it is much smaller and also the possibility of changing it is rather restricted.
3. In contrast to DME, the charging of the electric double layer makes no problem with RDE and UME.
4. The work with both DME and RDE requires the use of a *base* (*supporting or indifferent*) *electrolyte,* the concentration of which is at least twenty times higher than that of the electroactive species. With UME it is possible to work even in the absence of a base electrolyte. The ohmic potential difference represents no problem with UME while in the case of both other electrodes it must be accounted for in not sufficiently conductive media. The situation is particularly difficult with DME. Usually no potentiostat is needed for the work with UME.
5. The area of the electrode can easily be measured with DME and RDE while in the case of very small UME it is often difficult to determine.
6. The high overpotential of hydrogen is an advantage of DME while this property can be exploited only with mercury-covered RDE and UME. On the other hand, the dissolution of mercury at rather low positive potentials is a disadvantage of DME which is not shared by RDE and UME made of nobler metals than mercury.

In *differential pulse polarography,* the dropping mercury electrode is polarized by a sequence of pulses as shown in Fig. 5.18J. The difference between the currents flowing during several milliseconds before and at the end of a pulse is attenuated and recorded as function of the applied polarizing voltage, the charging current being thus eliminated. The resultant dependence of the current on the potential has the form of a peak, whose 'height' is proportional to the concentration. The high sensitivity of this method ensures it an important position in electroanalysis.

The *rotating ring disk electrode* can be used to study the products of the electrode reaction (Fig. 5.20). The products of the electrode reaction at the disk are carried by convective diffusion to the ring where they undergo a

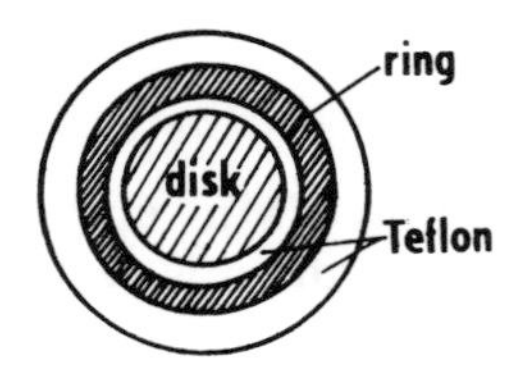

Fig. 5.20 The ring disk system. Sometimes the disk and the ring are made of different metals

further electrode reaction. The polarization curve at the ring is measured as a function of the current at the disk.

The simplest of the methods employing controlled current density is electrolysis at constant current density, in which the E–t dependence is measured (the *galvanostatic* or *chronopotentiometric method*). The instrumentation for this method is much less involved than for controlled-potential methods. The basic experimental arrangement for galvanostatic measurements is shown in Fig. 5.15, where a recording voltmeter or oscilloscope replaces the potentiometer. The theory of the simplest applications of this method to electrode processes was described in Section 5.4.1 (see Eqs 5.4.16 and 5.4.17).

Similar to potentiostatic methods, the main problem in the galvanostatic method is connected with charging of the electrode. This is especially true for high current densities where only part of the charge is consumed as the faradaic current and the rest is employed for charging the double layer, so that the main condition for this method is not fulfilled. This disadvantage can partly be removed by first applying a short impulse of strong current to charge the electrode double layer and then employing a weaker constant current (the *galvanostatic double-pulse method*).

The *potential-decay method* can be included in this group. Either a current is passed through the electrode for a certain period of time or the electrode is simply immersed in the solution and the dependence of the electrode potential on time is recorded in the currentless state. At a given electrolyte composition, various cathodic and anodic processes (e.g. anodic dissolution of the electrode) can proceed at the electrode simultaneously. The sum of their partial currents plus the charging current is equal to zero. As concentration changes thus occur in the electrolyte, the rates of the partial electrode reactions change along with the value of the electrode potential. The electrode potential has the character of a mixed potential (see Section 5.8.4).

In the closely related *coulostatic method* based on injection of a charge from a small condenser into an electrode in equilibrium with a redox system. The resulting time dependence of the electrode potential originates from the discharging of the electrical double layer by electrode reactions that were initially in equilibrium. The resultant E–t dependence is a function of the differential capacity of the electrode and the parameters of the electrode reaction. The advantage of this method is that no external current passes through the system so that no complicated instrumentation is required for elimination of the ohmic drop.

5.5.3 *Periodic methods*

If the working electrode–reference electrode–auxiliary electrode system is polarized in a potentiostatic circuit (see page 293) by alternating voltage with angular frequency ω and small amplitude ΔE ($\Delta E < 5$ mV), $E + \Delta E \sin \omega t$, then an alternating current, $I = \Delta I \sin(\omega t + \theta)$, is obtained, where θ is the phase angle. The electrode (or, in fact, the whole electrolytic cell) thus acts as an impedance (called the *faradaic or electrochemical impedance*) $Z = Z' - \mathrm{j}Z'' = R_s - \mathrm{j}/(\omega C_s)$, where $\mathrm{j} = \sqrt{-1}$, and R_s and C_s are the resistance and capacitance in series connection. For the phase angle θ we have

$$\theta = \arctan(Z''/Z') \tag{5.5.15}$$

In an analysis of an electrode process, it is useful to obtain the 'impedance spectrum'—the dependence of the impedance on the frequency in the complex plane, or the dependence of Z'' on Z', and to analyse it by using suitable equivalent circuits for the given electrode system and electrode process. Figure 5.21 depicts four basic types of impedance spectra and the corresponding equivalent circuits for the capacity of the electrical double layer alone (A), for the capacity of the electrical double layer when the electrolytic cell has an ohmic resistance R_E (B), for an electrode with a double-layer capacity C_D and simultaneous electrode reaction with polarization resistance R_P(C) and for the same case as C where the ohmic resistance of the cell R_E is also included (D). It is obvious from the diagram that the impedance for case A is

$$Z = \frac{-\mathrm{j}}{\omega C_D} \tag{5.5.16}$$

for case B,

$$Z = R_E - \frac{\mathrm{j}}{\omega C_D} \tag{5.5.17}$$

for case C,

$$Z = \left(\frac{1}{R_P} + \mathrm{j}\omega C_D\right)^{-1} = \frac{R_P}{1 + \omega^2 C_D^2 R_P^2} - \frac{\mathrm{j}\omega C_D R_P^2}{1 + \omega^2 C_D^2 R_P^2} \tag{5.5.18}$$

and, finally, for case D,

$$Z = R_E + \frac{R_P}{1 + \omega^2 C_D^2 R_P^2} - \frac{\mathrm{j}\omega C_D R_P^2}{1 + \omega^2 C_D^2 R_P^2} \tag{5.5.19}$$

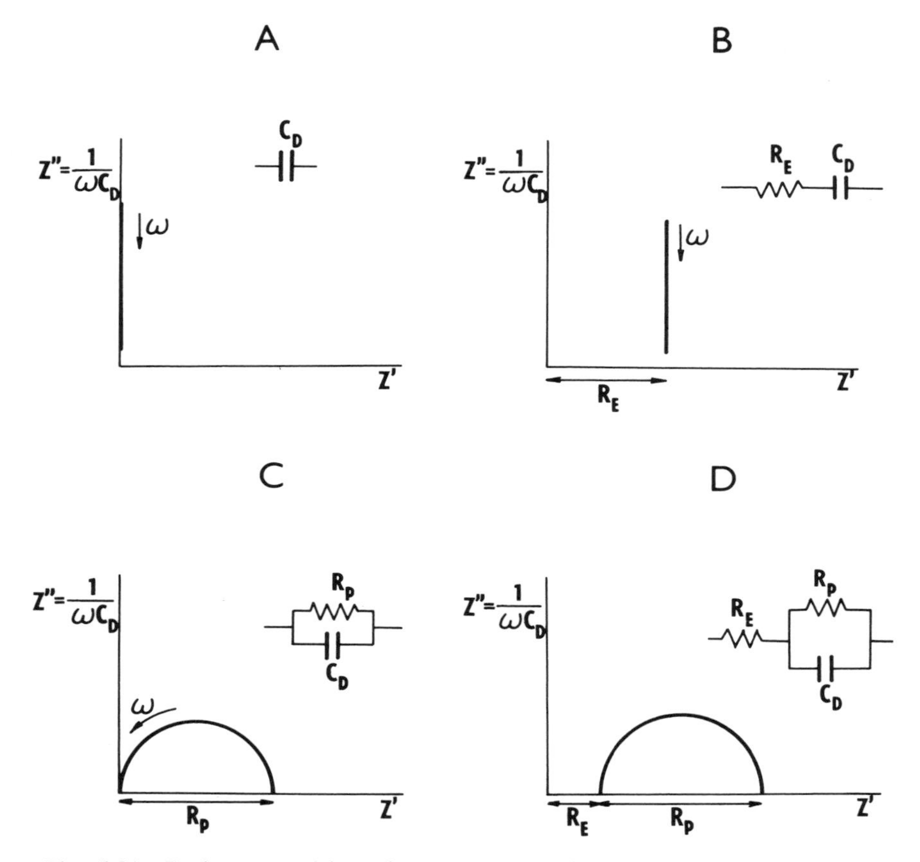

Fig. 5.21 Basic types of impedance spectra and of corresponding equivalent circuits (Eqs 5.5.16 to 5.5.19). (According to R. D. Armstrong *et al*)

The calculation becomes more difficult when the polarization resistance R_P is relatively small so that diffusion of the oxidized and reduced forms to and from the electrode becomes important. Solution of the partial differential equation for linear diffusion (2.5.3) with the boundary condition $D(\partial c_{Red}/\partial x) = -D(\partial_{Ox}/\partial x) = \Delta I \sin \omega t$ for a steady-state periodic process and a small deviation of the potential from equilibrium is

$$\Delta E = \frac{RT\Delta I}{n^2F^2\omega^{1/2}}\left(\frac{1}{c^0_{Ox}D^{1/2}_{Ox}} + \frac{1}{c^0_{Red}D^{1/2}_{Red}}\right)\sin\left(\omega t - \frac{\pi}{4}\right) \tag{5.5.20}$$

showing that the potential is behind the current by the phase angle $\theta = 45°$.

When the rate of the electrode reaction is measurable, being characterized by a definite polarization resistance R_P (Eq. 5.2.31), the electrode system can be characterized by the equivalent circuit shown in Fig. 5.22.

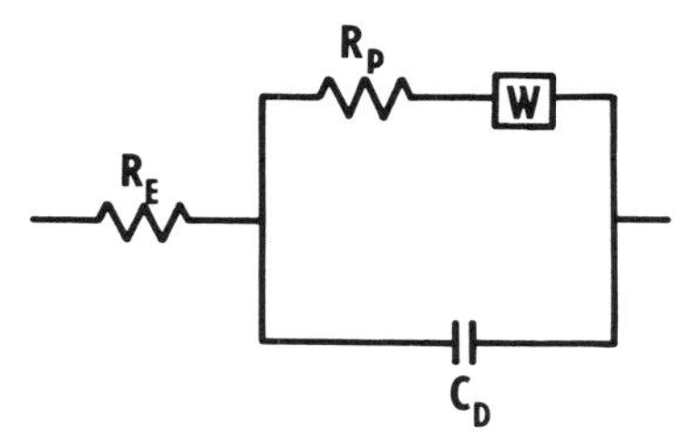

Fig. 5.22 Equivalent circuit of an electrode with diffusing electroactive substances. W is the Warburg impedance (Eq. 5.5.21)

Here W is the Warburg impedance corresponding to the diffusion process

$$W = \frac{RT}{n^2F^2\sqrt{2\omega}}\left(\frac{1}{c^0_{\mathrm{Ox}}D^{1/2}_{\mathrm{Ox}}} + \frac{1}{c^0_{\mathrm{Red}}D^{1/2}_{\mathrm{Red}}}\right)(1 - \mathrm{j}) = \sigma\omega^{-1/2}(1 - \mathrm{j}). \quad (5.5.21)$$

The impedance of the whole system is given as

$$Z = R_{\mathrm{E}} + \frac{1}{\mathrm{j}\omega C_{\mathrm{D}} + 1/(R_{\mathrm{P}} + W)} \quad (5.5.22)$$

The resulting dependence of Z'' on Z' (Nyquist diagram) is involved but for values of R_{p} that are not too small it has the form of a semicircle with diameter R_{p} which continues as a straight line with a slope of unity at lower frequencies (higher values of Z' and Z''). Analysis of the impedance diagram then yields the polarization resistance (and thus also the exchange current), the differential capacity of the electrode and the resistance of the electrolyte.

The impedance can be measured in two ways. Figure 5.23 shows an impedance bridge adapted for measuring the electrode impedance in a potentiostatic circuit. This device yields results that can be evaluated up to a frequency of 30 kHz. It is also useful for measuring the differential capacity of the electrode (Section 4.4). A phase-sensitive detector provides better results and yields (mostly automatically) the current amplitude and the phase angle directly without compensation.

A modification of faradaic impedance measurement is *a.c. polarography*, where a small a.c. voltage is superimposed on the voltage polarizing the dropping mercury electrode (Fig. 5.18I).

5.5.4 *Coulometry*

If all the constants in the expression for the limiting diffusion current are known, including the diffusion coefficient, then the number of electrons consumed in the reaction can be found from the limiting current data. However, this condition is often not fulfilled, or the limiting diffusion

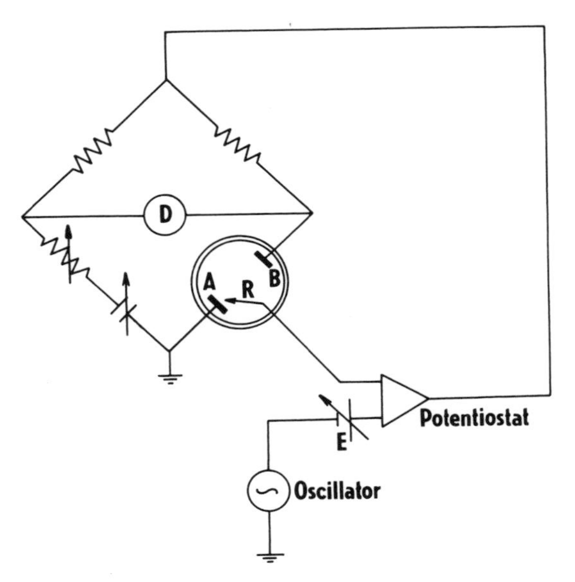

Fig. 5.23 A bridge for electrode impedance measurement in potentiostatic arrangement. D is the bridge balance detector and E the variable d.c. voltage source. For A, B and R see Fig. 5.15. (According to R. D. Armstrong *et al*)

current cannot be measured at all. The coulometric method is then used to determine the electrochemical equivalent of the electroactive substance. In this method, the change of the concentration of the electroactive substance is measured in dependence on the current passed through the electrolytic cell. The amount of charge is determined as the time integral of the current.

The electrolysis efficiency must be 100 per cent, i.e. the charge passed must be consumed only in the studied electrode reaction. There are four possible sources of error:

(a) Electrolysis of water. The range of useful potentials is determined by the potentials of oxygen and hydrogen evolution.
(b) Dissolution of the electrode material. This occurs particularly in complexing media and leads to an increase of the anodic or decrease of the cathodic current.
(c) Subsequent electrode reaction of the products of electrolysis.
(d) Electrode reduction of dissolved atmospheric oxygen.

Potentiostatic coulometry is mostly employed in the investigation of electrode processes. In this case, the current has to be directly proportional to the concentration of the electroactive substance:

$$I = A\kappa c = -\mathrm{d}m/\mathrm{d}t\, zF \tag{5.5.23}$$

where c is the instantaneous concentration of the electroactive species, $\mathrm{d}m/\mathrm{d}t$ is the amount of substance reduced per unit time, A is the area of the electrode and κ is a proportionality constant; I and c are functions of time because of the solution being exhausted by the electrode process. If the volume of the solution is V, then $c = m/V$ and a first-order differential equation

$$\mathrm{d}m/m = -A/(zFV)\,\mathrm{d}t \tag{5.5.24}$$

is obtained, the solution of which gives an exponential dependence of current on time. The plot of $\log I$ versus t has the slope $-\kappa A/zFV$, giving the value of z. A uniform concentration in the cell is maintained by stirring.

Investigation of intermediates of an electrode reaction and rapid determination of the electrochemical equivalents may be achieved by means of thin-layer electrolytic cell only about 10 μm thick, consisting of two platinum electrodes which are the opposing spindle faces of an ordinary micrometer.

5.5.5 *Electrode materials and surface treatment*

Metals are the most important electrode materials. Because of the readily renewable surface of mercury electrodes, they have been for several decades and, to a certain degree, still remain the most popular material for theoretical electrochemical research. The large-scale mercury electrode also plays a substantial role in technology (brine electrolysis) but the general tendency to replace it wherever possible is due to the environmental harmfulness of mercury.

At mercury electrodes the geometric and actual surface areas are identical. There is no site on the electrode that is energetically preferable to any other site (except for the deposition of heterogeneous patches). Except at rather positive potentials, the mercury electrode is usually not covered with a film of insoluble oxides or other compounds.

A liquid gallium electrode has similar properties to the mercury electrode (at temperatures above 29°C), but it has a far greater tendency to form surface oxides.

Mercury electrodes require far less maintenance than solid metal electrodes. Especially for the dropping mercury electrode, a noticeable amount of impurities present in the solution at low concentrations ($<10^{-5}\,\mathrm{mol\cdot dm^{-3}}$) cannot appreciably reach the surface of the electrode through diffusion during the drop-time (see Section 5.7.2).

Solid metal electrodes with a crystalline structure are different. The crystal faces forming the surface of these electrodes are not ideal planes but always contain *steps* (Fig. 5.24). Although equilibrium thermal roughening corresponds to temperatures relatively close to the melting point, steps are a common phenomenon, even at room temperature. A *kink* (*half-crystal position*—Fig. 5.24c) is formed at the point where one step ends and the

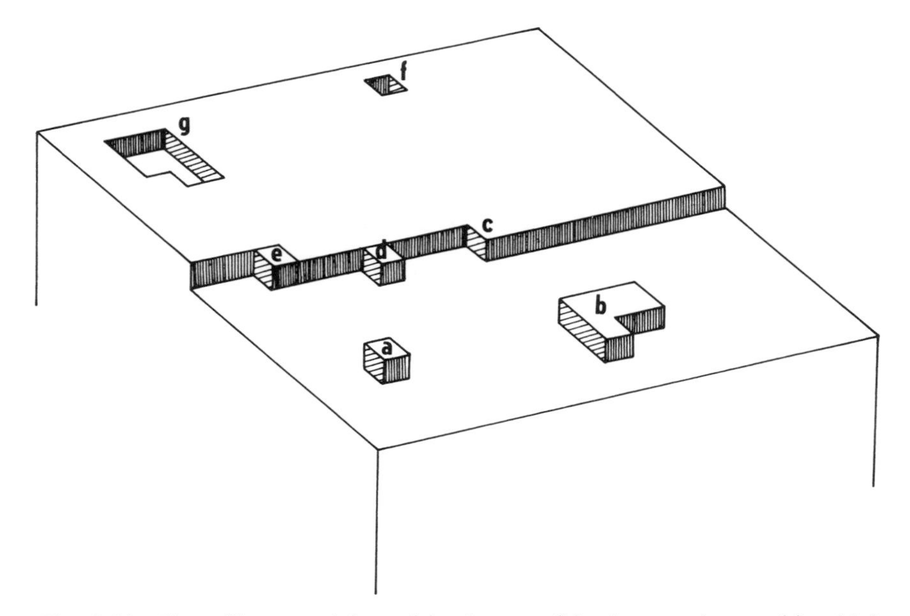

Fig. 5.24 Crystalline metal face: (a) ad-atom, (b) ad-atom cluster, (c) a kink (half-crystal position), (d) atom at the step, (e) atom in the kink, (f) vacancy, (g) vacancy cluster. (According to E. Budevski)

next begins, which is of great importance in crystal growth. In addition to steps, the crystal plane also contains individual metal atoms (*ad-atoms,* Fig. 5.24a) and *ad-atom clusters* (Fig. 5.24b). A. Kossel and I. Stranski have stated that the individual atoms in the crystal plane have different energies depending on their degree of contact with the other atoms of the crystal. The atoms inside the crystal have the lowest energy, as would be expected, followed by the atoms incorporated in the crystal plane and then by atoms in steps and in kinks and by ad-atoms.

In addition, in the great majority of cases metal crystals contain irregularities in the crystal lattice, termed dislocations. The rôle of *screw dislocations* will be considered in Section 5.8. In this type of dislocation, the individual atomic planes in the crystal are mutually translocated by a small angle, so that a wedge-shaped step arises from the crystal surface (Fig. 5.25). A single-crystal electrode without dislocations can be made by the method described by E. Budevski *et al.* A silver single crystal is fused in a glass tube with a capillary opening. This electrode is used as the cathode in the electrolysis of $AgNO_3$ by direct current with a small superimposed a.c. component. The front face then grows as a perfect, dislocation-free face, filling the cross-section of the capillary, usually approximately 0.2 mm in diameter.

Fundamental studies of the electrolyte solution/solid electrode interface

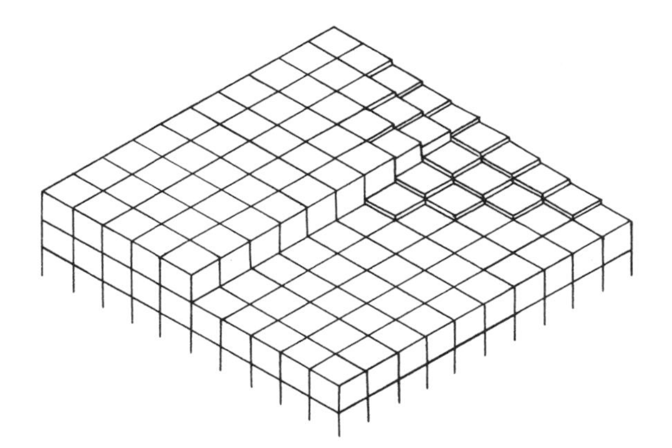

Fig. 5.25 A screw dislocation

can only be made by using well-defined single crystal electrodes. This additionally requires not only electrochemical methods of characterization, but also other methods supplying detailed structural information (LEED, STM, etc., see Section 5.5.6).

Solid metal electrodes are usually polished mechanically and are sometimes etched with nitric acid or aqua regia. Purification of platinum group metal electrodes is effectively achieved also by means of high-frequency plasma treatment. However, electrochemical preparation of the electrode immediately prior to the measurement is generally most effective. The simplest procedure is to polarize the electrode with a series of cyclic voltammetric pulses in the potential range from the formation of the oxide layer (or from the evolution of molecular oxygen) to the potential of hydrogen evolution (Fig. 5.18F).

The most important electrode material is platinum. Its widespread electrochemical use follows from its relatively high chemical inertness and electrocatalytic properties. The available potential window of platinum (the potential region where there is comparatively small faradaic current flow) is the interval from about 0 to +1.6 V versus SHE in acidic aqueous electrolytes; the negligible hydrogen overpotential of platinum makes it ideally suitable for the preparation of a standard hydrogen electrode (see Section 3.2.1). Platinum electrode is in aqueous electrolytes quickly covered, according to the polarization voltage, either with adsorbed hydrogen or surface oxide; only in a narrow potential interval (from 0.4 to 0.8 V versus SHE, the so-called 'double layer region') the Pt surface might be considered almost free from adsorbed or chemisorbed species. In practice, however, anions or organic impurities are always adsorbed even in this potential region. The Pt surface particularly interacts with double bonds in organic molecules. Chemisorption of electroactive olefinic molecules has also been employed for the preparation of *chemically modified electrodes* (see below).

Platinum electrodes are made usually from polycrystalline metal; the crystal planes at the surface include both the (111) and (100) faces in approximately equal proportions. The electrochemical properties of Pt(111) and Pt(100) faces are not identical. (Generally, the physical properties of individual metal crystal faces, such as work function, catalytic activity, etc., are different.)

Cyclic voltammetry studies of single-crystal platinum electrodes in acidic aqueous electrolytes showed that the two characteristic peaks of hydrogen adsorption/desorption on platinum (see Fig. 5.40) correspond in fact to reactions at two different crystal faces: the peak at lower potential to Pt(100) and the other one to Pt(111).

The opposite case to Pt single crystal are electrodes whose surface area has been increased by covering with platinum black. This so-called platinized platinum electrode (see page 173) shows up to 1000× higher physical surface area than the geometric surface area (the ratio of these two quantities is called *roughness factor,* see page 246). An increase in the physical surface area (roughness factor) of the electrode results in decreasing the relative coverage with surface-active impurities. The roughness factor of a single- crystal electrode is theoretically equal to one, but this value can hardly be achieved in view of microscopic imperfections of the real crystal face (see Fig. 5.24). A bright polycrystalline platinum electrode shows a roughness factor 1.3–3 depending on the level of polishing.

The second most widely used noble metal for preparation of electrodes is *gold.* Similar to Pt, the gold electrode, contacted with aqueous electrolyte, is covered in a broad range of anodic potentials with an oxide film. On the other hand, the hydrogen adsorption/desorption peaks are absent on the cyclic voltammogram of a gold electrode in aqueous electrolytes, and the electrocatalytic activity for most charge transfer reactions is considerably lower in comparison with that of platinum.

Semiconductors. In Sections 2.4.1, 4.5 and 5.10.4 basic physical and electrochemical properties of semiconductors are discussed so that the present paragraph only deals with practically important electrode materials. The most common semiconductors are Si, Ge, CdS, and GaAs. They can be doped to p- or n-state, and used as electrodes for various electrochemical and photoelectrochemical studies. Germanium has also found application as an infrared transparent electrode for the *in situ* infrared spectroelectrochemistry, where it is used either pure or coated with thin transparent films of Au or C (Section 5.5.6). The common disadvantage of Ge and other semiconductors mentioned is their relatively high chemical reactivity, which causes the practical electrodes to be almost always covered with an oxide (hydrated oxide) film.

The surface reactivity is especially pronounced under illumination of the semiconductor electrode with photons of an energy greater than the band gap. The photogenerated minority carriers (electrons or holes) may react with the electrode material to cause its photodecomposition, usually called *photocorrosion* (for electrochemical corrosion, see Section 5.8.4). It appears with both p-type semiconductors (surface reduction), and n-type semiconductors (surface oxidation), but it is practically more important in the latter case. The photocorrosion reactions can be demonstrated, e.g., on Si and CdS as follows:

(a) Cathodic photocorrosion of p-type materials

$$\begin{gathered} \text{p-Si} + 4e^- + 2H_2O \rightarrow SiH_4 + 4OH^- \\ \text{p-CdS} + 2e^- \rightarrow Cd + S^{2-} \end{gathered} \tag{5.5.25}$$

(b) Anodic photocorrosion of n-type materials

$$\begin{gathered} \text{n-Si} + 4h^+ + 2H_2O \rightarrow SiO_2 + 4H^+ \\ \text{n-CdS} + 2h^+ \rightarrow Cd^{2+} + S \end{gathered} \tag{5.5.26}$$

Here, the electrons are denoted e^- and the holes h^+.

The stability of n-Si, n-Ge, n-CdS, and n-GaAs photoanodes against photocorrosion increases in the presence of a solution species that reacts rapidly enough with photogenerated holes to compete with the surface oxidation. An alternative way is to cover the semiconductor electrode with an electronically conducting polymer film, such as polyvinylferrocene, polypyrrole, polyaniline, Nafion/tetrathiafulvalene, etc. These films are capable of efficiently capturing photogenerated holes and transport them to the solution, where they undergo a useful redox reaction (photoelectrosynthesis). The removal of photogenerated holes suppresses not only the photocorrosion, but also electron-hole recombination within the illuminated semiconductor, which increases the quantum yield of the photoelectrochemical process. Protection of p-type semiconductors by the mentioned conductive films is also possible; it enhances, for example, the rate of H_2 evolution at semiconductor photocathodes.

Various other semiconductor materials, such as CdSe, MoSe, WSe, and InP were also used in electrochemistry, mainly as n-type photoanodes. Stability against photoanodic corrosion is, naturally, much higher with semiconducting oxides (TiO_2, ZnO, $SrTiO_3$, $BaTiO_3$, WO_3, etc.). For this reason, they are the most important n-type semiconductors for photoanodes. The semiconducting metal oxide electrodes are discussed in more detail below.

Metal oxides. Noble metals are covered with a surface oxide film in a broad range of potentials. This is still more accentuated for common metals, and other materials of interest for electrode preparation, such as semiconductors and carbon. Since the electrochemical charge transfer reactions mostly occur at the surface oxide rather than at the pure surface, the study of electrical and electrochemical properties of oxides deserves special attention.

The electronic conductivity of metal oxides varies from values typical for insulators up to those for semiconductors and metals. Simple classification of solid electronic conductors is possible in terms of the band model, i.e. according to the relative positions of the Fermi level and the conduction/valence bands (see Section 2.4.1).

A transition behaviour between a semiconductor and a metal can also be observed with metal oxides. This occurs when the Fermi level of the semiconductor is shifted (usually by heavy doping) into the conduction or valence band. This system is called *degenerate semiconductor.* If there exists a sufficient density of electronic states above and below thus shifted Fermi energy, the degenerate semiconductor can behave in contact with an electrolyte as a quasimetal electrode (e.g., with the charge transfer coefficient α close to 0.5, cf. Section 5.2.8). The quasimetallic behaviour of a degenerate semiconductor electrode is also conditioned by a sufficient capacity of the electronic states near the Fermi energy. Assuming, for simplicity, the double layer at the semiconductor/electrolyte interface as a parallel plate capacitor, whose capacity is independent of the potential difference across the electric double layer, the charge σ_{sc} corresponding to a potential difference $\Delta_2^s\phi$ applied across the Helmholtz layer of thickness d is (cf. Section 4.5.1)

$$\sigma_{sc} = \Delta_2^s\phi\varepsilon\varepsilon_0/d \tag{5.5.27}$$

If the charge σ_{sc} is fully accommodated by electronic states near the Fermi energy, no space charge is formed in the electrode phase, and any voltage applied to the electrode appears exclusively across the Helmholtz layer, i.e. the system behaves like a metal.

Degeneracy can be introduced not only by heavy doping, but also by high density of surface states in a semiconductor electrode (pinning of the Fermi level by surface states) or by polarizing a semiconductor electrode to extreme potentials, when the bands are bent into the Fermi level region.

Another possibility of the appearance of a quasimetallic behaviour of metal oxides occurs with thin oxide films on metals. If the film thickness is sufficiently small (less than *ca.* 3 nm), the electrons can tunnel through the oxide film, and the charge transfer actually proceeds between the level of solution species and the Fermi level of the supporting metal (Fig. 4.12).

Although the band model explains well various electronic properties of metal oxides, there are also systems where it fails, presumably because of neglecting electronic correlations within the solid. Therefore, J. B. Goodenough presented alternative criteria derived from the crystal structure, symmetry of orbitals and type of chemical bonding between metal and oxygen. This 'semiempirical' model elucidates and predicts electrical properties of simple oxides and also of more complicated oxidic materials, such as bronzes, spinels, perowskites, etc.

There is a number of essentially non-conducting metal oxides acting as passive layers on electrodes; the best known example is Al_2O_3. Metals that form insulating oxide films by anodic bias (e.g. Al, V, Hf, Nb, Ta, Mo) are termed 'valve metals' (cf. Section 5.8.3).

Some insulating oxides become semiconducting by doping. This can be achieved either by inserting certain heteroatoms into the crystal lattice of the oxide, or more simply by its partial sub-stoichiometric reduction or oxidation, accompanied with a corresponding removal or addition of some oxygen anions from/into the crystal lattice. (Many metal oxides are, naturally, produced in these mixed-valence forms by common preparative techniques.) For instance, an oxide with partly reduced metal cations behaves as a n-doped semiconductor; a typical example is TiO_2.

Titanium dioxide is available as rutile (also in the form of sufficiently large single crystals) or anatase (only in polycrystalline form). Third modification, brookite, has no significance for electrode preparation. Pure rutile or anatase are practically insulators, with conductivities of the order of 10^{-13} S/cm. Doping to n-type semiconducting state is performed, for example, by partial reduction with gaseous hydrogen, or simply by heating at temperatures about 500°C. This is accompanied by a more or less deep coloration of the originally white material, and by an increase of the electronic conductivity by orders of magnitude. Semiconducting titanium dioxide is one of the most important electrode materials in photoelectrochemistry (see Section 5.9.2).

Another example of a non-stoichiometric, n-semiconducting oxide is PbO_2. This has been extensively studied as a positive electrode material for lead/acid batteries. Lead dioxide occurs in two crystal modifications, orthorhombic α-PbO_2, and more stable tetragonal β-PbO_2; both polymorphs show high electronic conductivity of the order of 10^2 S/cm, and a typical mixed-valence Pb(II)/Pb(IV) composition with the O/Pb ratio in the range about 1.83–1.96. If the lead dioxide is prepared in an acidic aqueous medium (as in a lead/acid battery), the oxygen deficiency is partly introduced by replacement of O by OH, i.e. a more correct formula of this material is

$$\mathrm{Pb(IV)}_{(1-x/2)}\mathrm{Pb(II)}_{x/2}\mathrm{H}_x\mathrm{O}_2$$

Both modifications of PbO_2 occur in the lead/acid battery in the approximate ratio of $\alpha/\beta \simeq 1/4$. The standard potential of the α-$PbO_2/PbSO_4$ redox couple in acidic medium is $E_0 = 1.697$ V (for the β modification by about 10 mV lower). Although the lead dioxide is thermodynamically unstable in contact with water, its relatively high oxygen overvoltage is responsible for a reasonable kinetic stability. A PbO_2 layer supported by Pb is suitable for anodic oxidations of various inorganic and organic substances, since it is stable against further oxidation and its ohmic resistance is negligible.

Another class of conducting oxides are degenerate semiconductors, obtained by heavy doping with suitable foreign atoms. Two oxides, n-SnO_2 (doped with Sb, F, In) and n-In_2O_3 (doped with Sn), are of particular interest. These are commercially available in the form of thin optically transparent films on glass (Nesa, Nesatron, Intrex-K, etc.). Typical layer thickness is several hundreds of nm, and resistivities of up to 5–20 Ω/square. (Note: the unit 'Ω/square' follows from the fact that the resistance of a uniform film on square support is independent of the size of the square, if we measure the resistance in direction parallel to the square.)

The electrochemical properties of conducting SnO_2 and related materials have been studied especially by T. Kuwana *et al.* Optically transparent SnO_2 electrodes have found interesting applications in spectroelectrochemistry (see Section 5.5.6). Similar to PbO_2, these electrodes exhibit relatively high oxygen overpotential and a good anodic stability, especially in acidic medium. Their thermal stability is also satisfactory, i.e. SnO_2 can be utilized in electrolytes such as eutectic melts. The available potential window is, nevertheless, limited at the cathodic side by irreversible reduction of the oxide to the metal.

High electrical conductivity is also attained in oxides with very narrow, partially filled conduction bands; the best known example is RuO_2. This material has a conductivity of about 2–3 10^4 S/cm at the room temperature, and metal-like variations with the temperature. Some authors consider RuO_2 and similar oxides as true metallic conductors, but others describe them rather as n-type semiconductors.

RuO_2 is an important electrode material for industrial anodic processes. Special attention is deserved by the so-called *dimensionally stable anodes* (DSA) invented by H. B. Beer in 1968. These are formed by a layer of a microcrystalline mixture of TiO_2 and RuO_2 (crystallite size less thn 0.1 μm) on a titanium support (Fig. 5.26). This material is suitable as anode for chlorine and oxygen evolution at high current densities. For industrial chlorine production, it replaced the previously used graphite anodes. These

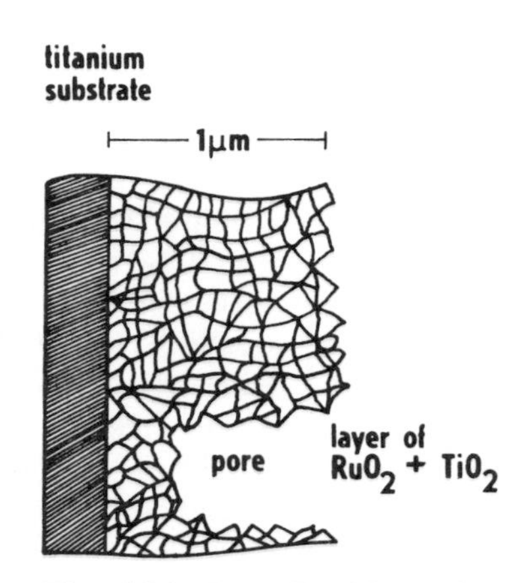

Fig. 5.26 Beer–De Nora dimensionally stable electrode with a RuO_2–TiO_2 layer

were rapidly 'burnt' by evolved O_2 and Cl_2, and thus the cathode–anode spacing gradually increased, which caused an unwanted increase of voltage during prolonged electrolysis.

Carbon. The electronically conductive carbons are derived from the hexagonal crystalline modification—*graphite.*

(Note: Electronic conductivity, and even superconductivity at *ca.* 10 K, was observed with a new carbon polymorph, buckminsterfullerene, synthesized in 1990. Electrochemical and photoelectrochemical properties of a semiconducting diamond electrode were studied by Pleskov.)

Graphite is a thermodynamically stable form of elemental carbon below approximately 2600 K at 10^2 MPa; its crystallographic description is D_{6h}^4–$P6_3/mmc$, unit cell constants: $a_0 = 245.6$ pm, $c_0 = 670.8$ pm. It shows a marked anisotropy of electronic conductivity: in the c-direction around 1 S/cm (increasing with temperature as with semiconductors), but in the a-direction around 10^5 S/cm (decreasing with temperature as with metals). The latter value can further be increased to a 10^6 S/cm level by intercalation with AsF_5 or SbF_5 (see the discussion of intercalation compounds below). At least one of the graphite intercalation compounds (C_8K) is a superconductor at low temperature.

Large single crystals of graphite are rare; therefore, graphite is employed in polycrystalline form (solid blocks, paste electrodes, suspensions, etc.). Nevertheless, some carbons prepared by chemical vapour deposition (pyrolytic carbons) match the properties of a graphite single crystal. The supported layer of pyrolytic graphite is actually polycrystalline, but the individual crystallites show a high degree of preferred orientation with carbon hexagons parallel to the surface of the substrate. This is especially true with the so-called *highly oriented pyrolytic graphite* (HOPG), prepared from ordinary pyrolytic graphite by heat treatment under simultaneous compressive stress (stress annealing) around 3500°C. A HOPG contains well-ordered crystallites, whose angular scatter is typically less than 1°.

The faces parallel to the plane of carbon hexagons (100), also called basal planes, show chemical and electrochemical properties significantly different from those of the perpendicular (edge) planes (110). The latter contain unsaturated carbon atoms (dangling bonds), which are highly active, for example, against oxidation with oxygen, both chemical and electrochemical. The structure of thus formed surface oxides on carbons is very complex, but it can be derived from the common oxygen functional groups known from organic chemistry: carboxyl, hydroxyl, and carbonyl. These occur on the oxidized carbon surface either individually or in various combinations: surface anhydride, ether, lactone, etc. (see Fig. 5.27). Surface carbonyls (quinones) are reversibly reducible; further anodic oxidation of graphite and carbons occurs through intermediates such as mellithic acid to finally give CO_2.

The basal plane of graphite (or HOPG) is essentially free from surface oxygen functionalities. This is manifested by almost flat, featureless

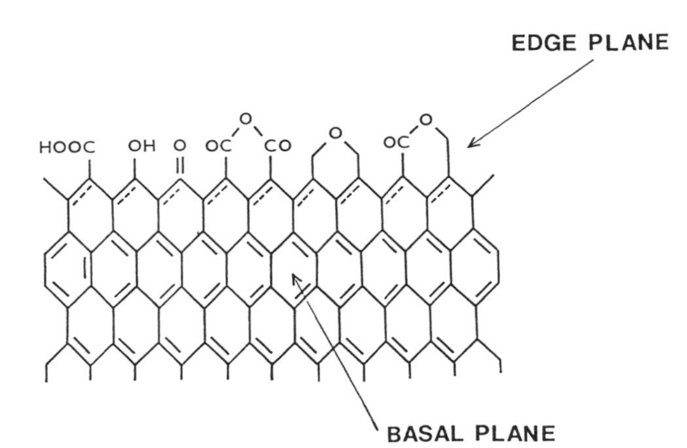

Fig. 5.27 Oxygen functional groups on the edge plane of graphite

voltammetric curves in aqueous electrolytes, and relatively low charging currents. Unlike platinum, the basal plane HOPG virtually does not adsorb anions and most organic substances. Some molecules, such as phtalocyanines or porphyrines, form, however, π–π complexes with graphite, and are thus strongly adsorbed, being oriented parallel to the basal plane.

The most popular carbon material for electrochemistry is probably the *glass-like carbon.* (This term is preferable to the common names 'glassy carbon' and 'vitreous carbon', which are actually trade marks and should not be used as terms.) Glass-like carbon is prepared by pyrolysis of suitable polymers, e.g. polyfurfurylalcohol or phenolic resins. If the heating programme is properly controlled, the gaseous by-products of pyrolysis evolve sufficiently slowly and the final product is a monolithic, hard carbon, showing very low porosity and permeability to liquids and gases.

Glass-like carbon is, in contrast to graphite, isotropic in electrical conductivity and other physico–chemical properties. This explains its structure, characterized by randomly oriented strips (lamellae) of pseudographitic layers of carbon hexagons. The lamellae occasionally approach the graphitic interlayer spacing (335 pm), forming extremely small graphitic regions with crystalline dimensions L_a, L_c of the order of 1–10 nm. These regions are interconnected by a non-graphitic carbon material as shown in Fig. 5.28. A similar three-dimensional network of graphitic microcrystallites can also be found in other forms of 'amorphous' carbon, and, according to the forces needed for breaking the network crosslinks (orientation of crystallites), we can classify almost all the solid carbons as hard (non-graphitizable) and soft (graphitizable) ones. Most polycrystalline carbons, including glass-like carbon, show an increase of electrical conductivity with increasing temperature; they can be regarded as intrinsic semiconductors with a band-gap energy of several tens to hundreds meV.

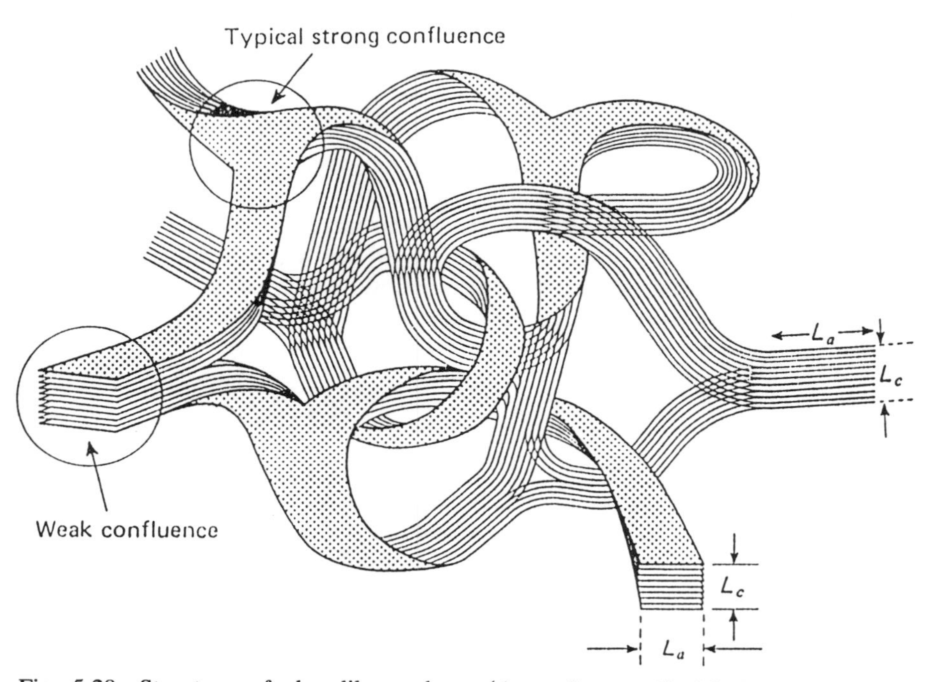

Fig. 5.28 Structure of glass-like carbon. (According to G. M. Jenkins and K. Kawamura)

Carbon electrodes are widely used in electrochemistry both in the laboratory and on the industrial scale. The latter includes production of aluminium, fluorine, and chlorine, organic electrosynthesis, electrochemical power sources, etc. Besides the use of graphite (carbons) as a virtually inert electode material, the electrochemical intercalation deserves special attention. This topic will be treated in the next paragraph.

In spite of the high effort focused on the carbon electrochemistry, very little is known about the electrochemical preparation of carbon itself. This challenging idea appeared in the early 1970s in connection with the cathodic reduction of poly(tetrafluoroethylene) (PTFE) and some other perfluorinated polymers. The standard potential of the hypothetical reduction of PTFE to elemental carbon:

$$(CF_2)_n + 2e^- + 2H^+ \rightarrow C_n + 2HF \qquad (5.5.28)$$

estimated from the Gibbs free energies of formation (cf. Sections 3.1.4 and 3.1.5) equals 1.00 V at 25°C. The reductive carbonization of PTFE is, therefore, thermodynamically highly favoured, and it indeed occurs at the carbon/PTFE interface in aprotic liquid or solid electrolytes. The produced elemental carbon is termed *electrochemical carbon*. It is reversibly dopable (reducible) under insertion of supporting electrolyte cations into the carbon skeleton.

It should be emphasized that the electrochemical carbonization proceeds, in contrast to all other common carbonization reactions (pyrolysis), already at the room temperature. This fact elucidates various surprising physico–chemical properties of electrochemical carbon, such as extreme chemical reactivity and adsorption capacity, time-dependent electronic conductivity and optical spectra, as well as its very peculiar structure which actually matches the structure of the starting fluorocarbon chain. The electrochemical carbon is, therefore, obtained primarily in the form of linear polymeric carbon chains (polycumulene, polyyne), generally termed carbyne. This can be schematically depicted by the reaction:

$$-\underset{\mathrm{F}}{\overset{\mathrm{F}}{\underset{|}{\overset{|}{\mathrm{C}}}}}-\underset{\mathrm{F}}{\overset{\mathrm{F}}{\underset{|}{\overset{|}{\mathrm{C}}}}}-\underset{\mathrm{F}}{\overset{\mathrm{F}}{\underset{|}{\overset{|}{\mathrm{C}}}}}-\underset{\mathrm{F}}{\overset{\mathrm{F}}{\underset{|}{\overset{|}{\mathrm{C}}}}}- \xrightarrow{8e^-} \underset{\mathrm{F^-}}{\overset{\mathrm{F^-}}{=\mathrm{C}}}\underset{\mathrm{F^-}}{\overset{\mathrm{F^-}}{=\mathrm{C}}}\underset{\mathrm{F^-}}{\overset{\mathrm{F^-}}{=\mathrm{C}}}\underset{\mathrm{F^-}}{\overset{\mathrm{F^-}}{=\mathrm{C}}}= \qquad (5.5.29)$$

The product of the reaction (5.5.29) is, however, unstable against subsequent interchain crosslinking and insertion of supporting electrolyte cations. The electrochemical carbonization of fluoropolymers was recently reviewed by L. Kavan.

Insertion (intercalation) compounds. Insertion compounds are defined as products of a reversible reaction of suitable crystalline host materials with guest molecules (ions). Guests are introduced into the host lattice, whose structure is virtually intact except for a possible increase of some lattice constants. This reaction is called *topotactic.* A special case of topotactic insertion is reaction with host crystals possessing stacked layered structure. In this case, we speak about ‘intercalation’ (from the Latin verb *intercalare,* used originally for inserting an extra month, *mensis intercalarius,* into the calendar).

The intercalated guest molecules have a tendency to fill up completely the particular interlayer space in the host crystal, rather than to occupy randomly distributed interstitial sites between the layers. A perculiarity of intercalation is that the sequence of filled and empty interlayer spaces is very often regular, i.e. the structure of intercalation compounds consists of stacked guest layers interspaced between a defined number of the host layers. The number of host layers corresponds to the so-called *stages* of intercalation compounds. Stage 1 denotes successively stacked host and guest layers, stage 2 denotes a system with two host layers between one guest layer, etc. This is schematically depicted in Fig. 5.29.

Graphite crystal is the best known host material for intercalation reactions. The host layers are clearly defined by planes of carbon hexagons, also called graphene layers. (The term ‘graphene’ has been derived by analogy with condensed aromatic hydrocarbons: naphthalene, anthracene, tetracene, perylene, etc.). In contrast to other host materials (e.g. inorganic sulphides), graphite also forms compounds with high stage numbers

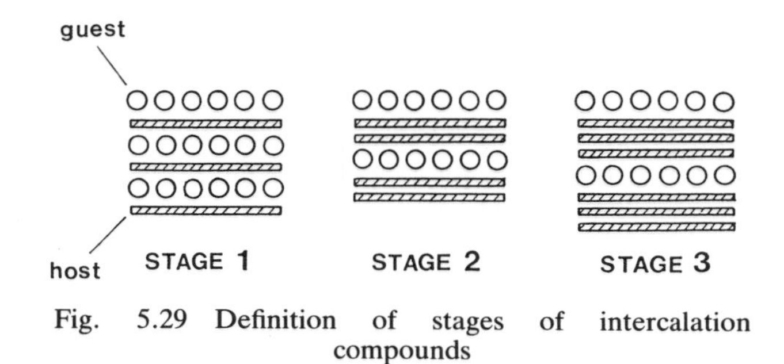

Fig. 5.29 Definition of stages of intercalation compounds

(10, 11, . . ,). Intercalation of some guest species, such as alkali metals, can simply be performed via a chemical reaction of a gaseous reactant with graphite.

Some other guest species (e.g. H_2SO_4, $HClO_4$ and other inorganic acids), however, do not react spontaneously with graphite, but the intercalation can be induced by an auxiliary oxidizing or reducing agent. The redox reaction promoting intercalation can also be performed electrochemically; the advantage of electrochemical intercalation is not only the absence of any foreign chemical agent and the corresponding reaction by-products, but also a precise control of potential, charge and kinetics of the process.

The electrochemical intercalation into graphite leads in the most simple case to binary compounds (graphite salts) according to the schematic equations:

$$C_n + M^+ + e^- \rightarrow M^+C_n^- \tag{5.5.30}$$

$$C_n + A^- - e^- \rightarrow C_n^+A^- \tag{5.5.31}$$

In the first case, the graphite lattice is negatively charged and the electrolyte cations, M^+ (alkali metals, NR_4^+, etc.) compensate this charge by being inserted between the graphene layers. In the second case, anions, A^- (ClO_4^-, AsF_6^-, BF_4^-, etc.) are analogously inserted to compensate the positive charge of the graphite host.

Reactions (5.5.30) and (5.5.31) proceed prevailingly during intercalation from solid or polymer electrolytes (cf. Section 2.6) or melts. When using common liquid electrolyte solutions, a co-insertion of solvent molecules (and/or intercalation of solvated ions) very often occurs. The usual products of electrochemical intercalation are therefore ternary compounds of a general composition:

$$M^+(\mathrm{Solv})_y C_n^- \quad \text{or} \quad C_n^+(\mathrm{Solv})_y A^-$$

The electrochemical intercalation of HSO_4^- anions together with H_2SO_4 was described by Thiele in 1934. The composition of the product of prolonged anodic oxidation of graphite in concentrated sulphuric acid is

$$C_{24}^+HSO_4^- \cdot 2H_2SO_4$$

and it is the stage 1 intercalation compound. By a very rapid heating of this compound, the intercalate vaporizes and forces the graphene layers apart, like in an accordion. The product is called *exfoliated graphite*; it exhibits various peculiar properties, for example, up to hundred times smaller density than graphite. It is used, for example, in the preparation of graphite foils.

The electrochemical intercalation/insertion has not only a preparative significance, but appears equally useful for charge storage devices, such as electrochemical power sources and capacitors. For this purpose, the co-insertion of solvent molecules is undesired, since it limits the accessible specific faradaic capacity.

The electrochemical intercalation/insertion is not a special property of graphite. It is apparent also with many other host/guest pairs, provided that the host lattice is a thermodynamically or kinetically stable system of interconnected vacant lattice sites for transport and location of guest species. Particularly useful are host lattices of inorganic oxides and sulphides with layer or chain-type structures. Figure 5.30 presents an example of the cathodic insertion of Li^+ into the TiS_2 host lattice, which is practically important in lithium batteries.

The concept of electrochemical intercalation/insertion of guest ions into the host material is further used in connection with redox processes in electronically conductive polymers (polyacetylene, polypyrrole, etc., see below). The product of the electrochemical insertion reaction should also be an electrical conductor. The latter condition is sometimes by-passed, in systems where the non-conducting host material (e.g. fluorographite) is finely mixed with a conductive binder. All the mentioned host materials (graphite, oxides, sulphides, polymers, fluorographite) are studied as prospective cathodic materials for Li batteries.

From this point of view, fluorographite, $(CF_x)_n$ with $x \leqq 1$, deserves

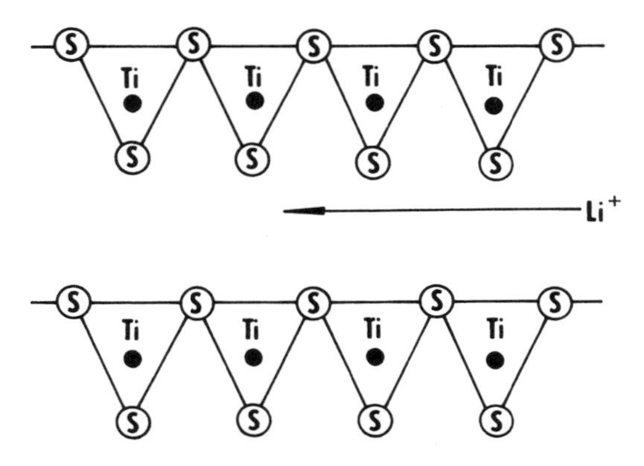

Fig. 5.30 Insertion of Li^+ between TiS_2 host layers. (According to C. A. Vincent *et al.*)

special attention. It is a covalent non-conductive compound whose structure is characterized by parallel planes of sp^3 carbon atoms in trans-linked perfluorinated cyclohexane chairs. The electrochemical reduction of fluorographite in a lithium cell was originally considered as a simple defluorination with the formation of LiF and elemental carbon. Further studies revealed, however, that the primary process, controlling the discharge rate and the cell voltage, is an electrochemical Li^+ intercalation between the layers in the fluorographite lattice. A typical product of this reaction is a ternary intercalation compound containing the co-inserted solvent molecules, Solv:

$$(CLi_xF_xSolv_y)_n$$

The electrochemical properties of fluorographite are also interesting in connection with the electrolysis of melted $KF \cdot 2HF$, which is used for industrial production of fluorine. Fluorine is here evolved at the carbon anode, which is spontaneously covered with a passivating layer of fluorographite; hence it causes an undesired energy loss during the electrolysis.

Chemically modified electrodes. Chemical and/or electrochemical surface pretreatments are always needed in order to get reproducible data on solid electrodes (see page 307); they are mostly focused on the removal of surface impurities (oxides). The 'activation' of solid electrodes by (electro)chemical pretreatment is, however, a complex and not well understood process. For instance, recent studies by scanning tunnelling microscopy demonstrated that the common pretreatment of a platinum electrode by a series of cyclic voltammetric pulses in the potential range of H_2/O_2 evolution (see page 307) cannot simply be regarded as chemical 'purification' from adsorbed species, since it is accompanied with marked changes in surface topography.

A qualitatively new approach to the surface pretreatment of solid electrodes is their chemical modification, which means a controlled attachment of suitable redox-active molecules to the electrode surface. The anchored surface molecules act as charge mediators between the elctrode and a substance in the electrolyte. A great effort in this respect was triggered in 1975 when Miller *et al.* attached the optically active methylester of phenylalanine by covalent bonding to a carbon electrode via the surface oxygen functionalities (cf. Fig. 5.27). Thus prepared, so-called 'chiral electrode' showed stereospecific reduction of 4-acetylpyridine and ethylphenylglyoxylate (but the product actually contained only a slight excess of one enantiomer).

A large selection of chemical modifications employing the surface oxygen groups on carbon and other electrode materials (metals, semiconductors, oxides) has been presented. For instance, surfaces with a high concentration of OH groups (typically SnO_2 and other oxides) are modified by organosilanes:

$$[S]\text{—}OH + Cl\text{—}SiR_3 \rightarrow [S]\text{—}O\text{—}SiR_3 \qquad (5.5.32)$$

where [S] stands for electrode surface, and the radical R might bear various electroactive species, such as ferrocene, viologenes, porphyrins, etc. Surfaces with OH groups can also be modified via 1,3,5-trichloro-2,4,6-triazine as a coupling agent. This further reacts with alcohols, amines, Grignard reagents, etc. Another method employes the surface carboxylic groups on carbon electrodes, usually activated by conversion to acylchloride with $SOCl_2$, followed by esterification or amidization in order to anchor the desired electroactive species.

In an ideal case the electroactive mediator is attached in a monolayer coverage to a flat surface. The immobilized redox couple shows a significantly different electrochemical behaviour in comparison with that transported to the electrode by diffusion from the electrolyte. For instance, the reversible charge transfer reaction of an immobilized mediator is characterized by a symmetrical cyclic voltammogram ($E_{pc} - E_{pa} = 0$; $j_{pa} = -j_{pc} = |j_p|$) depicted in Fig. 5.31. The peak current density, j_p, is directly proportional to the potential sweep rate, v;

$$j_p = n^2F^2\Gamma v/4RT \tag{5.5.33}$$

where Γ is the surface coverage of electroactive species, which might simply be determined from the voltammetric charge per unit electrode area (area of the peak, see Fig. 5.31):

$$\Gamma = Q/nF \tag{5.5.34}$$

or, alternatively, by non-electrochemical techniques of surface analysis (Section 5.5.6).

Besides the charge transfer reactions of the immobilized mediator, the

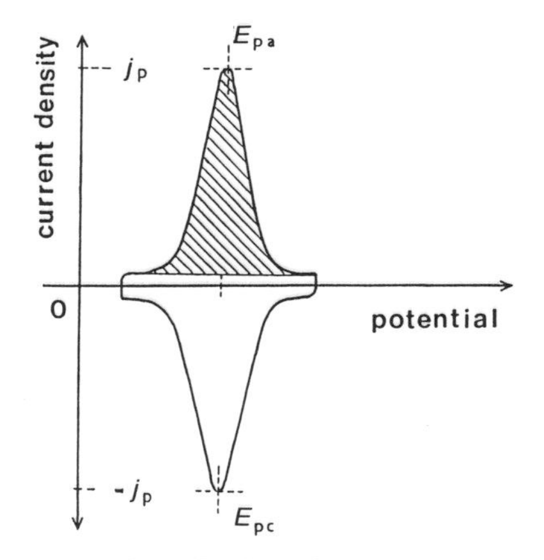

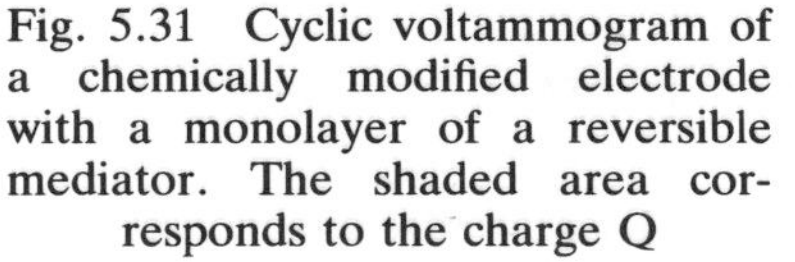
Fig. 5.31 Cyclic voltammogram of a chemically modified electrode with a monolayer of a reversible mediator. The shaded area corresponds to the charge Q

mediated electron transfer to a solution species has to be considered. A practically important situation occurs when the mediated electron transfer is faster than the charge transfer reaction of the same solution species on a bare electrode. If the charge transfer reaction at the electrode surface (which regenerates the original form of the mediator) is also faster, we speak about an *electron-transfer-mediated electrocatalysis*.

Since 1975, hundreds of papers dealing with chemically modified electrodes have appeared in the literature. The initial effort in this topic was reviewed by Murray in 1984. In subsequent years, the interest in surface modification gradually decreased. The reason might be that the rather demanding preparative techniques of chemical modification result, naturally, in a product possessing all the disadvantages of a two-dimensional (monolayer) system. The number of the redox-mediator turnovers, needed for sufficiently high electrode currents, is, therefore, relatively very large. For instance, the mediated current of 1 A/cm^2 would require (for a typical mediator coverage $\Gamma = 10^{-10}\,mol/cm^2$), the turnover rate n_t:

$$n_t = j/F\Gamma \simeq 10^5\,s^{-1} \tag{5.5.35}$$

which is beyond the practically available values. The surface modified electrodes may also be rapidly deactivated by the mediator desorption or decomposition as well as by adsorption of impurities from the electrolyte.

Therefore, recent interest has been focused prevailingly on electrodes modified by a multilayer coverage, which can easily be achieved by using polymer films on electrodes. In this case, the mediated electron transfer to solution species can proceed inside the whole film (which actually behaves as a system with a homogeneous catalyst), and the necessary turnover rate is relatively lower than in a monolayer.

A discussion of the charge transfer reaction on the polymer-modified electrode should consider not only the interaction of the mediator with the electrode and a solution species (as with chemically modified electrodes), but also the transport processes across the film. Let us assume that a solution species S reacts with the mediator Red/Ox couple as depicted in Fig. 5.32. Besides the simple charge transfer reaction with the mediator at the interface film/solution, we have also to include diffusion of species S in the polymer film (the diffusion coefficient D_{Sp}, which is usually much lower than in solution), and also charge propagation via immobilized redox centres in the film. This can formally be described by a diffusion coefficient D_p which is dependent on the concentration of the redox sites and their mutual distance (cf. Eq. (2.6.33).

The charge propagates in the film by electron hopping between the polymer Red/Ox couples. This is controlled by the electrode potential only in a close proximity of the electrode; in more distant sites, the charge transport is driven by a concentration gradient of reduced or oxidized mediators. The observed faradaic current density, j_F, is a superposition of

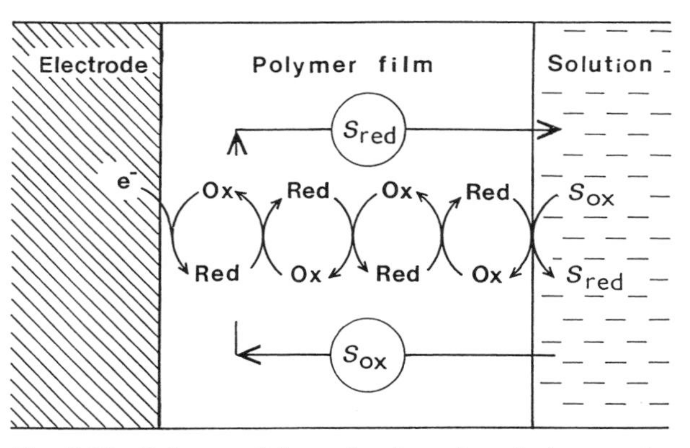

Fig. 5.32 Scheme of the reduction of a solution species S_{ox} within the polymer film. Electron transfer is mediated by the immobilized redox active sites Red/Ox

these two contributions: diffusion of S and electron hopping in the film:

$$j_F/nF = D_{Sp}(ds/dx)_0 - D_p(dr/dx)_0 \tag{5.5.36}$$

(s and r are the concentrations of species S and Red/Ox sites, respectively). The solution of Eq. (5.5.36) may differ for various particular situations and boundary conditions, but we can phenomenologically describe the mediated charge transfer process in the polymer film by an effective heterogeneous rate constant k_p and the concentration of S in solution, c_s:

$$j_F/nF = k_p c_s \tag{5.5.37}$$

The redox active polymer films might bear the mediator group attached either covalently to the polymer backbone (polyvinylferrocene, Ru(II) complexes of polyvinylpyridine, etc.) or electrostatically within the ion-exchange polymer (e.g. in Nafion, cf. Section 2.6).

Polymer films on electrodes have been prepared by various techniques; the most simple one is dip coating, i.e. adsorption of a preformed polymer from its solution. Another method consists in spreading and evaporating the polymer solution on the electrode. More uniform films are obtained by spin coating, which means evaporating on a rotating electrode. Polymer films have also been prepared by plasma polymerization, but the product is mostly ill-defined, owing to complicated side reactions introduced by high-energy particles in the plasma. Electrochemical polymerization deserves special attention and will be treated in the next section.

Modification of electrodes by electroactive polymers has several practical applications. The mediated electron transfer to solution species can be used in electrocatalysis (e.g. oxygen reduction) or electrochemical synthesis. For electroanalysis, preconcentration of analysed species in an ion-exchange film may remarkably increase the sensitivity (cf. Section 2.6.4). Various applications appeared also in corrosion protection, sensors, display devices, photoelectrochemistry, etc.

Electronically conducting polymers. The world-wide interest in electronically conducting polymers has been triggered in 1977/8, when Heeger, MacDiarmid, Shirakawa, and others have studied the oxidation and reduction reactions of polyacetylene,

$$(\text{—CH=CH—})_n$$

This polymer occurs as *trans*- or a less stable *cis*-isomer. (The *cis*-isomer can be converted to *trans*- by heating.) Both isomers are intrinsic semiconductors with conductivities around 10^{-9} S/cm (*cis*) and 10^{-6} S/cm (*trans*).

Polyacetylene can partly be oxidized or reduced with considerable increase in conductivity, up to 12 orders of magnitude. Oxidation or reduction of polyacetylene can be performed either chemically or electrochemically; this process is often termed p- or n-doping. In both cases, the polymer backbone is virtually intact except for accumulating positive (p-doping) or negative (n-doping) charge. This charge is compensated by ions of opposite sign, which are inserted into the polymer backbone either from a chemical dopant or from a supporting electrolyte.

The concept of p- and n-doping of polymers is similar to that of p- and n-doping of inorganic semiconductors, but it is, in fact, not equivalent. For instance, doping of silicon means that several atoms in the lattice of silicon are replaced by heteroatoms such as As or B (cf. Section 2.4, and Fig. 2.3). These sites are virtually electroneutral as long as they dissociate to form free electrons or holes and ionized heteroatoms (e.g. As^+ or B^-). On the other hand, doping of polyacetylene and other conjugated polymers means oxidation (p-doping) or reduction (n-doping) of the polymer skeleton associated with insertion of a corresponding number of anions (cations) into the polymer matrix. This can formally be described by Eqs (5.5.30) and (5.5.31) (where C_n stands for the polymer).

The individual macromolecular chains of conducting polymers agglomerate into more complicated structures, usually fibrous. The electronic conductivity of this system is a superposition of the conductivity of the individual fibres (chains) and that due to electron hopping between these domains. The latter is usually much lower, i.e. it controls the total conductivity of the system.

The electron hopping differs from the conventional solid-state conduction in that the charge transfer does not take place in the conduction band, but between localized states within the energy gap. This can be described by Mott's model of *variable range electron hopping*. This model was originally considered for non-crystalline inorganic semiconductors, such as amorphous germanium or chalcogenide glasses, but it is readily applicable also to conducting polymers and other amorphous or pseudoamorphous solids with microheterogeneous distribution of conductive and insulating regions (a typical example is glass-like carbon, see Fig. 5.28, and other 'amorphous carbons'). Mott's model follows from the preconception that the distance of electron hops varies with temperature and energy of the electron. Accordingly, the temperature dependence of conductivity of polymers (glass-like carbon, etc.) can be expressed by the equation:

$$\kappa = A \exp(\varepsilon / kT^n) \tag{5.5.38}$$

where A is a constant and ε is the activation energy of hopping. The exponent n is usually 1/4 or 1/3 (it is the reciprocal of the effective dimensionality of the space in which the charge transport occurs).

The charge transport in a conjugated chain and the interchain hopping is explained in terms of conjugation defects (radical or ionic sites), called *solitons* and *polarons*. Several possible conjugation defects are demonstrated in Fig. 5.33 on the example of *trans*-polyacetylene.

A soliton is an electroneutral radical site with a spin equal to 1/2 connected to the rest of the chain by somewhat longer bonds; the charge and the bond lengths are, however, disturbed also in the close surroundings of the defect, hence the soliton extends over about 15 (—CH—) units. If we remove or add an electron to the soliton, a zero spin and the corresponding charge appears (charged soliton). Two neutral solitons can recombine and the chain defect is cancelled. On the other hand, charged and neutral solitons in neighbouring sites form a stable pair called polaron. Both the solitons and polarons are mobile along the chain, but a soliton, unlike a polaron, moves virtually without overcoming energy barriers. If the number

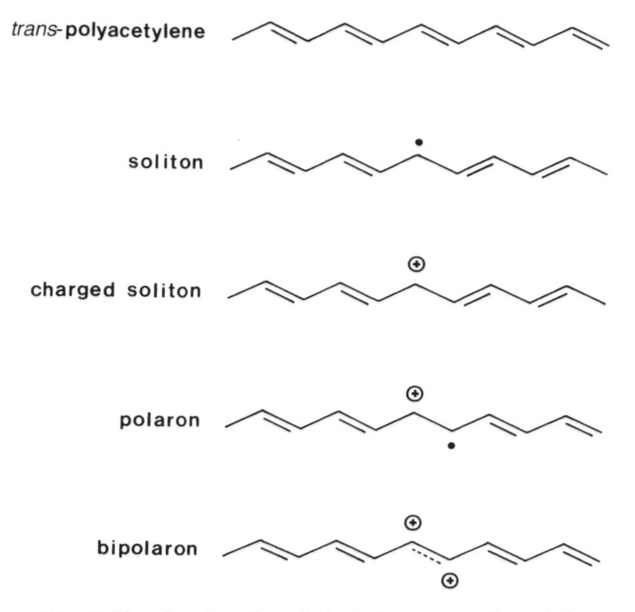

Fig. 5.33 Conjugation defects in *trans*-polyacetylene

of polarons in the chain increases, they can interact basically in two ways: two polarons can either form two charged solitons or a new stable zero-spin defect called bipolaron.

The electronic band structure of a neutral polyacetylene is characterized by an empty band gap, like in other intrinsic semiconductors. Defect sites (solitons, polarons, bipolarons) can be regarded as electronic states within the band gap. The conduction in low-doped polyacetylene is attributed mainly to the transport of solitons within and between chains, as described by the 'intersoliton-hopping model (IHM)'. Polarons and bipolarons are important charge carriers at higher doping levels and with polymers other than polyacetylene.

There exist numerous conjugated macromolecules showing high electronic conductivities after doping. Materials thus prepared are also termed 'synthetic metals' (their significance is documented, for example, by the existence of a journal of the same name). This section presents only a very brief summary of this constantly growing field. Further information can be found in numerous reviews and monographs; the most representative ones are listed below.

All important electronically conducting polymers, except perhaps for polyacetylene, can be prepared electrochemically by anodic oxidation of the monomers. The reaction is initiated by splitting off two hydrogen atoms from the monomer molecule (H—M—H), which subsequently polymerizes by interconnecting thus activated sites:

$$n\mathrm{H{-}M{-}H} \rightarrow \mathrm{H{-}(M)_{\mathit{n}}{-}H} + (2n-2)\mathrm{H}^+ + (2n-2)\mathrm{e}^- \quad (5.5.39)$$

According to this scheme, the consumption of positive charge should be close to 2F/mol of monomer. Experimental values for various monomers are considerably higher, in the range from 2.1 to 2.7. This is due to the fact that the dehydrogenation of the monomer generally occurs at more positive potentials than the p-doping of the polymer thus prepared. Therefore, the produced polymer is oxidized immediately after its formation (p-doped) with the consumption of a non-stoichiometric, additional positive charge:

$$\mathrm{H{-}(M)_{\mathit{n}}{-}H} + ny\mathrm{A}^- \rightarrow \mathrm{H{-}(M)_{\mathit{n}}^{\mathit{ny}+}{-}H}\cdot ny\mathrm{A}^- + ny\mathrm{e}^- \quad (5.5.40)$$

where y is the so-called doping level ($y \simeq 0.1$–0.7).

Both the mentioned reactions proceed in parallel. The mechanism of electrochemical polymerization is not completely elucidated, but obviously the first step of the polymerization, i.e. the formation of a dimer, H—M—M—H, is rate-determining. As soon as the dimer is formed, it further oxidizes more easily to the trimer and higher oligomers, since the oxidation potential of the oligomer gradually decreases with increasing polymerization degree, n. The polymerization reaction is, therefore, accelerated at a constant electrode potential as the polymer chain grows. Several examples of conducting polymers that can be electrochemically doped are quoted in Table 5.3.

Table 5.3 Examples of electronically conducting polymers. y is the level of electrochemical doping and κ is the maximum electrical conductivity. Except for polyacetylene and polyparaphenylene, only p-doping is considered

Polymer	Repeat unit	y	κ (S/cm)
Polyacetylene	—CH=CH—		
n-doped		0.09	100
p-doped		0.05	1000
Polyparaphenylene	—C_6H_4— (p-phenylene ring)		
n-doped		0.44	100
p-doped		0.1	100
Polypyrrole	pyrrole ring (N–H), linked at 2,5	0.4	500
Polyaniline	—C_6H_4—NH—	0.2	5
Polythiophene	thiophene ring (S), linked at 2,5	0.2	190

The monomer and lower oligomers are soluble in the electrolyte, but with increasing polymerization degree the solubility decreases. After attaining some critical value, an insoluble film is formed on the anode. Lower (soluble) oligomers can also diffuse from the electrode into the bulk of the electrolyte, hence the faradaic yield of electrochemical polymerization is, at least in the primary stages, substantially lower than 100 per cent.

The ideal electropolymerization scheme (Eq. (5.5.39)) is further complicated by the fact that lower oligomers can react with nucleophilic substances (impurities, electrolyte anions, and solvent) and are thus deactivated for subsequent polymerization. The rate of these undesired side reactions apparently increases with increasing oxidation potential of the monomer, for example, in the series:

$$\text{aniline} < \text{pyrrole} < \text{thiophene} < \text{furan} < \text{benzene}$$

The nucleophilic reaction with the solvent is of crucial importance. Monomers with lower oxidation potentials (aniline and pyrrole) can easily be polymerized even in aqueous electrolytes. For monomers with higher oxidation potentials, aprotic solvents must be used, such as acetonitrile propylene carbonate, dichloromethane, etc. (Benzene can be polymerized only in rather unusual electrolytes.) Side reactions with the solvent as well as its own oxidation, however, cannot be excluded even in aprotic electrolytes where they impair the electrochemical properties of the product. A more elegant way to polythiophene and similar polymers is therefore polymerization starting from oligomers, such as 2,2′-bithiophene (whose oxidation potential is by about 0.4 V lower than that of thiophene).

The best known electrochemically prepared polymer is *polypyrrole*. The oxidation potential of pyrrole is about 0.5 V versus SHE; that of polypyrrole is about −0.3 V versus SHE. The oxidative polymerization and p-doping proceeds, therefore, rather easy. On the other hand, the n-doping of polypyrrole has not been so far realized. The electrochemical polymerization has been performed with a large selection of electrodes (Pt, carbon, metal oxides) and electrolytes ranging from aqueous solutions to molten salts. Under suitable experimental conditions, a free-standing film of polypyrrole can be stripped off from the anode. This forms a basis of an industrial production of polypyrrole foils, invented by the BASF company. Polypyrrole foils have been studied as charge storing materials for batteries; further prospective applications range from polymeric sensors and electrochemic devices to antistatic foils for packing and other use.

Pyrrole derivatives substituted in positions 1-, 3-, or 4- have also been electrochemically polymerized (positions 2- and 5- must be free for polymerization). Besides homopolymers, copolymers can also be prepared in this way. Other nitrogen heterocycles that have been polymerized by anodic oxidation include carbazole, pyridazine, indole, and their various substitution derivatives.

In contrast to pyrrole and carbazole, the electrochemical polymerization of aniline takes place both at the ring and nitrogen positions. The polymerization is initiated in acidic medium by the formation of an aniline radical cation, whose charge is localized mainly on the nitrogen atom. The monomer units are then connected by C—N bonds in para positions (head-to-tail coupling) as depicted in Fig. 5.34A. This form is termed leucoemeraldine (or leucoemeraldine base) and it is essentially non-conducting in both the base and protonized forms (Fig. 5.34B). Strong oxidation of leucoemeraldine base leads to a non-conductive semiquinonic form called pernigraniline base (Fig. 5.34C). Pernigraniline is a rather unusual form of polyaniline; it was synthesized in a pure form only recently by chemical oxidation.

The common form of polyaniline is a 1:1 combination of alternating reduced (A) and oxidized (C) units; it is termed emeraldine (or emeraldine base). The emeraldine base is essentially non-conductive, but its conductivity increases by 9–10 orders of magnitude by treating with aqueous protonic acids. The conductive form of polyaniline can therefore be roughly depicted as a 1:1 combination of alternating A and D units.

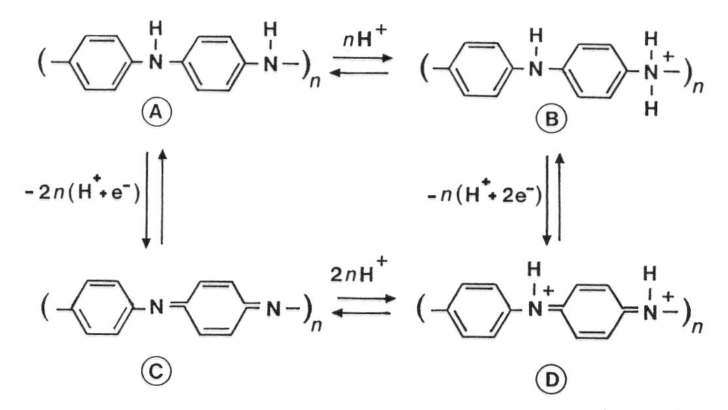

Fig. 5.34 Schematic representation of the oxidation/reduction and acidobasic reactions or polyaniline

5.5.6 *Non-electrochemical methods*

While methods employing *radiaoactive tracer techniques* have become a classical tool for the study of adsorption on electrodes, *optical methods* for the study of electrodes and processes occurring on them at an atomic or molecular level have undergone enormously rapid progress, which is characteristic for the contemporary development of electrochemistry.

Most suitable are methods studying the interphase between the electrode and the electrolyte solution under the same conditions as those employed in electrochemical methods, i.e. *in situ*. In the first place methods employing electromagnetic radiation in the microwave, infrared, visible, ultraviolet, X-ray and γ-radiation regions belong to this group. The radiation can pass more or less efficiently through the electrolyte, through the material of the electrochemical cell (window), and even through the electrode without a decrease in its intensity below the detection limit.

The methods of electron and ion optics (Table 5.4) employing a beam of slow electrons or ions (with energies up to 10^4 eV) for excitation of the system have found many applications in the study of the surfaces of solid electrodes. The very small mean free path of a beam of slow electrons or ions, of the order of 1–10 nm in solids and 100 nm in gases at atmospheric pressure, results in the very high surface sensitivity of these methods. On the other hand, however, its application is limited to an *ex situ* study of the electrode surface in an ultra high vacuum, i.e. at pressures below 10^{-6} Pa. Various experimental procedures have been proposed for anaerobic transfer of the electrode from the electrolyte solution to an ultra high vacuum, if necessary with an intact double layer, but without the bulk electrolyte (electrode emersion). In spite of substantial success, e.g. in the study of the adsorbed anions on an immersed gold electrode polarized to various potentials, it is apparent that transfer of the electrode from the electrolyte

Table 5.4 Basic methods of surface analysis (e = electron, I = ion, $h\nu$ = photon, other abbreviations are explained in the text)

Method	Exciting particle	Detected particle	Information depth (nm)	Detection limit (at.%)	Lateral resolution (μm)	Range of elements	Bond information	Destructibility
XPS	$h\nu$	e	1–3	0.1–1	1000	≧Li	+ +	–
AES	e ($h\nu$, I)	e	0.4–3	0.1–1	0.1–10	≧Li	+	(+)
EMP	e	$h\nu$	10^3	0.1	1	>Na	–	(+)
HIXE PIXE	I	$h\nu$	10^3	10^{-5}–10^{-3}	3–1000	>Na	–	(+)
SIMS	I	I	0.3–1	10^{-5}	100	≧H	+	+ +
ISS	I	I	0.3–1	0.1	100	≧Li	–	–
RBS	I	I	3–10^3	10^{-2}	1–3	≧C	–	–

solution to an ultra high vacuum spectrometer is the most sensitive operation in the whole procedure of a surface analysis of electrodes.

Methods employing electromagnetic radiation. The interaction of electromagnetic waves with the electrode, the electrolyte or the interphase between them can be connected with a change in the amplitude, phase, frequency and direction of wave propagation. Evaluation of these changes often yields very detailed information on the studied object. Some methods (e.g. ESR, absorption or luminescence spectroscopy in the ultraviolet and visible regions) are more suitable studying electrochemically generated substances in solution, while others (e.g. reflection spectroscopy, ellipsometry, Mössbauer spectroscopy and X-ray diffraction) are applied to the study of changes of the electrode surface or in the immediate vicinity of the surface. The boundary between the two types of methods is not a sharp one, as procedures have been developed that are suitable for obtaining both types of information (e.g. Raman spectroscopy).

A change in the amplitude of the excitation radiation occurs especially during the non-elastic interaction of photons with a substance, connected with the transfer of a defined quantum of energy; methods measuring this change are termed spectroscopic methods. Relatively, the smallest energy quantum corresponds to the energy difference between two states differing in the spin of the unpaired electron, whose degeneracy has been removed by the action of an external magnetic field. These processes are studied by methods termed *electron spin* (*paramagnetic*) *resonance,* ESR (or EPR). Of substances with an unpaired electron, transition metal compounds and free radicals can be studied. Electrochemistry is concerned mainly with the study of organic radicals formed during redox processes at electrodes or in subsequent chemical reactions, often in non-aqueous solvents. With the ESR method, radicals in electrolytes at concentrations as low as $10^{-8}\,\text{mol}\cdot\text{dm}^{-3}$ can be detected in a wide lifetime range. Very unstable radicals can also be studied indirectly, by conversion to a more stable radical by the spin-trapping method. ESR measurements can readily be carried out *in situ,* and the term SEESR (*simultaneous electrochemical ESR*) is sometimes used. The shape and hyperfine structure of the ESR lines also yields information on the structure of radicals, ion association, solvation, etc. In practice, even the information on the mere fact that the product or intermediate in an electrode process is a radical is valuable and can be of great importance in the discussion of the mechanism of the electrode process. The ESR method has been quite extensively used in electrochemistry in the study of substances in the electrolyte phase, while less attention has been paid to the study of paramagnetic sites on the electrode itself.

Electrochemically generated products can be readily characterized by *in situ* measurement of their absorption spectra in the ultraviolet and visible regions. Optically transparent electrodes (OTEs) prepared from thin layers of metals or semiconductors on glass or polymer supports play a key role in these investigations. Alternatively, very fine, transparent metal mesh can be used as an electrode. Absorption spectra can be advantageously studied by means of a working electrode in a thin-layer system (with a thickness of about 0.05–0.5 mm), where electrolysis of all the bulk electrolyte lasts only a few seconds (thin-layer spectroelectrochemistry).

Under certain conditions, the product of the electrode process can be studied by using *electrochemically generated luminescence*. This phenomenon occurs when the reactants generated at the electrode or in a subsequent chemical reaction have such high free energy that they can interact to form products in electronically excited states. If these states are capable of radiative relaxation, then they can be identified according to their luminescence spectra. Reactants for luminescence studies can be generated in two ways, either subsequently on a single electrode polarized, e.g. by a rectangular or sinusoidal voltage, or simultaneously, e.g. at a rotating ring disk electrode. Electrochemically generated luminescence finds wide application in the study of organic substances and radicals and also of some complexes of transition metals (for example, $Ru(bpy)_3^{2+}$, where bpy = bipyridine). This method is especially useful in determining the mechanisms of electrochemical processes.

The most important methods used in *in-situ* studies of electrode surfaces are various modifications of *reflection spectroscopy* in the ultraviolet through infrared regions. For electrochemical applications, the specular reflection (at smooth electrode surfaces) is much more important than the diffuse reflection from matt surfaces. The reflectivity, R, of the electrode/electrolyte interface is defined by:

$$R = I/I_0 \tag{5.5.41}$$

where I_0 is the intensiy of incident radiant energy, and I is the intensity of that reflected from the interface (some authors distinguish the term 'reflectivity' for specular reflection and 'reflectance' for diffuse reflection.)

Specular reflection of electromagnetic radiation at the (electrochemical) interface is generally described by Fresnel equations. Supposing the most simple case that both the electrolyte and electrode are transparent and differ only in their refractive indexes, n_1 and n_2, the reflectivity for normal incidence of the radiation equals:

$$R = [(n_2 - n_1)/(n_2 + n_1)]^2 \tag{5.5.42}$$

and $R = 1$ for grazing incidence. At other angles of incidence $0 < \theta < 90°$, the reflectivity of radiation polarized normal and parallel to the plane of reflection is different, the former increases monotonously from value given by Eq. (5.5.42) to unity, but the latter goes through minimum at angle θ_b (so called Brewster angle):

$$\theta_b = \arctan(n_2/n_1). \tag{5.5.43}$$

The approximate picture of a transparent electrochemical interface is rather far from the actual situation. Nevertheless, the Fresnel equations are valid also for absorbing media, if one uses the complex refractive indexes, N_k:

$$N_k = n_k - i(\alpha\lambda/4\pi), \qquad k = 1, 2 \tag{5.5.44}$$

(α is absorption coefficient of the medium and λ wavelength). In this case, Eq. (5.5.42) is replaced by

$$R = [(n-1)^2 + n^2\alpha^2\lambda^2/8\pi^2]/[(n+1)^2 + n^2\alpha^2\lambda^2/8\pi^2], \tag{5.5.45}$$

where $n = n_1/n_2$ is the relative refractive index. Equation (5.5.45) demonstrates the fact that strongly absorbing materials, like metals, are also strongly reflecting, with $R \to 1$ even at the normal incidence.

The most important situation occurs when a film of different optical properties is formed at the electrode surface. In this case, theory predicts that the R value can be changed, even for non-absorbing films, as a result of existence of a third phase with different refractive index interspaced between the electrode and electrolyte. Therefore, the entire observed decrease in reflectivity R is not necessarily caused by the absorption of radiation in the film. This approximation, is, however, reasonably acceptable when the film is supported by a highly reflective phase, such as smooth metal electrode.

In general, two basic experimental arrangements of reflective spectroelectrochemistry are possible:

(a) The excitation beam passes first through a window in the electrochemical cell, then through the electrolyte and is finally reflected from the surface of the electrode back into the electrolyte phase (*specular external reflectance*). There is a sufficient number of suitable substances for the preparation of electrodes, electrolytes and windows for working in the ultraviolet and visible regions. More problems are encountered in the infrared region, as practically all electrolytes absorb infrared radiation so strongly that external reflection spectra can be studied only in cells with a very thin (1–10 μm) layer of electrolyte between the window (usually Si, Ge, ZnSe, CaF_2) and the electrode.

(b) The excitation beam passes through the electrode material and is reflected once or twice from the interface with the electrolyte back into the electrode material (*internal* or *attenuated total reflection*). This method requires the use of an optically transparent electrode. *Internal reflectance of infrared radiation* can be studied, for example, on a germanium electrode; other materials such as carbon and various metals have such high absorption coefficients for infrared radiation that they can be used in only very thin layers (5–30 nm) fixed to transparent support materials. Vacuum deposited gold layer on a Ge electrode is the most common arrangement. In practice, a compromise between the electrical conductivity and infrared absorbance must always be found.

The relative changes in reflectance at the electrode are usually very small, typically of the order of 10^{-4}–10^{-2} per cent, and thus a special technique must be used to measure these differences; this is based primarily on signal modulation and lock-in detection. The simplest modulation method is based on mechanical interruption of the excitation light beam; this approach has a disadvantage in that it in no way increases the surface sensitivity of reflectance spectroscopy. Thus it is preferable to employ signal modulation by periodic changes in the electrode potential (electrochemically modulated reflectance spectroscopy). A further approach to modulation involves application of polarized excitation radiation. This modulation depends on periodic changes in the polarization plane of the radiation between the directions perpendicular to and parallel to the plane of reflection of the beam. This method also increases the surface sensitivity of reflectance spectroscopy as a result of application of specific selection rules for the reflectance of polarized light from the electrode surface. In addition, the measurement is carried out at constant potential and the information on the electrode surface is not distorted by changes that could occur during potential modulation.

The third approach to reflectance studies at the electrochemical interface presents *Fourier transformation spectroscopy*. This method is used in the infrared spectral region, where standard commercial instruments (with only minor laboratory modifications) are readily applicable. The most common experimental technique is known as SNIFTIRS (subtractively normalized interfacial Fourier transformation infrared spectroscopy). It is based on the measurement of external reflectance, R_w at a working potential of the electrode, E_w, and is compared to external reflectance, R, at a potential E, where no significant charge transfer reaction takes place. The plot of $(R_w - R)/R = \Delta R/R$ versus wavenumber contains therefore only the characteristic features introduced by the potential change, whereas the potential-independent infrared absorptions (electrolyte window, and vibrational structures intact by the charge transfer reaction) are numerically cancelled. Negative-going peaks correspond to the increase in concentration of species generated at the working potential E_w and vice versa.

Reflectance spectroscopy in the infrared and visible ultraviolet regions provides information on electronic states in the interphase. The external reflectance spectroscopy of the pure metal electrode at a variable potential (in the region of the minimal faradaic current) is also termed ‘electroreflectance’. Its importance at present is decreased by the fact that no satisfactory theory has so far been developed. The application of reflectance spectroscopy in the ultraviolet and visible regions is based on a study of the electronic spectra of adsorbed substances and oxide films on electrodes.

Infrared reflectance spectroscopy provides information on the vibrational states in the interphase. It can be interpreted in terms of molecular symmetry, force constants and chemical bond lengths. The intensity of the spectral peaks of the adsorbed molecules is determined both by standard selection rules and also by specific surface selection rules, differentiating various directions of the polarization of radiation and molecular orientation with respect to the surface of the electrode. Infrared reflectance spectroscopy has been used, for example, to study the structure of water in the electrical double layer, adsorption of H_2 and CO on a platinum electrode and the adsorption of intermediates in the oxidation of organic substances on Pt, Au, Rh and Pd electrodes. The spectra can be interpreted in terms of the symmetry changes in the molecules in the electrochemical interface, bonding to the electrode surface, aggregation, etc.

An example of the SNIFTIRS study of a Pt electrode in contact with 0.5 M $LiClO_4$ in propylene carbonate (PC) is demonstrated in Fig. 5.35. The reference potential, E was 2 V versus Li/Li^+, the working potential, E_w was 3.2 V versus Li/Li^+. Anodic oxidation of PC is clearly apparent at these conditions; it is accompanied by the relative intensity drop of the C=O stretching vibration at 1785 cm^{-1} (from the intact PC), and increase of the C=O stretching vibration at 1766 cm^{-1} (from the oxidized PC). The oxidation products are further characterized by several new negative going peaks in the 1000–1800 cm^{-1} region, whereas the content of CH_3 groups remains unchanged by oxidation. The most characteristic oxidation product of PC is molecular CO_2 identified at 2342 cm^{-1}. The onset of PC oxidation (CO_2 formation) was detected by SNIFTIRS already at 2.1 V versus Li/Li^+, which indicates that propylene carbonate is much less stable against oxidative decomposition than it was formerly believed on the basis of electrochemical methods (anodic currents at the platinum electrode in the mentioned electrolyte are practically negligible in the interval 2–4 V versus Li/Li^+).

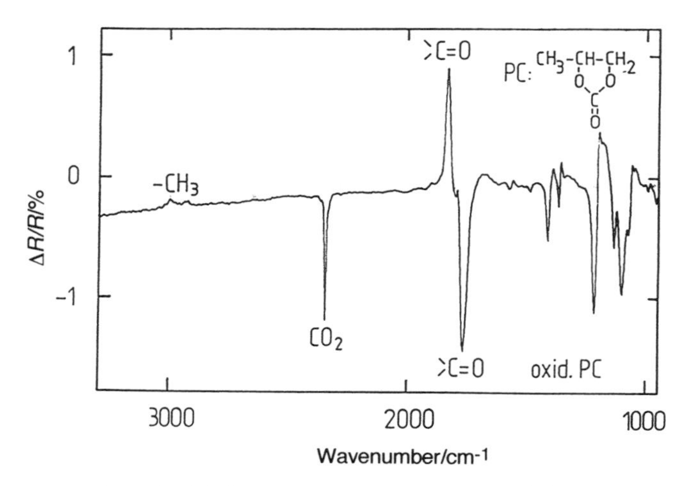

Fig. 5.35 SNIFTIRS spectrum from a polished Pt electrode in 0.5 M $LiClO_4$ in propylene carbonate. Reference potential: 2.00 V versus Li/Li^+ electrode; working potential 3.20 V versus Li/Li^+ electrode. According to P. Novák *et al.*

The absorption of radiation at the electrochemical interface can be studied by a number of methods—not only optically by the reflection or transmission technique but also by direct measurement of the energy released during radiationless relaxation of excited states. The simplest version is based on measuring temperature changes in the region around an intermittently irradiated electrode—called *photothermal spectroscopy*. Alternatively, periodic temperature changes of the electrode can be monitored in terms of periodic changes in its dimensions; these can be measured with a piezoelectric detector. The most commonly employed method involves measurement of the intensity of an acoustic signal formed by periodic pressure changes in the region around the intermittently irradiated electrode—termed *photoacoustic* (or *optoacoustic*) *spectroscopy* or PAS (OAS). In this method, the excitation radiation is modulated with an acoustic frequency of 10^1–10^3 Hz, and detection is carried out with a microphone. The photothermal and photoacoustic spectroscopic methods have a great advantage over reflectance and transmission techniques in their minimal requirements on electrode preparation; the electrode can be opaque, can have a rough surface, etc. Photoacoustic spectroscopy and related techniques can be used from the ultraviolet to infrared regions; only the ultraviolet and visible regions have so far been used in electrochemical applications. Interpretation of the spectra is similar to that for absorption or reflectance spectroscopy in the same region. Specific applications include a study of the efficiency of photoelectrochemical conversion of energy at semiconductor electrodes.

The reflectance of light waves from the electrode surface involves a change not only in the amplitude but also in the phase of the reflected radiation. The method termed *ellipsometry* is based on the fact that, during reflection from the electrode, linearly polarized light is changed into elliptically polarized light. The polarized light incident on the electrode surface can be considered as consisting of two superimposed light waves with the same frequency and phase but polarized perpendicularly and parallel to the plane of incidence. After reflection from the surface, these two waves are, in general, out of phase and their amplitudes are changed compared to those of the incident radiation. The interference of these two reflected waves produces elliptically polarized light or, in special cases (waves with the same amplitude and phase-shifted by $\pi/2$), circularly polarized light. The phase shift and the ratio of the amplitudes and thus the corresponding optical parameters of the surface can be found experimentally. The capability to determine the thickness of surface oxide films on electrodes and the kinetics of their growth *in situ* is most important in electrochemical applications of this method. Surface films can be studied with sufficient sensitivity in a wide range of thicknesses, starting with submonolayers. Ellipsometry was also applied to the study of the electrical double layer.

Raman spectroscopy is certainly one of the most important optical methods employed in electrochemistry. This method is based on an analysis of non-elastically scattered radiation in the ultraviolet and visible regions. Interaction with vibrational states in the scattering medium leads to a slight decrease (or increase) in the frequency of the scattered radiation compared to the incident radiation. As a result of specific selection rules, this method yields structural bonding data complementary to the data obtained by other vibrational spectroscopic methods, especially infrared. The low intensity of the scattered light and the low shift of the frequency make it essential that the excitation radiation have very precisely defined frequency and high intensity; at present, lasers are exclusively used for this purpose. When the frequency of the excitation radiation is identical with that of the electronic absorption band of the studied molecule, a marked increase in the intensity of the Raman effect is observed, exploited in *Raman resonance spectroscopy*. Compared to infrared spectroscopy, the use of Raman spectroscopy in electrochemistry is favoured by the high transparence of the electrolyte in the ultraviolet and visible regions, enabling measurements *in situ*. Compared to electronic spectra in the ultraviolet and visible regions, Raman spectra of electrochemically generated substances are characterized by high chemical selectivity and thus far greater capabilities in the study of mechanistic and kinetic problems. An optical multichannel analyser can be used to study spectra with a high time resolution, i.e. at various stages during the change of the electrode potential.

A study of the adsorption of pyridine on a silver electrode first demonstrated that, under suitable conditions, Raman spectroscopy can even be sufficiently sensitive for the study of substances adsorbed on electrodes, where a modification called *surface enhanced Raman spectroscopy* (SERS) was used. The adsorption of various molecules on a silver electrode is accompanied by an enormous increase in the intensity of the Raman scattered radiation—by up to 10^6-fold compared to the intensity in the homogeneous phase. The surface enhancement effect led to great interest during the 1970s among theoreticians and experimentalists, including non-electrochemists. In spite of considerable effort, no satisfactory theoretical basis for this effect has yet been found. The amplification factor depends on the wavelength of the exciting radiation and also very strongly on the electrode material. In addition to the silver electrode, which is mostly used, copper (an increase of about 10^5) and gold (about 10^4) electrodes are also suitable for these studies. In addition, some silver alloys or other materials can also be used, whose surface has been modified by a small amount of silver. Increases in the Raman scattering radiation have also been reported for Ni, Pd, Cd, Hg, Pt and Al electrodes.

Methods employing X-rays and γ-radiation are used less often in electrochemistry. The possibility of using X-ray diffraction for *in situ* study of the electrode surface was first demonstrated in 1980. This technique has long been used widely as a method for the structural analysis of crystalline substances. Diffraction patterns that are characteristic for the electrochemical interface can be obtained by using special electrochemical cells and electrodes. The electrode can also be studied *in situ* by measuring absorption of X-ray radiation. The *extended X-ray absorption fine structure* (EXAFS) method is based on the excitation of core-level electrons and, in addition to elemental analysis, permits study of the electronic structure of surfaces. The EXAFS method requires a quite intense source of X-ray radiation, preferably a synchrotron.

Mössbauer spectroscopy can be used for *in situ* study of electrodes containing nuclei capable of resonance absorption of γ radiation; for practical systems, primarily the ^{57}Fe isotope is used (passivation layers on iron electrodes, adsorbed iron complexes, etc.). It yields valuable information on the electron density on the iron atom, on the composition and symmetry of the coordination sphere around the iron atom and on its oxidation state.

Methods employing electron and ion beams. Of the methods employing a beam of electrons or ions for excitation of the sample and/or detection of the signal, various surface analysis methods are of greatest importance. Qualitative and semiquantitative elemental analysis of the electrode surface is more or less routine for practically all the elements in the periodic system. Nonetheless, the individual methods vary, among other factors, in their sensitivity for various elements, information depth (i.e. the thickness of the sample layer yielding a detectable signal) and lateral resolution. The latter is determined by the surface area of the sample irradiated by the excitation beam. With a beam of charged species, sufficiently intense radiation can be obtained even for focusing to very small cross-sections, which is the principle of microanalysis (microprobes). Scanning of the surface of the sample by the excitation beam yields the distribution of the elements over the sample surface (scanning microanalysis).

The basic methods of surface analysis are listed in Table 5.4. The *X-ray photoelectron spectroscopic* (XPS) *method,* also termed *electron spectroscopy for chemical analysis* (ESCA), one of the most frequently used methods, is based on the determination of the kinetic energy of the electrons emitted from the core levels in a photoelectric process. The energy of the core levels is assessed with the help of the kinetic energy of the photoelectrons. These energies unambiguously characterize the given atom. The measurement is quite precise (with an error of about 10^{-1} eV) and thus not only can elemental analysis be carried out but also fine changes in the core level energies produced by changes in the electron densities due to the formation of chemical bonds, i.e. chemical shifts, can be studied. Photoelectron spectroscopy can indicate, for example, various oxidation states of a metal in oxide films on the electrode; it is widely used in the study of chemically modified electrodes and substances adsorbed on electrodes. The XPS method yields information on several monolayers of atoms in the surface. In practice, it is often necessary to analyse deeper layers in the sample; the standard procedure involves a destructive technique of ion sputtering, i.e. depth profiling.

Auger electron spectroscopy (AES) is a further important method of surface analysis. It is based on the non-radiative relaxation of vacant core levels, connected with the emission of electrons (the Auger phenomenon). Measurement of the kinetic energy of emitted Auger electrons permits determination of the core level energies including chemical shifts. The spectra are more difficult to interpret than XPS spectra, as the electron reorganization occurs among three energy levels. Thus, the energy of the Auger electron (in contrast to the energy of a photoelectron) does not depend on the excitation energy employed. In the AES method, the core level can be excited by any type of radiation with sufficient energy; excitation using electrons is used most often and in general permits much better lateral resolution than excitation using photons. Auger electron spectroscopy is also often combined with depth profiling. It is employed in electrochemistry for similar purposes as XPS, e.g. to study oxide films, adsorption, corrosion, etc.

A complementary phenomenon to the emission of Auger electrons is the emission of characteristic X-ray radiation (radiative relaxation of the vacant core level). The probability of emission of X-ray radiation is greater for heavier elements and, on the other hand, the probability of the Auger phenomenon increases for lighter elements. The *electron microprobe* (EMP) employs excitation by characteristic X-ray radiation using a focused beam of electrons. This is not a surface method in the strictest sense, as the information depth is much greater than for XPS and AES. A disadvantage of electron excitation of X-ray spectra is the relatively high background of X-ray bremsstrahlung, decreasing the sensitivity of the analysis. The bremsstrahlung background can be considerably suppressed by excitation of X-ray spectra using an ion beam (*proton-induced X-ray emission,* PIXE, or *helion-induced X-ray emission,* HIXE).

The impact of an ion beam on the electrode surface can result in the transfer of the kinetic energy of the ions to the surface atoms and their release into the vacuum as a wide range of species—atoms, molecules, ions, atomic aggregates (clusters), and molecular fragments. This is the effect of ion sputtering. The SIMS (*secondary ion mass spectrometry*) method deals with the mass spectrometry of sputtered ions. The SIMS method has high analytical sensitivity and, in contrast to other methods of surface analysis, permits a study of isotopes. In materials science, the SIMS method is the third most often used method of surface analysis (after AES and XPS); it has so far been used only rarely in electrochemistry.

Methods based on the study of the scattering of an ion beam also belong among techniques for analysis of surfaces that have been used only occasionally in electrochemistry. The *ion-scattering spectroscopic* (ISS) *method* studies the scattering of slow ions (with energy up to 1 keV). The scattering of fast ions (with energy of about 1 MeV) has been studied by the *Rutherford backscattering spectroscopic* (RBS) *method.* The scattering of ions from a surface can be described over the whole energy range as an elastic binary collision. Determination of the energy and angle of the scattered ions yields, in the first place, information on the mass of the surface atoms, i.e. elemental analysis. The RBS method is especially useful for the detection of heavy atoms at an electrode made of a light element. It is employed for non-destructive analysis of depth concentration profiles.

In addition to the above methods, yielding primarily semiquantitative elemental analysis of the surface, further methods of electron and ion optics are suitable for the study of electrochemical systems. The diffraction of slow electrons (*low energy electron diffraction,* LEED) is widely employed for structural investigation of single-crystal electrodes. In contrast to X-ray diffraction, the LEED method yields information only on the two-dimensional surface structure. The LEED method is used to study clean surfaces and substances adsorbed on crystalline surfaces. An interesting structural change ('surface reconstruction') is often observed after contact of the electrode with the electrolyte. A disadvantage of the LEED method compared to X-ray diffraction is the necessity of working *ex situ.*

The electrochemistry of modified electrodes occasionally employs a rather special method of electron spectroscopy—*inelastic electron tunnelling spectroscopy* (IETS)—based on precise measurement of the electrical conductivity of an electrode covered with an oxide layer and possibly with other adsorbed substances, in dependence on the voltage. The electrons 'see' the layer of oxide or adsorbed molecules as a potential barrier that can be overcome by tunnelling. At a definite potential, a characteristic change in the conductivity is observed, resulting from non-elastic interaction of the tunnelling electrons with the vibrational levels in the studied layer of oxide or adsorbates on the electrode. The IETS method thus yields analogous data to those obtained from the other vibrational spectroscopic methods. The non-elastic tunnelling phenomenon is not limited by the selection rules of vibrational spectroscopy and thus bands can be observed that are active in both the Raman and infrared spectra. The IETS method has an advantage in its ability to study very thin surface layers on the electrodes and, last but not least, in the quite simple instrumentation (e.g. it is not necessary to work in a vacuum). Measurements can be carried out, however, only *ex situ* and on a limited number of materials. The aluminium oxide-covered electrode is most advantageous.

As mentioned above, the methods based on detection of electrons or ions or probing the electrode surface by these particles are generally handicapped by the necessity to move the studied electrode into vacuum, i.e. to work *ex-situ.* There are, however, two important exceptions to this rule: electrochemical mass spectrometry and electrochemical scanning tunnelling microscopy.

Electrochemical mass spectrometry was recently introduced for direct study of the products of electrochemical processes. For this purpose, an electrochemical cell is connected to the vacuum chamber of a mass spectrometer by a porous Teflon membrane. This membrane is at the electrolyte side covered by a layer of porous platinum serving as a working electrode. When the pore size is suitably selected, the solvent molecules do not pass through the hydrophobic Teflon membrane, but some molecules generated at the platinum electrode (e.g. H_2, O_2, NO, NO_2, CO_2) can penetrate through the membrane, and are subsequently detected by the mass spectrometer. The analytical response is rather fast, typically < 0.5 s, which allows the measurement of mass–potential curves in parallel with usual current–potential diagrams. According to the used potential regime, we can distinguish the *mass spectrometric cyclic voltammetry,* MSCV (with triangular potential changes) or *differential electrochemical mass spectrometry,* DEMS (with square-wave potential changes). The electrochemical mass spectrometry seems promising, especially for the study of anodic oxidation of organic substances. For instance, the detection of CO_2 during anodic oxidation of propylene carbonate is a suitable complementary method to SNIFTIRS (see the example discussed on page 334).

Electrochemical scanning tunnelling microscopy (ESTM) has recently been recognized as a powerful technique for *in-situ* characterization of the topography of electrode surfaces extending to the atomic level. This method was originally developed for the study of conducting surfaces in vacuum, but is readily applicable also to surfaces in gaseous or liquid environment. The latter application was first demonstrated in 1986 on graphite in aqueous solutions. The principle of ESTM is very simple: it is based on electron tunnelling between the studied electrode surface and a tip of sharply pointed wire, which is placed in close vicinity above the studied surface. The tunnelling current decreases exponentially with the distance of the tip from the surface (roughly by an order of magnitude for every distance increase by 0.1 nm). A significant tunnelling current flows therefore only between several atoms at the apex of the tip and the nearest atoms in the examined surface. The exponential decay of the tunnelling current with distance enables one to measure the vertical position of the tip with a precision in the order of 0.01 nm. By sweeping the tip across the studied surface, an image of its topography with a comparable vertical resolution can be obtained (the lateral resolution is about 0.2 nm).

The experimental set-up usually utilizes a piezoelectric tripod as a support of the tip (Fig. 5.36). This is movable vertically and laterally over the examined surface; the vertical distance is fixed by a feedback loop to a constant tunnelling current at each point of the scan. The contours of the surface are thus visualized by voltage changes needed to move the piezoelectric tripod to a desired position.

The *in-situ* ESTM was utilized for studying adsorbed or underpotential-

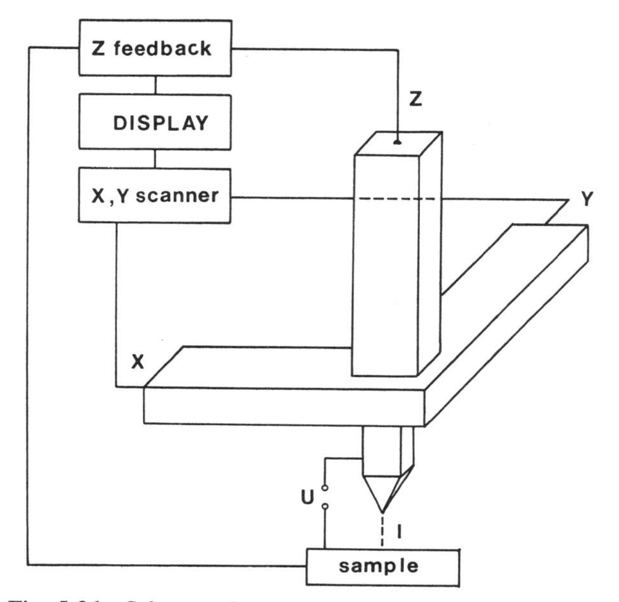

Fig. 5.36 Scheme of a scanning tunnelling microscope

deposited atoms or molecules on metal and semiconductor electrodes, effects of surface reconstruction and electrochemical roughening, etc. The latter was demonstrated, for example after successive oxidation and reduction of the electrode surface in aqueous electrolytes, which is widely used for 'activation' of Au or Pt electrodes. Fundamental ESTM studies are performed prevailingly with single-crystal electrodes.

The ESTM experiment provides actually five measurable quantities: tunnelling current, I at the applied voltage, U, and three dimensions, x, y, z. The standard STM can therefore easily be modified by recording the local I-U, U-z, or I-z characteristics (z is vertical distance of the tip from the electrode surface). Plot of dI/dU or dI/dz versus x and y brings additional information on the electronic and chemical surface properties (local work functions, density-of-states effects, etc.), since these manifest themselves primarily as I-U dependences. The mentioned plots are basis of the *scanning tunnelling spectroscopy* (STS).

Other techniques derived from tunnelling microscopy are based on the measurement of small repulsive forces (down to 10^{-9} N) acting on a diamond tip attached on the gold foil. These forces originate from repulsion of individual atoms between the tip and the examined surface (*atomic force microscope,* AFM). Unlike standard tunnelling microscopy, the AFM is not confined only to electrically conducting samples. Other methods based on the concept of a force microscopy are *laser force microscope* (LFN) which works with attractive rather than repulsive forces, and *magnetic force* microscope (MFM). The use of these methods in electrochemistry is so far only marginal. Electrochemical background has, however, a recently developed *scanning ion conductance microscope* (SICM). In this case, the sample is immersed in an electrolyte bath and scanned by a fine electrolyte-filled micropipette while measuring the ion current between two silver electrodes placed in the bath and inside the micropipette. The obtained resolution (*ca.* 10 nm) makes possible to monitor, for example, ion currents flowing through individual channels in biological membranes.

Radioactive tracer techniques. In electrochemistry, the procedure is essentially the same as in studies of chemical reactions: the electroactive substance or medium (solvent, electrolyte) is labelled, the product of the electrode reaction is isolated and its activity is determined, indicating which part of the electroactive substance was incorporated into a given product or which other component of the electrolysed system participated in product formation. Measurement of the exchange current at an amalgam electrode by means of a labelled metal in the amalgam (see page 262) is based on a similar principle.

These techniques are especially useful for studies of the adsorption of reactants, intermediates and products of electrode reactions. The simplest case corresponds to adsorption that is so strong that the electrode can be removed from the solution, rinsed and its activity measured without interference from desorption. When this procedure is impossible, the activity of the adsorbate can be measured by the 'electrode lowering method'. The radioactive counter is placed under the bottom of the cell, which is made of a plastic foil. The electrode can be located at large distances from the bottom or can be placed so close to the bottom that only a thin layer of solution remains beneath it. The radioactivity values at the two electrode positions permit determination of the adsorbate activity. This procedure can be repeated many times, thus supplying data on the kinetics of the adsorption process.

References

Adams, R. N., *Electrochemistry at Solid Electrodes,* M. Dekker, New York, 1969.

Albery, W. J., *Electrode Kinetics,* Oxford University Press, 1975.

Albery, W. J., and M. E. Hitchman, *Ring Disc Electrodes,* Clarendon Press, Oxford, 1970.

Aruchamy, A. (Ed.), *Photochemistry and Photovoltaics of Layered Semiconductors,* Kluwer Academic Publishers, Dordrecht, 1992.

Augustinski, J., and L. Balsenc, Application of Auger and photoelectron spectroscopy to electrochemical problems, *MAE,* **13,** 243 (1978).

Baker, B. G., Surface analysis by electron spectroscopy, *MAE,* **10,** 93 (1975).

Bard, J. A., and L. R. Faulkner, see page 290.

Besenhard, J. O., and H. P. Fritz, Elektrochemie schwarzer Kohlenstoffe, *Angew. Chem.* **95,** 954 (1983).

Bond, A. M., *Modern Polarographic Methods in Analytical Chemistry,* M. Dekker, New York, 1980.
Bond, A. M., K. B. Oldham and C. G. Zoski, Ultramicroelectrode, *Anal. Chim. Acta,* **216,** 177 (1989).
Chien, J. C. W., *Polyacetylene: Chemistry, Physics and Materials Science,* Academic Press, New York, 1984.
Clavilier, J., R. Faure, G. Guinet, and R. Durand, Preparation of monocrystalline Pt microelectrodes and electrochemical study of the plane surfaces cut in the direction of (111) and (110) planes, *J. Electroanal. Chem.,* **107,** 205 (1980).
Electrodics—Experimental techniques, CTE, **9** (1984).
Evans, D. H., Doing chemistry with electrodes, *Acc. Chem. Res.* **10,** 313 (1977).
Experimental methods of electrochemistry, CTE, **8** (1984).
Faulkner, L. R., Chemical microstructures on electrodes, *Chem. Eng. News,* 27 February 1984, p. 28.
Finklea, H. O. (Ed.), *Semiconductor Electrodes, Studies in Physical and Theoretical Chemistry, 55,* Elsevier, Amsterdam, 1988.
Fleischmann, M., and I. R. Hill, Raman spectroscopy, *CTE,* **8,** 373 (1984).
Fleischmann, M., S. Pons, D. R. Rolinson, and P. P. Schmidt, *Ultramicroelectrodes,* Datatech Systems, Marganton, N. C., 1987.
Furtak, T. E., K. L. Kliewer, and D. W. Lynch (Ed), Non-traditional approaches to the study of the solid-electrolyte interface, *Surface Sci.,* **101,** 1 (1980).
Galus, Z., *Fundamentals of Electrochemical Analysis,* Ellis Horwood, Chichester, 1976.
Greef, R. Ellipsometry, *CTE,* **8,** 339 (1984).
Hammond, J. S., and N. Winograd, ESCA and electrode surface chemistry, *CTE,* **8,** 445 (1984).
Hansen, W. N., Internal reflection spectroscopy in electrochemistry, *AE,* **9,** 1 (1973).
Heineman, W. R., F. M. Hawkridge, and H. N. Blount, Spectroelectrochemistry at optically transparent electrodes, in *Electroanalytical Chemistry* Vol. 13 (Ed. A. J. Bard), M. Dekker, New York, 1984.
Heyrovský, J., and J. Kůta, *Principles of Polarography,* Academic Press, New York, 1966.
Hubbard, A. T., Electrochemistry of well-defined surfaces, *Acc. Chem. Res.,* **13,** 177 (1980).
Justice, J. B. (Ed.), *Voltammetry in Neurosciences,* Humana Press, Clifton, 1987.
Kastening, B., Electron spin resonance, *CTE,* **8,** 433 (1984).
Kazarinov, V. E., and V. N. Andreev, Tracer methods in electrochemical studies, *CTE,* **8,** 393 (1984).
Kavan, L., Electrochemical carbonization of fluoropolymers, in *Chemistry and Physics of Carbon* (Ed. P. A. Thrower), Vol. 23, M. Dekker, New York, 1991, p. 69.
Kinoshita, K., *Carbon, Electrochemical and Physicochemical Properties,* John Wiley & Sons, New York, 1988.
Kissinger, P., and W. Heinemann (Eds), *Laboratory Techniques in Electroanalytical Chemistry,* M. Dekker, New York, 1984.
Kuwana, T., and N. Winograd, Spectroelectrochemistry at optically transparent electrodes, in *Electroanalytical Chemistry* (Ed. A. J. Bard), Vol. 7, p. 1, M. Dekker, New York, 1974.
Kuzmany H., M. Mehring, and S. Roth (Eds), *Electronic Properties of Polymers (and Related Compounds), Springer Series of Solid State Sciences,* Vols 63, 76, 91, 107, Springer-Verlag, Heidelberg, 1985–1992.
Linford, R. G., *Electrochemical Science and Technology of Polymers,* Vols 1 and 2, Elsevier, London, 1987 and 1990.

Macdonald, D. D., *Transient Techniques in Electrochemistry,* Plenum Press, New York, 1977.
McIntyre, J. D. E., Specular reflection spectroscopy of the electrode/solution interface, *AE,* **9,** 61 (1973).
McIntyre, J. D. E., New spectroscopic methods for electrochemical research, in *Trends in Electrochemistry* (Eds J. O'M. Bockris, D. A. J. Rand and B. J. Welch), p. 203, Plenum Press, New York, 1977.
McKiney, T. M., Electron spin resonance in electrochemistry, in *Electroanalytical Chemistry* (Ed. A. J. Bard), Vol. 10, p. 97, M. Dekker, New York, 1977, p. 97.
Morrison, S. R., see page 244.
Muller, R. H., Principles of ellipsometry, *AE,* **9,** 167 (1973).
Muller, R. H., Recent advances in some optical experimental methods, *Electrochim. Acta,* **22,** 951 (1977).
Murray, R. W., Chemically modified electrodes, in *Electroanalytical Chemistry* (Ed. A. J. Bard), Vol. 13, M. Dekker, New York, 1984.
Novák, P., P. A. Christensen, T. Iwashita, and W. Vielstich, Anodic oxidation of propylene carbonate on platinum, glassy carbon and polypyrrole, *J. Electroanal. Chem.,* **263,** 37 (1989).
Osteryoung, J., Developments in electroanalytical instrumentation, *Science,* **218,** 261 (1982).
Pleskov, Yu. V., and V. Yu. Filinovskii, *The Rotating Disc Electrode,* Consultants Bureau, New York, 1976.
Santhanam, K. S. V., and M. Sharon (Eds), *Photochemical Solar Cells,* Elsevier, Amsterdam, 1988.
Sarangapani, S., J. R. Akridge and B. Schumm (Eds), *The Electrochemistry of Carbon,* The Electrochemical Society, Pennington (1983).
Sawyer, D. T., W. R. Heineman, and J. M. Bube, *Chemistry, Experiments for Instrumental Methods,* John Wiley & Sons, New York, 1984.
Sawyer, D. T., and J. L. Roberts, *Experimental Electrochemistry for Chemists,* Wiley-Interscience, New York, 1974.
Skotheim, T. A. (Ed.), *Handbook of Conducting Polymers,* Vols 1 and 2, M. Dekker, New York, 1986.
Snell, K. D., and A. G. Keenan, Surface modified electrodes, *Chem. Soc. Rev.,* **8,** 259 (1979).
Southampton Electrochemistry Group, see page 253.
Trasatti, S. (Ed.), *Electrodes of Conductive Oxides,* Elsevier, Amsterdam, 1980.
Vincent, C. A., F. Bonino, M. Lazari, and B. Scrosati, *Modern Batteries, An Introduction to Electrochemical Power Sources,* E. Arnold, London, 1984.
Whittingham M. S., and N. G. Jacobson, *Intercalation Chemistry,* Academic Press, Orlando, 1982.
Wightman, R. M., and D. O. Wipf, *Electroanalytical Chemistry* (Ed. A. J. Bard) Vol. 16, M. Dekker, New York, 1990.
Yeager, E., and A. J. Salkind (Eds), *Techniques of Electrochemistry,* Vols 1–3, Wiley-Interscience, New York, 1972–1978.
Zak, J., and T. Kuwana, Chemically modified electrodes and electrocatalysis, *J. Electroanal. Chem.,* **150,** 645 (1983).

5.6 Chemical Reactions in Electrode Processes

Chemical reactions accompanying the electrode reaction can be classified from three points of view. They can be sufficiently fast so that they proceed in equilibrium or they are slow enough to constitute a rate-controlling step. Another classification is based on their position with respect to the electrode reaction.

5.6.1 *Classification*

There are an infinite number of combinations here, only some of which will be considered (the individual steps are denoted by symbols as described on page 252):

$$
\begin{gathered}
I_1 + I_2 \underset{k_-}{\overset{k_+}{\rightleftarrows}} R \overset{\text{el}}{\rightleftarrows} P \qquad \text{(CE)} \quad \text{(a)} \\
\text{C(a)} \qquad \text{E(a)} \\
R \overset{\text{el}}{\rightleftarrows} M \underset{k_-}{\overset{k_+}{\rightleftarrows}} P \qquad \text{(EC)} \quad \text{(b)} \\
\text{E(b)} \qquad\qquad \text{C(b)} \\
R \underset{k_+}{\overset{\text{el}}{\rightleftarrows}} P_1 + P_2 \qquad \text{(EC)} \quad \text{(c)} \\
\text{C(c)}
\end{gathered}
\tag{5.6.1}
$$

Case (a) If the chemical reaction preceding the electrode reaction, C(a), and the electrode reaction itself, E(a), are sufficiently fast compared to the transport processes, then both of these reactions can be considered as equilibrium processes and the overall electrode process is reversible (see page 290). If reaction C(a) is sufficiently fast and E(a) is slow, then C(a) affects the electrode reaction as an equilibrium process. If C(a) is slow, then it becomes the rate-controlling step (a detailed discussion is given in Section 5.6.3).

Case (b) Reaction E(b) is sufficiently fast; then a chemical reaction subsequent to the electrode reaction, C(b), is either an equilibrium or rate-controlling step. If E(b) is sufficiently slow, then C(b) has no effects whatever its rate.

Case (c) A chemical reaction C(c) parallel to the electrode reaction has an effect only when E(c) is sufficiently fast.

Cases (a) and (c) for relatively slow chemical reactions are particularly interesting when the electrode reaction is fast and unidirectional, so that the concentration of substance R at the surface of the electrode approaches zero. Then characteristic *limiting kinetic currents* are formed on the polarization curves.

The third approach to classification considers the interaction of substances undergoing chemical reactions with the electrode. The reactions considered in this section are either volume reactions or heterogeneous reactions with weak interaction with the electrode surface; such reactions occur often at mercury electrodes. Reactions with strong chemisorption interaction will be considered in Section 5.7.

5.6.2 *Equilibrium of chemical reactions*

A typical example is protonation preceding the electrode reaction

$$A + H_3O^+ \rightleftarrows AH^+ + H_2O \tag{5.6.2}$$

characterized by the dissociation constant K':

$$K' = \frac{[A][H_3O^+]}{[AH^+]} \tag{5.6.3}$$

If substance AH^+ participates in an irreversible electrode reaction

$$AH^+ + e \rightarrow AH \tag{5.6.4}$$

the current density is given by the equation

$$\begin{aligned} j &= -Fk_{\text{conv}} \exp\left(-\frac{\alpha FE}{RT}\right) c_{AH^+} \\ &= -Fk_{\text{conv}} \exp\left(-\frac{\alpha FE}{RT}\right) \frac{c_{H_3O^+} c}{K' + c_{H_3O^+}} \end{aligned} \tag{5.6.5}$$

where $c = c_A + c_{AH^+}$. If $a_{H_3O^+} \ll K'$ and therefore $c_{AH^+} \ll c_A$, then Eq. (5.6.5) assumes the form

$$j = -Fk_{\text{conv}} \exp\left(-\frac{\alpha FE}{RT}\right) c_{H_3O^+} \frac{c}{K'} \tag{5.6.6}$$

with an overall reaction order of 2.

The deposition of a metal from a complex also involves equilibria in solution between the free metal ions, the complexed ions and the complexing agent. The formation of the complex MX^{z+} from the metal ion M^{z+} and complexing agent X is characterized by the consecutive stability constants (Section 1.4.3)

$$K_i' = \frac{[MX_i^{z+}]}{[MX_{i-1}^{z+}][X]} \tag{5.6.7}$$

so that the solution contains various proportions of the complexes MX^{z+}, $MX_2^{z+}, \cdots, MX_n^{z+}$ in addition to the free metal ions. Both the free and the complexed ions can participate in the electrode reaction. Under equilibrium conditions described by Eqs (5.6.7), the following electrode reactions can occur according to H. Gerischer:

$$\begin{aligned} &M^{z+} + ze \rightleftarrows \\ &MX^{z+} + ze \rightleftarrows M + X \\ &\cdots\cdots\cdots\cdots\cdots\cdots \\ &MX_n^{z+} + ze \rightleftarrows M + nX \end{aligned} \tag{5.6.8}$$

Each sort of complex including the free ions has a characteristic electrode reaction rate constant (for example, k_{c,MX_i}, $i = 0, 1, \ldots, n$).

Consider the relatively simple case where the chemical reactions of formation and dissociation of the complexes are sufficiently fast so that equilibrium among the complexes, free ions and complexing agent is maintained everywhere in the electrolyte during the electrolysis. The rate of deposition of the metal is then determined by the electrode reaction of the complex MX_i, for which the product $k_{c,MX_i}c_{MX_i}$ is, at the given potential, the largest for all the complexes.

5.6.3 *Chemical volume reactions*

According to the theory developed by R. Brdička and K. Wiesner, a chemical reaction prior to the electrode reaction acts as the rate-controlling step in the electrode process when the electrode reaction disturbs the chemical equilibrium in solution and renewal of this equilibrium is not sufficiently fast. This effect extends to a certain depth in the solution as concentration changes resulting from the electrode reaction are propagated by diffusion. According to J. Koutecký and R. Brdička, the differential equation describing diffusion to an electrode with a simultaneous chemical reaction is given by combination of the second Fick law and the equation for the reaction rate (this idea first appeared in a paper by A. Eucken). Consider a system of two substances participating in a reversible reaction

$$B \underset{k_1}{\overset{k_2}{\rightleftarrows}} A \tag{5.6.9}$$

with the equilibrium constant $K = k_2/k_1$. The substance A is present in a marked excess over substance B ($K \gg 1$), so that its concentration is practically not changed by the electrode process. The substance B undergoes an irreversible electrode reaction with the rate constant

$$k_c = k^{\ominus} \exp\left[-\alpha nF(E - E^{0\prime})/RT\right]$$

(see Eq. 5.2.23). The differential equation and boundary and initial conditions for substance B are

$$\frac{\partial c_B}{\partial t} = D\frac{\partial^2 c_B}{\partial x^2} + k_1(c_A - Kc_B) \tag{5.6.10}$$

$$\begin{aligned} t = 0, x > 0: &\quad c_B = c_A/K \\ t > 0, x = 0: &\quad D(\partial c_B/\partial x) = k_c c_B \\ t > 0, x \to \infty: &\quad c_B = c_A/K \end{aligned}$$

After a sufficiently long time interval ($t \gg 1/k_1K$), a steady state is established ($\partial c_B/\partial t = 0$), giving

$$D\frac{d^2 c_B}{dx^2} + k_1(c_A - Kc_B) = 0 \tag{5.6.11}$$

with the solution

$$c_{\mathrm{B}} = \frac{c_{\mathrm{A}}}{K}\left[1 - \frac{1}{1+\sqrt{k_1 KD}/k_{\mathrm{c}}}\exp\left(-x\sqrt{\frac{k_1 K}{D}}\right)\right] \tag{5.6.12}$$

Figure 5.37 depicts the stationary distribution of the electroactive substance (the reaction layer) for $k_{\mathrm{c}} \rightarrow \infty$. The thickness of the reaction layer is defined in an analogous way as the effective diffusion layer thickness (Fig. 2.12). It equals the distance μ of the intersection of the tangent drawn to the concentration curve in the point $x = 0$ with the line $c = c_{\mathrm{A}}/K$,

$$\mu = \sqrt{\frac{D}{k_1 K}} \tag{5.6.13}$$

The current density is given by the equation

$$j = nFD\left(\frac{\mathrm{d}c_{\mathrm{B}}}{\mathrm{d}x}\right)_{x=0} = \frac{nFc_{\mathrm{A}}}{K}\frac{\sqrt{k_1 KD}}{1+\sqrt{k_1 KD}/k_{\mathrm{c}}} \tag{5.6.14}$$

which describes the polarization curve of a kinetic current, controlled by the electrode reaction and the steady state of the chemical volume reaction and diffusion. For $k_{\mathrm{c}} \gg \sqrt{k_1 KD}$, the limiting current density is obtained:

$$j_1 = zFc_{\mathrm{A}}\sqrt{\frac{k_1 D}{K}} \tag{5.6.15}$$

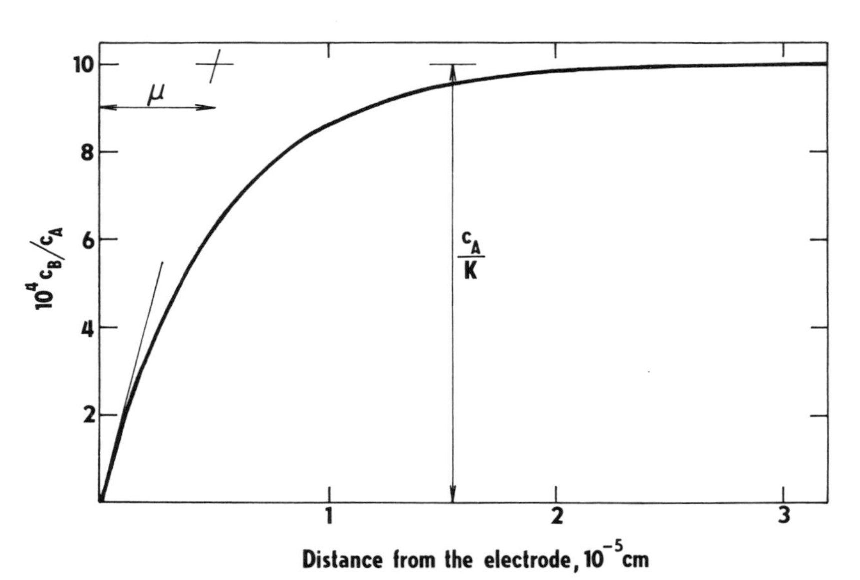

Fig. 5.37 Steady-state concentration distribution (reaction layer) in the case of a chemical volume reaction preceding an electrode reaction (Eq. (5.6.12)); $K = 10^3$, $k_{\mathrm{c}} \rightarrow \infty$, $k_1 = 0.04\ \mathrm{s}^{-1}$, $D = 10^{-5}\ \mathrm{cm \cdot s^{-1}}$, μ is the effective reaction layer thickness

This relationship forms the basis for the method of determining the rate constants of fast chemical reactions from the kinetic current.

A similar procedure can be employed for subsequent and parallel chemical reactions and for higher-order reactions.

The reduction of formaldehyde at a mercury electrode is an example of a system in which a chemical reaction precedes the electrode reaction. Formaldehyde is present in aqueous solution as the hydrated form (as dihydroxymethane), which cannot be reduced at a mercury electrode. This form is in equilibrium with the carbonyl form

According to R. Brdička and K. Veselý the carbonyl form of formaldehyde is reduced and the limiting kinetic current is given by the rate of the chemical volume reaction of dehydration. An analogous situation occurs for the equilibria among complexes, metal ions and complexing agents if the rates of complex formation and decomposition are insufficient to preserve the equilibrium. A simple example is the deposition of cadmium at a mercury electrode from its complex with nitrilotriacetic acid $HN^+\!\!<\!\!\begin{smallmatrix}COO^-\\COOH\\COOH\end{smallmatrix} = H_3X$. Here, a single X^{3-} species is bound in the complex. According to J. Koryta, equilibrium in solution in the pH range from 3 to 5 can be described by the equation

$$Cd^{2+} + HX^{2-} \underset{k_d}{\overset{k_f}{\rightleftharpoons}} CdX^- + H^+ \tag{5.6.17}$$

The symbols k_f and k_d are the rate constants for the formation and dissociation of the complex. The equilibrium constant of reaction (5.6.17) is given by the relationship

$$K_{CdX}K_3 = \frac{k_f}{k_d} \tag{5.6.18}$$

where K_{CdX} is the stability constant of the complex and K_3 is the third dissociation constant of the acid. If only free metal ions react in the given potential range at the electrode and the rate of the electrode reaction of the complex is negligible, then the constants in Eq. (5.6.15) for the limiting kinetic current are given as

$$k_1 = k_d c_{H_3O^+}$$
$$K = K_{CdX}K_3 \frac{c_{HX^{2-}}}{c_{H_3O^+}} \tag{5.6.19}$$

where $c_{HX^{2-}}$ is the concentration of the complexing agent, present in excess so that its concentration in the electrolyte does not change during the electrolysis.

A chemical reaction subsequent to a fast (reversible) electrode reaction (Eq. 5.6.1, case b) can consume the product of the electrode reaction, whose concentration in solution thus decreases. This decreases the overpotential of the overall electrode process. This mechanism was proposed by R. Brdička and D. H. M. Kern for the oxidation of ascorbic acid, converted by a fast electrode reaction at the mercury electrode to form dehydroascorbic acid. An equilibrium described by the Nernst equation is established at the electrode between the initial substance and this intermediate product. Dehydroascorbic acid is then deactivated by a fast chemical reaction with water to form diketogulonic acid, which is electroinactive.

If a chemical reaction regenerates the initial substance completely or partially from the products of the electrode reaction, such case is termed a chemical reaction parallel to the electrode reaction (see Eq. 5.6.1, case c). An example of this process is the catalytic reduction of hydroxylamine in the presence of the oxalate complex of Ti^{IV}, found by A. Blažek and J. Koryta. At the electrode, the complex of tetravalent titanium is reduced to the complex of trivalent titanium, which is oxidized by the hydroxylamine during diffusion from the electrode, regenerating tetravalent titanium, which is again reduced. The electrode process obeys the equations

$$\begin{gathered} Ti^{IV} + e \xrightarrow{\text{el}} Ti^{III} \\ Ti^{III} + NH_2OH \xrightarrow{\text{solution}} Ti^{IV} + OH^- + \dot{N}H_2 \end{gathered} \tag{5.6.20}$$

The $\dot{N}H_2$ radical rapidly reacts with excess oxalic acid required to form the oxalate complexes of titanium.

Partial regeneration of the product of the electrode reaction by a disproportionation reaction is characteristic for the reduction of the uranyl ion in acid medium, which, according to D. H. M. Kern and E. F. Orleman and J. Koutecký and J. Koryta, occurs according to the scheme

$$\begin{gathered} UO_2^{2+} + e \rightarrow UO_2^{+} \\ 2UO_2^{+} \xrightarrow{k} UO_2^{2+} + U^{IV} \text{ (products)} \end{gathered} \tag{5.6.21}$$

5.6.4 *Surface reactions*

Heterogeneous chemical reactions in which adsorbed species participate are not 'pure' chemical reactions, as the surface concentrations of these substances depend on the electrode potential (see Section 4.3.3), and thus the reaction rates are also functions of the potential. Formulation of the relationship between the current density in the stationary state and the concentrations of the adsorbing species in solution is very simple for a linear adsorption isotherm. Assume that the adsorbed substance B undergoes an irreversible electrode reaction with the rate constant k_c and is simultaneously formed at the electrode by a reversible heterogeneous reaction from the electroinactive substance A (see Eq. 5.6.9). The equilibrium constant for this reaction is $K = k_2/k_1$. In contrast to the system considered above, the quantities k_1 and k_2 are the rate constants of surface reactions. The substance A is present in solution and is adsorbed on the electrode according to the linear adsorption isotherm

$$\Gamma_A = \beta c_A \tag{5.6.22}$$

where Γ_A is the surface concentration and c_A is the bulk concentration of substance A, and β is the adsorption coefficient. At steady state,

$$\begin{aligned}\frac{d\Gamma_B}{dt} &= -k_c\Gamma_B + k_1(\Gamma_A - K\Gamma_B)\\ &= -k_c\Gamma_B + k_1(\beta c_A - K\Gamma_B) = 0\end{aligned} \tag{5.6.23}$$

so that the current density j and limiting current density j_l for $k_c \to \infty$ follow the equations

$$\begin{aligned}\frac{j}{nF} &= k_c\Gamma_B = \frac{k_c k_1 \beta c_A}{k_c + k_1 K}\\ j_l/nF &= k_1\beta c_A\end{aligned} \tag{5.6.24}$$

If relationship (5.2.23) is used for k_c and relationship (4.3.49) for β, then a

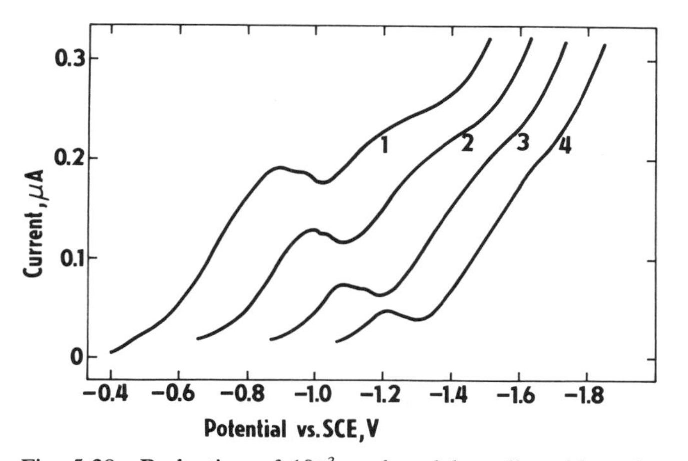

Fig. 5.38 Reduction of 10^{-3} M phenylglyoxylic acid at the mercury streaming electrode in acetate and phosphate buffers containing 1 M KNO_3: (1) pH 5.02, (2) pH 5.45, (3) pH 5.85, (4) pH 6.25. The curves 2, 3 and 4 are shifted by 0.2 V, 0.4 V and 0.6 V with respect to curve 1. The first wave is controlled by the surface protonation reaction while the second is a direct reduction of the acid anion. (According to J. Koryta)

j–E curve with a peak is obtained. This phenomenon is typical for a number of cases where the protonized electroactive form is formed by a surface reaction with protons (see Fig. 5.38).

References

Bard, A. J., and L. R. Faulkner, see page 290.

Brdička, R., V. Hanuš, and J. Koutecký, General theoretical treatment of polarographic kinetic currents, in *Progress in Polarography* (Eds P. Zuman and I. M. Kolthoff), Vol. 1, p. 145, Interscience, New York, 1962.

Heyrovský, J., and Kůta, see page 343.

Gerischer, H., and K. J. Vetter, Reaction overpotential (in German), *Z. Physik. Chemie,* **197,** 92 (1951).

Koryta, J., Diffusion and kinetic currents at the streaming mercury electrode, *Coll. Czech. Chem. Comm.,* **19,** 433 (1954).

Koryta, J., Electrochemical kinetics of metal complexes, *AE,* **6,** 289 (1967).

Koutecký, J., and J. Koryta, The general theory of polarographic kinetic currents, *Electrochim. Acta,* **3,** 318 (1961).

Southampton Electrochemistry Group, see page 253.

5.7 Adsorption and Electrode Processes

As mentioned in Section 5.1, adsorption of components of the electrolysed solution plays an essential role in electrode processes. Adsorption of reagents or products or of the intermediates of the electrode reaction or other components of the solution that do not participate directly in the electrode reaction can sometimes lead to acceleration of the electrode reaction or to a change in its mechanism. This phenomenon is termed *electrocatalysis.* It is typical of electrocatalytic electrode reactions that they depend strongly on the electrode material, on the composition of the electrode–solution interphase, and, in the case of single-crystal electrodes, on the crystallographic index of the face in contact with the solution.

Adsorption can also have the opposite effect on the rate of the electrode reaction, i.e. it can retard it. This is termed *inhibition of the electrode reaction.*

5.7.1 *Electrocatalysis*

A typical adsorption process in electrocatalysis is chemisorption, characteristic primarily for solid metal electrodes. The chemisorbed substance is often chemically modified during the adsorption process. Then either the substance itself or some fragment of it is bonded chemically to the electrode. As electrodes mostly have physically heterogeneous surfaces (see Sections 4.3.3 and 5.5.5), the Temkin adsorption isotherm (Eq. 4.3.46) is suitable for characterizing the adsorption.

The basic characteristics of electrocatalysis will be demonstrated on several examples, in the first place on the electrode processes of hydrogen, i.e. reduction of protons present in a proton donor, primarily oxonium ion or water, and oxidation of molecular hydrogen to form a proton that subsequently reacts with the solvent or with the lyate ion (in water forming the oxonium ion or the water molecule).

The electrochemical evolution of hydrogen has long been one of the most studied electrochemical processes. The mechanism of this process and the overpotential involved depend on the electrode material; here the overpotential is defined as the difference between the electrode potential at which hydrogen is formed at the given current density and the equilibrium potential of the hydrogen electrode in the given solution. Indeed Eq. (5.2.32) was formulated by J. Tafel as an empirical relationship for the evolution of hydrogen. The quantity a can also be a function of the hydrogen ion activity and of the composition of the solution in general. The electrocatalytic character of this process follows from the marked dependence of the constants of the Tafel equation for the cathodic process on the electrode material (Table 5.5). The cathodic process (and in the opposite direction, the anodic process) has either an EC or an EE mechanism (two subsequent electrode reactions, one of the original reactant and the other of the intermediate). It consists of two steps, in the first of which adsorbed hydrogen is formed from the reactants (H_3O^+ or H_2O):

$$\begin{aligned} H_3O^+ + e &\rightleftarrows H_{ads} + H_2O \\ H_2O + e &\rightleftarrows H_{ads} + OH^- \end{aligned} \tag{5.7.1}$$

This process is often called the *Volmer reaction* (I). In the second step, adsorbed hydrogen is removed from the electrode, either in a chemical reaction

$$2H_{ads} \rightleftarrows H_2 \tag{5.7.2}$$

(the *Tafel reaction* (II)) or in an electrode reaction ('electrochemical desorption'),

$$H_{ads} + H_3O^+ + e \rightleftarrows H_2 + H_2O \tag{5.7.3}$$

(the *Heyrovský reaction* (III)).

First, we shall discuss reaction (5.7.1), which is more involved than simple electron transfer. While the frequency of polarization vibration of the media where electron transfer occurs lies in the range 3×10^{10} to 3×10^{11} Hz, the frequency of the vibrations of proton-containing groups in proton donors (e.g. in the oxonium ion or in the molecules of weak acids) is of the order of 3×10^{12} to 3×10^{13} Hz. Then for the transfer proper of the proton from the proton donor to the electrode the classical approximation cannot be employed without modification. This step has indeed a quantum mechanical character, but, in simple cases, proton transfer can be described in terms of concepts of reorganization of the medium and thus of the exponential relationship in Eq. (5.3.14). The quantum character of proton transfer occurring through the tunnel mechanism is expressed in terms of the

Table 5.5 Constants a and b of the Tafel equation and the probable mechanism of the hydrogen evolution reaction at various electrodes with H_3O^+ as electroactive species ($a_{H_3O^+} \approx 1$). (According to L. I. Krishtalik)

Cathode material	$-a$	b	Mechanism
Pb	1.52–1.56	0.11–0.12	I(s),III
Tl	1.55	0.14	I(s),III
Hg	1.415	0.116	I(s),III
Cd	1.40–1.45	0.12–0.13	I(s),III
In	1.33–1.36	0.12–0.14	I(s),III
Sn	1.25	0.12	I(s),III
Zn	1.24	1.12	I(s),III
Bi	1.1	0.11	I(s),III
Ga (l)	1.05	0.11	I(s),III
(s)	0.90	0.10	I(s),III
Ag	0.95	0.12	I(s),III
Au	0.65–0.71	0.10–0.14	I(s),III
Cu	0.77–0.82	0.10–0.12	I(s),III
Fe	0.66–0.72	0.12–0.13	I(s),III
Co	0.67	0.15	I(s),III
Ni	0.55–0.72	0.10–0.14	I(s),III
Pt (anodically activated)	0.05–0.10	0.03	I,II(s)
(large j)	0.25–0.35	0.10–0.14	I,III(s)
(poisoned)	0.47–0.72	0.12–0.13	I(s),?
Rh (anodically activated)	0.05–0.10	0.03	I,II(s)
Ir (anodically activated)	0.05–0.10	0.03	I,II(s)
Re	0.15–0.21	0.03–0.04	I,II(s)
W	0.58–0.70	0.10–0.12	I,III(s)
Mo	0.58–0.68	0.10	I,III(s)
Nb (unsaturated with H)	0.92	0.11	I,III(s)
(saturated with H)	0.78	0.11	I,III(s)
Ta (unsaturated with H)	1.2	0.19	I,III(s)
(saturated with H)	1.04	0.15	I,III(s)
Ti	0.82–1.01	0.12–0.018	I,III(s)

I indicates the Volmer mechanism (Eq. 5.7.1), II the Tafel mechanism (Eq. 5.7.2) and III the Heyrovský mechanism (Eq. 5.7.3). The slowest step of the overall process is denoted (s).

transmission coefficient,

$$\kappa \approx \exp\left(-\frac{2\pi m \omega_i \omega_f}{h(\omega_i + \omega_f)} r^2\right) \tag{5.7.4}$$

where m is the mass of the proton or its isotope, h is Planck's constant, ω_i is the vibrational frequency of the bond between the proton and the rest of the molecule of the proton donor, ω_f is the vibrational frequency of the bond between the hydrogen atom and the metal and r is the proton tunnelling distance. When the hydrogen atom is weakly adsorbed, the vibrational frequency of the hydrogen–metal bond ω_f is small and the proton tunnelling distance r is large and thus κ is very small. Consequently, in the evolution of hydrogen at metals that adsorb hydrogen atoms very weakly, such as Hg, Pb, Tl, Cd, Zn, Ga and Ag, reaction (5.7.1) is the rate-controlling step. At high overpotentials, Eq. (5.3.17) is valid for the dependence of the rate constant on the potential. At lower overpotentials (at very small current densities) barrierless charge transfer occurs (see page 274), as indicated in Fig. 5.39.

Distinct adsorption of hydrogen can be observed with electrodes with a lower hydrogen overpotential, such as the platinum electrode. This phenomenon can be studied by cyclic voltammetry, as shown in Fig. 5.40 for a polycrystalline electrode. The potential pulse begins at $E = 0.0$ V, where the electrode is covered with a layer of adsorbed hydrogen. When the potential is shifted to a more positive value, the adsorbed hydrogen is oxidized in two anodic peaks in the potential range from 0.1 to 0.4 V. At even more positive potentials, no electrode process occurs and only the current for electrode charging flows through the system. This is especially noticeable at high polarization rates. The potential range from 0.4 to 0.8 V is termed the double-layer region. At potentials of $E > 0.8$ V, 'adsorbed oxygen' begins to form, i.e. a surface oxide or a layer of adsorbed OH radicals. This process is characterized by a drawn-out wave. Evolution of molecular oxygen starts at a potential of 1.8 V. When the direction of polarization is reversed, the oxide layer is first gradually reduced. This process has a certain activation energy and occurs at more negative potentials than the anodic process. The reduction of oxonium ions, accompanied by adsorption, occurs at the same potentials as the opposite anodic process. If this experiment is carried out on the individual crystal faces of a single-crystal electrode (see Fig. 5.41),

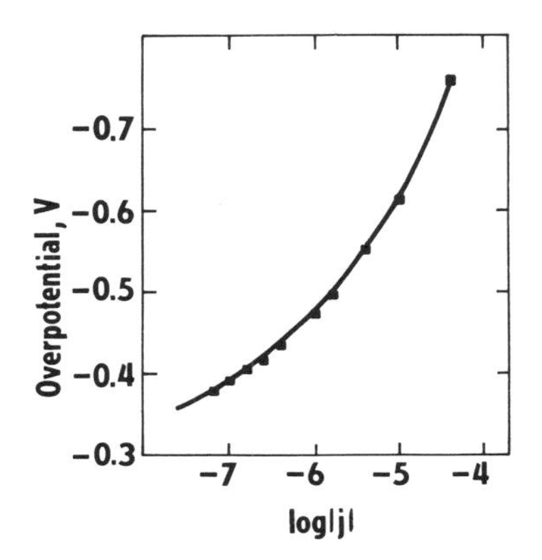

Fig. 5.39 Tafel plot of hydrogen evolution at a mercury cathode in 0.15 M HCl, 3.2 M KI electrolyte at 25°C. (According to L. I. Krishtalik)

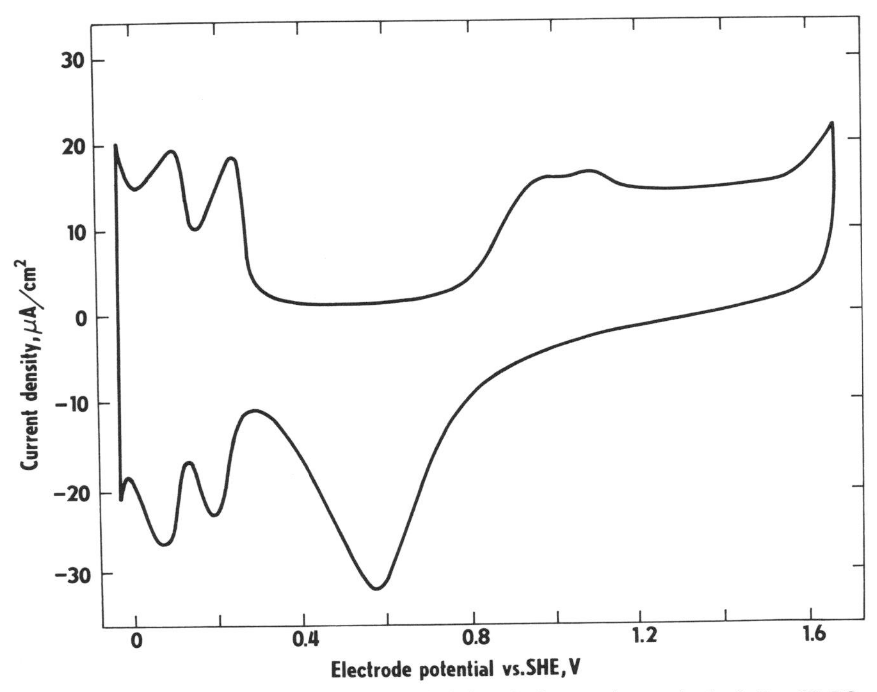

Fig. 5.40 Cyclic voltammogram of a bright platinum electrode in 0.5 M H_2SO_4. Geometrical area of the electrode 1.25×10^{-3} cm², periodical triangular potential sweep ($dE/dt = 30$ V · s^{-1}), temperature 20°C, the solution was bubbled with argon. (By courtesy of J. Weber)

then the picture changes quite markedly. The shape of the voltammetric curve is also affected strongly by the procedure of annealing the electrode prior to the experiment. There are also very marked differences between the first voltammetric curve and the curve obtained after repeated pulsing. All these features are typical for electrocatalytic phenomena.

It was demonstrated by R. Parsons and H. Gerischer that the adsorption energy of the hydrogen atom determines not only the rate of the Volmer reaction (5.7.1) but also the relative rates of all three reactions (5.7.1) to (5.7.3). The relative rates of these three reactions decide over the mechanism of the overall process of evolution or ionization of hydrogen and decide between possible rate-determining steps at electrodes from different materials.

The effect of adsorption on the electroreduction of hydrogen ions, i.e. the Volmer reaction, is strongly affected by the potential difference in the diffuse electrical layer (Eq. 5.3.20).

In the presence of iodide ions, the overpotential at a mercury electrode decreases, although the adsorption of iodide is minimal in the potential region corresponding to hydrogen evolution. The adsorption of iodide

Fig. 5.41 A single-crystal sphere of platinum (magnification 100×). It is prepared from a Pt wire by annealing in an oxygen–hydrogen flame or by electric current and by subsequent etching. The sphere is then cut parallel to the face with a required Miller index to obtain a single-crystal electrode. (By courtesy of E. Budevski)

retards the electrode process at other electrodes with large hydrogen overpotentials, such as the lead electrode.

Hydrogen is evolved from water molecules in alkaline media at mercury and some other electrodes.

As the adsorption energy increases, the rate of the Volmer reaction can increase until equilibrium is attained and the rate of the process is determined by either the Tafel or the Heyrovský reaction. However, it is more probable for kinetic reasons that the Tafel reaction will occur at electrodes that form a moderately strong bond with adsorbed hydrogen (e.g. at platinum electrodes, at least in some cases). Electrodes that adsorb hydrogen strongly such as tungsten electrodes, are in practice completely covered with adsorbed hydrogen over a wide range of electrode potentials.

The Tafel reaction would require breaking the adsorption bonds to two hydrogen atoms strongly bound to the electrode, while the Heyrovský reaction requires breaking only one such bond; this reaction then determines the rate of the electrode process.

The isotope effect (i.e. the difference in the rates of evolution of hydrogen from H_2O and D_2O) on hydrogen evolution is very important for theoretical and practical reasons. The electrolysis of a mixture of H_2O and D_2O is characterized, like in other separation methods, by a separation factor

$$S = \frac{(c_H/c_D)_{gas}}{(c_H/c_D)_{sol}} \tag{5.7.5}$$

where c_H/c_D is the ratio of the atomic concentrations of the two isotopes. The separation factor is a function of the overpotential and of the electrode material. The reacting species are H_3O^+ and H_2DO^+ in solutions with low deuterium concentrations. The S values for mercury electrodes lie between 2.5 and 4, for platinum electrodes with low overpotentials between 3 and 4 and, at large overpotentials, between 7 and 8.

The overpotential of hydrogen at a mercury electrode decreases sharply in the presence of readily adsorbed, weak organic bases (especially nitrogen-containing heterocyclic compounds). A peak appears on the polarization curves of these catalytic currents. The hydrogen overpotential is decreased as oxonium ions are replaced in the electrode reaction by the adsorbed cations of these compounds, $BH_{ads}{}^+$. The product of the reduction is the BH_{ads} radical. Recombination of these radicals yields molecular hydrogen and the original base. The evolution of hydrogen through this mechanism occurs more readily than through oxonium ions. The decrease in the catalytic current at negative potentials is a result of the desorption of organic compounds from the electrode surface.

The *electrode processes of oxygen* represent a further important group of electrocatalytic processes. The reduction of oxygen to water

$$O_2 + 4H^+ + 4e \rightleftarrows 2H_2O \tag{5.7.6}$$

has a standard potential determined by calculation from thermodynamic data as +1.227 V. The extremely low exchange current density prevents direct determination of this value. The simultaneous transfer of four electrons in reaction (5.7.6) is highly improbable; thus the reaction must consist of several partial processes. The non-catalytic electroreduction of oxygen at a mercury electrode will now be compared to the catalytic reduction at a silver electrode. J. Heyrovský demonstrated that the stable intermediate in the reduction at mercury is hydrogen peroxide. Figure 5.42 depicts the voltammogram (polarographic curve) for the reduction of oxygen at a dropping mercury electrode. The first wave corresponds to the reduction of oxygen to hydrogen peroxide and the second to the reduction

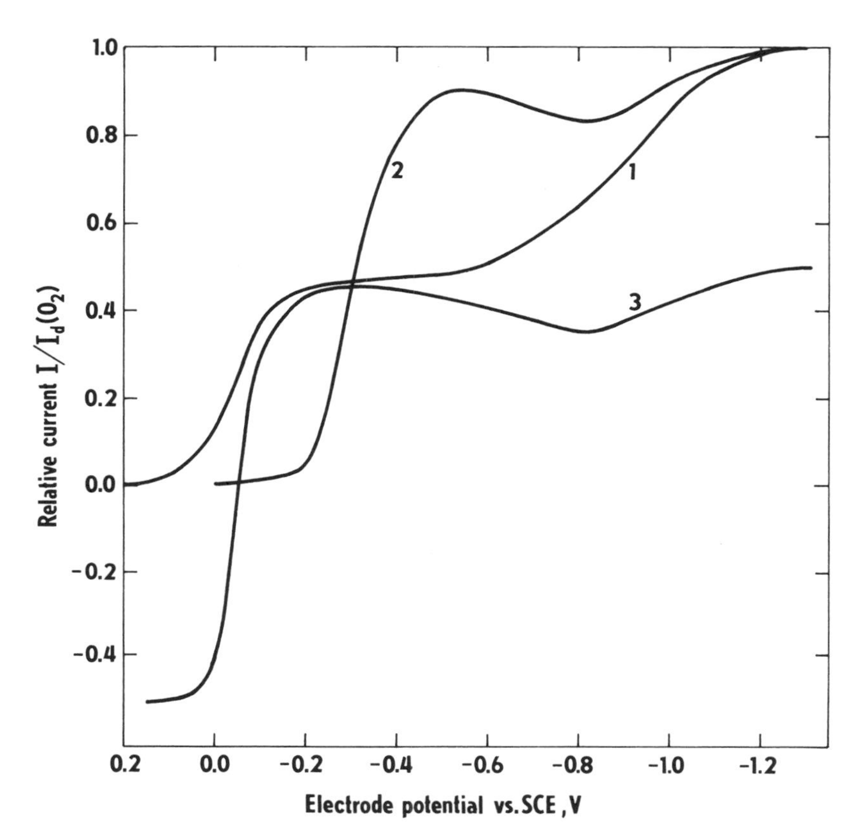

Fig. 5.42 Electrode reactions of oxygen and hydrogen peroxide at mercury (dropping electrode) and silver (rotating disk electrode): (1) reduction of O_2 to H_2O_2 (first wave) and of H_2O_2 to H_2O (second wave) in 1 M CH_3COOH, 1 M CH_3COONa electrolyte at Hg electrode, (2) reduction of O_2 in 0.1 M NaOH at Ag electrode, (3) oxidation of H_2O_2 to O_2 and reduction of H_2O_2 to H_2O in 0.1 M NaOH at rotating Ag electrode. (According to M. Březina, J. Koryta and M. Musilová)

of hydrogen peroxide to water. The first of these processes

$$O_2 + 2H^+ + 2e \rightleftarrows H_2O_2 \tag{5.7.7}$$

has a standard electrode potential of 0.68 V and the second reaction

$$H_2O_2 + 2H^+ + 2e \rightleftarrows 2H_2O \tag{5.7.8}$$

has a potential of +1.77 V. Reaction (5.7.7) occurs at a mercury electrode with an overpotential in acidic and neutral media and reaction (5.7.8) is always associated with very high overpotentials. A more detailed analysis revealed that reaction (5.7.7) corresponds to a multistep process, described by Eqs (5.2.39) to (5.2.48). The rate is determined by the transfer of the first electron, forming the superoxide anion:

$$O_2 + e \rightleftarrows O_2^- \tag{5.7.9}$$

The next reactions are connected with the transfer of an additional electron and of hydrogen ions:

$$\begin{aligned} O_2^- + H^+ &\rightleftarrows HO_2 \\ HO_2 + e &\rightleftarrows HO_2^- \\ HO_2^- + H^+ &\rightleftarrows H_2O_2 \end{aligned} \tag{5.7.10}$$

These reactions proceed very rapidly, so that the overall reaction corresponds to the transfer of two electrons. As reaction (5.7.9) is very slow in acid and neutral media, the electrode reaction is irreversible and the polarization curve does not depend on the concentration of hydrogen ions. In weakly alkaline media, reoxidation of HO_2^- begins to occur. At $pH > 11$, the polarization curve at a dropping mercury electrode becomes reversible. In this way, the process proceeds in water and water-like solvents. On the other hand, for example in carbonate melts, the step following after the reaction (5.7.9) is the slow reaction $O_2^- + e = O_2^{2-}$.

The reduction of hydrogen peroxide at a mercury electrode is also a stepwise electrode reaction, where the second step once again is fast, so that the observed electrode reaction corresponds to the transfer of two electrons:

$$\begin{aligned} H_2O_2 + e &\xrightarrow{k_c} \dot{O}H + OH^- \\ \dot{O}H + e &\longrightarrow OH^- \end{aligned} \tag{5.7.11}$$

The first step in this reaction is connected with breaking of the O—O bond, with a markedly unsymmetrical energy profile ($\alpha \approx 0.2$). Other reactions connected with a similar bond breakage have an analogous course, e.g. the reduction of cystine at a mercury electrode.

Curve 2 in Fig. 5.42 depicts the reduction of oxygen at a silver rotating disk electrode. The first wave, corresponding to almost four electrons, can be attributed to reaction (5.7.6), being a result of the catalytic reduction of hydrogen peroxide formed as an intermediate in the presence of surface silver oxide. This process occurs quite rapidly at potentials more positive than the reduction potential of oxygen alone. The peak on the voltammetric curve is formed because of the reduction and thus disappearance of the catalytically active surface oxide. A further increase in the current corresponds to the non-catalytic reduction of hydrogen peroxide, which occurs at high overpotentials such as that on mercury.

The behaviour of hydrogen peroxide alone (Fig. 5.42, curve 3) is in agreement with this explanation; the catalytic reduction obeys Eq. (5.7.8) at potentials more positive than the non-catalytic oxidation. The voltammetric curve obtained is characterized by a continuous transition from the anodic to the cathodic region. The process occurring at negative potentials is then only the reduction of H_2O_2, proceeding analogously as in the reduction of oxygen (curve 2).

The anodic evolution of oxygen takes place at platinum and other noble metal electrodes at high overpotentials. The polarization curve obeys the Tafel equation in the potential range from 1.2 to 2.0 V with a b value between 0.10 and 0.13. Under these conditions, the rate-controlling process is probably the oxidation of hydroxide ions or water molecules on the surface of the electrode covered with surface oxide:

$$OH^-(ads) \rightarrow OH(ads) + e \tag{5.7.12}$$

or

$$H_2O(ads) \rightarrow OH(ads) + H^+ + e \tag{5.7.13}$$

This is followed by a fast reaction, either

$$2OH(ads) \rightarrow O(ads) + H_2O \tag{5.7.14}$$

or

$$OH(ads) \rightarrow O(ads) + H^+ + e \tag{5.7.15}$$

and, finally,

$$2O(ads) \rightarrow O_2 \tag{5.7.16}$$

At more positive potentials, processes occur that depend on the composition of the electrolyte, such as the formation of $H_2S_2O_8$ and HSO_5 in sulphuric acid solutions, while the $ClO_4^{\cdot}$ radical is formed in perchloric acid solutions, decomposing to form ClO_2 and O_2. The formation of ozone has been observed at high current densities in solutions of rather concentrated acids.

5.7.2 *Inhibition of electrode processes*

Electrode processes can be retarded (i.e. their overpotential is increased) by the adsorption of the components of the electrolysed solution, of the products of the actual electrode reaction and of other substances formed at the electrode. Figure 5.43 depicts the effect of the adsorption of methanol on the adsorption of hydrogen at a platinum electrode (see page 353).

The integral of the anodic current over the range of potentials including both peaks of the adsorption of hydrogen, as well as the integral corresponding to the cathodic current, is equal to the charge $Q_m = -F\Gamma_m$, where Γ_m is the maximal surface concentration of hydrogen atoms. When another substance is also adsorbed on the electrode (here, methanol), with a surface concentration that does not change during the potential pulse, this substance prevents adsorption of hydrogen at sites that it occupies. If the curve of the hydrogen peaks is integrated under these circumstances, a different Q_m' value is obtained. If one molecule of adsorbate occupies one site for hydrogen adsorption, then the relative coverage of the electrode by

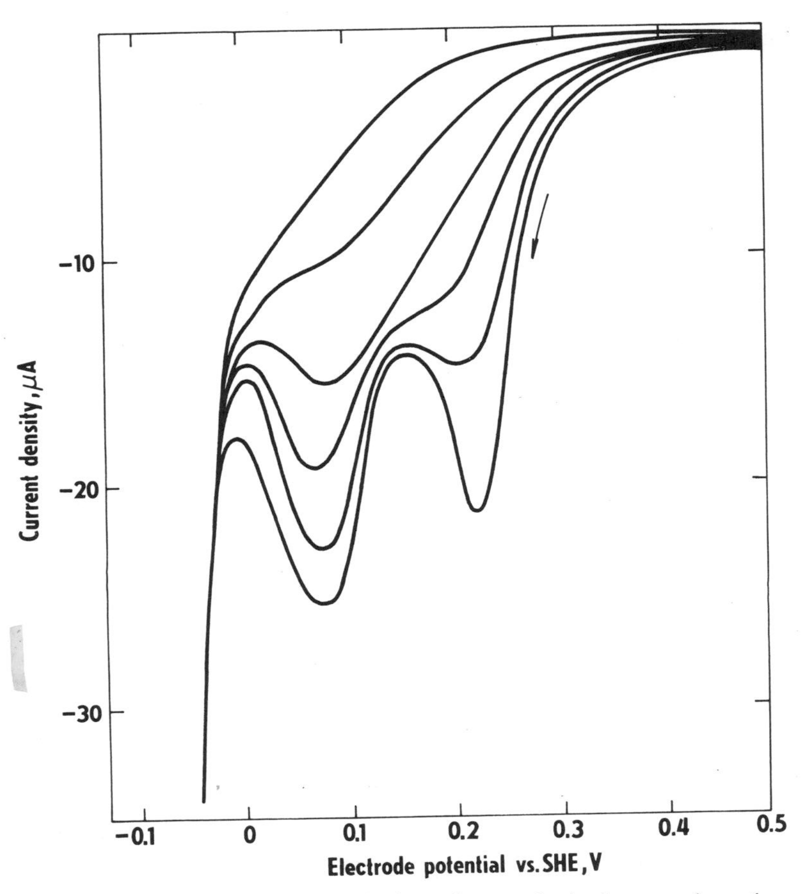

Fig. 5.43 Influence of methanol adsorption on the hydrogen adsorption at a bright platinum electrode in 0.5 M H_2SO_4. This is the same electrode as in Fig. 5.40 polarized by a single potential sweep ($dE/dt = 30\ V \cdot s^{-1}$) from $E = 0.5$ V to $E = -0.1$ V. Concentration of methanol (in $mol \cdot dm^{-3}$ from bottom to top): 0, 10^{-4}, 10^{-3}, 10^{-2}, 10^{-1}, 5. Methanol was adsorbed at $E = 0.5$ V during 2 min (the equilibrium coverage was achieved), $t = 20°C$. (By courtesy of J. Weber)

the adsorbed substance is

$$\Theta = \frac{Q_m - Q'_m}{Q_m} \tag{5.7.17}$$

The dependence of the relative coverage of the platinum electrode with methanol on its concentration in solution indicates that the adsorption of methanol obeys the Temkin isotherm (4.3.46).

The anodic oxidation of molecular hydrogen dissolved in the electrolyte at a Pt electrode consists of several steps: the transport of hydrogen molecules to the electrode, their dissociation with simultaneous adsorption of atomic hydrogen on the electrode (reversed Tafel reaction (5.7.2)) and its subsequent oxidation. The shape of the polarization curve depends on which of these processes is slowest. The adsorption of hydrogen is affected by the adsorption of anions, occurring at positive potentials. The adsorbed anions, especially iodide, decrease both the adsorption energy and the maximal amount of hydrogen adsorbed. The effect of iodide is analogous to that of substances termed cathodic poisons, such as mercury ions, trivalent arsenic or cyanide ions. The retarding effect of adsorbed anions is especially apparent for the anodic ionization of hydrogen dissolved in solution at a platinum electrode. Figure 5.44 shows the voltammograms for this process at a rotating disk electrode. In the vicinity of the equilibrium potential, this process is controlled by convective diffusion (see Eqs 5.4.20 and 5.4.23); at slightly more positive potentials a limiting current appears that is then independent of the potential. At much more positive potentials, the surface process is retarded by the adsorption of sulphate anions and becomes the

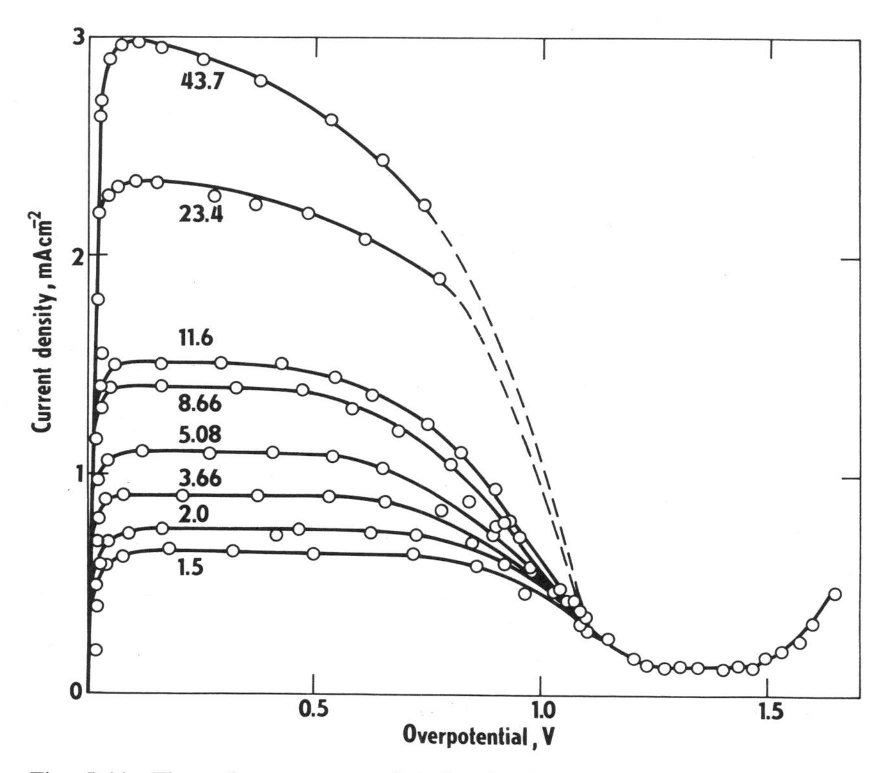

Fig. 5.44 The voltammogram of molecular hydrogen at a rotating bright platinum disk electrode in 0.5 M H_2SO_4, $p_{H_2} = 10^5$ Pa, 25°C. The rotation speed $\omega(s^{-1})$ is indicated at each curve. (According to E. A. Aykazyan and A. I. Fedorova)

rate-controlling step in place of diffusion. The higher the rotation rate of the electrode, the sooner the transition occurs between the transport-controlled process and the surface-controlled process. The current increase at 1.6 V corresponds to platinum 'surface oxide' formation.

The electrocatalytic oxidation of lower aliphatic compounds such as alcohols, aldehydes, hydrocarbons and formic and oxalic acid at platinum-group electrodes has a number of typical features. The chemisorption process usually involves the dissociation of several hydrogen atoms from the original molecule and a so-far not completely identified intermediate (often termed 'reduced carbon dioxide' for the reactions of methanol, formaldehyde and formic acid) remains adsorbed on the electrode. At electrode potentials higher than 0.35 V (versus a hydrogen electrode located in the same solution and saturated with hydrogen at standard pressure) all the hydrogen atoms formed during chemisorption are oxidized anodically. The adsorbed intermediate can be oxidized at potentials higher than +0.4 V; the larger the coverage by the adsorbed intermediate, the higher the threshold potential of the anodic oxidation. This phenomenon can be explained by the 'pair mechanism' of the anodic process, proposed by S. Gilman. Water molecules in the vicinity of the adsorbed intermediate are highly polarized, facilitating their oxidation. The OH radicals formed then immediately attack this intermediate in a surface reaction. At a constant potential, the rate of this process decreases because the number of sites available for oxidation of water gradually diminishes. However, if a pulse of increasing positive potential is applied to the electrode, then the oxidation of water resulting in OH radical formation is accelerated by increasing the potential and the rate of the overall electrode process is enhanced, as shown for methanol in Fig. 5.45. Clearly, the rate of the process at low potentials decreases with increasing concentration of methanol owing to the retarding effect of electrode coverage with the intermediate. In contrast, at higher potentials, the rate of the process increases with increasing concentration (the oxidation of water is so fast that the overall rate increases with increasing surface coverage).

So far, several examples have been given of the inhibition of electrocatalytic processes. This retardation is a result of occupation of the catalytically more active sites by electroinactive components of the electrolyte, preventing interaction of the electroactive substances with these sites. The electrode process can also be inhibited by the formation of oxide layers on the surface and by the adsorption of less active intermediates and also of the products of the electrode process.

The inhibition of electrode processes as a result of the *adsorption of electroinactive surfactants* has been studied in detail at catalytically inactive mercury electrodes. In contrast to solid metal electrodes where knowledge of the structure of the electrical double layer is small, it is often possible to determine whether the effect of adsorption on the electrode process at mercury electrodes is solely due to electrostatics (a change in potential ϕ_2)

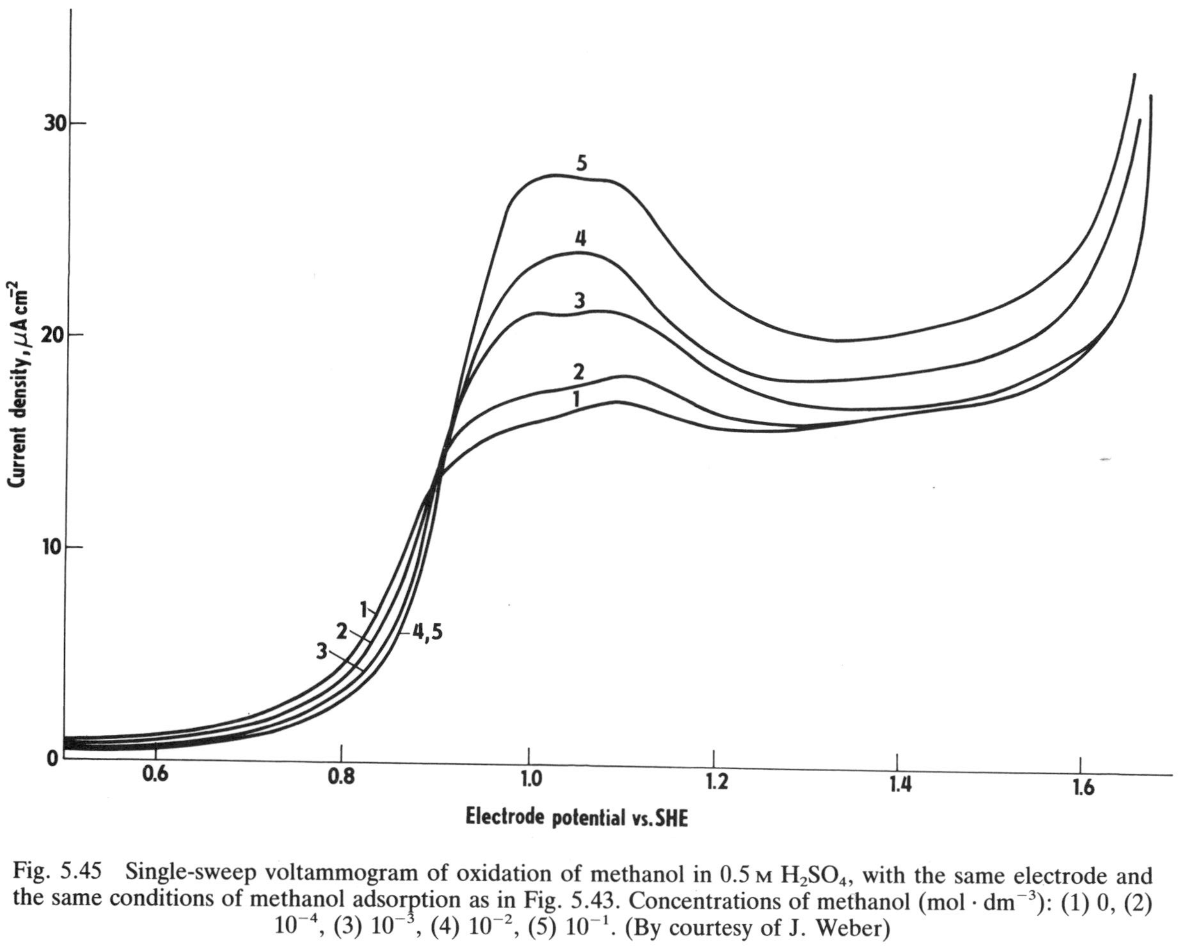

Fig. 5.45 Single-sweep voltammogram of oxidation of methanol in 0.5 M H_2SO_4, with the same electrode and the same conditions of methanol adsorption as in Fig. 5.43. Concentrations of methanol (mol · dm^{-3}): (1) 0, (2) 10^{-4}, (3) 10^{-3}, (4) 10^{-2}, (5) 10^{-1}. (By courtesy of J. Weber)

or is the result of some other type of inhibition. The steric effect where the sufficiently close approach of the electroactive substance to the electrode is inhibited is typical for large surface-active molecules. The adsorbed layer of surfactant replacing the layer of water molecules covering the surface of the electrode prevents the reacting species and their solvation shells from forming a configuration suitable for the electrode reaction (see Section 5.3). This leads to a decrease in the value of the rate constant of the electrode reaction. For example, in the case of an irreversible electrode reaction with a rate constant k_c and adsorption of an uncharged surfactant, we have

$$k_c = k_0(1 - \Theta) + k_1 \tag{5.7.18}$$

where k_0 and k_1 are the rate constants for the electrode reaction at unoccupied and occupied sites of the electrode surface, respectively. For the adsorption of surface-active ions, these quantities are also a function of the surface coverage Θ.

It is very simple to determine the value of $\Theta = \Gamma/\Gamma_m$ for a strongly adsorbed substance in electrolysis with a dropping mercury electrode. If a much smaller amount of substance is sufficient for complete electrode coverage than available in the test solution, then the surface concentration of the surface-active substance Γ is determined by its diffusion to the electrode.

The total amount M adsorbed on the electrode in time t (A is the surface of the electrode according to Eq. (4.4.4) and Γ is the surface excess in moles per square centimetre) is given as

$$M = 0.85\Gamma m^{2/3}t^{2/3} = \int_0^t D_p\left(\frac{\partial c_p}{\partial x}\right)_{x=0} A\,\mathrm{d}t \tag{5.7.19}$$

where $D_p(\partial c_p/\partial x)_{x=0}$ is the material flux of the surface-active substance to the electrode, corresponding to the limiting current density and given according to Eq. (2.7.17) by the relationship

$$D_p\left(\frac{\partial c_p}{\partial x}\right)_{x=0} = c_p^0\sqrt{\frac{7D_p}{3\pi t}} \tag{5.7.20}$$

where c_p is the concentration of the surface-active substance, D_p is its diffusion coefficient and c_p^0 is its concentration in the bulk of the solution. The surface concentration at time t is given by the equation deduced by J. Koryta:

$$\Gamma = 0.74c_p^0(D_p t)^{1/2} \tag{5.7.21}$$

If complete coverage of the electrode Γ_m is practically attained at time θ, then

$$\Theta = \frac{\Gamma}{\Gamma_m} = \left(\frac{t}{\theta}\right)^{1/2} \tag{5.7.22}$$

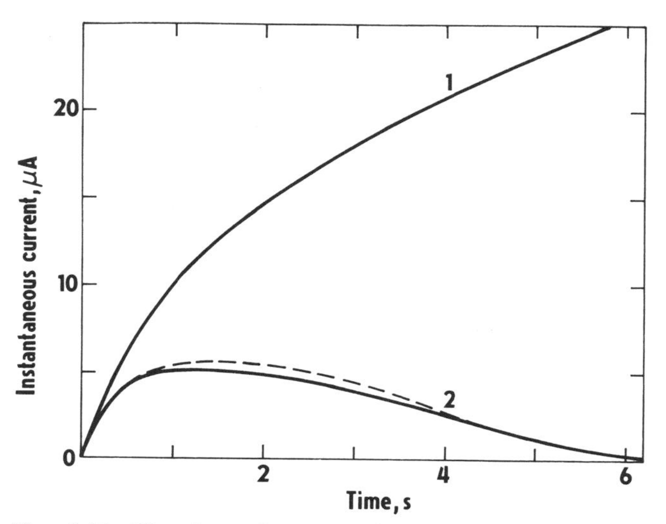

Fig. 5.46 The dependence on time of the instantaneous current I at a dropping mercury electrode in a solution of $0.08\,\mathrm{M}$ $Co(NH_3)_6Cl_3 + 0.1\,\mathrm{M}$ $H_2SO_4 + 0.5\,\mathrm{M}$ K_2SO_4 at the electrode potential where $-I \ll -I_d$ (i.e. the influence of diffusion of the electroactive substance is negligible): (1) in the absence of surfactant; (2) after addition of 0.08% polyvinyl alcohol. The dashed curve has been calculated according to Eq. (5.7.23). (According to J. Kůta and I. Smoler)

The value of Γ_m can be calculated from Eq. (5.7.21) where Γ approaches Γ_m at time $t = \theta$. The instantaneous polarographic current I_{ads} controlled by the rate of the irreversible electrode reaction of substance A is as follows:

$$I_{ads} = zFk_cAc_{Ox} = 0.85zFk_0m^{2/3}t^{2/3}\left[1 - \left(\frac{t}{\theta}\right)^{1/2}\right]c_{Ox}$$
$$= I\left[1 - \left(\frac{t}{\theta}\right)^{1/2}\right] \qquad (5.7.23)$$

(see Eq. 5.7.18), assuming that $k_1 = 0$ and that the current is so small that diffusion of the electroactive substance can be neglected; I is the current in the absence of the surface-active substance. Figure 5.46 shows the time dependence of the instantaneous current in the absence and presence of the surface-active substance.

References

Anson, F. C., Patterns of ionic and molecular adsorption at electrodes, *Acc. Chem. Res.*, **8,** 400 (1975).

Breiter, M. W., *Electrochemical Processes in Fuel Cells,* Springer-Verlag, New York, 1969.

Kinetics and mechanisms of electrode processes, *CTE,* **7** (1983)
Frumkin, A. N., Hydrogen overvoltage and adsorption phenomena, Part I, Mercury, *AE,* **1,** 65 (1961); Part II, Solid metals, *AE,* **3,** 267 (1963).
Gilman, S., The anodic film on platinum electrode, in *Electroanalytical Chemistry* (Ed. A. J. Bard), Vol. 2, p. 111, M. Dekker, New York, 1967.
Heyrovský, J., and J. Kůta, see page 343.
Hoare, J., *The Electrochemistry of Oxygen,* Interscience, New York, 1968.
Krishtalik, L. I., Hydrogen overvoltage and adsorption phenomena, Part III, Effect of the adsorption energy of hydrogen on overvoltage and the mechanism of the cathodic process, *AE,* **7,** 283 (1970).
Krishtalik, L. I., The mechanism of the elementary act of proton transfer in homogeneous and electrode reactions, *J. Electroanal. Chem.,* **100,** 547 (1979).
Lamy, C., Electrocatalytic oxidation of organic compounds on noble metals in aqueous solutions, *Electrochim. Acta,* **29,** 1581 (1984).
Marcus, R. A., Similarities and differences between electron and proton transfers at electrodes and in solution, Theory of hydrogen evolution reaction, *Proc. Electrochem. Soc.,* **80-3,** 1 (1979).
Schulze, J. W., and M. A. Habib, Principles of electrocatalysis and inhibition by electroadsorbates and protective layer, *J. Appl. Electrochem.,* **9,** 255 (1979).
Vielstich, W., *Fuel Cells,* John Wiley & Sons, New York, 1970.

5.8 Deposition and Oxidation of Metals

Sections 5.6.2 and 5.6.3 dealt with the deposition of metals from complexes; these processes follow the simple laws dealt with in Sections 5.2 and 5.3, particularly if they take place at mercury electrodes. The deposition of metals at solid electrodes (electrocrystallization) and their oxidation is connected with the kinetics of transformation of the solid phase, which has a specific character. A total of five different cases can be distinguished in these processes:

1. The deposition occurs at an electrode of a different metal or other conductive material (e.g. graphite).
2. The deposition occurs at an electrode of the same metal.
3. The metal is ionized in the anodic process to produce soluble ions.
4. The metal is anodically oxidized to ions that react with the components of the solution to yield an insoluble compound forming a surface film on the electrode.
5. A component of the solution participates in the anodic oxidation of the metal, so that the metal is converted directly into a surface film.

These processes either lead to the formation of a new solid phase or the original solid phase grows or disappears. In addition to the electrochemical laws discussed earlier, this group of phenomena must be explained on the basis of the theory of new phase formation (crystallization, condensation, etc.).

The basic properties of electrocrystallization can best be illustrated by the example of the deposition of a metal on an electrode of a different material (case 1).

5.8.1 *Deposition of a metal on a foreign substrate*

The formation of a new phase requires that the system be metastable, such as a supersaturated vapour prior to condensation, a supersaturated solution prior to crystallization, etc. The metastable situation of the system required for the formation of a new phase through an electrode reaction is brought about by the overpotential. At a suitable value of the excess energy in the system *nuclei* of the new phase are formed. If these nuclei are too small, then they become unstable as a result of the relatively high surface energy and decompose spontaneously. If their dimensions exceed a certain critical value, which is a function of the excess energy of the system, then they begin to grow spontaneously. Their formation requires a sufficiently large fluctuation of the variables of state of the system or the presence of sites at the interface at which nuclei can be formed with an overall decrease in the surface energy. The nucleation belongs to those processes where the individual events (of stochastic nature) occurring on the molecular level can be observed because of amplifying by subsequent macroscopic processes.

Assume that current is passed either through the total nucleus surface area or through part thereof, such as the edge of a two-dimensional nucleus of monoatomic thickness. The transition of the ion M^{z+} to the metallic state obeys the equation for an irreversible electrode reaction, i.e. Eqs (5.2.12), (5.2.23) and (5.2.37). The effect of transport processes is neglected. The current density at time t thus depends on the number of nuclei and their active surface area. If there is a large number of nuclei, then the dependence of their number on time can be considered to be a continuous function. For the overall current density at time t we have

$$j = \lambda \int_0^t A(t-\tau) v_n d\tau \qquad (5.8.1)$$

where $\lambda = zFk \exp(-\alpha zF\eta/RT)c_{M^{z+}}$ and A is the active surface area of a single nucleus. The rate of nucleation $v_n = dN_n/dt$ then gives the increase in the number of nuclei per unit time.

The simplest case occurs when a number of nuclei N_n^0 is formed at the beginning of the process and does not change with time (*instantaneous nucleation*). This is the case when the electrolysis is carried out by the double-pulse potentiostatic method (Fig. 5.18B), where the crystallization nuclei are formed in the first high, short pulse and the electrode reaction then occurs only at these nuclei during the second, lower pulse. A second situation in which instantaneous nucleation can occur is when the nuclei occupy all the active sites on the electrode at the beginning of the electrolysis.

For instantaneous nucleation Eq. (5.8.1) assumes the form

$$j = \lambda A(t) N_n^0 \qquad (5.8.2)$$

Hemispherical nuclei will be considered; this case is important for the deposition of liquid metal such as Hg on a platinum or graphite surface. The subsequent derivation can be used as an approximation for the general case of three-dimensional nuclei. The growth of the volume of the nucleus $V(t)$ is described by the equation

$$\frac{\mathrm{d}V(t)}{\mathrm{d}t} = \frac{\lambda V_{\mathrm{m}}}{zF} A(t) \tag{5.8.3}$$

where V_{m} is the molar volume of the deposited metal. Simple geometric considerations and Eq. (5.8.2) yield

$$j = \frac{2\pi V_{\mathrm{m}}^2 \lambda^3 N_{\mathrm{n}}^0 t^2}{z^2 F^2} \tag{5.8.4}$$

This relationship is valid when the growing nuclei do not overlap. Otherwise, the rate of growth of the surface gradually decreases. However, usually the dependence of the current on time in the initial stage is used as a diagnostic criterium for the type of nucleus and nucleation. The simplest case of *progressive nucleation* with a constant rate κ, Eq. (5.8.1) then gives

$$j = \frac{2\pi V_m^2 \kappa \lambda^3 t^3}{3z^2 F^2} \tag{5.8.5}$$

The calculation for the important case of two-dimensional nuclei growing only in the plane of the substrate will be based on the assumption that these are circular and that the electrode reaction occurs only at their edges, i.e. on the surface, $2\pi rh$, where r is the nucleus radius and h is its height (i.e. the crystallographic diameter of the metal atom). The same procedure as that employed for a three-dimensional nucleus yields the following relationship for instantaneous nucleation:

$$j = \frac{2\pi h \lambda^2 N_{\mathrm{n}}^0 t}{zF\Gamma_{\mathrm{m}}} \tag{5.8.6}$$

and for progressive nucleation with constant rate κ,

$$j = \frac{\pi h \lambda^2 \kappa t^2}{zF\Gamma_{\mathrm{m}}} \tag{5.8.7}$$

where Γ_{m} is the surface concentration of the metal in the compact monolayer ($\mathrm{mol \cdot cm^{-2}}$).

In this case the effect of overlapping of the individual nuclei can be simply expressed in terms of the statistical Kolmogorov–Avrami theory, where the right-hand side of Eq. (5.8.6) is multiplied by the factor $\exp(-N_{\mathrm{n}}^0 \lambda^2 t^2 / z^2 F^2 \Gamma_{\mathrm{m}}^2)$ and that of Eq. (5.8.7) by the factor $(-\pi\kappa\lambda^2 t^3 / 3z^2 F^2 \Gamma_{\mathrm{m}}^2)$.

The properties of the crystal nucleus are a function of its surface energy, its dimensions and the energy of adhesion to the substrate. The equilibrium form of a crystal with the nth plane lying on the substrate is described by the Gibbs–Wulff–Kaischew equation

$$\gamma_1 h_1 = \gamma_2 h_2 = \cdots = (\gamma_n - \beta)h_n \tag{5.8.8}$$

where γ_i is the specific surface energy, numerically equal to the surface tension of the ith crystal plane, h_1 is the distance of this plane from the centre of the crystal and β is the standard Gibbs energy of adhesion of the nth crystal plane to the substrate. The equilibrium thickness of the crystal (i.e. the distance of the highest point or upper plane of the crystal from the substrate) decreases with increasing adhesion energy, until the three-dimensional crystal is finally replaced by a monolayer (i.e. a two-dimensional crystal). If $\beta > 2\gamma_n$, then the surface layer of adsorbed atoms (Fig. 5.24), termed by M. Volmer *ad-atoms* ($\Gamma < \Gamma_m$), is more stable than the compact surface of the given metal. Consequently, ad-atoms of the metal can be deposited at a potential that is lower than the electrode potential of the corresponding metal electrode in the given medium (*underpotential deposition*). Underpotential-deposited ad-atoms exhibit remarkable electrocatalytic properties in the oxidation of simple organic substances.

The nucleation rate constant (Eq. (5.8.5)) κ depends on the size of the critical crystallization nucleus according to the Volmer–Weber equation

$$\kappa = k_\mathrm{n} \exp\left(-\frac{\Delta G_\mathrm{n}^\ddagger}{RT}\right) \tag{5.8.9}$$

where the preexponential factor k_n is a function of the number of sites on the substrate where the deposited atom can be accommodated and of the frequency of collisions of the ion M^{z+} with this site. The Gibbs energy for the formation of a critical spherical nucleus with negligible adhesion energy can be described using the Kelvin equation for the Gibbs energy of a critical nucleus in the condensation of a supersaturated vapour:

$$G_\mathrm{n}^\ddagger = \frac{16\pi V_\mathrm{m}^2 \gamma^3}{3(RT \ln p/p_0)^2} \tag{5.8.10}$$

where γ is the surface energy, p is the pressure of the supersaturated vapour and p_0 is that of the saturated vapour. The Gibbs energy of supersaturation $RT \ln p/p_0$ can be replaced by the electrical overpotential energy $zF\eta$, yielding

$$G_\mathrm{n}^\ddagger = \frac{16\pi V_\mathrm{m}^2 \gamma^3}{3z^2F^2\eta^2} \tag{5.8.11}$$

The rate of three-dimensional nucleation is described in general by the experimentally verified dependence

$$\log \kappa = \frac{-k_1}{\eta^2} + k_2 \tag{5.8.12}$$

while that of two-dimensional nucleation by

$$\log \kappa = \frac{-k_1}{\eta} + k_2 \tag{5.8.13}$$

5.8.2 *Electrocrystallization on an identical metal substrate*

It was mentioned on page 306 (see Fig. 5.24) that, even at room temperature, a crystal plane contains steps and kinks (half-crystal positions). Kinks occur quite often—about one in ten atoms on a step is in the half-crystal position. Ad-atoms are also present in a certain concentration on the surface of the crystal; as they are uncharged species, their equilibrium concentration is independent of the electrode potential. The half-crystal position is of basic importance for the kinetics of metal deposition on an identical metal substrate. Two mechanisms can be present in the incorporation of atoms in steps, and thus for step propagation:

(a) The ion M^{z+} is reduced to an ad-atom that is transported to the step by surface diffusion and then rapidly incorporated in the half-crystal position.
(b) The electrode reaction occurs directly between the metal ion and the kink, without intermediate formation of ad-atoms.

In the first case, the rate of deposition depends on the equilibrium concentration of ad-atoms, on their diffusion coefficient, on the exchange current density and on the overpotential. In the second case, the rate of deposition is a function, besides of the geometric factors of the surface, of the exchange current and the overpotential. This mechanism is valid, for example, in the deposition of silver from a $AgNO_3$ solution.

Formation of subsequent layers can occur either in the way of formation of a single nucleus which then spreads undisturbed over the whole face or by formation of other nuclei before the face is completely covered (multinuclear multilayer deposition). Now let us discuss the first case in more detail.

The basic condition for experimental study of nucleation on an identical surface requires that this surface be a single crystal face without screw dislocations (page 306). Such a surface was obtained by Budevski *et al.* when silver was deposited in a narrow capillary. During subsequent deposition of silver layers the screw dislocations 'died out' so that finally a surface of required properties was obtained.

Further deposition of silver on such a surface is connected with a cathodic current randomly oscillating around the mean value

$$\langle \mathrm{I} \rangle = zF\kappa N_m A \tag{5.8.14}$$

where N_m is the number of atoms per unit surface and A the surface area of the electrode. The individual current pulses are shown in Fig. 5.47.

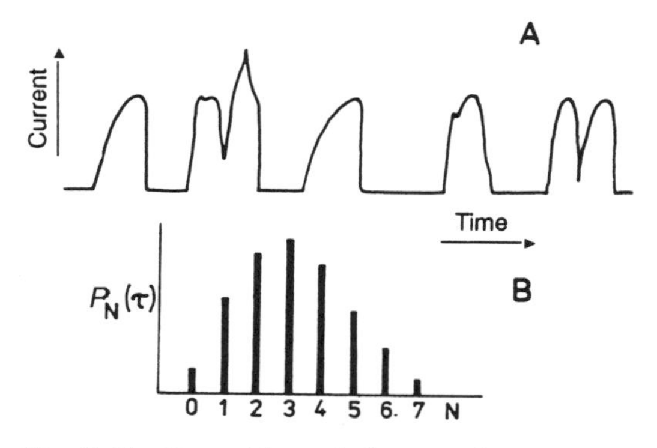

Fig. 5.47 Deposition of silver on (100) face of a silver electrode in absence of screw dislocations. A—time dependence of current pulses, B—probability distribution of the occurrences of pulses in the time interval τ

Obviously, the nucleation is a randon process which is amplified by subsequent deposition of many thousands of silver atoms before the surface is completely covered (if integrated over the time interval of monolayer formation the current in each pulse corresponds to an identical charge). Such an amplification of random processes is the only way they can be observed. This situation is quite analogous, for example, to radioactive decay where a single disintegration is followed, in a Geiger tube, by the flow of millions of electrons.†

In electrochemistry similar phenomena are observed, for example, with the formation of insoluble films on electrodes or with ion selective channel formation in bilayer lipid membranes or nerve cell membranes (pages 377 and 458).

At low overpotentials, the silver electrode prepared according to Budevski *et al.* behaves as an ideal polarized electrode. However, at an overpotential higher than -6 mV the already mentioned current pulses are observed (Fig. 5.48A). Their distribution in the time interval τ follows the Poisson relation for the probability that N nuclei are formed during the time interval τ

$$P(\tau) = [(\kappa\tau)^N/N!] \exp(-\kappa\tau) \tag{5.8.15}$$

The validity of the Poisson distribution for silver nucleation is demonstrated in Fig. 5.48B. The assumption for this kind of treatment is that the nucleus formation is irreversible and that the event is binary consisting of a discontinuous process (nucleus formation) and a continuous process (flow of

† Because of discontinuous microstructure of all natural systems all processes occurring in nature are essentially stochastic (random) and what we observe are only outlines of the phenomena like the average current value in Eq. (5.8.14).

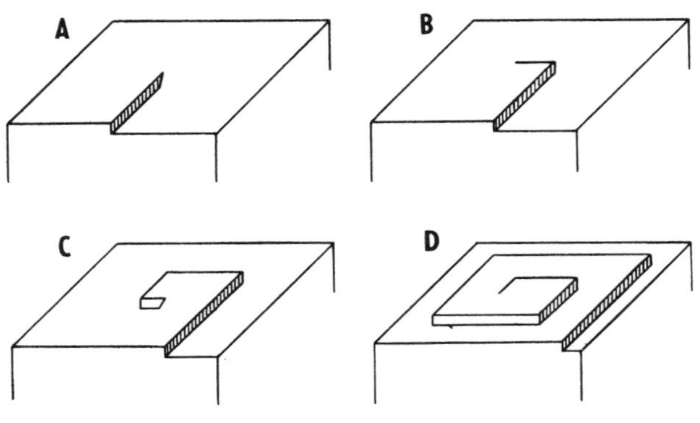

Fig. 5.48 A scheme of the spiral growth of a crystal. (According to R. Kaischew, E. Budevski and J. Malimovski)

time). In the case of silver deposition, the dependence of the nucleation rate constant κ on the overpotential η follows Eq. (5.8.13).

Another approach to stochastic processes of this kind is the observation of the system in equilibrium (for example, of an electrode in equilibrium with the potential determining system). The value of the electrode potential is not completely constant but it shows irregular small deviations from the average value E,

$$E(t) = E + x(t) \tag{5.8.16}$$

where E is the constant average value and $x(t)$ the fluctuating part, the *noise*. For the experimentalist the noise seems to be an unwelcome nuisance as he is interested in the 'deterministic' average value; however, the 'stochastic' data $x(t)$ can surprisingly bring a lot of useful information.

The noise can be described by means of two approaches, namely, in the time domain and in the frequency domain. For the time domain approach the characteristic value is the well-known mean square deviation from the value over the time T,

$$D = 1/T\int_0^T [x(t) - X]^2\, \mathrm{d}t \tag{5.8.17}$$

with $D^{1/2}$ = root mean square (rms).

The frequency domain description is based on the Fourier integral transformation of the signal in the time domain into the frequency domain,

$$X(\omega) = \int_{\infty}^{\infty} x(t)\exp(-j\omega t)\, \mathrm{d}t \tag{5.8.18}$$

By this transformation, an important set of values, the *power spectrum*,

$$S(\omega) = |X(\omega)|^2 \tag{5.8.19}$$

is obtained. The analysis of the power spectrum, the aim of which is, for example, to determine the characteristic constants of the processes from which the noise stems will not be discussed here and the reader is recommended to inspect the relevant papers listed on page 384.

The electrocrystallization on an identical metal substrate is the slowest process of this type. Faster processes which are also much more frequent, are connected with ubiquitous defects in the crystal lattice, in particular with the *screw dislocations* (Fig. 5.25). As a result of the helical structure of the defect, a monoatomic step originates from the point where the new dislocation line intersects the surface of the crystal face. It can be seen in Fig. 5.48 that the wedge-shaped step gradually fills up during electrocrystallization; after completion it slowly moves across the crystal face and winds up into a spiral. The resultant progressive spiral cannot disappear from the crystal surface and thus provides a sufficient number of growth

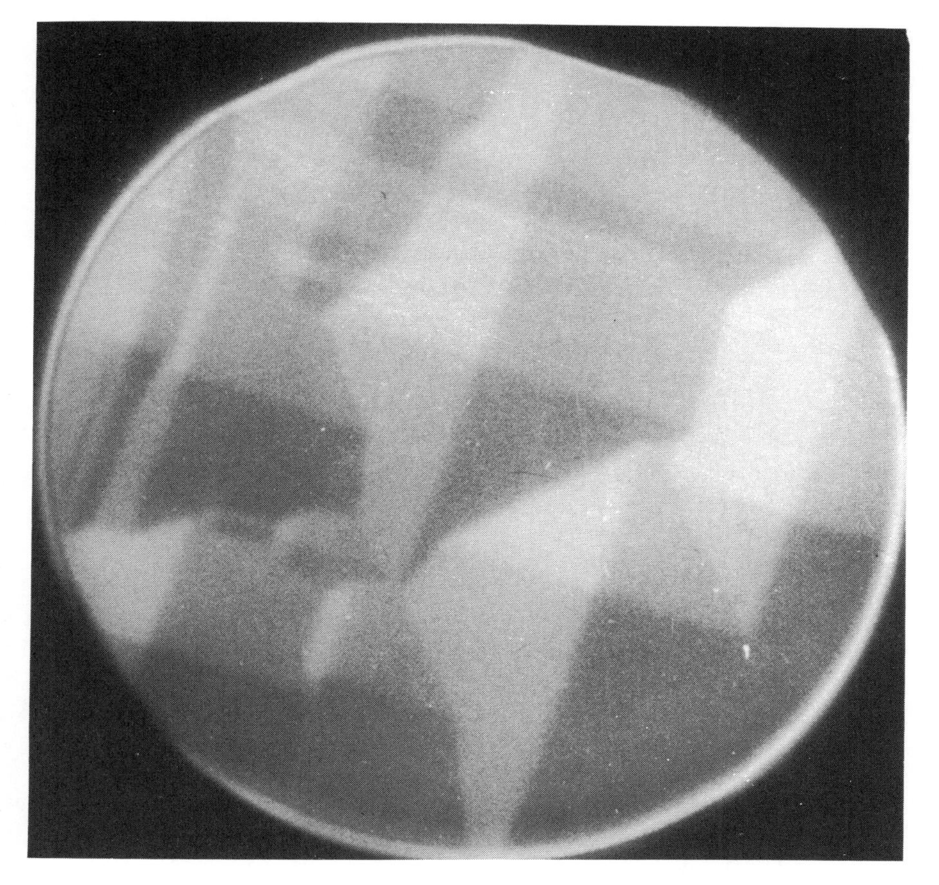

Fig. 5.49 A microphotograph of low pyramid formation during electrocrystallization of Ag on a single crystal Ag surface. (By courtesy of E. Budevski)

sites. Consequently, growth no longer requires nucleation and continues at a low overpotential. Repeated spiral growth forms low pyramids, with a steepness increasing with increasing overpotential. Figure 5.49 depicts a quadratic pyramid on an Ag(100) face. Macroscopic growth on a single crystal substrate is called *epitaxy*.

The kinetics of electrocrystallization conforms to the above description only under precisely defined conditions. The deposition of metals on polycrystalline materials again yields products with polycrystalline structure, consisting of *crystallites*. These are microscopic formations with the structure of a single crystal.

Whiskers are sometimes formed in solutions with high concentrations of surface-active substances. These are long single crystals, growing in only one direction, while growth in the remaining directions is retarded by adsorption of surface-active substances. Whiskers are characterized by quite

Fig. 5.50 Spiral growth of Cu from a 0.5 M $CuSO_4$ + 0.5 M H_2SO_4 electrolyte, 25°C, current density 15 $mA \cdot cm^{-2}$, magnification 1250×. (From H. Seiter, H. Fischer, and L. Albert, *Electrochim. Acta*, **2**, 97, 1960)

high strength. From solutions of low concentration metals are often deposited as *dendrites* at high current densities.

Macroscopic growth during electrocrystallization occurs through fast movements of steps, 10^{-4}–10^{-5} cm high, across the crystal face. Under certain conditions, spirals also appear, formed of steps with a height of a thousand or more atomic layers, so that they can be studied optically (Fig. 5.50).

5.8.3 *Anodic oxidation of metals*

The anodic dissolution of metals on surfaces without defects occurs in the half-crystal positions. Similarly to nucleation, the dissolution of metals involves the formation of empty nuclei (atomic vacancies). Screw dislocations have the same significance. Dissolution often leads to the formation of continuous crystal faces with lower Miller indices on the metal. This process, termed *facetting,* forms the basis of metallographic etching.

Anodic oxidation often involves the formation of films on the surface, i.e. of a solid phase formed of salts or complexes of the metals with solution components. They often appear in the potential region where the electrode, covered with the oxidation product, can function as an electrode of the second kind. Under these conditions the films are thermodynamically stable. On the other hand, films are sometimes formed which in view of their solubility product and the pH of the solution should not be stable. These films are stabilized by their structure or by the influence of surface forces at the interface.

Thus films can be divided into two groups according to their morphology. *Discontinuous films* are porous, have a low resistance and are formed at potentials close to the equilibrium potential of the corresponding electrode of the second kind. They often have substantial thickness (up to 1 mm). Films of this kind include halide films on copper, silver, lead and mercury, sulphate films on lead, iron and nickel; oxide films on cadmium, zinc and magnesium, etc. Because of their low resistance and the reversible electrode reactions of their formation and dissolution, these films are often very important for electrode systems in storage batteries.

Continuous (barrier, passivation) *films* have a high resistivity ($10^6 \Omega \cdot$ cm or more), with a maximum thickness of 10^{-4} cm. During their formation, the metal cation does not enter the solution, but rather oxidation occurs at the metal–film interface. Oxide films at tantalum, zirconium, aluminium and niobium are examples of these films.

If the electrode is covered with a film, then anodic oxidation of the metal does not involve facetting. The surface of the metal either becomes more rough (if the film is discontinuous) or becomes very lustrous (continuous films). Films, especially continuous films, retard the electrode reaction of metal oxidation. The metal is said to be in its *passive state.*

Film growth (and its cathodic dissolution) is controlled by similar laws as

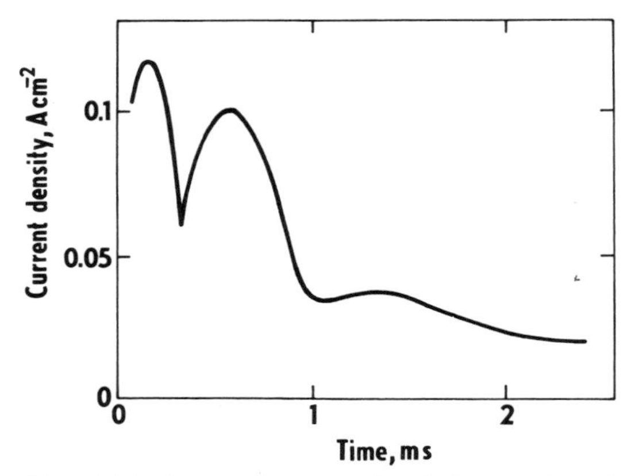

Fig. 5.51 Successive growth of three calomel layers at an overpotential of 40 mV on a mercury electrode. Electrolyte 1 M HCl. (According to A. Bewick and M. Fleischmann)

metal deposition, i.e. oxidation in half-crystal positions, nucleation and spiral growth, all playing analogous roles. An example is the kinetics of formation of a calomel film on mercury (Fig. 5.51). The time dependence of the current obeys approximately the relationship for the progressive formation of two-dimensional nuclei (see Eq. (5.8.7) and the following text on page 370). However, further nucleation and the formation of a second and a third layer occur on a partly formed monolayer. Dendrites are sometimes formed during macroscopic growth of oxide layers (Fig. 5.52). This phenomenon is particularly unfavourable in batteries.

Nucleation in continuous films sometimes occurs at less positive potentials than those for metal dissolution (Al, Ta) and metal is thus always in a passive state. Again, on other metals (Fe, Ni) nuclei are formed at more positive potentials than those corresponding to anodic dissolution. The potential curve for an iron electrode in 1 M H_2SO_4 at an increasingly positive potential has the shape depicted in Fig. 5.53 (the Fe–Fe^{3+} curve). The electrode is first in an active state where dissolution occurs. An oxide film begins to form at the potentials of the peak on the polarization curve (called the Flade potential, E_F) and the electrode becomes passivated; a small corrosion current, j_{corr}, then passes through the electrode. Oxygen is formed at even more positive potentials. The Flade potential depends on the pH of the solution and on admixtures in the iron electrode. With increasing chromium content, the Flade potential is shifted to more negative values and the current at the peak decreases. This retardation of the oxidation of the electrode leads to the anticorrosion properties of alloys of iron and chromium.

The rate-controlling step in the growth of a multimolecular film is the transport of metal ions from the solution through the film. Diffusion and

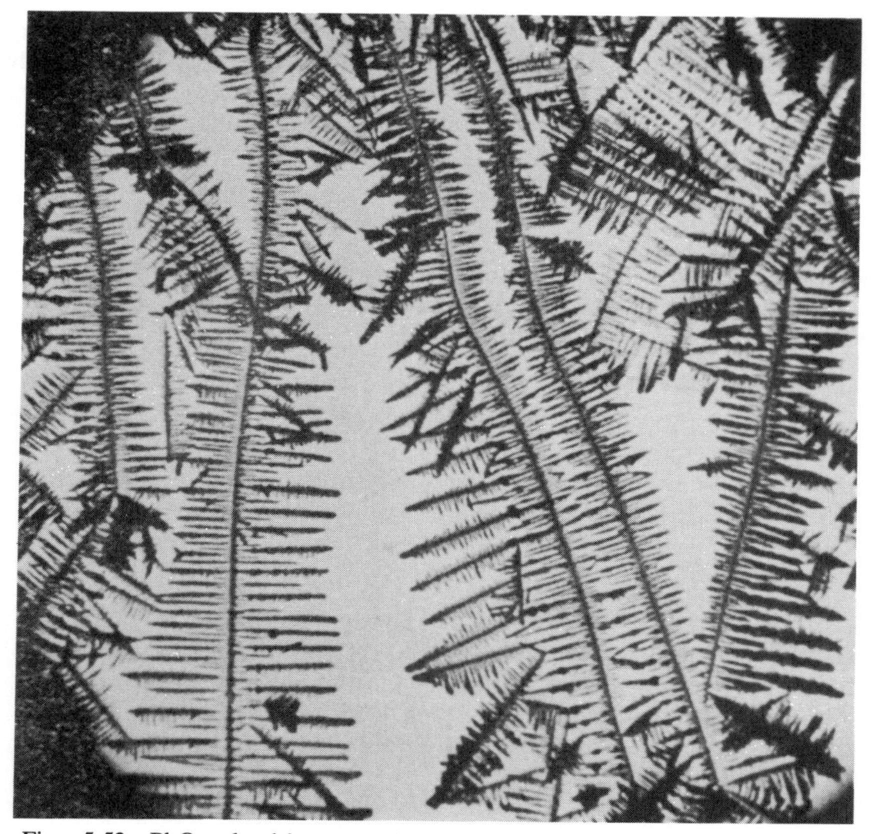

Fig. 5.52 PbO_2 dendrite growth at a lead electrode. Current density $0.3\ mA \cdot cm^{-2}$. (From G. Wranglén, *Electrochim. Acta*, **2**, 130, 1960)

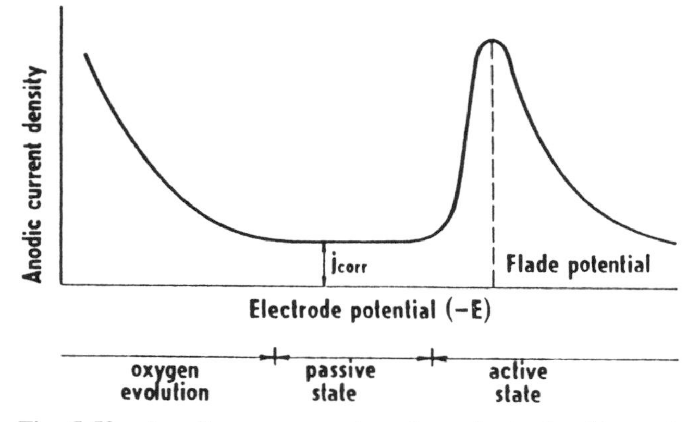

Fig. 5.53 Anodic processes at an iron electrode. (According to K. J. Vetter)

migration control this transport when the electric field in the film is not very strong. If the electrochemical potential of an ion in the film in the steady state decreases with increasing distance from the surface of the electrode, then the current passing through the film (cf. Eq. 2.5.23) is given as

$$j = k_1 \frac{\tilde{\mu}_1 - \tilde{\mu}_2}{l} \tag{5.8.20}$$

where $\tilde{\mu}_1$ is the electrochemical potential of the ion at the film–electrolyte interface, $\tilde{\mu}_2$ is the value at the metal–film interface and l is the thickness of the film. If the film is insoluble then

$$j = k_2 \frac{\mathrm{d}l}{\mathrm{d}t} \tag{5.8.21}$$

Solution of the differential equation obtained by combining Eqs (5.8.20) and (5.8.21) yields the *Wagner parabolic law* of film growth,

$$l^2 = kt \tag{5.8.22}$$

A strong electric field is formed in very thin films (with a thickness of about 10^{-5} cm) during current flow. If the average electrochemical potential difference between two neighbouring ions in the lattice is comparable with their energy of thermal motion, kT, then Ohm's law is no longer valid for charge transport in the film. Verwey, Cabrera, and Mott developed a theory of ion transport for this case.

If the energy of an ion in the lattice is a periodic function of the interionic distance, then the minima of this function correspond to stable positions in the lattice and the maxima to the most unstable positions. The maxima are situated approximately half-way between two neighbouring stable positions. If there is an electric field in the film, $\Delta\phi/l$, where $\Delta\phi$ is the electric potential difference between the two edges of the film, then a potential energy difference of $zFa\Delta\phi/l$ is formed between two neighbouring ions. In these expressions, z is the charge number and a is the distance between two neighbouring ions. The energies corresponding to the transfer of an ion from two neighbouring rest positions to the energy maximum are the activation energies for transport of an ion in the direction and against the direction of the field. Analogous considerations to those employed in the derivation of the basic relationship for electrode kinetics (Section 5.2) yield the relationship for the current density, given by the difference between the rates of these two processes:

$$j = k\left\{\exp\left(\frac{\alpha zFa}{RTl}\Delta\phi\right) - \exp\left[-\frac{(1-\alpha)zFa}{RTl}\Delta\phi\right]\right\} \tag{5.8.23}$$

In very strong fields, where the electric energy of the ion is much larger than the energy of thermal motion ($zFa\Delta\phi/l \gg RT$), then

$$j = k \exp\left(\frac{\alpha zFa}{RTl}\Delta\phi\right) \tag{5.8.24}$$

which is the *exponential law of film growth*. On the other hand, if $zFa\Delta\phi/l \ll RT$, then, on expanding the exponentials, we obtain the equation

$$j = k\frac{zFa}{RTl}\Delta\phi \tag{5.8.25}$$

i.e. Ohm's law. The significance of the constant k follows directly from Eq. (5.8.25):

$$k = \frac{RT\kappa}{zFa} \tag{5.8.26}$$

where κ is the conductivity of the film.

5.8.4 *Mixed potentials and corrosion phenomena*

As demonstrated in Section 5.2, the electrode potential is determined by the rates of two opposing electrode reactions. The reactant in one of these reactions is always identical with the product of the other. However, the electrode potential can be determined by two electrode reactions that have nothing in common. For example, the dissolution of zinc in a mineral acid involves the evolution of hydrogen on the zinc surface with simultaneous ionization of zinc, where the divalent zinc ions diffuse away from the electrode. The sum of the partial currents corresponding to these two processes must equal zero (if the charging current for a change in the electrode potential is neglected). The potential attained by the metal under these conditions is termed the *mixed potential*, E_{mix}. If the polarization curves for both processes are known, then conditions can be determined such that the absolute values of the cathodic and anodic currents are identical (see Fig. 5.54A). The rate of dissolution of zinc is proportional to the partial anodic current.

In contrast to the equilibrium electrode potential, the mixed potential is given by a non-equilibrium state of two different electrode processes and is accompanied by a spontaneous change in the system. Besides an electrode reaction, the rate-controlling step of one of these processes can be a transport process. For example, in the dissolution of mercury in nitric acid, the cathodic process is the reduction of nitric acid to nitrous acid and the anodic process is the ionization of mercury. The anodic process is controlled by the transport of mercuric ions from the electrode; this process is accelerated, for example, by stirring (see Fig. 5.54B), resulting in a shift of the mixed potential to a more negative value, E'_{mix}.

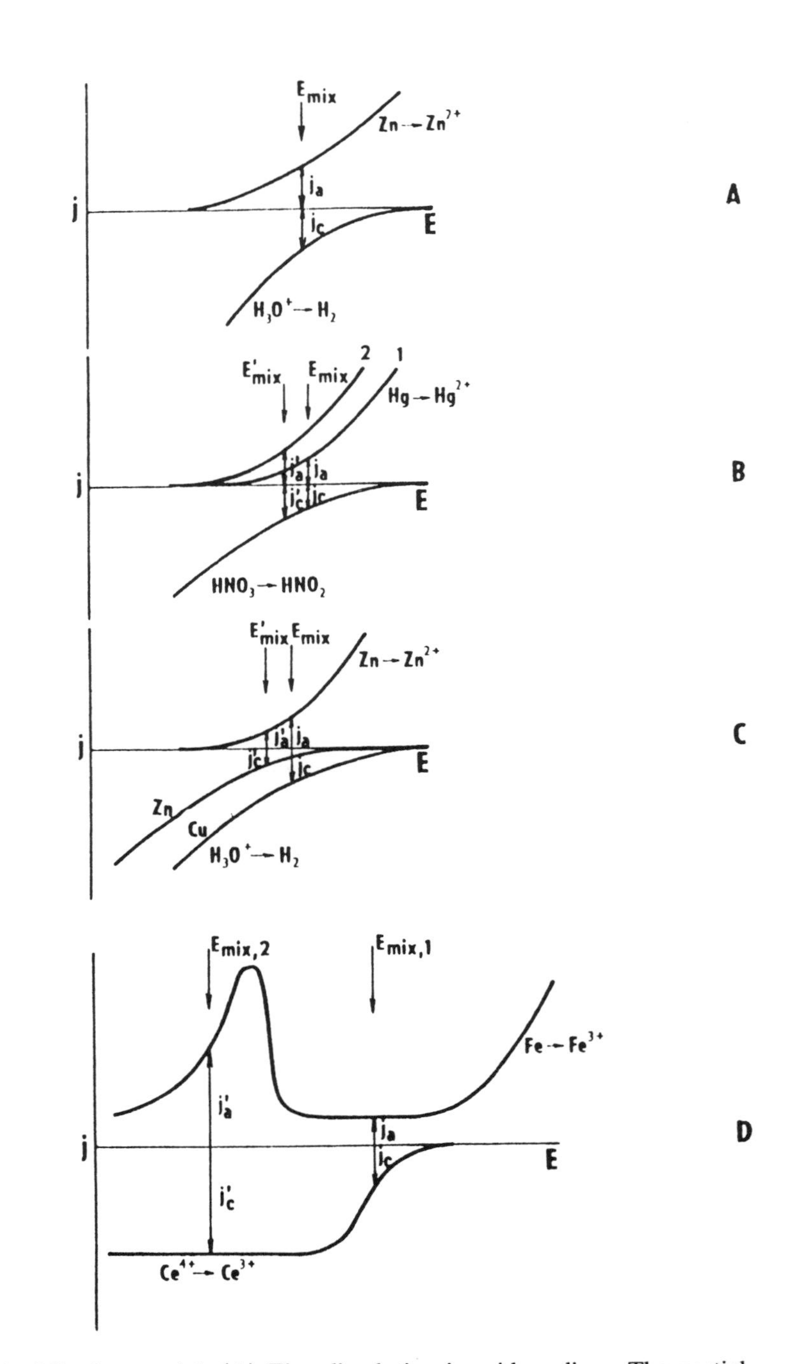

Fig. 5.54 Mixed potential. (A) Zinc dissolution in acid medium. The partial processes are indicated at the corresponding voltammograms. (B) Dissolution of mercury in nitric acid solution. The original dissolution rate characterized by (1) the corrosion current j_a is enhanced by (2) stirring which causes an

Processes associated with two opposing electrode processes of a different nature, where the anodic process is the oxidation of a metal, are termed *electrochemical corrosion processes.* In the two above-mentioned cases, the surface of the metal phase is formed of a single metal, i.e. corrosion occurs on a chemically homogeneous surface. The fact that, for example, the surface of zinc is physically heterogeneous and that dissolution occurs according to the mechanism described in Section 5.8.3 is of secondary importance.

The dissolution of zinc in a mineral acid is much faster when the zinc contains an admixture of copper. This is because the surface of the metal contains copper crystallites at which hydrogen evolution occurs with a much lower overpotential than at zinc (see Fig. 5.54C). The mixed potential is shifted to a more positive value, E'_{mix}, and the corrosion current increases. In this case the cathodic and anodic processes occur on separate surfaces. This phenomenon is termed corrosion of a chemically heterogeneous surface. In the solution an electric current flows between the cathodic and anodic domains which represent short-circuited electrodes of a galvanic cell. A. de la Rive assumed this to be the only kind of corrosion, calling these systems *local cells.*

An example of the process of a passivating metal is the reaction of tetravalent cerium with iron (see Fig. 5.54D). Iron that has not been previously passivated dissolves in an acid solution containing tetravalent cerium ions, in an active state at a potential of $E_{mix,2}$. After previous passivation, the rate of corrosion is governed by the corrosion current j_a and the potential assumes a value of $E_{mix,1}$.

For the corrosion phenomena which are of practical interest, the cathodic processes of reduction of oxygen and hydrogen ions are of fundamental importance, together with the structure of the metallic material, which is often covered by oxide layers whose composition and thickness depend on time. The latter factor especially often prevents a quantitative prediction of the rate of corrosion of a tested material.

Electrochemical corrosion processes also include a number of processes in organic chemistry, involving the reduction of various compounds by metals or metal amalgams. A typical example is the electrochemical carbonization of fluoropolymers mentioned on p. 316. These processes, that are often described as 'purely chemical' reductions, can be explained relatively easily on the basis of diagrams of the anodic and cathodic polarization curves of the type shown in Fig. 5.54.

Fig. 5.54 (*caption continued*)
increase of the transport rate of mercuric ions from the metal surface. (C) Effect of copper admixture on zinc dissolution. The presence of copper patches on zinc surface increases the rate of hydrogen evolution. (D) Oxidation of iron in the solution of Ce^{4+}. On passivated iron the oxidation rate is low (j_a) while iron in active state dissolves rapidly (j'_a). According to K. J. Vetter

A mixed potential can also be established in processes that do not involve metal dissolution. For example, the potential of a platinum electrode in a solution of permanganate and manganese(II) ions does not depend on the concentration of divalent manganese. Anodic oxidation of manganese(II) ions does not occur, but rather anodic evolution of oxygen. The kinetics of this process and the reduction of permanganate ions determine the value of the resulting mixed potential.

References

Bewick, A., and M. Fleischmann, Formation of surface compounds on electrodes, in *Topics in Surface Chemistry* (Eds E. Kay and P. S. Bagres), p. 45, Plenum Press, New York, 1978.

Bezegh, A., and J. Janata, Information from noise, *Anal. Chem.,* **59,** 494A (1987).

Budevski, E., Deposition and dissolution of metals and alloys, Part A, Electrocrystallization, *CTE,* **7,** 399 (1983).

De Levie, R., Electrochemical observation of single molecular events, *AE,* **13,** 1 (1985).

Despić, A. R., Deposition and dissolution of metals and alloys, Part B, Mechanism, kinetics, texture and morphology, *CTE,* **7,** 451 (1983).

Fleischmann, M., M. Labram, C. Gabrielli, and A. Sattar, The measurement and interpretation of stochastic effects in electrochemistry and bioelectrochemistry, *Surface Science,* **101,** 583 (1980).

Fleischmann, M., and H. R. Thirsk, Electrocrystallization and metal deposition, *AE,* **3,** 123 (1963).

Froment, M. (Ed.), *Passivity of Metals and Semiconductors,* Elsevier, Amsterdam, 1983.

Gilman, S., see page 368.

Kokkinidis, G., Underpotential deposition and electrocatalysis, *J. Electroanal. Chem.,* **201,** 217 (1986).

McCafferty, E., and R. J. Broid (Eds), *Surfaces, Inhibition and Passivation,* Ellis Horwood, Chichester, 1989.

Pospíšil, L., Random processes in electrochemistry, in J. Koryta *et al., Contemporaneous Trends in Electrochemistry* (in Czech), Academia, Prague, 1986, p. 14.

Štefec, R., *Corrosion Data from Polarization Measurements,* Ellis Horwood, Chichester, 1990.

Uhlig, H. H., *Corrosion and Corrosion Control,* John Wiley & Sons, New York, 1973.

Vermilyea, D. A., Anodic films, *AE,* **3,** 211 (1963).

Vetter, K. J., see page 353.

Young, L., *Anodic Films,* Academic Press, New York, 1961.

5.9 Organic Electrochemistry

As this broad subject has been treated in a great many monographs and surveys and falls largely in the domain of organic chemistry, the treatment here will be confined to several examples.

The tendency of an organic substance to undergo electrochemical reduction or oxidation depends on the presence of electrochemically active groups in the molecule, on the solvent, on the type of electrolyte and especially on the nature of the electrode. While oxidation processes at the mercury electrode are limited to hydroquinones and related substances, radical intermediates formed by cathodic reduction (e.g. in the potentiostatic double-pulse method) and some substances forming complexes with mercury, almost all organic substances can be oxidized under suitable conditions at a platinum electrode (e.g. saturated aliphatic hydrocarbons in quite concentrated solutions of sulphuric or phosphoric acid at elevated temperatures).

The reduction of organic substances requires the presence of a π-electron system, a suitable substituent (e.g. a halogen atom), an electron gap or an unpaired electron in the molecule.

The electrode reaction of an organic substance that does not occur through electrocatalysis begins with the acceptance of a single electron (for reduction) or the loss of an electron (for oxidation). However, the substance need not react in the form predominating in solution, but, for example, in a protonated form. The radical formed can further accept or lose another electron or can react with the solvent, with the base electrolyte (this term is used here rather than the term 'indifferent' electrolyte) or with another molecule of the electroactive substance or a radical product. These processes include substitution, addition, elimination, or dimerization reactions. In the reactions of the intermediates in an anodic process, the reaction partner is usually nucleophilic in nature, while the intermediate in a cathodic process reacts with an electrophilic partner.

According to G. J. Hoijtink, aromatic hydrocarbons are usually reduced in aprotic solvents in two steps:

$$\begin{aligned} R + e &\rightleftarrows \dot{R}^- \\ \dot{R}^- + e &\rightleftarrows R^{2-} \end{aligned} \tag{5.9.1}$$

The first step is so fast that it can hardly be measured experimentally, while the second step is much slower (probably as a result of the repulsion of negatively charged species, R^- and R^{2-}, in the negatively charged diffuse electric layer). The reduction of an isolated benzene ring is very difficult and can occur only indirectly with solvated electrons formed by emission from the electrode into solvents such as some amines (see Section 1.2.3). This is a completely different mechanism than the usual interaction of electrons from the electrode with an electroactive substance.

In the presence of a proton donor, such as water, radical anions accept a proton and the reaction scheme assumes the form

$$\begin{aligned} R + e &\rightleftarrows \dot{R}^- \\ \dot{R}^- + H^+ &\rightleftarrows \dot{R}H \\ \dot{R}H + e &\rightleftarrows RH^- \\ RH^- + H^+ &\rightleftarrows RH_2 \end{aligned} \tag{5.9.2}$$

In the oxidation of aromatic substances at the anode, radical cations or dications are formed as intermediates and subsequently react with the solvent or with anions of the base electrolyte. For example, depending on the conditions, 1,4-dimethoxybenzene is cyanized after the substitution of one methoxy group, methoxylated after addition of two methoxy groups or acetoxylated after substitution of one hydrogen on the aromatic ring, as shown in Fig. 5.55, where the solvent is indicated over the arrow and the base electrolyte and electrode under the arrow for each reaction; HAc denotes acetic acid.

J. Heyrovský and K. Holleck and B. Kastening pointed out that the reduction of aromatic nitrocompounds is characterized by a fast one-electron step, e.g.

$$\text{—NO}_2 + e \rightleftarrows \text{—}\dot{\text{N}}\text{O}_2^- \tag{5.9.3}$$

This reduction step can be readily observed at a mercury electrode in an aprotic solvent or even in aqueous medium at an electrode covered with a suitable surfactant. However, in the absence of a surface-active substance, nitrobenzene is reduced in aqueous media in a four-electron wave, as the first step (Eq. 5.9.3) is followed by fast electrochemical and chemical reactions yielding phenylhydroxylamine. At even more negative potentials phenylhydroxylamine is further reduced to aniline. The same process occurs at lead and zinc electrodes, where phenylhydroxylamine can even be oxidized to yield nitrobenzene again. At electrodes such as platinum, nickel or iron, where chemisorption bonds can be formed with the products of the

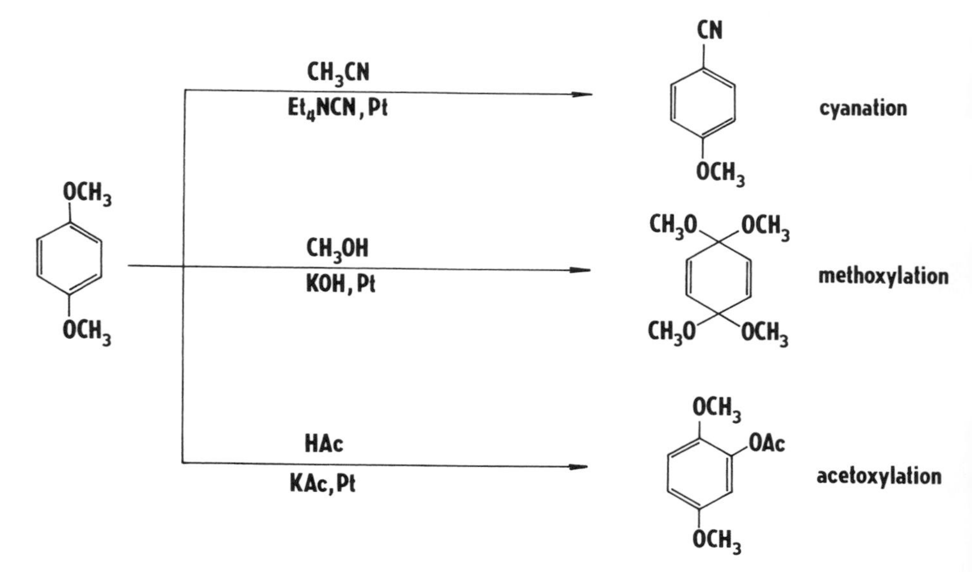

Fig. 5.55 Different paths of anodic oxidation dependent on electrolyte composition. (According to L. E. Eberson and N. L. Weinberg)

reduction of nitrobenzene, azoxybenzene and hydrazobenzene are produced as a result of interactions among the adsorbed intermediates. If azobenzene is reduced in acid medium to hydrazobenzene, then the product rearranges in a subsequent chemical reaction in solution to give benzidine.

An example of dimerization of the intermediates of an electrode reaction is provided by the reduction of acrylonitrile in a sufficiently concentrated aqueous solution of tetraethylammonium *p*-toluene sulphonate at a mercury or lead electrode. The intermediate in the reaction is probably the dianion

$$CH_2{=}CHCN + 2e \rightarrow \overset{(-)}{C}H_2{-}\overset{(-)}{C}HCN \tag{5.9.4}$$

which reacts in the vicinity of the electrode with another acrylonitrile molecule

$$\begin{aligned} &\overset{(-)}{C}H_2{-}\overset{(-)}{C}HCN + CH_2{=}CHCN \rightarrow NC\overset{(-)}{C}HCH_2CH_2\overset{(-)}{C}HCN \\ &NC\overset{(-)}{C}HCH_2CH_2\overset{(-)}{C}HCN + 2H_2O \rightarrow NC(CH_2)_4CN + 2OH^- \end{aligned} \tag{5.9.5}$$

finally to form adiponitrile (an important material in the synthesis of Nylon 606).

Carbonyl compounds are reduced to alcohols, hydrocarbons or pinacols (cf., for example, Eq. 5.1.8), where the result of the electrode process depends on the electrode potential.

The electrocatalytic oxidation of methanol was discussed on page 364. The extensively studied oxidation of simple organic substances is markedly dependent on the type of crystal face of the electrode material, as indicated in Fig. 5.56 for the oxidation of formic acid at a platinum electrode.

M. Faraday was the first to observe an electrocatalytic process, in 1834, when he discovered that a new compound, ethane, is formed in the electrolysis of alkali metal acetates (this is probably the first example of electrochemical synthesis). This process was later named the *Kolbe reaction,* as Kolbe discovered in 1849 that this is a general phenomenon for fatty acids (except for formic acid) and their salts at higher concentrations. If these electrolytes are electrolysed with a platinum or irridium anode, oxygen evolution ceases in the potential interval between +2.1 and +2.2 V and a hydrocarbon is formed according to the equation

$$2RCOO^- \rightarrow R_2 + 2CO_2 + 2e \tag{5.9.6}$$

In addition to hydrocarbons, other products have also been found, especially in the reactions of the higher fatty acids. In steady state, the current density obeys the Tafel equation with a high value of constant $b \approx 0.5$. At a constant potential the current usually does not depend very much on the sort of acid. The fact that the evolution of oxygen ceases in the

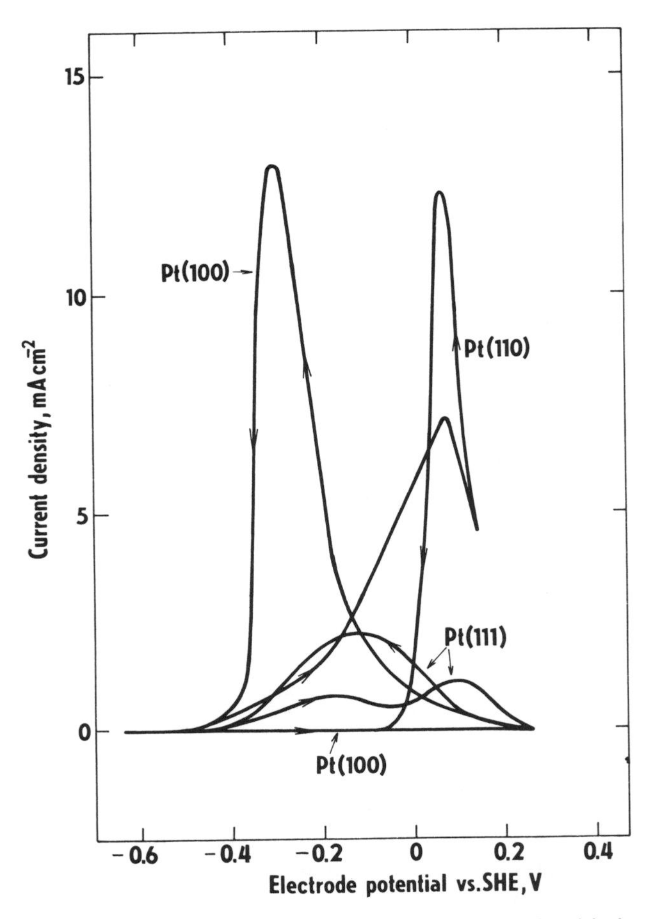

Fig. 5.56 Triangular-pulse voltammograms of oxidation of 0.1 M formic acid in 0.5 M H_2SO_4 at various faces of a single-crystal platinum electrode, 22°C, $dE/dt = 50\ mV \cdot s^{-1}$. (According to C. Lamy)

potential range where the Kolbe reaction proceeds can be explained in two ways. The anion of the fatty acid or an oxidation intermediate may block the surface of the electrode (cf. page 361), or as a second alternative, the acid may react with an intermediate of anodic oxygen evolution, which then cannot occur. The mechanism of this interesting process has not yet been completely elucidated; the two main hypotheses are illustrated in the following schemes:

$$\begin{aligned} RCOO^- &\rightarrow RCOO_{ads}^- \\ RCOO_{ads}^- &\rightarrow R\dot{C}OO_{ads} + e \\ R\dot{C}OO_{ads} &\rightarrow \dot{R}_{ads} + CO_2 \\ \dot{R}_{ads} &\rightarrow \text{products, for example } R_2 \end{aligned} \tag{5.9.7}$$

and

$$
\begin{array}{ccc}
 & & \text{products, for example } R_2 \\
 & & \uparrow \\
RCOO^- & \longrightarrow & \dot{R} + CO_2 + e \\
\downarrow & & \\
RCOO^+ + 2e & \longrightarrow & R^+ + CO_2 \\
 & & \downarrow \\
 & & \text{further products}
\end{array}
\tag{5.9.8}
$$

Similarly as for other chemical reactions, the activation energy for the electrode reactions of organic substances depends on the nature and position of the substituents on the aromatic ring of an organic molecule. As the standard potentials are not known for many of these mostly irreversible processes and as they were mainly investigated by polarography, their basic data is the half-wave potential (Eq. 5.4.27). The difference of the half-wave potential of the tested substance, *i,* and of the reference substance (which is, for example, non-substituted) for the same type of electrode reaction (e.g. nitrogroup reduction) is given by the equation

$$E_{1/2,i} - E_{1/2,\mathrm{ref}} = \Delta E_{1/2} = \Delta E^{0\prime} + \frac{RT}{\alpha nF} \ln \frac{k_i^{\ominus}}{k_{\mathrm{ref}}^{\ominus}} \tag{5.9.9}$$

We assume that neither the preexponential factor of the conditional electrode reaction rate constant nor the charge transfer coefficient changes markedly in a series of substituted derivatives and that the diffusion coefficients are approximately equal. In view of (5.2.52) and (5.2.53),

$$\Delta E_{1,2} = \Delta E^{0\prime} - \frac{\Delta\Delta H_0^{\ddagger}}{\alpha nF} \tag{5.9.10}$$

where $\Delta\Delta H_0^{\ddagger}$ is the difference between the activation energies of the electrode reactions of substance *i* and of the reference substance. R. W. Brockman and D. E. Pearson, S. Koide and R. Motoyama and P. Zuman applied the *Hammett equation,* originally deduced for chemical reactions, to electrode processes. In this concept the activation energy or standard Gibbs energy of a reaction where members of a homologous series participate is given by the relationship

$$
\begin{aligned}
\Delta\Delta H_0^{\ddagger} &= \rho_i \sigma_k \\
\Delta\Delta G^{0\prime} &= zFE^{0\prime} = \rho_j \sigma_k
\end{aligned}
\tag{5.9.11}
$$

where ρ_j and ρ_j' characterize the type of the reaction and σ_k the substituent. The values of σ_k which are the same for all reaction types have been listed by L. Hammett. Thus, for the series of half-wave potentials of the same electrode reaction type we have

$$\Delta E_{1/2} = \rho_j'' \sigma_k \tag{5.9.12}$$

where ρ_j'' characterizes the electrode reaction type (reduction of the nitro, aldehydic, chlorine, etc., group) and σ_k the substituent (mainly in the para or meta position on the benzene ring).

References

Baizer, M. and H. Lund (Eds), *Organic Electrochemistry,* M. Dekker, New York, 1983.

Fleischmann, M., and D. Pletcher, Organic electrosynthesis, *Roy. Inst. Chem. Rev.,* **2,** 87 (1969).

Fry, A. J., *Synthetic Organic Electrochemistry,* John Wiley & Sons, Chichester, 1989.

Fry, A. J., and W. E. Britton (Eds), *Topics in Organic Electrochemistry,* Plenum Press, New York, 1986.

Kyriacou, D. K., and D. A. Jannakoudakis, *Electrocatalysis for Organic Synthesis,* Wiley–Interscience, New York, 1986.

Lamy, C., see page 368.

Peover, M. E., Electrochemistry of aromatic hydrocarbons and related substances, in *Electroanalytical Chemistry* (Ed. A. J. Bard), Vol. 2, p. 1, M. Dekker, New York, 1967.

Shono, T., *Electroorganic Chemistry as a New Tool in Organic Synthesis,* Springer-Verlag, Berlin, 1984.

Torii, S., *Electroorganic Syntheses, Methods and Applications,* Kodansha, Tokyo, 1985.

Weinberg, N. L. (Ed.), *Techniques of Electroorganic Synthesis,* Wiley–Interscience, New York, Part 1, 1974, Part 2, 1975.

Weinberg, N. L., and H. R. Weinberg, Oxidation of organic substances, *Chem. Rev.,* **68,** 449 (1968).

Yoshida, K., *Electrooxidation in Organic Chemistry, The Role of Cation Radicals as Synthetic Intermediates,* John Wiley & Sons, New York, 1984.

Zuman, P., *Substitution Effects in Organic Polarography,* Plenum Press, New York, 1967.

5.10 Photoelectrochemistry

5.10.1 *Classification of photoelectrochemical phenomena*

Photoelectrochemistry studies the effects occurring in electrochemical systems under the influence of light in the visible through ultraviolet region. Light quanta supply an extra energy to the system, hence the electrochemical reactions, which are thermodynamically or kinetically suppressed in the dark, may proceed at a high rate under illumination. (There also exists an opposite effect, where the (dark) electrochemical reactions lead to highly energetic products which are able to emit electromagnetic radiation. This is the principle of 'electrochemically generated luminescence', mentioned in Section 5.5.6.) Two groups of photoelectrochemical effects are traditionally distinguished: *photogalvanic* and *photovoltaic.*

The *photogalvanic effect* is based on light absorption by a suitable photoactive redox species (dye) in the electrolyte solution. The photoexcited dye subsequently reacts with an electron donor or acceptor process, taking place in the vicinity of an electrode, is linked to the electrode reaction which restores the original form of the dye. The cycle is terminated by a counterelectrode reaction in the non-illuminated compartment of the cell. Both electrode reactions may (but need not) regenerate the reactants consumed at the opposite electrodes. The photopotential and photocurrent appear essentially as a result of concentration gradients introduced by an asymmetric illumination of the photoactive electrolyte solution between two inert electrodes. Therefore, any photogalvanic cell can, in principle, be considered as a concentration cell.

The *photovoltaic effect* is initiated by light absorption in the electrode material. This is practically important only with semiconductor electrodes, where the photogenerated, excited electrons or holes may, under certain conditions, react with electrolyte redox systems. The photoredox reaction at the illuminated semiconductor thus drives the complementary (dark) reaction at the counterelectrode, which again may (but need not) regenerate the reactant consumed at the photoelectrode. The regenerative mode of operation is, according to the IUPAC recommendation, denoted as 'photovoltaic cell' and the second one as 'photoelectrolytic cell'. Alternative classification and terms will be discussed below.

The term 'photovoltaic effect' is further used to denote non-electrochemical photoprocesses in solid-state metal/semiconductor interfaces (Schottky barrier contacts) and semiconductor/semiconductor (p/n) junctions. Analogously, the term 'photogalvanic effect' is used more generally to denote any photoexcitation of the d.c. current in a material (e.g. in solid ferroelectrics). Although confusion is not usual, electrochemical reactions initiated by light absorption in electrolyte solutions should be termed 'electrochemical photogalvanic effect', and reactions at photoexcited semiconductor electodes 'electrochemical photovoltaic effect'.

The boundary between effects thus defined is, however, not sharp. If, for instance, light is absorbed by a layer of a photoactive adsorbate attached to the semiconductor electrode, it is apparently difficult to identify the light-absorbing medium as a 'solution' or 'electrode material'. Photoexcited solution molecules may sometimes also react at the photoexcited semiconductor electrode; this process is labelled *photogalvanovoltaic effect.*

The electrochemical photovoltaic effect was discovered in 1839 by A. E. Becquerel† when a silver/silver halide electrode was irradiated in a solution of diluted HNO_3. Becquerel also first described the photogalvanic effect in a cell consisting of two Pt electrodes, one immersed in aqueous and the other in ethanolic solution of $Fe(ClO_4)_3$. This discovery was made about the same time as the observation of the photovoltaic effect at the Ag/AgX electrodes. The term 'Becquerel effect' often appears in the old literature, even for denoting the vacuum photoelectric effect which was discovered almost 50 years later. The electrochemical photovoltaic effect was elucidated in 1955 by W. H. Brattain and G. G. B. Garrett; the theory was further developed

† Father of Antoine-Henri Becquerel, the famous French physicist and discoverer of radioactivity.

in detail by H. Gerischer. The photogalvanic effect was first interpreted in 1940 by E. Rabinowitch, and more recently studied by W. J. Albery, M. D. Archer and others.

Besides the classification of photoelectrochemical effects (cells) according to the nature of the light-absorbing system, we can also distinguish them according to the effective Gibbs free energy change in the electrolyte, ΔG:

1. $\Delta G = 0$: *Regenerative photoelectrochemical cell.* The reactants consumed during the photoredox process (either in homogeneous phase or at the semiconductor photoelectrode) are immediately regenerated by subsequent electrode reactions. The net output of the cell is an electrical current flowing through the cell and an external circuit, but no permanent chemical change in the electrolyte takes place.
2. $\Delta G \neq 0$: *Photoelectrosynthetic cell.* The products of the homogeneous or heterogeneous photoredox process and the corresponding counter-electrode reactions are sufficiently stable to be collected for later use. Practically interesting is the endoergic photoelectrosynthesis ($\Delta G > 0$), producing species with stored chemical energy (fuel). The most important example is the splitting of water to oxygen and hydrogen. The exoergic process ($\Delta G < 0$) is also called *photocatalytic,* since the light provides only the activation energy for the reaction, which is otherwise thermodynamically possible even in the dark. Although no chemical energy is stored, useful products can sometimes result; an example is the synthesis of NH_3 from N_2 and H_2.

5.10.2 *Electrochemical photoemission*

In contrast to semiconductors, metals are usually not active as photoelectrodes owing to the fast thermal recombination of photogenerated charge carriers in the metal. The only light-stimulated charge transfer process at the electrolyte/metal electrode interface is *photoemission* of electrons. It is, in principle, similar to the conventional photoelectric effect in vacuum, i.e. it takes place under illumination with photons of sufficiently short wavelengths to overcome the binding energy of electron in the metal. Photoelectrons, emitted from the metal electrode into the electrolyte solution, are rapidly solvated and subsequently captured by a reducible solution species. The electrochemical photoemission was studied mostly at mercury electrode, first by M. Heyrovský. The dependence of the photoemission current on the radiation frequency differs from that of a conventional vacuum photoemission. Gurevich *et al.* deduced the relationship:

$$I_c = A(h\nu - h\nu_0 - eE)^{5/2} \tag{5.10.1}$$

where I_c is the photocurrent, A is a constant, $h\nu$ is the energy of the light quantum, $h\nu_0$ is the threshold energy for photoemission at zero charge potential, e is the electron charge, and E is the electrode potential.

5.10.3 *Homogeneous photoredox reactions and photogalvanic effects*

The photogalvanic effects are initiated by a homogeneous photoredox reaction of an electrolyte redox system with a suitable photoexcited organic or organometallic substance (dye), S. The photon absorption produces a short-lived, electronically excited dye molecule, S^*:

$$S + h\nu \rightarrow S^*(\xrightarrow{k_b} S)$$

whose structure and physicochemical properties are different. The conversion of S^* back to S proceeds by various radiative and non-radiative relaxation processes, which are characterized by a pseudo-first-order rate constant, k_b. Its reciprocal value defines the *lifetime of the excited state,* τ_0:

$$\tau_0 = 1/k_b \qquad (5.10.2)$$

(the fraction of the S^* molecules, deactivated during the time τ_0, equals $1 - e^{-1} \simeq 63\%$). Only excitations of electrons from the highest occupied molecular orbital (HOMO) to the lowest unoccupied molecular orbital (LUMO), which produce reasonably long-lived S^* molecules are practically important. This is schematically depicted in Fig. 5.57.

Most of the ground-state molecules contain all the electrons paired. (The O_2 molecule is an important exception to this rule.) The excited molecule may occur in two states, either with none or two unpaired electrons; these states are called excited singlet, S_1, or excited triplet, T_1. The latter state is, according to the Pauli rule, energetically favoured and its lifetime is higher by orders of magnitude, since the reactions with singlet molecules as well as radiative relaxation (phosphorescence) are spin-forbidden.

The photoexcited $S^*(T_1)$ molecule may have a sufficiently long lifetime to diffuse some distance in the electrolyte solution and to take part in a

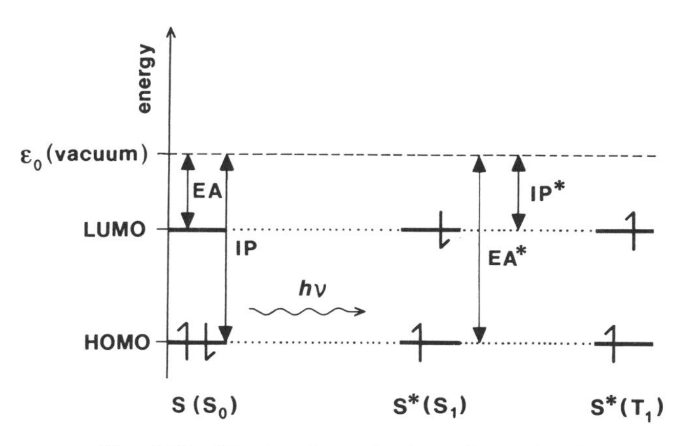

Fig. 5.57 Electronic excitation of a molecule

bimolecular homogeneous reaction with a molecule acting as electron donor (D) or acceptor (A). In the first case, we speak about *reductive quenching*:

$$S^* + D \rightarrow S^- + D^+ \tag{5.10.3}$$

and in the second case, about *oxidative quenching*:

$$S^* + A \rightarrow S^+ + A^- \tag{5.10.4}$$

Photoredox reactions involving excited singlet molecules, $S^*(S_1)$ are relatively rare owing to short lifetimes τ_0; one exception is the molecule of oxonine (oxygen analogue of thionine, see below) possessing relatively long singlet lifetime (2.4 ns) and also small quantum yield of a non-radiative $S_1 \rightarrow T_1$ conversion (intersystem crossing), about $3 \cdot 10^{-3}$. Oxonin, therefore, reacts photochemically, e.g. with Fe^{2+} *via* the excited singlet state.

Figure 5.57 demonstrates that the photoexcited molecule S^* exhibits, in both spin states, a lower ionization potential (IP*) and a higher electron affinity (EA*) in comparison with that of the ground state S molecule (IP, EA). Consequently, the photoexcited molecule S^* is a better oxidizing as well as reducing agent than the ground state molecule, S. Neglecting the entropy and pressure changes, we can express the relative change of the standard redox potential of the S/S^- couple as follows:

$$E^0(S/S^-) - E^0(S^*/S^-) = -(\Delta G - \Delta G^*)/F \simeq -\varepsilon_{ex}/F \tag{5.10.5}$$

and analogously:

$$E^0(S^+/S) - E^0(S^+/S^*) \simeq \varepsilon_{ex}/F \tag{5.10.6}$$

ε_{ex} is the excitation energy of S (mostly of the first triplet state, $S^*(T_1)$). The actual changes of the redox potentials introduced by photoexcitation are dramatic (cf. Table 5.6). Thus, the photoexcited molecule S^* is a powerful redox reactant to drive various unusual redox reactions in the electrolyte.

Table 5.6 Properties of three typical photoredox-active molecules. bpy denotes 2,2′ bipyridine, TMPP is tetra N-methylpyridine porphyrin; λ_{max} is the wavelength of the absorption maximum, ε is the extinction coefficient at λ_{max}, φ_T is the quantum yield of the formation of the excited triplet state, τ_0 is its lifetime, and E_0 are standard redox potentials

	Thionine[a]	$Ru(bpy)_3^{2+}$	$ZnTMPP^+$
λ_{max}	599	452	560
ε [cm^2 μmol^{-1}]	56	14	16
φ_T	0.55	1.0	0.90
τ_0 [μs]	7.7	0.6	655
E_0 (S/S^-) [V]	0.192	−1.28	—
E_0 (S^+/S) [V]	—	1.26	1.2
E_0 (S^*/S^-) [V]	1.4	0.84	—
E_0 (S^+/S^*) [V]	—	−0.86	−0.6

[a] pH = 2 (redox potentials of the system thionine/semithionine).

The participation of S^* molecules in homogeneous redox processes is mainly controlled by their short lifetimes. The lifetime of S^* is shortened by a bimolecular redox reaction (5.10.3) or (5.10.4) as follows:

$$\tau = 1/(k_b + k_r c) \tag{5.10.7}$$

(k_r is a rate constant of the quenching reaction and c is a concentration of the reactant D or A). Equation (5.10.7) can be rearranged to a more practical form:

$$\tau_0/\tau = 1 - K_{SV}c \tag{5.10.8}$$

where $K_{SV} = \tau_0 k_r$ is the so-called Stern–Volmer constant.

Figure 5.58 presents a general scheme of photogalvanic cells based on reductive (a) or oxidative (b) quenching of S^*. These cells are regenerative if D and A are two forms of the same redox couple ($D^+ = A$ and $D = A^-$), otherwise the products D^+ and A^- accumulate in the electrolyte and the cell is photoelectrosynthetic.

Some properties of three typical photoredox-active dyes are summarized in Table 5.6. The most important photogalvanic system is based on the photoredox reaction of thionine with Fe^{2+}. It was first studied by E. Rabinowitch in 1940 and recently optimized by J. W. Albery *et al.* The structural formula of thionine, Thi^+ and its reduction products (Sem^+, Leu^{2+}) are as follows:

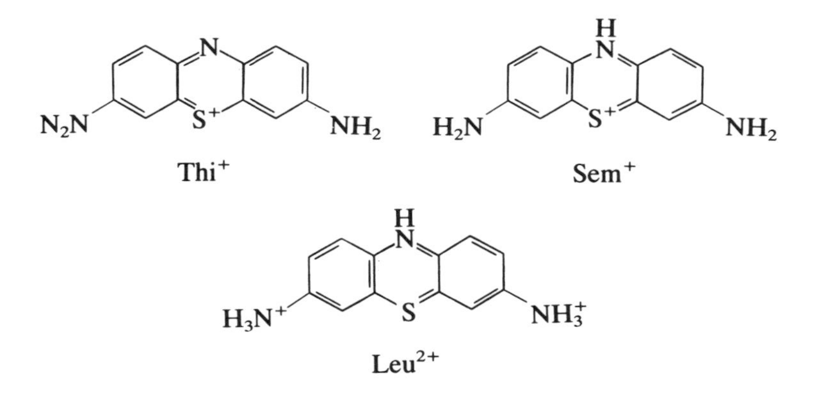

The photoexcited thionone, $^*Thi^+$ is reduced by Fe^{2+} to the unstable semithionine (Sem^+) which further disproportionates to Thi^+ and the doubly reduced form of thionine, leucothionine (Leu^{2+}):

$$^*Thi + Fe^{2+} + H^+ \rightarrow Sem^+ + Fe^{3+} \tag{5.10.9}$$

$$2Sem^+ + H^+ \rightarrow Thi^+ + Leu^{2+} \tag{5.10.10}$$

The anodic reaction (taking place in the illuminated cell compartment) is a two-electron process regenerating the initial form of Thi^+:

$$Leu^{2+} \rightarrow Thi^+ + 2e^- + 3H^+ \tag{5.10.11}$$

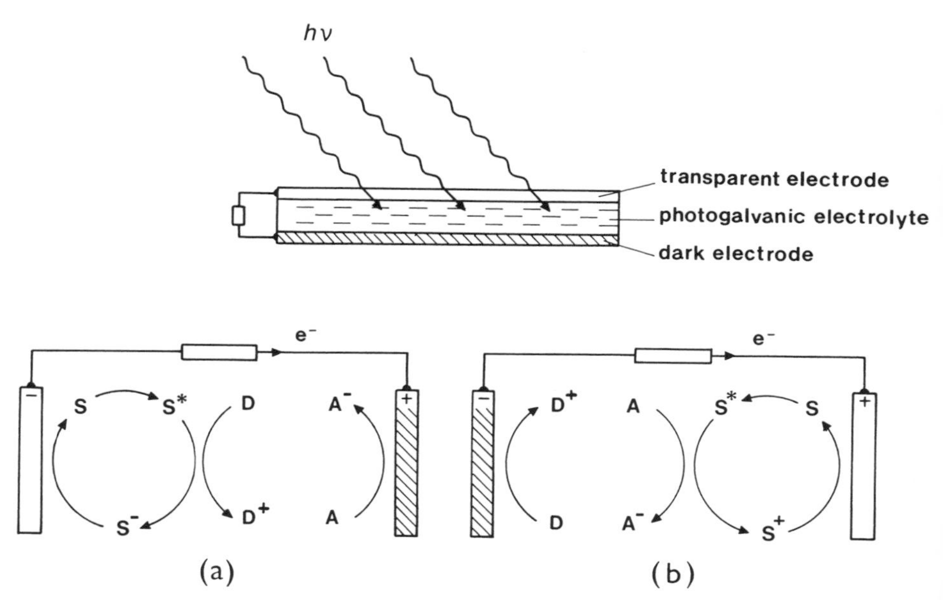

Fig. 5.58 Scheme of a photogalvanic cell. The homogeneous photoredox process takes place in the vicinity of the optically transparent anode (a) or cathode (b)

The two-electron oxidation of the dye is not very common; other dyes usually undergo one-electron redox reactions. The cathodic reaction (taking place in the non-illuminated cell compartment) regenerates the complementary redox system:

$$Fe^{3+} + e^- \rightarrow Fe^{2+} \tag{5.10.12}$$

According to the reactions (5.10.9)–(5.10.12), we can classify this cell as a regenerative ($\Delta G = 0$) photogalvanic cell based on reductive quenching (Fig. 5.58a). Good performance of the iron–thionine cell is conditioned by the generation of $^*Thi^+$ (and Leu^{2+}) in close vicinity (below *ca.* 1 μm) of the anode; this can be achieved by using an optically transparent (e.g. SnO_2) anode as a window (Fig. 5.58). The selectivity of the anode and cathode for the desired reactions (5.9.24) and (5.9.25) may be further increased by selecting the proper electrode materials or by chemical modification of the anode.

A regenerative photogalvanic cell with oxidative quenching (Fig. 5.58b) is based, for example, on the Fe^{3+}–$Ru(bpy)_3^{2+}$ system. In contrast to the iron–thionine cell, the homogeneous photoredox process takes place near the (optically transparent) cathode. The photoexcited $^*Ru(bpy)_3^{2+}$ ion reduces Fe^{3+} and the formed $Ru(bpy)_3^{3+}$ and Fe^{2+} are converted at the opposite electrodes to the initial state.

More complicated photogalvanic cells may employ two light absorbing systems: one in the cathodic process and the other in the anodic process. An interesting example is a cell with two different chlorophyll-based absorbers. The cathodic process starts by photoexcitation of a modified natural chloroplast with deactivated photosystem II (PS II—for terminology of photosynthetic processes see page 469). The remaining PS I reduces, for example, anthraquinone-2-sulphonate. The regenerative cycle of PS I is linked to tetramethyl-*p*-phenylenediamine which is cycled at the cathode. The complementary anodic process is O_2 evolution by the action of a second chloroplast system with active PS II; electrons from photoexcited PS II are transferred to the tetramethyl-*p*-phenylenediamine and the cycle is terminated by its oxidation at the anode. The whole cell resembles photosynthesis, the net output being O_2 and reduced anthraquinone-2-sulphonate (cf. Section 6.5.2).

The efficiency of the photocurrent generation in practical photogalvanic cells is generally low, since it is limited by short lifetimes of the excited dyes, parasitic electron transfer reactions, etc.

The importance of the photogalvanic effect rests in basic electrochemical research of homogeneous photoredox reactions. To this purpose, original experimental techniques have been developed, e.g. voltammetry with an optically transparent rotating disk electrode. The photoexcitation of reactants makes the study of redox reactions possible in a considerably broader region of the Gibbs free energy changes than with the ground-state reactants. This brings interesting conclusions, for example, for the formulation of a more general theory of electron transfer, preparation of compounds in less-common oxidation states, etc.

5.10.4 *Semiconductor photoelectrochemistry and photovoltaic effects*

Basic properties of semiconductors and phenomena occurring at the semiconductor/electrolyte interface in the dark have already been discussed in Sections 2.4.1 and 4.5.1. The crucial effect after immersing the semiconductor electrode into an electrolyte solution is the equilibration of electrochemical potentials of electrons in both phases. In order to quantify the dark- and photoeffects at the semiconductor/electrolyte interface, a common reference level of electron energies in both phases has to be defined.

Let us choose, as an arbitrary reference level, the energy of an electron at rest in vacuum, ε_0 (cf. Section 3.1.2). This reference energy is obvious in studies of the solid phase, but for the liquid phase, the Trasatti's conception of 'absolute electrode potentials' (Section 3.1.5) has to be adopted. The formal energy levels of the electrolyte redox systems, $\varepsilon_{\mathrm{REDOX}}$, referred to ε_0, are given by the relationship:

$$\varepsilon_{\mathrm{REDOX}} = -eE^0_{\mathrm{REDOX}} - kT \ln (c_{\mathrm{ox}}/c_{\mathrm{red}}) + \varepsilon_{\mathrm{SHE}} \qquad (5.10.13)$$

where E^0_{REDOX} is the standard redox potential of the particular redox system and c_{ox}, c_{red} the concentrations of the oxidized and reduced forms, respectively. ε_{SHE} is the energy level of the standard hydrogen electrode referred to ε_0 (see Section 3.1.5).

The energy levels of several redox couples as well as the positions of the band edges of typical semiconductors in equilibrium with aqueous electrolyte (pH 7) are compared in Fig. 5.59.

The electrochemical potential of an electron in a solid defines the Fermi energy (cf. Eq. 3.1.9). The Fermi energy of a semiconductor electrode (ε_F^0) and the electrolyte energy level (ε_{REDOX}^0) are generally different before contact of both phases (Fig. 5.60a). After immersing the semiconductor electrode into the electrolyte, an equilibrium is attained:

$$\varepsilon_F = \varepsilon_{REDOX} \tag{5.10.14}$$

The equilibration proceeds by electron transfer between the semiconductor and the electrolyte. The solution levels are almost intact ($\varepsilon_{REDOX} \simeq \varepsilon_{REDOX}^0$), since the number of transferred electrons is negligible relative to the number of the redox system molecules (c_{ox} and c_{red}). On the other hand, the energy levels of the semiconductor phase may shift considerably. The region close to the interface is depleted of majority charge carriers and the energy bands are bent upwards or downwards as depicted in Fig. 5.60b.

The equilibrium potential between the bulk of the semiconductor and the bulk of the electrolyte (Galvani potential, $\Delta_s^{sc}\phi$) is expressed by Eq. (4.5.6).

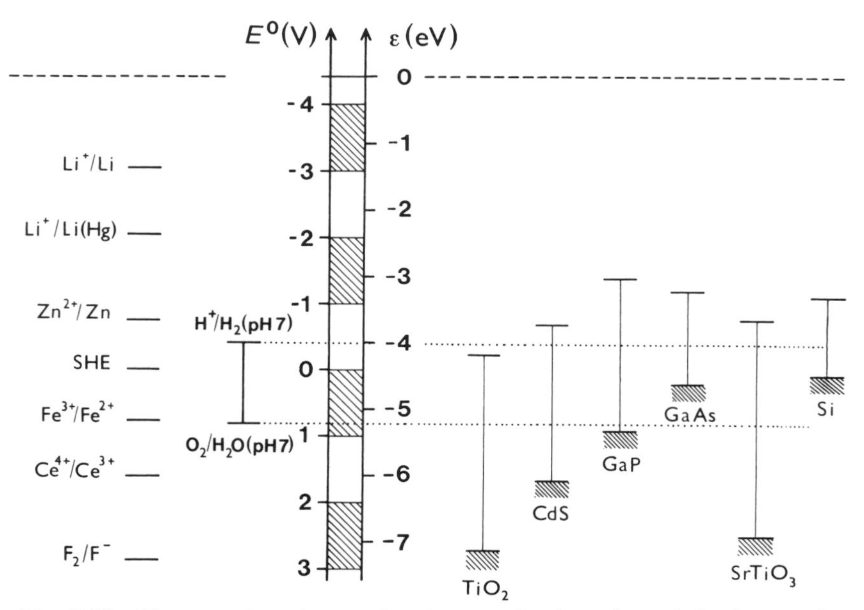

Fig. 5.59 The energies of several redox couples (ε_{redox}), and the lower edge of the conduction band (ε_c) and the upper edge of the valence band (ε_v) of typical semiconductors. The positions of band edges are given for semiconductors in equilibrium with aqueous electrolyte at pH 7

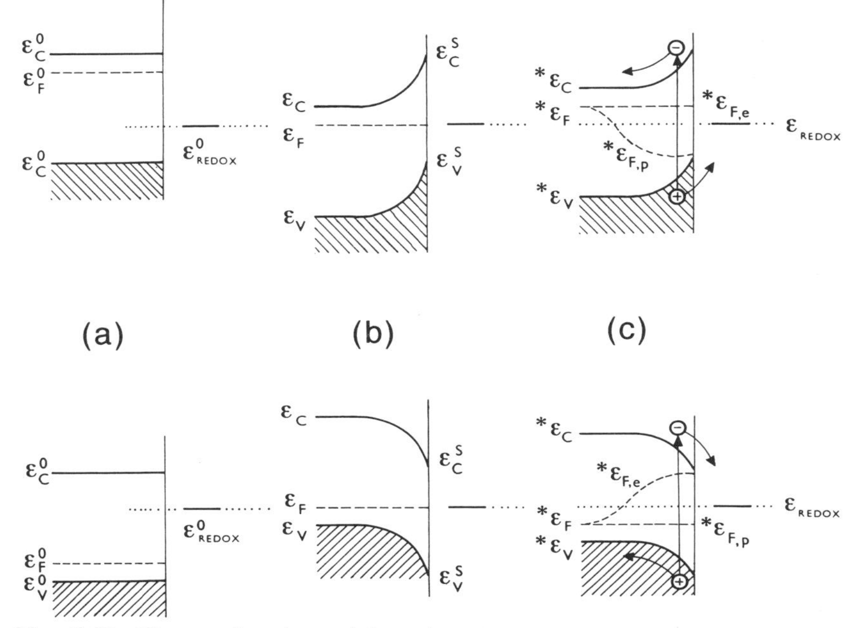

Fig. 5.60 The semiconductor/electrolyte interface (a) before equilibration with the electrolyte, (b) after equilibration with the electrolyte in the dark, and (c) after illumination. The upper part depicts the n-semiconductor and the lower the p-semiconductor

This potential is related to the electrode potential, E (apart from an additive constant depending on the definition of the reference potential). The first term in Eq. (4.5.6), $\Delta_s^{sc}\phi$ characterizes the band bending

$$e\Delta_s^{sc}\phi = \varepsilon_c^s - \varepsilon_c = \varepsilon_v^s - \varepsilon_v \tag{5.10.15}$$

where ε_c^s and ε_c are the energies of the lower edge of the conduction band at the surface and in the bulk, respectively, and ε_v^s and ε_v are analogous energies of the upper edge of the valence band. The positions of ε_c^s and ε_v^s are determined by the Helmholtz electric layer, hence they remain fixed as long as $\Delta_2^s\phi$ is not changed. External changes of the electrode potential, E, manifest themselves as changes of $\Delta_s^{sc}\phi$, at a certain electrode potential (the flatband potential, E_{fb}), the band bending disappears, $\Delta_s^{sc}\phi = 0$.

For a more detailed description of the semiconductor/electrolyte interface, it is convenient to define the quasi-Fermi levels of electrons, $\varepsilon_{F,e}$ and holes, $\varepsilon_{F,p}$,

$$\varepsilon_{F,e}(x) = \varepsilon_c(x) + kT \ln [n_e(x)/(N_c - n_e(x))] \tag{5.10.16}$$

and analogously

$$\varepsilon_{F,p}(x) = \varepsilon_v(x) - kT \ln [(n_p(x)/(N_v - n_p(x))] \tag{5.10.17}$$

N_c and N_v are the effective densities of electronic states in the corresponding band edges which can be expressed as follows:

$$N_c = (2\pi m_e kT/h^2)^{3/2}; \qquad N_v = (2\pi m_p kT/h^2)^{3/2} \tag{5.10.18}$$

m_e and m_p are the effective masses of electron and hole, respectively; $n_e(x)$ and $n_p(x)$ are the carrier concentrations defined by Eqs (4.5.1), (4.5.2) and (4.5.8). In equilibrium, both quasi-Fermi levels are equal:

$$\varepsilon_{F,e}(x) = \varepsilon_{F,p}(x) = \varepsilon_F \tag{5.10.19}$$

for any $x > 0$. Combining Eqs (5.10.16)–(5.10.19) and assuming $n_e \ll N_c$ and $n_p \ll N_v$, we can easily derive the equilibrium equations (4.5.8) and (4.5.9).

Upon illumination, photons having energy higher than the band gap ($\varepsilon_g = \varepsilon_c - \varepsilon_v$) are absorbed in the semiconductor phase and the electron-hole-pairs (e^-/h^+) are generated. This effect can be considered equivalent to the photoexcitation of a molecule (Fig. 5.57) if we formally identify the HOMO with the ε_c level and LUMO with the ε_v level. The lifetime of excited e^-/h^+ pairs (in the bulk semiconductor) is defined analogously as the lifetime of the excited molecule in terms of a pseudo-first-order relaxation (Eq. 5.10.2).

The undesired recombination of the photogenerated e^-/h^+ pairs is considerably suppressed if the charge carriers are photogenerated in the space charge region or in its close vicinity accessible by diffusion of excited electrons or holes during their lifetime. In this case, electrons and holes are separated by the electric field of the space charge as depicted in Fig. 5.60c. The charge separation is apparently conditioned by the band bending, hence the potential $\Delta_1^{sc}\phi$ must be more positive (n-semiconductors) or more negative (p-semiconductors) than the flatband potential E_{fb}.

Under illumination, the electrons and holes within the space charge layer are no more in equilibrium defined by Eq. (5.10.19), i.e. the levels $^*\varepsilon_{F,e}(x)$ and $^*\varepsilon_{F,p}(x)$ are different in this region. The concentration of the minority carriers near the surface is increased dramatically in comparison with the equilibrium value, hence the quasi-Fermi level of the minority carriers shows complicated dependence on x (Fig. 5.60c). On the other hand, the quasi-Fermi level of the majority charge carriers remains practically flat and is only shifted relative to ε_F. This shift occurs generally towards the initial value ε_F^0, i.e. the band bending decreases (Fig. 5.60c). The potential change (ΔE_{ph}) corresponding to the shift of the bulk Fermi level $^*\varepsilon_F$ relative to ε_F is called *photopotential.* It can simply be determined as a difference of the potentials measured by using an ohmic contact at the backside of the illuminated (E_{LIGHT}) and non-illuminated (E_{DARK}) semiconductor electrode:

$$\Delta E_{ph} = (1/e)\,|\varepsilon_F^* - \varepsilon_F| = |E_{LIGHT} - E_{DARK}| \tag{5.10.20}$$

The photopotential can also be expressed in terms of the relative change in the minority carrier density, Δn_{min}, introduced by photoexcitation. By using

Eqns (5.10.16), (5.10.17) and (5.10.20) we obtain

$$\Delta E_{\mathrm{ph}} \simeq (kT/e) \ln (1 + \Delta n_{\mathrm{min}}/n^0_{\mathrm{min}}) \qquad (5.10.21)$$

The magnitude of the photopotential is also related to light intensity and to the value of E_{DARK}. The value of E_{LIGHT} cannot obviously exceed the flatband potential. At $E_{\mathrm{DARK}} = E_{\mathrm{fb}}$, the *photopotential* drops to zero which can be used for a simple measurement of the flatband potential.

The photopontential also approaches to zero when the semiconductor photoelectrode is short-circuited to a metal counterelectrode at which a fast reaction (injection of the majority carriers into the electrolyte) takes place. The corresponding photocurrent density is defined as a difference between the current densities under illumination, j_{LIGHT} and in the dark, j_{DARK}:

$$j_{\mathrm{ph}} = |j_{\mathrm{LIGHT}} - j_{\mathrm{DARK}}| \qquad (5.10.22)$$

The photocurrent density (j_{ph}) is proportional to the light intensity, but almost independent of the electrode potential, provided that the band bending is sufficiently large to prevent recombination. At potentials close to the flatband potential, the photocurrent density again drops to zero. A typical current density-voltage characteristics of an n-semiconductor electrode in the dark and upon illumination is shown in Fig. 5.61. If the electrode reactions are slow, and/or if the e^-/h^+ recombination *via* impurities or surface states takes place, more complicated curves for j_{LIGHT} result.

The photocurrent is cathodic or anodic depending on the sign of the minority charge carriers injected from the semiconductor electrode into the electrolyte, i.e. the n-semiconductor electrode behaves as a photoanode and

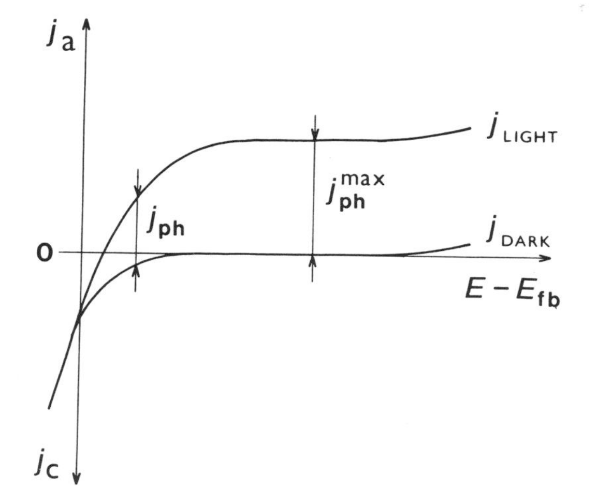

Fig. 5.61 Current density–potential characteristics of n-semiconductor electrode in the dark and upon illumination

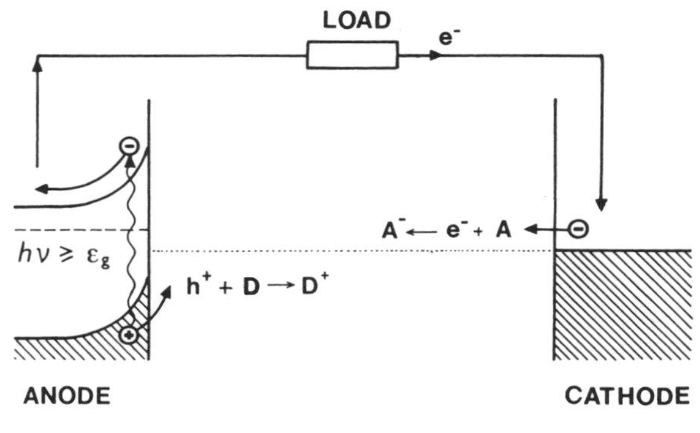

Fig. 5.62 Scheme of a photovoltaic cell with n-semiconductor photoanode

the p-semiconductor electrode as a photocathode. The most common photovoltaic cells are based on the combination of a semiconductor photoanode and a metal cathode (Fig. 5.62). Cells with a metal anode and a semiconductor photocathode or with a semiconductor photoanode and a semiconductor photocathode have, however, also been described.

The potential which controls the photoelectrochemical reaction is generally not the photopotential defined by Eqs (5.10.20) and (5.10.21) (except for the very special case where the values of $^*\varepsilon_v$, ε_{REDOX} and the initial Fermi energy of the counterelectrode are equal). The energy which drives the photoelectrochemical reaction, ε_R can be expressed, for example, for an n-semiconductor electrode as

$$\varepsilon_R = \varepsilon_g - e\Delta_s^{sc}\phi - (\varepsilon_c - \varepsilon_F) \tag{5.10.23}$$

This energy is consumed to overcome the overpotentials of the electrode reactions (η_a, η_c), the IR drop in the external circuit and electrolyte, and it might partly be converted into the free energy, ΔG, of stable products (if any) of the endoergic electrochemical reaction:

$$\varepsilon_R = e(\eta_a + \eta_c + IR) + \Delta G/N_A \tag{5.10.24}$$

The electrochemical cell can again be of the regenerative or electrosynthetic type, as with the photogalvanic cells described above. In the regenerative photovoltaic cell, the electron donor (D) and acceptor (A) (see Fig. 5.62) are two redox forms of one reversible redox couple, e.g. $Fe(CN)_6^{3-/4-}$, I_2/I^-, Br_2/Br^-, S^{2-}/S_x^{2-}, etc. the cell reaction is cyclic ($\Delta G = 0$, cf. Eq. (5.10.24) since $D^+ = A$ and $D = A^-$). On the other hand, in the electrosynthetic cell, the half-cell reactions are irreversible and the products (D^+ and A^-) accumulate in the electrolyte. The most carefully studied reaction of this type is photoelectrolysis of water ($D^+ = O_2$ and $A^- = H_2$). Other photoelectrosynthetic studies include the preparation of $S_2O_8^{2-}$, the reduction of CO_2 to formic acid, N_2 to NH_3, etc.

The photoelectrolysis of water was first studied in 1971 by A. Fujisima and K. Honda, who used a n-TiO_2 photoanode and a Pt counterelectrode. From the band positions and the formal energy levels of the H^+/H_2 and O_2/H_2O couples (Fig. 5.59), it is, nevertheless, apparent that the cell:

$$\text{n-TiO}_2 \,|\text{aqueous electrolyte}|\, \text{Pt} \tag{5.10.25}$$

does not provide sufficient voltage for photoelectrolysis of water (1.229 V) since the energy level of hydrogen electrode at pH 7 is about 0.2 V above the conduction band edge (this difference remains constant also at other pH values since both energy levels show the same Nernstian shift with pH). The missing voltage has, therefore, to be supplied either from an external source (*photoassisted electrolysis* of water) or by immersing the electrodes into electrolytes of different pH values (photoanode in alkaline and cathode in acidic electrolyte, the electrolytes being separated by a suitable membrane).

The direct photoelectrolysis of water requires that the ε_v level be below the O_2/H_2O level and the ε_c level be above the H^+/H_2 level. This condition is satisfied, e.g. for CdS, GaP, and several large-band gap semiconductors, such as $SrTiO_3$, $KTaO_3$, Nb_2O_5 and ZrO_2 (cf. also Fig. 5.59). From the practical points of view, these materials show, however, other specific problems, e.g. low electrocatalytic activity, sensitivity to photocorrosion (CdS, GaP), and inconvenient absorption spectrum (oxides).

The world-wide effort in photo(electro)chemical splitting of water started in the 1970s. It was motivated by two factors: (1) the oil crisis which stimulated the research in the solar energy conversion, and (2) the optimism connected with the 'hydrogen economy'. At present, however, the conversion of solar energy *via* photo(electro)lysis of water as well as the use of hydrogen as a medium for the energy storage and transport do not seem to compete seriously with the currently available energy resources.

5.10.5 *Sensitization of semiconductor electrodes*

The band-gap excitation of semiconductor electrodes brings two practical problems for photoelectrochemical solar energy conversion: (1) Most of the useful semiconductors have relatively wide band gaps, hence they can be excited only by ultraviolet radiation, whose proportion in the solar spectrum is rather low. (2) the photogenerated minority charge carriers in these semiconductors possess a high oxidative or reductive power to cause a rapid photocorrosion.

Both these disadvantages can be overcome by sensitization. In this case, the light is not absorbed in the semiconductor phase, but in an organic or organometallic photoredox active molecule (sensitizer, S), whose excitation energy is lower than ε_g. The sensitizer might either be dissolved in the electrolyte or adsorbed at the semiconductor surface. If the redox levels of the sensitizer in the ground and excited states are properly positioned relative to the ε_c and ε_v levels, a charge injection from the photoexcited sensitizer (S*) might occur. The photoexcited molecule reacts at the

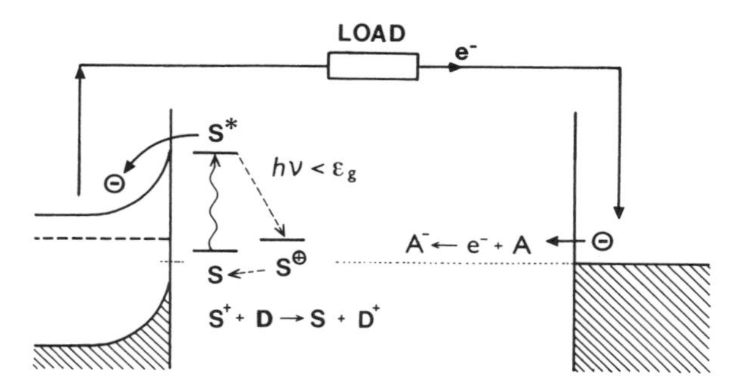

Fig. 5.63 Scheme of a photoelectrochemical cell with sensitized semiconductor anode

interface to form the oxidized, S^+ (reduced, S^-) ground-state form *via* the majority charge carrier injection into the conduction (valence) band of the n- (p-) type semiconductor electrode.

Practically more important is the sensitization of the n-type semiconductor electrode (Fig. 5.63). The depicted scheme is virtually equivalent to that in Fig. 5.62; the only exception is that the 'hole' is not created in the valence band but formally in the sensitizer molecule.

The charge injection from the sensitizer, S^*, dissolved in the electrolyte solution, might formally be considered as a photogalvanic process. The S^* molecules must be able to diffuse to the semiconductor surface during their lifetime, τ_0. Assuming the usual lifetime, $\tau_0 \simeq 10^{-6}$ to 10^{-4} s and the diffusion coefficient, D of 10^{-5} cm^2/s, the effective diffusion distance of S^*, δ_D (see Eq. 2.5.9) is very small:

$$\delta_D = (\pi D \tau)^{1/2} \simeq 56 \text{ nm} \tag{5.10.26}$$

Let us assume that the electrolyte is illuminated through the semiconductor electrode as shown in Fig. 5.58 (this is, in principle, possible since the semiconductor is transparent for wavelengths $\lambda > hc/\varepsilon_g$ at which the sensitizer absorbs the radiation). The relative intensity of radiation transmitted to the distance δ_D is given by the Lambert–Beer law:

$$\log (I/I_0) = -\varepsilon c \delta_D \tag{5.10.27}$$

where ε is the extinction coefficient of S and c its concentration. The light-harvesting efficiency ϕ_L is defined as:

$$\varphi_L = (1 - I/I_0) = 1 - 10^{-\varepsilon c \delta_D} \tag{5.10.28}$$

For a typical extinction coefficient $\varepsilon \simeq 10^7$ cm^2/mol (cf. Table 5.6) and $c = 1$ M, the calculated, φ_L values (Eqs (5.9.42) and (5.9.44)) are only *ca.* 12 per cent. The number of active S^* molecules is further reduced by the quantum yield of the formation of the triplet state φ_T (cf. Table 5.6) and the quantum yield of the injection of the majority charge carriers from S^* into the semiconductor electrode (φ_I). Thus, the overall *quantum yield of photocurrent* (IPCE = *incident photon to current efficiency*) is a product of all the three quantities:

$$\text{IPCE} = \varphi_L \cdot \varphi_T \cdot \varphi_I \qquad (5.10.29)$$

It actually denotes the number of electrons flowing through the interface per one photon impinging on it, IPCE is usually considered for monochromatic light (radiant power, P, in W/cm^2) whose wavelength (λ) matches the wavelength of the absorption maximum of the sensitizer, λ_{max}. A more practical equation for IPCE can easily be derived by combining the photocurrent density, j_{ph} with the radiant power and wavelength:

$$\text{IPCE} = j_{ph}hc/\lambda Pe \qquad (5.10.30)$$

where c is the velocity of light.

The light-harvesting efficiency of an electrode covered with an adsorbed sensitizer can be expressed in the form similar to Eq. (5.10.28):

$$\varphi_L = 1 - 10^{-\varepsilon_{ads}\Gamma r} \qquad (5.10.31)$$

where ε_{ads} is the extinction coefficient of adsorbed sensitizer, Γ is the surface concentration of the adsorbate (in mol/cm^2 of the physical surface area) and r is the roughness factor of the electrode. The latter is defined as the ratio of physical to geometrical surface areas (polycrystalline materials, such as semiconducting oxides may achieve the r values of about 10^2–10^3). The extinction coefficient, ε_{ads}, can be approximated by the usual extinction coefficient of the solution, ε (in cm^2/mol).

For a typical monomolecular coverage, $\Gamma \simeq 10^{-10}\ mol/cm^2$, an electrode roughness factor $r = 1000$ and an extinction coefficient $\varepsilon_{ads} = 10^7\ cm^2/mol$, the light-harvesting efficiency is, in comparison to the preceding case, very high, $\varphi_L = 90$ per cent. Moreover, the adsorbed sensitizer is in intimate contact with the semiconductor surface, hence the conditions for charge injection from S* into the semiconductor are almost ideal ($\varphi_I \rightarrow 100$ per cent).

The most successful sensitizers so far tested are complexes of Ru(II) with various derivatives of 2,2′ bipyridine, e.g. 2,2′-bipyridine 4,4′-dicarboxylic acid (L). The Ru(II)L_3 complex is adsorbed from an aqueous solution of suitable pH value to oxidic semiconductors *via* electrostatic bonds between $—COO^-$ groups of the ligands and the positively charged (protonized) semiconductor surface.

The sensitization of n-TiO_2 by RuL_3 was studied in detail by M. Grätzel *et al.* The photoelectrochemical reaction is initiated by an electron injection into the conduction band of TiO_2:

$$^*\text{Ru(II)L}_3 \rightarrow \text{Ru(III)L}_3 + e^-_{cb}(\text{TiO}_2) \qquad (5.10.32)$$

The standard redox potential of Ru(III)L_3/Ru(II)L_3 equals 1.56 V(protonated form). Even energetically highly demanding processes, such as oxidation of water to oxygen ($E_0 = 1.229$ V) and bromide to Br_2 ($E_0 = 1.0873$ V) are thermodynamically possible. Two reactions deserve special attention, since they enable the construction of highly efficient regenerative cells

$$2\text{Ru(III)L}_3^{3-} + 2\text{I}^- \rightarrow 2\text{Ru(II)L}_3^{4-} + \text{I}_2, \qquad \text{IPCE (450 nm)} \simeq 70\%$$

$$2\text{Ru(III)L}_3^{3-} + 2\text{Br}^- \rightarrow 2\text{Ru(II)L}_3^{4-} + \text{Br}_2, \qquad \text{IPCE (450 nm)} \simeq 55\%$$

The best parameters of the I^-/I_2 cell were achieved by using a non-aqueous electrolyte and the anatase electrode of nanometer-sized particles, sensitized with $RuL_2(\mu - (NC)Ru(CN)(bpy)_2)_2$. In 1991, B. O'Regan and M. Grätzel described a cell attaining parameters competitive to commercial, solid-state photovoltaic devices.

5.10.6 *Photoelectrochemical solar energy conversion*

The solar spectrum corresponds roughly to the radiation of a black body at the temperature 5750 K. If we denote the photon flux in the wavelength interval from λ to $(\lambda + d\lambda)$ (in photons/m^2 s) as $J(\lambda)$, the total solar power flux (in W/m^2) is given as:

$$P_s = hc \int_0^\infty J(\lambda)\, d\lambda \qquad (5.10.33)$$

The power flux corrected for atmospheric absorptions equals about 1 kW/m^2 if the Sun stands at zenith and the sky is clear. This situation is customarily labelled AM 1 (one standard air mass).

The total world consumption of energy (*ca.* 3.10^{20} J/year) can be met by converting the solar energy impinging on about 0.01 per cent of the Earth's surface, which is about 0.3 per cent of the desert areas. The research on solar energy conversion is further encouraged by the existence of natural photosynthesis, which provides all the fossile energy resources of the Earth. The natural photosynthesis transforms about 0.1 per cent of the solar energy impinging on the Earth, but the energy conversion in green plants is not as efficient as that attained by currently available photovoltaic cells and other artificial devices.

The solar energy conversion efficiency is defined as:

$$\text{CE} = P_{out}/P_s \qquad (5.10.34)$$

where P_{out} is the maximum power output of the cell per unit illuminated area. For a regenerative photoelectrochemical cell, the maximum value of P_{out} has to be derived from the photocurrent/potential characteristic of the cell (Fig. 5.64). The quality of the cell is also described by the so-called fill factor, f:

$$f = P_{out}/(j_{ph}^{max} \Delta E_{ph}^{max}) \qquad (5.10.35)$$

where j_{ph}^{max} is the maximum photocurrent density (at short circuit) and ΔE_{ph}^{max} is the maximum photovoltage (at open circuit).

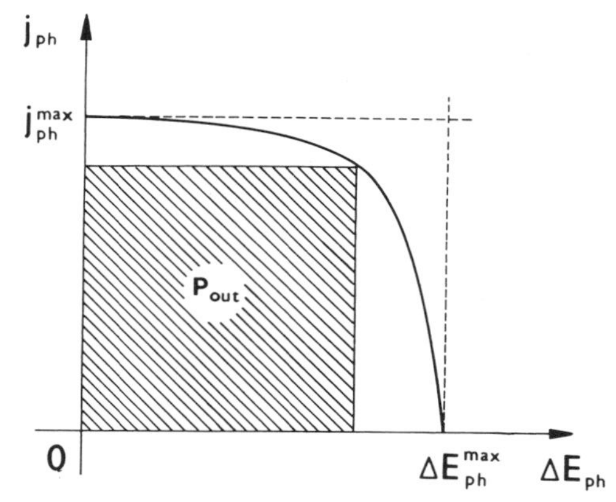

Fig. 5.64 Photocurrent density–voltage characteristics of the regenerative photoelectrochemical cell. The shaded area is the maximum output power, P_{out}

Any quantum device, including photoelectrochemical cell, is in principle not able to absorb (and convert) light whose wavelength is longer than a certain threshold, λ_g. This wavelength corresponds to the photon energy equal to the excitation energy of the photoactive molecule (with photogalvanic effects or sensitized semiconductors) or to the band gap of the semiconductor photoelectrode:

$$\lambda_g = \frac{hc}{\varepsilon_{ex}}, \qquad \lambda_g = \frac{hc}{\varepsilon_g} \tag{5.10.36}$$

The theoretically convertible fraction of the solar power equals:

$$P_{out} = \frac{hc}{\lambda_g}\int_0^{\lambda_g} J(\lambda)\, d\lambda \tag{5.10.37}$$

The plot of $CE = P_{out}/P_s$ (from Eqs (5.10.33) and (5.10.37)) versus λ_g for AM 1.2 is shown in Fig. 5.65 (curve 1). It has a maximum of 47 per cent at 1100 nm. Thermodynamic considerations, however, show that there are additional energy losses following from the fact that the system is in a thermal equilibrium with the surroundings and also with the radiation of a black body at the same temperature. This causes partial re-emission of the absorbed radiation (principle of detailed balance). If we take into account the equilibrium conditions and also the unavoidable entropy production, the maximum CE drops to 33 per cent at 840 nm (curve 2, Fig. 5.65).

Besides the mentioned thermodynamic losses, there always exist kinetic losses arising from the competitive non-radiative quenching of the excited state. For instance in photovoltaic devices, the undesired thermal recom-

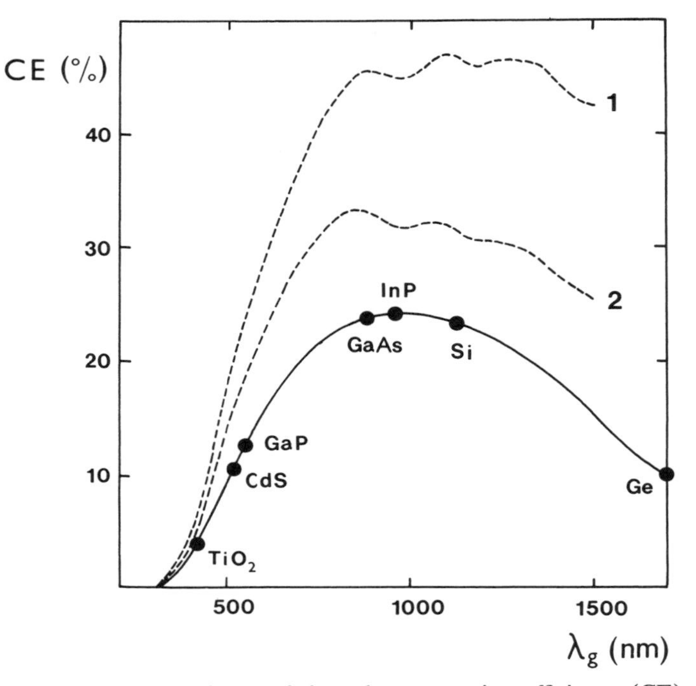

Fig. 5.65 Dependence of the solar conversion efficiency (CE) on the threshold wavelength (λ_g) for a quantum converter at AM 1.2. Curve 1: Fraction of the total solar power convertible by an ideal equilibrium converter with no thermodynamic and kinetic losses. Curve 2: As 1 but the inherent thermodynamic losses (detailed balance and entropy production) are considered. Continuous line: Efficiency of a regenerative photovoltaic cell, where the thermodynamic and kinetic losses are considered. The values of λ_g for some semiconductors are also shown (according to J. R. Bolton *et al.*)

bination of photogenerated e^-/h^+ pairs is prevented by a band bending. This leads to the necessary charge separation, but is also paid by additional energy losses (cf. Eq. 5.10.23), which can be estimated as about 0.6–0.8 eV per photon.

The theoretical solar conversion efficiency of a regenerative photovoltaic cell with a semiconductor photoelectrode therefore depends on the model used to describe the thermodynamic and kinetic energy losses. The CE values, which consider all the mentioned losses can generally only be estimated; the full line in Fig. 5.65 represents such an approximation. Unfortunately, the materials possessing nearly the optimum absorption properties (Si, InP, and GaAs) are handicapped by their photocorrosion sensitivity and high price.

References

Aruchamy, A., see p. 342.

Bard, A. J., R. Memming, and B. Miller, Terminology in semiconductor electrochemistry and photoelectrochemical energy conversion, *Pure Appl. Chem.*, **63,** 569 (1991).

Brodsky, A. M., and Yu. V. Pleskov, Electron photoemission at a metal–electrolyte solution interface, *Progr. Surf. Sci.*, **2,** 1 (1972).

Cardon, F., W. P. Gomes, and W. Dekeyser (Eds), *Photovoltaic and Photoelectrochemical Solar Energy Conversion,* Plenum Press, New York, 1981.

Connolly, J. S., *Photochemical Conversion and Storage of Solar Energy,* Academic Press, New York, 1981.

Conway, B. E., *Solvated Electrons in Field- and Photo-Assisted Processes at Electrodes, MAE* **7,** 83 (1972).

Finklea, H. O. see p. 343.

Fox, M. A., and M. Chanon, *Photoinduced Electron Transfer,* Elsevier, Amsterdam (1988).

Gerischer, H., The impact of semiconductors on the concepts of electrochemistry, *Electrochim. Acta,* **35,** 1677 (1990).

Gerischer, H., and J. Katz (Eds), *Light-Induced Charge Separation in Biology and Chemistry,* Verlag Chemie, Weinheim, 1979.

Grätzel, M., *Heterogeneous Photochemical Electron Transfer Reactions,* CRDC Press, Baton Rouge, Florida, USA (1987).

Hautala, R. R., R. B. King, and C. Kutel, *Solar Energy,* The Humana Press, Clifton, 1979.

Memming, R., Charge transfer processes at semiconductor electrodes, in *Electroanalytical Chemistry,* Vol. 11, p. 1 (Ed. A. J. Bard), M. Dekker, New York, 1979; *CTE,* **7,** 529 (1983).

Nozik, H. J., Photoelectrochemistry: applications to solar energy conversion, *Ann. Rev. Phys. Chem.,* **29,** 184 (1978).

O'Regan, B., and M. Grätzel, *Nature,* **353,** 737 (1991).

Pelizetti, E., and N. Serponne (Eds) *Homogeneous and Heterogeneous Photocatalysis,* D. Reidel, Dordrecht, 1986.

Photoelectrochemistry. Faraday Discussions of the Royal Society of Chemistry No. 70 (1980).

Pleskov, Yu. V., *Solar Energy Conversion. A Photoelectrochemical Approach,* Springer-Verlag, Berlin, 1990.

Pleskov, Yu. V., and Yu. Ya. Gurevich, *Semiconductor Photoelectrochemistry,* Plenum Press, New York, 1986.

Santhanam, K. S. V., and M. Sharon, see p. 344.

Sciavello, M. (Ed.), *Photochemistry, Photocatalysis and Photoreactors,* NATO ASI Series C, Vol. 146, D. Reidel, Dordrecht (1985).

Vlachopoulos, N., P. Liska, J. Augustynski, and M. Grätzel, *J. Am. Chem. Soc.,* **110,** 1216 (1988).

Chapter 6

Membrane Electrochemistry and Bioelectrochemistry

Not only electric currents in muscles and nerves but particularly even the mysterious effects occurring with electric fish can be explained by means of the properties of semipermeable membranes.

W. Ostwald, 1891

6.1 Basic Concepts and Definitions

The discovery of galvanic electricity (i.e. electrical phenomena connected with the passage of electric current) by L. Galvani in 1786 occurred simultaneously with his study of a bioelectrochemical phenomenon which was the response of excitable tissue to an electric impulse. E. du Bois-Reymond found in 1849 that such electrical phenomena occur at the surface of the tissue, but it was not until almost half a century later that W. Ostwald demonstrated that the site of these processes are electrochemical semipermeable membranes. In the next decade, research on semipermeable membranes progressed in two directions—in the search for models of biological membranes and in the study of actual biological membranes.

The search for models of biological membranes led to the formation of a separate branch of electrochemistry, i.e. membrane electrochemistry. The most important results obtained in this field include the theory and application of ion-exchanger membranes and the discovery of ion-selective electrodes (including glass electrodes) and bilayer lipid membranes.

The study of biological membranes led to the conclusion that the great majority of the processes in biological systems occur at cell and organelle membranes. The electrochemical aspects of this subject form the basis of *bioelectrochemistry,* dealing with the processes of charge separation and transport in biological membranes and their models, including electron and proton transfer in cell respiration and photosynthesis as well as ion transport in the channels of excitable cells. The electrokinetic phenomena (electrical double layer, interfacial tension of cells and organelles, cell membrane extension and contraction, etc.) also belong to this field. Bioelectrochemistry includes the more classical subjects of the thermodynamics and kinetics of redox processes of components of biological systems at electrodes.

This chapter will deal with the basic properties of electrochemical membranes in general and the membrane aspects of bioelectrochemistry in particular. A number of bioelectrochemical topics was discussed in Sections 1.5.3 and 3.2.5.

6.1.1 *Classification of membranes*

In contrast to mechanics, where the term *membrane* (Lat. *membrana* = parchment) designates an elastic, two-dimensional plate, this term is used in chemistry, biophysics and biology to designate a solid or liquid phase (usually, but not always, with a thickness substantially smaller than its other dimensions) separating two, usually liquid, phases. The transport (permeation) of the various components of both phases through the membrane occurs at different rates relative to those in the homogeneous phases with which the membrane is in contact. The membrane is consequently called *semipermeable.*

The thickness of the membrane phase can be either macroscopic ('thick')—membranes with a thickness greater than micrometres—or microscopic ('thin'), i.e. with thicknesses comparable to molecular dimensions (biological membranes and their models, bilayer lipid films). Thick membranes are crystalline, glassy or liquid, while thin membranes possess the properties of liquid crystals (fluid) or gels (crystalline).

Depending on their structure, membranes can be separated into *porous,* where matter is transported through pores in the membrane, and *compact,* where the substance is transported either through the entire homogeneous membrane phase or its homogeneous parts.

Membranes can be *homogeneous,* where the whole membrane participates in the permeation of a substance, or *heterogeneous,* where the active component is anchored in a suitable support (for solid membranes) or absorbed in a suitable diaphragm or acts as a plasticizer in a polymeric film. Both of the latter cases are connected with liquid membranes. Biological membranes show heterogeneity at a molecular level.

Membranes exhibiting selectivity for ion permeation are termed *electrochemical membranes*. These membranes must be distinguished from simple liquid junctions that are often formed in porous diaphragms (see Section 2.5.3) where they only prevent mixing of the two solutions by convection and have no effect on the mobility of the transported ions. It will be seen in Sections 6.2 and 6.3 that the interior of some thick membranes has properties analogous to those of liquid junctions, but that the mobilities of the transported ions are changed.

6.1.2 *Membrane potentials*

A characteristic property of electrochemical membranes is the formation of an electric potential difference between the two sides of the membrane, termed the membrane potential $\Delta\phi_M$:

$$\begin{array}{ccc} \phi(1) & d & \phi(2) \\ \text{solution 1} \,|\, & \text{membrane} & |\, \text{solution 2} \\ & x = \mathrm{p} \quad x = \mathrm{q} & \end{array} \tag{6.1.1}$$

Similar to galvanic cells, the membrane potential is determined by subtracting the electric potential of the phase on the left from that of the phase on the right, i.e.

$$\Delta\phi_M \equiv \Delta_1^2\phi = \phi(2) - \phi(1) \tag{6.1.2}$$

For cell membranes, the intracellular liquid is usually denoted as solution 2, while solution 1 is the extracellular liquid.

The formation of a membrane potential is connected with the presence of an electrical double layer at the surface of the membrane. For a thick, compact membrane, an electrical double layer is formed at both interfaces. The electrical double layer at a porous membrane is formed primarily in the membrane pores (see Section 6.2). The electrical double layer at thin membranes is formed on both membrane surfaces. It is formed by fixed ions on the surface of the membrane and the diffuse layer in the electrolyte.

Consider the simple case where both sides of the membrane are in contact with a solution of symmetrical electrolyte BA in a single solvent and the membrane is permeable for only one ionic species. In equilibrium its electrochemical potential (Eq. (3.1.5)) in both solutions adjacent to the membrane has the same value. Thus,

$$\Delta\phi_M = \frac{RT}{zF} \ln \frac{a_B(1)}{a_B(2)} \tag{6.1.3}$$

if the membrane is permeable for cation B^{z+} and

$$\Delta\phi_M = \frac{RT}{|z|\,F} \ln \frac{a_A(2)}{a_A(1)} \tag{6.1.4}$$

if the membrane is permeable for anion A^{z-}.

The membrane potential expressed by Eqs (6.1.3) and (6.1.4) is termed the Nernst membrane potential as it originates from the analogous ideas as the Nernst equation of the electrode potential (p. 165) and the equation of the Nernst potential at ITIES (Eq. (3.3.50)).

Consider a system in which both solutions contain various ions for which the membrane is permeable (diffusible ions) and one type of ion that, for some reason (e.g. a macromolecular ion for a porous membrane), cannot pass through the membrane (non-diffusible ion). The membrane is permeable for the solvent.

The equilibrium conditions for the diffusible ions are

$$\tilde{\mu}_i(1, p_1) = \tilde{\mu}_i(2, p_2) \tag{6.1.5}$$

This condition expresses the fact that the two solutions are under different pressures, p_1 and p_2, as a result of their, in general, different osmotic pressures. An analogous equation cannot be written for the non-diffusible ion as it cannot pass through the membrane and the 'equilibrium' concentrations cannot be established.

First, let us consider dilute solutions, where it is possible to set $p_1 = p_2$. Then the electrochemical potentials in Eq. (6.1.5) are expanded in the usual manner, yielding

$$RT \ln \frac{a_+(2)}{a_+(1)} + z_+ F \Delta\phi_M = 0 \tag{6.1.6}$$

for the diffusible cation and

$$RT \ln \frac{a_-(2)}{a_-(1)} + z_- F \Delta\phi_M = 0 \tag{6.1.7}$$

for the diffusible anion. Elimination of the terms containing the electric potentials in these equations yields

$$\left[\frac{a_+(2)}{a_+(1)}\right]^{1/z_+} = \left[\frac{a_-(2)}{a_-(1)}\right]^{1/z_-} = \lambda \tag{6.1.8}$$

i.e. for univalent, divalent, etc., cations and anions

$$\begin{aligned} \frac{a_+(2)}{a_+(1)} &= \left[\frac{a_{2+}(2)}{a_{2+}(1)}\right]^{1/2} = \left[\frac{a_{3+}(2)}{a_{3+}(1)}\right]^{1/3} = \cdots = \lambda \\ \frac{a_-(1)}{a_-(2)} &= \left[\frac{a_{2-}(1)}{a_{2-}(2)}\right]^{1/2} = \left[\frac{a_{3-}(1)}{a_{3-}(2)}\right]^{1/3} = \cdots = \lambda \end{aligned} \tag{6.1.9}$$

The constant λ is termed the Donnan distribution coefficient.

In the simple case of a diffusible, univalent cation B^+ and anion A^- and non-diffusible anion X^- present in phase 2, the condition of electroneutrality gives

$$\begin{aligned} c_{B^+}(2) &= c_{A^-}(2) + c_{X^-}(2) \\ c_{B^+}(1) &= c_{A^-}(1) \end{aligned} \tag{6.1.10}$$

For dilute solutions, activities can be replaced by concentrations in Eqs. (6.1.9), yielding

$$c_{B^+}(1) c_{A^-}(1) = c_{B^+}(2) c_{A^-}(2) \tag{6.1.11}$$

Equations (6.1.10) and (6.1.11) give

$$\lambda = \left[\frac{c_{A^-}(2) + c_{X^-}(2)}{c_{A^-}(2)}\right]^{1/2} \tag{6.1.12}$$

A more exact solution of the equilibrium conditions (Eq. 6.1.5) must consider that the standard term μ^0 depends on the pressure, which is different in the two solutions:

$$\mu^0(T, p) = \mu^*(T, p) = \int_0^p v \, \mathrm{d}p \tag{6.1.13}$$

where μ^* is the limiting value of μ^0 at $p \rightarrow 0$ and v is the molar volume of the component at pressure p. If the volume v is considered to be independent of the pressure (more accurate calculations employ a linear dependence) then

$$\mu^0(T, p) = \mu^*(T) + vp \tag{6.1.14}$$

It can be readily seen that this procedure yields an equation whose left-hand side is the same as that for Eqs (6.1.6) and (6.1.7) and whose right-hand side is not equal to zero, but rather to $v_+(p_1 - p_2)$ or $v_-(p_1 - p_2)$.

Equations (6.1.6) and (6.1.7) yield the Donnan potential $\Delta\phi_D \equiv \Delta\phi_M$ in the form

$$\Delta\phi_D = -\frac{RT}{F} \ln \lambda \tag{6.1.15}$$

The Donnan potentials contain the individual ionic activities and cannot be measured by using a purely thermodynamic procedure. In the concentration range where the Debye–Hückel limiting law is valid, the ionic activities can be replaced by the mean activities.

The membrane potentials are measured by constructing a cell with a semipermeable membrane separating solutions 1 and 2:

$$\mathrm{Ag}\,|\,\mathrm{AgCl, sat.\ KCl}\,|\,\text{solution 1}\,|\,\text{solution 2}\,|\,\mathrm{sat.\ KCl, AgCl}\,|\,\mathrm{Ag} \tag{6.1.16}$$

whose EMF is given by the sum of the membrane potential and the two liquid junction potentials, which can usually be neglected.

Equations (6.1.3) and (6.1.4) could be derived from the relationships for the liquid junction potential formed by a single salt (Eq. 2.5.31), assuming that the anion mobility in Eq. (6.1.3) and the cation mobility in Eq. (6.1.4) equal zero. The general case, i.e. that the membrane only changes the transport number of the cation and anion, was solved by J. Bernstein at the beginning of this century using a relationship analogous to Eq. (2.5.31), which assumes the following form for a uni-univalent electrolyte:

$$\Delta\phi_M = (\tau_- - \tau_+) \frac{RT}{F} \ln \frac{c(2)}{c(1)} \tag{6.1.17}$$

where τ_+ and τ_- are the transport numbers of the cation and anion in the membrane, respectively, which are different from the corresponding values in solution.

References

Blank, M. (Ed.), *Bioelectrochemistry: Ions, Surfaces, Membranes,* American Chemical Society, Washington, 1980.

Bioelectrochemistry, CTE, **10** (1985).

Butterfield, D. A., *Biological and Synthetic Membranes,* John Wiley & Sons, New York, 1989.

Eisenman, G. (Ed.), *Membranes—A Series of Advances,* M. Dekker, New York, Vol. 1, 1972; Vol. 2, 1973; Vol. 3, 1975.
Galvani, L., *De Viribus Electricitatis in Motu Musculari,* Typographia Instituti Scientiarum, Bologna, 1791, Translation: Green, R. M., *Commentary on the Effect of Electricity on Muscular Motion,* p. 66, E. Licht, Cambridge, Mass., 1953.
Jain, M. K., *Introduction to Biological Membranes,* John Wiley & Sons, New York, 1988.
Keynes, R. D., The generation of electricity by fishes, *Endeavour,* **15,** 215 (1956).
Koryta, J., *Ions, Electrodes and Membranes,* 2nd ed. John Wiley & Sons, Chichester, 1991.
Koryta, J., What is bioelectrochemistry?, *Electrochim. Acta,* **29,** 1291 (1984).
Membrane phenomena, *Discussion Faraday Soc.,* **21,** 1956.
Milazzo, G., and M. Blank (Eds), *Bioelectrochemistry III—Charge Separation Across Biomembranes,* Plenum, New York, 1988.
Overbeck, J. T., The Donnan equilibrium, in *Progress in Biophysics and Biophysical Chemistry,* Vol. 6, p. 57, Pergamon Press, London, 1965.
Schlögl, R., *Stofftransport durch Membranen,* D. Steinkopff, Darmstadt, 1964.
Sollner, K., The early developments of the electrochemistry of polymer membranes, in *Charged Gells and Membranes* (Ed. E. Sélégny), Vol. 1, Reidel, Dordrecht, 1976.
Williams, R. J. P., A general approach to bioelectrics, in *Charge and Field Effects in Biosystems* (Eds M. J. Allen and P. N. R. Usherwood), Abacus Press, London, 1984.
Wu, C. H., Electric fish and the discovery of animal electricity, *Am. Scientist,* **72,** 598 (1984).

6.2 Ion-exchanger Membranes

Ion-exchanger membranes with fixed ion-exchanger sites contain ion conductive polymers (*ionomers*) the properties of which have already been described on p. 128. These membranes are either homogeneous, consisting only of a polyelectrolyte that may be chemically bonded to an un-ionized polymer matrix, and heterogeneous, where the grains of polyelectrolyte are incorporated into an un-ionized polymer membrane. The electrochemical behaviour of these two groups does not differ substantially.

All ion-exchanger membranes with fixed ion-exchanger sites are porous to a certain degree (in contrast to liquid membranes and to membranes of ion-selective electrodes based on solid or glassy electrolytes, such as a single crystal of lanthanum fluoride).

6.2.1 *Classification of porous membranes*

Depending on the pore size, porous membranes can be divided into three groups:

(a) Membranes with wide pores are simply diaphragms limiting flow and diffusion of the solutions with which the membrane is in contact. When the diaphragm contains cylindrical pores with identical radii r and a density N per unit area, then, in the ideal case, the material flux of the

ith component of the solution through the diaphragm is given as

$$J_i = c_i J_V - Nr^2 \pi D_i (RT)^{-1} c_i \frac{\mathrm{d}\tilde{\mu}_i}{\mathrm{d}x} \tag{6.2.1}$$

where c_i is the concentration of the ith component, D_i is its diffusion coefficient, $\tilde{\mu}_i$ is its electrochemical potential and J_V is the volume flux of the solution:

$$J_V = d_h \frac{\mathrm{d}P}{\mathrm{d}x} = N\pi r^2 (8\eta d)^{-1} \frac{\mathrm{d}P}{\mathrm{d}x} \tag{6.2.2}$$

where d_h is the mechanical permeability (permeability per unit pressure gradient), η is the viscosity of the solution, d is the thickness of the membrane and P is the hydrostatic pressure. Equation (6.2.1) follows directly from Eq. (2.3.23) and (2.5.23).

(b) *Fine-pore membranes* ($r = 1$–100 nm) selectively affect the character of transport. These are often called semipermeable membranes—the permeability of the membrane is different for different components of the solution as a result of the properties of the membrane itself (rather than as a result of different mobilities of the components of the solution). The walls of the pores of ion-exchanger membranes are electrically charged and contain an aqueous solution. The contents of the electrolyte components in the pores depends on the electric charge on the pore walls and concentration of the electrolyte in contact with the membrane. The electrical double layer formed inside the pores results in the *counterions* (= gegenions, ions with opposite charge to the fixed ions on the membrane wall) having a greater concentration in the pores than ions with the same charge as the fixed ions (*coions*). When the solution is dilute and the pores so narrow that the diffuse electrical layer has an effective thickness comparable with the pore radius, then the gegenions are present in a clear excess over the coions (see Fig. 6.1). In the extreme case, the electrical diffuse layers completely fill the pores, which then contain only counterions and the membrane is permeable only for these ions, the transport number of counterion being $\tau_i = 1$. Such a membrane is termed *permselective*. On the other hand, if the electrolyte is more concentrated and the pores wider, then the excess of the counterions over the coions eventually becomes negligible.

(c) *Microporous membranes* have such small pore radii that mass is transported by an exchange process between the dissolved species and the solvent particles. This structure is characteristic, for example, for amorphous polymer films below the glass transition temperature. The porosity is a result of irregular coiling of segments of the polymer chains. These membranes are used to separate mixtures of gases and liquids (these are not electrochemical membranes) and to desalinate

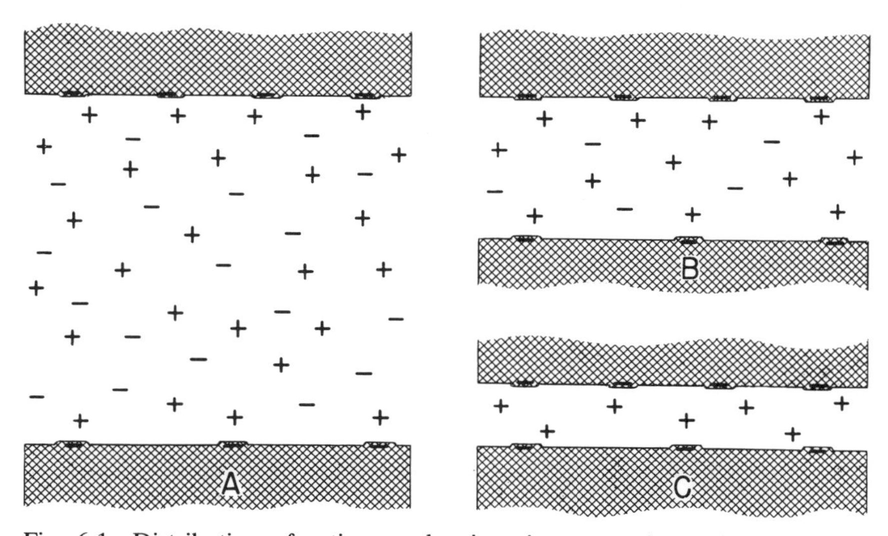

Fig. 6.1 Distribution of cations and anions in pores of a cation-exchanger membrane depends on pore radius which decreases in the sequence A→B→C. In case C the membrane becomes permselective. (According to K. Sollner)

water by *hyperfiltration*. They must be very thin for transport to occur at all and must simultaneously be mechanically strong.

6.2.2 *The potential of ion-exchanger membranes*

The distribution of electric potential across the membrane and the dependence of the membrane potential on the concentration of fixed ions in the membrane and of the electrolyte in the solutions in contact with the membrane is described in the model of an ion-exchanger membrane worked out by T. Teorell, and K. H. Meyer and J. F. Sievers.

This theory will be demonstrated on a membrane with fixed univalent negative charges, with a concentration in the membrane, c_x. The pores of the membrane are filled with the same solvent as the solutions with which the membrane is in contact that contain the same uni-univalent electrolyte with concentrations c_1 and c_2. Conditions at the membrane–solution interface are analogous to those described by the Donnan equilibrium theory, where the fixed ion X^- acts as a non-diffusible ion. The Donnan potentials $\Delta\phi_{D,1} = \phi_p - \phi(1)$ and $\Delta\phi_{D,2} = \phi(2) - \phi_q$ are established at both surfaces of the membranes ($x = p$ and $x = q$). A liquid junction potential, $\Delta\phi_L = \phi_q - \phi_p$, due to ion diffusion is formed within the membrane. Thus

$$\Delta\phi_M = \Delta\phi_{D,2} + \Delta\phi_L + \Delta\phi_{D,1} \qquad (6.2.3)$$

The Donnan equilibrium condition (6.1.9) gives

$$\begin{aligned} c_{+,p}c_{-,p} &= c_{+,1}c_{-,1} = c_1^2 \\ c_{+,q}c_{-,q} &= c_{+,2}c_{-,2} = c_2^2 \end{aligned} \qquad (6.2.4)$$

(For simplicity, the effect of the activity coefficients is neglected; it should be recalled that the membrane is not completely impermeable for anions so that neither $c_{-,p}$ nor $c_{-,q}$ is equal to zero.) These relationships and the electroneutrality condition in the membrane

$$c_{+,p} = c_{-,p} + c_x$$
$$c_{+,q} = c_{-,q} + c_x$$

yield the expressions for the surface concentrations at the membrane:

$$\begin{aligned} c_{+,p} &= (\tfrac{1}{4}c_x^2 + c_1^2)^{1/2} + \tfrac{1}{2}c_x \\ c_{+,q} &= (\tfrac{1}{4}c_x^2 + c_2^2)^{1/2} + \tfrac{1}{2}c_x \\ c_{-,p} &= (\tfrac{1}{4}c_x^2 + c_1^2)^{1/2} - \tfrac{1}{2}c_x \\ c_{-,q} &= (\tfrac{1}{4}c_x^2 + c_2^2)^{1/2} - \tfrac{1}{2}c_x \end{aligned} \tag{6.2.5}$$

When the expressions (6.2.5) are substituted into the Henderson equation (2.5.34) $\Delta\phi_L$ is obtained. Both contributions $\Delta\phi_D$ are calculated from the Donnan equation. From Eq. (6.2.3) we obtain, for the membrane potential,

$$\Delta\phi_M = \frac{RT}{F}\left[\ln\frac{c_{+,q}c_1}{c_{-,p}c_2} - \frac{U_+ - U_-}{U_+ + U_-}\ln\frac{(U_+ + U_-)c_{+,q} - U_-c_x}{(U_+ + U_-)c_{+,p} - U_-c_x}\right] \tag{6.2.6}$$

Figure 6.2 schematically depicts the concentration distribution in this ion-exchanger membrane.

In the limiting case, Eq. (6.2.6) is converted to the relationships derived above: if the membrane is completely impermeable for the anion, then $c_{-,p} = c_{-,q} = 0$ and also $U_- = 0$ and the membrane potential is given by the Nernst term $(RT/F)\ln(c_1/c_2)$ (cf. Eq. 6.1.3). If, on the other hand, the

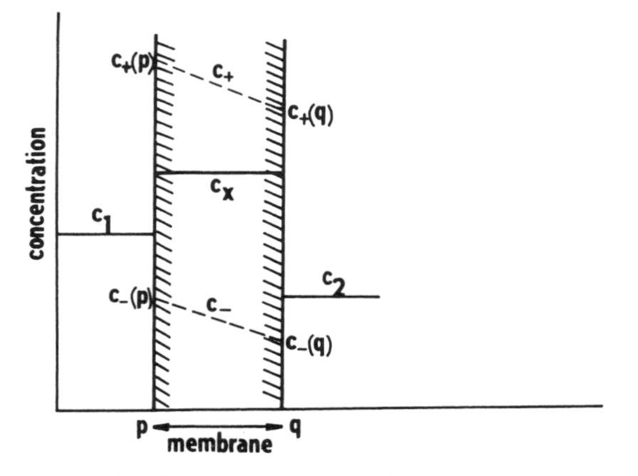

Fig. 6.2 Concentration distribution of cations c_+, anions c_- and fixed anions in a cation-exchanging membrane. The bathing solutions have electrolyte concentrations c_1 and c_2. (According to K. Sollner)

membrane is completely permeable for both the cation and the anion, then $c_x = 0$ as well as $c_{+,q} = c_{-,q} = c_2$ and $c_{+,p} = c_{-,p} = c_1$ and $\Delta\phi_M$ is given by the Henderson term alone, $-(RT/F)(t_+ - t_-) \ln(c_2/c_1)$ (cf. Eq. 2.5.31).

6.2.3 *Transport through a fine-pore membrane*

Transport processes in ion-exchanger membranes are important in most applications. This section will first deal with interaction of the material flux through the membrane with the field of the diffuse double layer in its pores and also with mutual interaction of the fluxes of various thermodynamic quantities, treated by using the thermodynamics of irreversible processes (the phenomenological description of membrane processes in terms of the thermodynamics of irreversible processes is the principal application of this field in electrochemistry). First, for example, a cation-exchanger membrane will be considered, containing an excess of mobile cations (counterions) over anions (coions). On the transport of electric current, the cations give a greater electrical impulse to the solvent molecules than the anions, leading to formation of an electroosmotic flux in the direction of movement of the cations in the electric field.

An *electroosmotic flux* is formed as a result of the effect of the electric field in the direction normal to the pores in the membrane, $\partial\phi/\partial x$, (assumed to be constant) on the charge in the electric diffuse layer in the pore with a charge density ρ. The charges move in the direction of the x axis (i.e. in the direction of the field), together with the whole solution with velocity v. At steady state

$$\frac{\partial\phi}{\partial x}\rho = \eta\frac{\partial^2 v}{\partial y^2} \tag{6.2.7}$$

where η is the viscosity of the liquid in the pores. According to this equation the force acting on the charges is compensated by internal friction of the liquid. The coordinate in the direction perpendicular to the pore surface is denoted by y. The Poisson equation (Eq. 1.3.8) holds for the space charge

$$-\frac{\rho}{\varepsilon} = \frac{\partial^2\phi}{\partial y^2} \tag{6.2.8}$$

so that substitution into Eq. (6.2.7) gives

$$\eta\frac{d^2 v}{dy^2} = -\frac{\partial\phi}{\partial x}\varepsilon\frac{\partial^2\phi}{\partial y^2} \tag{6.2.9}$$

The electrical potential in the layer of the liquid moving only negligibly at the surface of the pores is equal to the electrokinetic potential ζ (see page 242). At the middle of the pore, the electric potential and its gradient $\partial\phi/\partial y$ equal zero and also $\partial v/\partial y = 0$. Integration of Eq. (6.2.9) yields the Helmholtz–Smoluchowski equation

$$v = \frac{\partial\phi}{\partial x}\frac{\varepsilon\zeta}{\eta} \tag{6.2.10}$$

For a simple model of the membrane it is assumed that N cylindrical pores with radius r are located perpendicular to a membrane of thickness d. The resistance of such a membrane is given by the relationship

$$R = \frac{d}{N\kappa\pi r^2} = \frac{\partial\phi}{\partial x}\frac{d}{j} \tag{6.2.11}$$

where κ is the conductivity of the pores of the membrane and j is the current density (referred to the unit surface area of the membrane). The volume flux of the liquid through the membrane (electroosmotic flux) then is

$$J_V = v\pi r^2 N = \frac{\varepsilon\zeta j}{\kappa\eta} \tag{6.2.12}$$

Because of the electroosmotic flux, a cation-exchanger membrane is less resistant to cation flux than to anion flux.

Some of the elements of thermodynamics of irreversible processes were described in Sections 2.1 and 2.3. Consider the system represented by n fluxes of thermodynamic quantities and n driving forces; it follows from Eqs (2.1.3) and (2.1.4) that $\frac{1}{2}n(n+1)$ independent experiments are needed for determination of all phenomenological coefficients (e.g. by gradual elimination of all the driving forces except one, by gradual elimination of all the fluxes except one, etc.). Suitable selection of the driving forces restricted by relationship (2.3.4) leads to considerable simplification in the determination of the phenomenological coefficients and thus to a complete description of the transport process.

The theory of transport in membranes is based on the thermodynamics of irreversible processes (Sections 2.1 and 2.3). If n species are involved in the transport, we have to consider n independent fluxes (including that of the solvent), each of them characterized by n phenomenological coefficients. Owing to the Onsager reciprocity relations, however, there are only $n(n+1)/2$ independent phenomenological coefficients, which can be determined by as many independent experiments. As usual in the theory of transport in electrolyte solutions, the driving forces are gradients of electrochemical potentials (in dilute solutions gradients of concentrations and of the electric potential can also be used).

Consider the system shown in Fig. 6.3. The ion-exchanger membrane separates solutions of a single, completely dissociated, uni-univalent electrolyte. Two pistons can be employed to form a pressure difference between the two compartments. The two electrodes W_1 and W_2 are

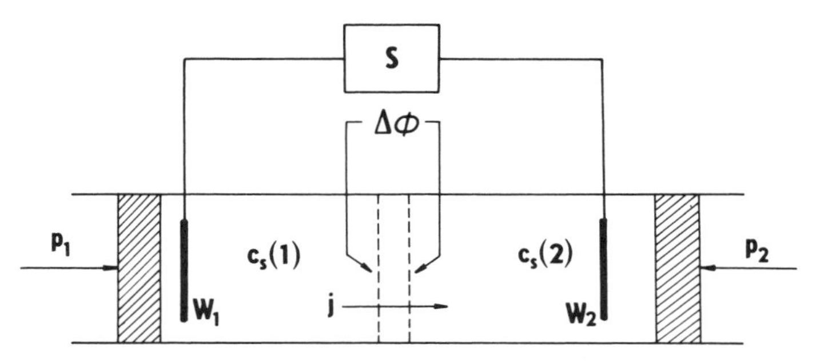

Fig. 6.3 A membrane system during flow of electric current. $\Delta p = p_2 - p_1$, $\Delta c_s = c_s(2) - c_s(1)$. S denotes the current source

connected to an external source S to produce a flow of electric current through the membrane. Two usually identical reference electrodes are placed in the two compartments (e.g. two saturated calomel electrodes) and are used to measure the membrane potential.

Since the ionic fluxes cannot be measured individually, it is preferable to introduce the salt flux, besides solvent flux and charge flux (current density). The driving forces would then be the gradients or differences of the chemical potentials in media with different salt concentrations and different pressures, multiplied by -1. These differences must be relatively small to remain within the framework of linear irreversible thermodynamics, so that

$$\Delta\mu_s \approx \frac{RT\Delta c_s}{\bar{c}_s} + v_s\Delta p$$
$$\Delta\mu_w \approx \frac{RT\Delta c_w}{\bar{c}_w} + v_w\Delta p \tag{6.2.13}$$

where c_s is the salt concentration, c_w is the solvent concentration, $\bar{c}_s$ and $\bar{c}_w$ are the average salt and solvent concentrations between the two compartments, respectively, v_s and v_w are the molar volumes of the salt and the solvent, respectively, and Δp is the pressure difference between the two compartments.†

Even then, the formulation of the system of fluxes and driving forces is not yet satisfactory, because of the quantity Δc_w. Thus, a new definition of the fluxes will be introduced, i.e. the volume flux of the solution, defined by the equation

$$J_V = v_w J_w + v_s J_s \tag{6.2.14}$$

† An approximate expression for $\Delta\mu_i$ follows from the relationship

$$\Delta \ln c_i = \ln \frac{\bar{c}_i + \Delta c_i/2}{\bar{c}_i - \Delta c_i/2} = \ln \frac{1 + \Delta c_i/2\bar{c}_i}{1 - \Delta c_i/2\bar{c}_i} \approx \frac{\Delta c_i}{\bar{c}_i}$$

The conjugate driving force is the pressure gradient Δp multiplied by -1. Further, the relative flux of the dissolved substance compared to the flux of the solvent, i.e. the 'exchange' flux J_D, is defined by the relationship

$$J_D = \frac{J_s}{\bar{c}_s} - \frac{J_w}{\bar{c}_w} \tag{6.2.15}$$

where J_s and J_w are the fluxes of the salt and the solvent, respectively. The conjugate driving force is then the osmotic pressure gradient (multiplied by -1), since (cf. Eq. 1.1.17)

$$\Delta\pi = 2RT\Delta c_s \tag{6.2.16}$$

The dissipation function is given as

$$\Phi = -J_V\Delta p - J_D\Delta\pi - j\Delta\phi < 0 \tag{6.2.17}$$

The present example of membrane transport is described by the following set of equations:

$$\begin{aligned} J_V &= -L_{vv}\Delta p - L_{v\pi}\Delta\pi - L_{v\phi}\Delta\phi \\ J_D &= -L_{\pi v}\Delta p - L_{\pi\pi}\Delta\pi - L_{\pi\phi}\Delta\phi \\ j &= -L_{\phi v}\Delta p - L_{\phi\pi}\Delta\pi - L_{\phi\phi}\Delta\phi \end{aligned} \tag{6.2.18}$$

containing six phenomenological coefficients, which must be determined in independent experiments.

The coefficient L_{vv} can be determined from the mechanical permeability of the membrane (cf. Eq. 6.2.2) in the presence of a pressure difference and absence of a solute concentration difference ($\Delta\pi = 0$) and of an electric potential difference $\Delta\phi$ (electrodes W_1 and W_2 are connected in short, that is $\Delta\phi = 0$). The electrical conductivity of the membrane g_M in the absence of a pressure and solute concentration difference gives the coefficient $L_{\phi\phi}$:

$$j = -g_M\Delta\phi = -L_{\phi\phi}\Delta\phi \tag{6.2.19}$$

Of the remaining three cross phenomenological coefficients, $L_{v\pi}$ can be found from the volume flux during dialysis ($\Delta p = 0$, $\Delta\phi = 0$, because electrodes W_1 and W_2 are short-circuited):

$$L_{v\pi} = -\frac{J_V}{\Delta\pi} \tag{6.2.20}$$

Equations (6.2.18) yield the electroosmotic flux ($\Delta p = 0$, $\Delta\pi = 0$) in the form

$$\frac{J_V}{j} = \frac{L_{v\phi}}{L_{\phi\phi}} \tag{6.2.21}$$

The sign of the cross coefficient $L_{v\phi}$ determines the direction of the electroosmotic flux and of the cation flux. From Eq. (6.2.12) we have

$$\frac{L_{v\phi}}{L_{\phi\phi}} = \frac{\varepsilon\zeta j}{\kappa\eta} \tag{6.2.22}$$

where, for a cation-exchanger membrane, $\zeta > 0$ and, for an anion-exchanger membrane, $\zeta < 0$. Thus, the electroosmotic flux for an anion-exchanger membrane will go in the opposite direction.

When a change in pressure Δp (the *electroosmotic pressure*) stops the electroosmotic flux ($J_V = 0$, $\Delta\pi = 0$), then the first of Eqs (6.2.18) gives

$$\Delta p = -\frac{L_{v\phi}}{L_{vv}}\Delta\phi \tag{6.2.23}$$

Figure 6.4 depicts the formation of the electroosmotic pressure.

The last experiment involves measurement of the membrane potential (in the last of Eqs (6.2.18), $j = 0$ and $\Delta p = 0$), for which it follows from Eq. (6.2.16) that

$$\begin{aligned}\Delta\phi_M &= -\frac{L_{\phi\pi}}{L_{\phi\phi}}\Delta\pi \\ &= -\frac{L_{\phi\pi}}{L_{\phi\phi}}2RT\Delta c_s \\ &\approx -\frac{L_{\phi\pi}}{L_{\phi\phi}}2RT\bar{c}_s \ln\frac{1+\Delta c_s/2\bar{c}_s}{1-\Delta c_s/2\bar{c}_s} \\ &= -\frac{L_{\phi\pi}}{L_{\phi\phi}}2RT\bar{c}_s \ln\frac{c_s(2)}{c_s(1)}\end{aligned} \tag{6.2.24}$$

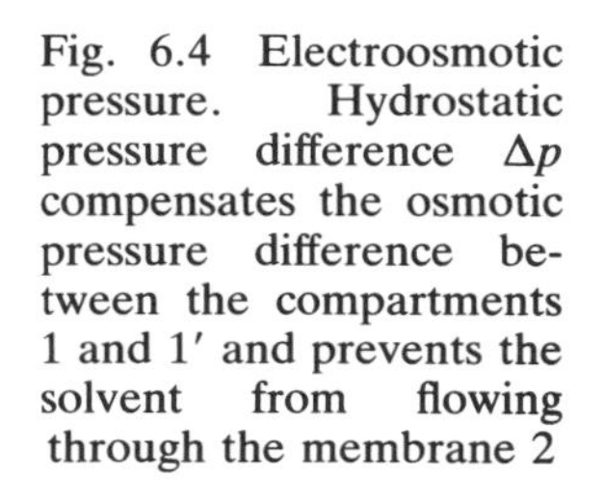

Fig. 6.4 Electroosmotic pressure. Hydrostatic pressure difference Δp compensates the osmotic pressure difference between the compartments 1 and 1′ and prevents the solvent from flowing through the membrane 2

From Eqs (6.1.17) and (6.2.19) we then obtain

$$L_{\phi\pi} = -\frac{(\tau_+ - \tau_-)g_M}{2\bar{c}_s F} \tag{6.2.25}$$

It will now be demonstrated how Eqs (6.2.18) assist in the orientation in the transport system, e.g. in the determination of the process that is suitable for separation and accumulation of the components, etc. According to Eqs (6.2.14) and (6.2.15) the solute flux with respect to the membrane, J_s, is given by the expression

$$J_s = \frac{v_w \bar{c}_w \bar{c}_s}{v_w \bar{c}_w + v_s \bar{c}_s} J_D + \bar{c}_s J_V \tag{6.2.26}$$

As the denominator in the first term on the right-hand side is equal to 1, we obtain with respect to Eqs (6.2.18) ($\Delta p = 0$, $\Delta\phi = 0$)

$$J_s = -(L_{\pi\pi} v_w \bar{c}_w + L_{V\pi})\bar{c}_s \Delta\pi \tag{6.2.27}$$

By definition, $L_{\pi\pi} > 0$. The sign of J_s is determined by the sign of the cross coefficient $L_{V\pi}$ and its absolute value. If $L_{V\pi} < 0$, the volume flux of the solvent occurs in the direction from more dilute to more concentrated solutions (i.e. in the direction of the osmotic pressure gradient). If $|L_{V\pi}|$ is smaller than $L_{\pi\pi} v_w \bar{c}_w$, then the solute flows in the direction of the drop of concentration (in the opposite direction to the concentration or osmotic pressure gradient). This case is termed *congruent flux*. In the opposite case ($-L_{V\pi} > L_{\pi\pi} v_w \bar{c}_w$), *non-congruent flux* is involved. For $L_{V\pi} > 0$, the solvent flux goes in the same direction as the salt flux, i.e. in the direction of the concentration gradient of the salt. This interesting phenomenon, termed *negative osmosis,* has not yet been utilized in practice.

Electrodialysis is a process for the separation of an electrolyte from the solvent and is used, for example, in desalination. This process occurs in a system with at least three compartments (in practice, a large number is often used). The terminal compartments contain the electrodes and the middle compartment is separated from the terminal compartments by ion-exchanger membranes, of which one membrane (1) is preferentially permeable for the cations and the other one (2) for the anions. Such a situation occurs when the concentration of the electrolyte in the compartments is less than the concentration of bonded ionic groups in the membrane. During current flow in the direction from membrane 1 to membrane 2, cations pass through membrane 1 in the same direction and anions pass through membrane 2 in the opposite direction. In order for the electrolyte to be accumulated in the central compartment, i.e. between membranes 1 and 2 (it is assumed for simplicity that a uni-univalent electrolyte is involved), the relative flux of the cations with respect to the flux of the solvent, $J_{D,+}$, and the relative flux of the anions with respect to that of the solvent, $J_{D,-}$, must be related by the expression

$$J_{D,+} = -J_{D,-} > 0 \tag{6.2.28}$$

When $\Delta p = 0$ (there is no pressure gradient in the system), $L_{\pi\pi} = 0$ (diffusion of ions through the membrane can be neglected), and if the conductivities of the two membranes are identical, then in view of the second of Eqs (6.2.18) this condition can be expressed as

$$L_{\pi\phi,+} = -L_{\pi\phi,-} > 0 \tag{6.2.29}$$

where $L_{\pi\phi,+}$ and $L_{\pi\phi,-}$ are the phenomenological coefficients appearing in Eqs (6.2.18) adjusted for the fluxes of the individual ionic species.

References

Bretag, A. H. (Ed.), *Membrane Permeability: Experiments and Models,* Techsearch Inc., Adelaide, 1983.
Flett, D. S. (Ed.), *Ion Exchange Membranes,* Ellis Horwood, Chichester, 1983.
Kedem, O., and A. Katchalsky, Permeability of composite membranes I, II, III, *Trans. Faraday Soc.,* **59,** 1918, 1931, 1941 (1963).
Lacey, R. E., and S. Loeb, *Industrial Processes with Membranes,* Wiley–Interscience, New York, 1972.
Lakshminarayanaiah, N., *Equations of Membrane Biophysics,* Academic Press, Orlando, 1984.
Meares, P. (Ed.), *Membrane Separation Processes,* Elsevier, Amsterdam, 1976.
Meares, P., J. F. Thain, and D. G. Dawson, Transport across ion-exchange membranes: The frictional model of transport, in *Membranes—A Series of Advances* (Ed. G. Eisenman), Vol. 1, p. 55, M. Dekker, New York, 1972.
Schlögl, R., see page 415.
Spiegler, K. S., *Principles of Desalination,* Academic Press, New York, 1969.
Strathmann, M., Membrane separation processes, *J. Membrane Sci.,* **9,** 121 (1981).
Woermann, D., Selektiver Stofftransport durch Membranen, *Ber. Bunseges.,* **83,** 1075 (1979).

6.3 Ion-selective Electrodes

Ion-selective electrodes are membrane systems used as potentiometric sensors for various ions. In contrast to ion-exchanger membranes, they contain a compact (homogeneous or heterogeneous) membrane with either fixed (solid or glassy) or mobile (liquid) ion-exchanger sites.

6.3.1 *Liquid-membrane ion-selective electrodes*

Liquid membranes in this type of ion-selective electrodes are usually heterogeneous systems consisting of a plastic film (polyvinyl chloride, silicon rubber, etc.), whose matrix contains an ion-exchanger solution as a plasticizer (see Fig. 6.5).

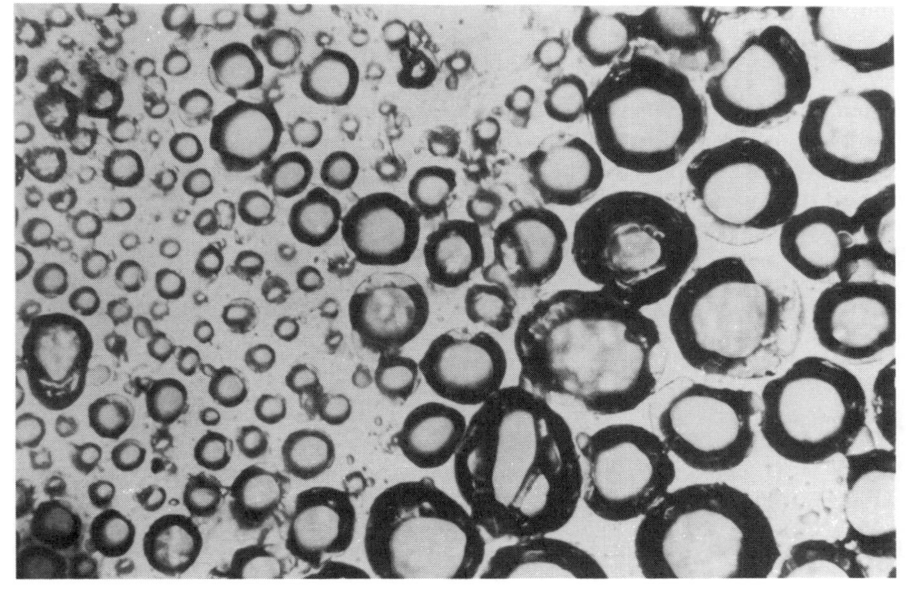

Fig. 6.5 Visible-light microscope photomicrograph showing pores (dark circles) in a polyvinyl chloride matrix membrane incorporating the ion-exchanger solution. (From G. H. Griffiths, G. J. Moody and D. R. Thomas)

In electrophysiology *ion-selective microelectrodes* are employed. These electrodes, resembling micropipettes (see Fig. 3.8), consist of glass capillaries drawn out to a point with a diameter of several micrometres, hydrophobized and filled with an ion-exchanger solution, forming the membrane in the ion-selective microelectrode.

The ion-selective electrode–test solution (analyte)–reference electrode system is shown in the scheme

$$\begin{array}{c} \quad\quad 1 \quad\quad\quad m \quad\quad\quad 2 \\ \text{ref. el. 1} \mid B_1A_2 \mid B_1A_1 \mid B_1A_2 \mid \text{ref. el. 2} \\ \Delta\phi_L(1) \quad\quad\quad\quad \Delta\phi_L(2) \end{array} \tag{6.3.1}$$

The membrane phase m is a solution of hydrophobic anion A_1^- (ion-exchanger ion) and cation B_1^+ in an organic solvent that is immiscible with water. Solution 1 (the test aqueous solution) contains the salt of cation B_1^+ with the hydrophilic anion A_2^-. The Gibbs transfer energy of anions A_1^- and A_2^- is such that transport of these anions into the second phase is negligible. Solution 2 (the internal solution of the ion-selective electrode) contains the salt of cation B_1^+ with anion A_2^- (or some other similar hydrophilic anion). The reference electrodes are identical and the liquid junction potentials $\Delta\phi_L(1)$ and $\Delta\phi_L(2)$ will be neglected.

The EMF of cell (6.3.1) is given by the relationship

$$\begin{aligned} E &= E_2 + \Delta\phi_L(2) + \Delta_m^2\phi + \Delta\phi_L(m) + \Delta_1^m\phi + \Delta\phi_L(1) - E_1 \\ &\approx \Delta_m^2\phi + \Delta_1^m\phi = \Delta\phi_M \end{aligned} \tag{6.3.2}$$

when the diffusion potential inside of the membrane, $\Delta\phi_L(m)$ is neglected. With respect to Eq. (3.2.48),

$$E = \frac{RT}{F} \ln \frac{a_{B_1^+}(1)}{a_{B_1^+}(2)} \tag{6.3.3}$$

As the concentration of the internal solution of the ion-selective electrode is constant, this type of electrode indicates the cation activity in the same way as a cation electrode (or as an anion electrode if the ion-exchanger ion is a hydrophobic cation).

Consider that the test solution 1 contains an additional cation B_2^+ that is identically or less hydrophilic than cation B_1^+. Then the exchange reaction

$$B_1^+(m) + B_2^+(w) \rightleftarrows B_1^+(w) + B_2^+(m) \tag{6.3.4}$$

occurs. The equilibrium constant of this reaction is (cf. page 63)

$$K_{exch} = \exp\left(-\frac{\Delta G^{0,o\to w}_{tr,B_1^+} - \Delta G^{0,o\to w}_{tr,B_2^+}}{RT}\right) \tag{6.3.5}$$

As the concentration of the ion-exchanger ion is constant in the membrane phase, it holds that

$$c_{B_1^+}(m) + c_{B_2^+}(m) = c_{A_1^-}(m) \tag{6.3.6}$$

For the potential difference $\Delta_1^m\phi$ (cf. Eqs 3.2.48 and 6.3.6) we have

$$\begin{aligned}\Delta_1^m\phi &= \frac{\Delta G^{0,o\to w}_{tr,B_1^+}}{F} + \frac{RT}{F}\ln\frac{a_{B_1^+}(1)}{c_{B_1^+}(m)\gamma_{B_1^+}(m)} \\ &= \frac{\Delta G^{0,o\to w}_{tr,B_2^+}}{F} + \frac{RT}{F}\ln\frac{a_{B_2^+}(1)}{c_{B_2^+}(m)\gamma_{B_2^+}(m)} \\ &= \frac{RT}{F}\ln\frac{a_{B_1^+}(1) + [\gamma_{B_1^+}(m)/\gamma_{B_2^+}(m)]K_{exch}a_{B_2^+}(1)}{\gamma_{B_1^+}c_{A_1^-}(m)\exp(\Delta G^{0,o\to w}_{tr,B_1^+}/RT)}\end{aligned} \tag{6.3.7}$$

Similarly, $\Delta_m^2\phi$ is given as

$$\Delta_m^2\phi = \frac{RT}{F}\ln\frac{\gamma_{B_1^+}c_{A_1-}(m)\exp(\Delta G^{0,o\to w}/RT)}{a_{B_1^+}(2)} \tag{6.3.8}$$

The EMF of cell (6.3.1) can be written as

$$E = E_{ISE} - E_{ref} \tag{6.3.9}$$

where the 'potential of the ion-selective electrode, E_{ISE}' is expressed by collecting the variables related to test solution 1 in one term and the variables related to the membrane, the internal solution of the ion-selective electrode and the internal reference electrode in another constant term so that in view of (6.3.2), (6.3.7) and (6.3.8)

$$\begin{aligned}E_{ISE} &= E_{0,ISE} + \frac{RT}{F}\ln\left[a_{B_1^+}(1) + \frac{\gamma_{B_1^+}(m)}{\gamma_{B_2^+}(m)}K_{exch}a_{B_2^+}(1)\right] \\ &= E_{0,ISE} + \frac{RT}{F}\ln[a_{B_1^+}(1) + K^{pot}_{B_1^+,B_2^+}a_{B_2^+}(1)]\end{aligned} \tag{6.3.10}$$

The quantity $K^{\text{pot}}_{B_1^+,B_2^+}$ is termed the *selectivity coefficient* for the determinand B_1^+ with respect to the interferent B_2^+. Obviously, for $a_{B_1^+}(1) \gg K^{\text{pot}}_{B_1^+,B_2^+} a_{B_2^+}(1)$, Eq. (6.3.9) is converted to Eq. (6.3.3), often expressed by stating that the potential of the ion-selective electrode depends on the logarithm of the activity of the determinand ion with Nernstian slope. A similar dependence is obtained for the ion B_2^+ when $a_{B_1^+}(1) \ll K^{\text{pot}}_{B_1^+,B_2^+} a_{B_2^+}(1)$. Equation (6.3.10) is termed the Nikolsky equation.

Among cations, potassium, acetylcholine, some cationic surfactants (where the ion-exchanger ion is the *p*-chlorotetraphenylborate or tetraphenylborate), calcium (long-chain alkyl esters of phosphoric acid as ion-exchanger ions), among anions, nitrate, perchlorate and tetrafluoroborate (long-chain tetraalkylammonium cations in the membrane), etc., are determined with this type of ion-selective electrodes.

Especially sensitive and selective potassium and some other ion-selective electrodes employ special complexing agents in their membranes, termed *ionophores* (discussed in detail on page 445). These substances, which often have cyclic structures, bind alkali metal ions and some other cations in complexes with widely varying stability constants. The membrane of an ion-selective electrode contains the salt of the determined cation with a hydrophobic anion (usually tetraphenylborate) and excess ionophore, so that the cation is mostly bound in the complex in the membrane. It can readily be demonstrated that the membrane potential obeys Eq. (6.3.3). In the presence of interferents, the selectivity coefficient is given approximately by the ratio of the stability constants of the complexes of the two ions with the ionophore. For the determination of potassium ions in the presence of interfering sodium ions, where the ionophore is the cyclic depsipeptide, valinomycin, the selectivity coefficient is $K^{\text{pot}}_{K^+,Na^+} \approx 10^{-4}$, so that this electrode can be used to determine potassium ions in the presence of a 10^4-fold excess of sodium ions.

6.3.2 *Ion-selective electrodes with fixed ion-exchanger sites*

This type of membrane consists of a water-insoluble solid or glassy electrolyte. One ionic sort in this electrolyte is bound in the membrane structure, while the other, usually but not always the determinand ion, is mobile in the membrane (see Section 2.6). The theory of these ion-selective electrodes will be explained using the glass electrode as an example; this is the oldest and best known sensor in the whole field of ion-selective electrodes.

The membrane of the glass electrode is blown on the end of a glass tube. This tube is filled with a solution with a constant pH (acetate buffer, hydrochloric acid) and a reference electrode is placed in this solution (silver chloride or calomel electrodes). During the measurement, this whole system is immersed with another reference electrode into the test solution. The membrane potential of the glass electrode, when the internal and analysed solutions are not too alkaline, is given by the equation

$$\Delta\phi_M = \frac{RT}{F}\ln\left[\frac{a_{H_3O^+}(1)}{a_{H_3O^+}(2)}\right] \qquad (6.3.11)$$

The theory of the function of the glass electrode is based on the concept of exchange reactions at the surface of the glass. The glass consists of a solid silicate matrix in which the alkali metal cations are quite mobile. The glass membrane is hydrated to a depth of about 100 nm at the surface of contact with the solution and the alkali metal cations can be exchanged for other cations in the solution, especially hydrogen ions. For example, the following reaction occurs at the surface of sodium glass:

$$Na^+(\text{glass}) + H^+(\text{solution}) \rightleftarrows H^+(\text{glass}) + Na^+(\text{solution}) \qquad (6.3.12)$$

characterized by an ion-exchange equilibrium constant (cf. Eq. (6.3.5)).

These concepts were developed kinetically by M. Dole and thermodynamically by B. P. Nikolsky. The last-named author deduced a relationship that is analogous to Eq. (6.3.10). It is approximated by Eq. (6.3.11) for pH 1–10 and corresponds well to deviations from this equation in the alkaline region. The potentials measured in the alkaline region are, of course, lower than those corresponding to Eq. (6.3.11). The difference between the measured potential and that calculated from Eq. (6.3.11) is termed the alkaline or 'sodium' error of the glass electrode, and depends on the type of glass and the cation in solution (see Fig. 6.6). This dependence is understandable on the basis of the concepts given above. The cation of the glass can be replaced by hydrogen or by some other cation that is of the same size or smaller than the original cation. The smaller the cation in the glass, the fewer the ionic sorts other than hydrogen ion that can replace it and the greater the concentration of these ions must be in solution for them to enter the surface to a significant degree. The smallest alkaline error is thus exhibited by lithium glass electrodes. For a given type of glass, the error is greatest in LiOH solutions, smaller in NaOH solutions, etc.

Glass for glass electrodes must have rather low resistance, small alkaline error (so that the electrodes can be used in as wide a pH range as possible) and low solubility (so that the pH in the solution layer around the electrode is not different from that of the analyte). These requirements are contradictory to a certain degree. For example, lithium glasses have a small alkaline error but are rather soluble. Examples of glasses suitable for glass electrodes are, for example, Corning 015 with a composition of 72% SiO_2, 6% CaO and 22% Na_2O, and lithium glass with 72% SiO_2, 6% CaO and 22% Li_2O.

Glasses containing aluminium oxide or oxides of other trivalent metals exhibit high selectivity for the alkali metal ions, often well into the acid pH region. A glass electrode with a glass composition of 11 mol.% Na_2O, 18 mol.% Al_2O_3 and 71 mol.% SiO_2, which is sensitive for sodium ions and

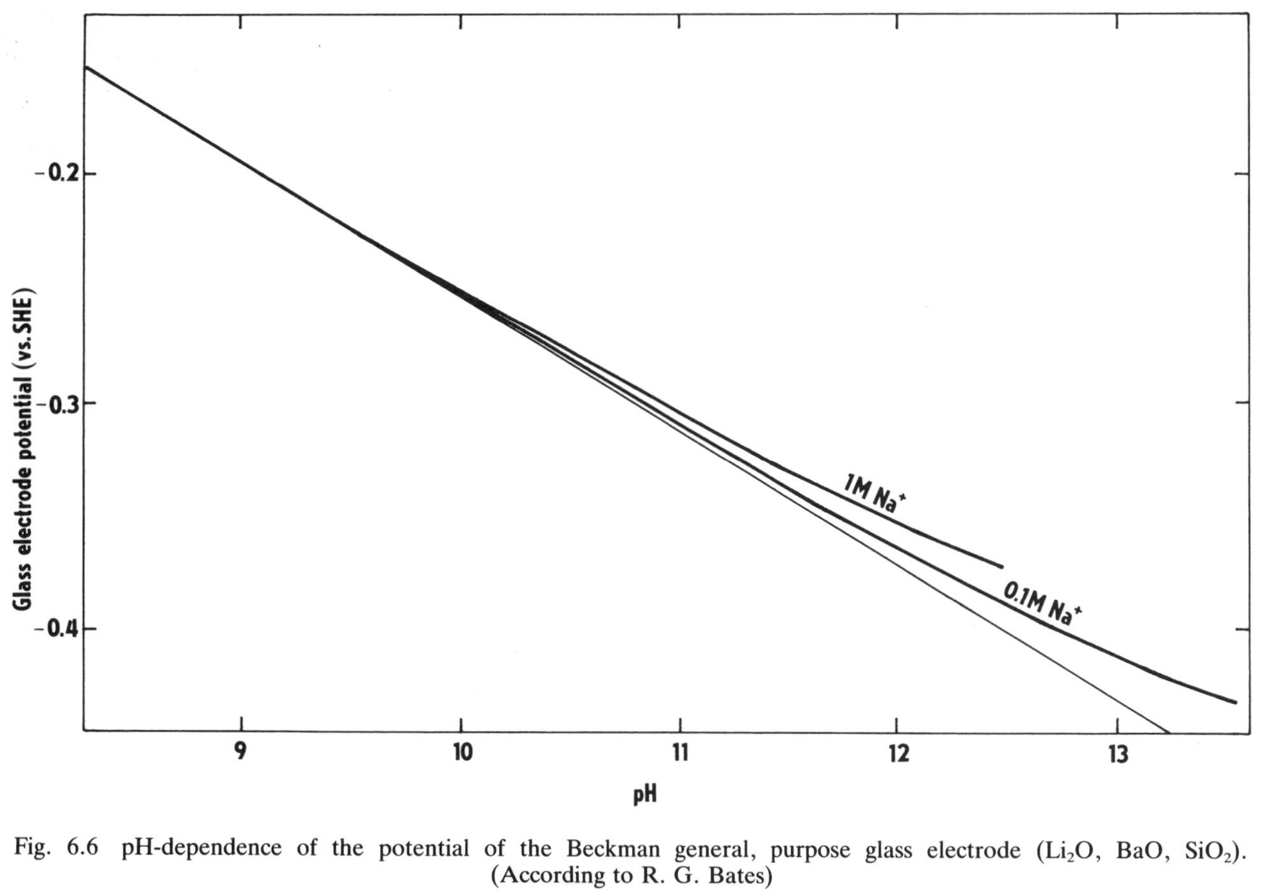

Fig. 6.6 pH-dependence of the potential of the Beckman general, purpose glass electrode (Li_2O, BaO, SiO_2). (According to R. G. Bates)

poorly sensitive for both hydrogen and potassium ions, has found a wide application. At pH 11 its selectivity constant for potassium ions with respect to sodium ions is $K^{pot}_{Na^+,K^+} = 3.6 \times 10^{-3}$.

The membranes of the other ion-selective electrodes can be either homogeneous (a single crystal, a pressed polycrystalline pellet) or heterogeneous, where the crystalline substance is incorporated in the matrix of a suitable polymer (e.g. silicon rubber or Teflon). The equation controlling the potential is analogous to Eq. (6.3.9).

Silver halide electrodes (with properties similar to electrodes of the second kind) are made of AgCl, AgBr and AgI. These electrodes, containing also Ag_2S, are used for the determination of Cl^-, Br^-, I^- and CN^- ions in various inorganic and biological materials.

The lanthanum fluoride electrode (discussed in Section 2.6) is used to determine F^- ions in neutral and acid media. After the pH–glass electrode, this is the most important of this group of electrodes.

The silver sulphide electrode is the most reliable electrode of this kind and is used to determine S^{2-}, Ag^+ and Hg^{2+} ions.

Electrodes containing a mixture of divalent metal sulphides and Ag_2S are used to determine Pb^{2+}, Cu^{2+} and Cd^{2+}.

6.3.3 *Calibration of ion-selective electrodes*

It has been emphasized repeatedly that the individual activity coefficients cannot be measured experimentally. However, these values are required for a number of purposes, e.g. for calibration of ion-selective electrodes. Thus, a conventional scale of ionic activities must be defined on the basis of suitably selected standards. In addition, this definition must be consistent with the definition of the conventional activity scale for the oxonium ion, i.e. the definition of the practical pH scale. Similarly, the individual scales for the various ions must be mutually consistent, i.e. they must satisfy the relationship between the experimentally measurable mean activity of the electrolyte and the defined activities of the cation and anion in view of Eq. (1.1.11). Thus, by using galvanic cells without transport, e.g. a sodium-ion-selective glass electrode and a Cl^--selective electrode in a NaCl solution, a series of $a_\pm$(NaCl) is obtained from which the individual ion activity a_{Na^+} is determined on the basis of the Bates–Guggenheim convention for a_{Cl^-} (page 37). Table 6.1 lists three such standard solutions, where pNa = $-\log a_{Na^+}$, etc.

6.3.4 *Biosensors and other composite systems*

Devices based on the glass electrode can be used to determine certain gases present in gaseous or liquid phase. Such a *gas probe* consists of a glass electrode covered by a thin film of a plastic material with very small pores,

Table 6.1 Conventional standards of ion activities. (According to R. G. Bates and M. Alfenaar)

Electrolyte	Molality ($mol \cdot kg^{-1}$)	pNa	pCa	pCl	pF
NaCl	0.001	3.015		3.015	
	0.01	2.044		2.044	
	0.1	1.108		1.110	
	1.0	0.160		0.204	
NaF	0.001	3.015			3.105
	0.01	2.044			2.048
	0.1	1.108			1.124
$CaCl_2$	0.000 333		3.530	3.191	
	0.003 33		2.653	2.220	
	0.033 3		1.883	1.286	
	0.333		1.105	0.381	

which are hydrophobic, so that the solution cannot penetrate the pores. A thin layer of an indifferent electrolyte is present between the surface of the film and the glass electrode and is in contact with a reference electrode. The gas (ammonia, carbon dioxide etc.) permeates through the pores of the film, dissolves in the solution at the surface of the glass electrode and induces a change of pH. For the determination of the concentration of the gas a calibration plot is used.

The *Clark oxygen sensor* is based on a similar principle. It contains an amperometric Pt electrode indicating the concentration of oxygen permeating through the pores of the membrane. In *enzyme electrodes* a hydrophilic polymer layer contains an enzyme which transforms the determinand into a substance that is sensed by a suitable electrode. There are potentiometric enzyme electrodes based mainly on glass electrodes as, for example, the urea electrode, but the most important is the *amperometric glucose sensor*. It contains β-glucose oxidase immobilized in a polyacrylamide gel and the Clark oxygen sensor. The enzyme reaction is

$$\text{glucose} + O_2 + H_2O \xrightarrow[\text{oxidase}]{\text{glucose}} H_2O_2 + \text{gluconic acid}$$

The decrease of oxygen reduction current measured with the Clark oxygen sensor indicates the concentration of glucose.

Enzyme electrodes belong to the family of *biosensors*. These also include systems with tissue sections or immobilized microorganism suspensions playing an analogous role as immobilized enzyme layers in enzyme electrodes. While the stability of enzyme electrode systems is the most difficult problem connected with their practical application, this is still more true with the *bacteria* and *tissue electrodes*.

References

Ammann, D., *Ion-Selective Microelectrodes,* Springer-Verlag, Berlin, 1986.
Baucke, F. G. K., The glass electrode—Applied electrochemistry of glass surfaces, *J. Non-Crystall. Solids,* **73,** 215 (1985).
Berman, H. J., and N. C. Hebert (Eds), *Ion-Selective Microelectrodes,* Plenum Press, New York, 1974.
Cammann, K., *Working with Ion-Selective Electrodes,* Springer-Verlag, Berlin, 1979.
Durst, R. A. (Ed.), *Ion-Selective Electrodes,* National Bureau of Standards, Washington, 1969.
Eisenman, G. (Ed.), *Glass Electrodes for Hydrogen and Other Cations,* M. Dekker, New York, 1967.
Freiser, H. (Ed.), *Analytical Application of Ion-Selective Electrodes,* Plenum Press, New York, Vol. 1, 1978; Vol. 2, 1980.
Janata, J., *The Principles of Chemical Sensors.* Plenum Press, New York, 1989.
Koryta, J. (Ed.), *Medical and Biological Applications of Electrochemical Devices,* John Wiley & Sons, Chichester, 1980.
Koryta, J., and K. Štulík, *Ion-Selective Electrodes,* 2nd ed., Cambridge University Press, Cambridge, 1983.
Morf, W. E., *The Principles of Ion-Selective Electrodes and of Membrane Transport,* Akadémiai Kiadó, Budapest and Elsevier, Amsterdam, 1981.
Scheller, F., and F. Schubert, *Biosensors,* Elsevier, Amsterdam, 1992.
Syková, E., P. Hník, and L. Vyklický (Eds), *Ion-Selective Microelectrodes and Their Use in Excitable Tissues,* Plenum Press, New York, 1981.
Thomas, R. C., *Ion-Sensitive Intracellular Microelectrodes. How To Make and Use Them,* Academic Press, London, 1978.
Turner, A. P. F., I. Karube and G. S. Wilson, *Biosensors, Fundamentals and Applications,* Oxford University Press, Oxford, 1989.
Zeuthen, I. (Ed.), *The Application of Ion-Selective Microelectrodes,* Elsevier, Amsterdam, 1981.

6.4 Biological Membranes

All living organisms consist of cells, and the higher organisms have a multicellular structure. In addition, the individual cells contain specialized formations called organelles like the nucleus, the mitochondria, the endoplasmatic reticulum, etc. (see Fig. 6.7). A great many life functions occur at the surfaces of cells and organelles, such as conversion of energy, the perception of external stimuli and the transfer of information, the basic elements of locomotion and metabolism. The large 'internal' surface of the organism at which these processes occur leads to their immense variety. The surfaces of cells and cellular organelles consist of biological membranes. In contrast to those discussed in the previous text these membranes are incomparably thinner. Electron microscope studies show them to consist of three layers with a thickness of 7–15 nm (see Fig. 6.8). In 1925, E. Gorter and F. Grendel isolated phospholipids from red blood cells and spread them out as a monolayer on a water–air interface. The overall surface of this monolayer was approximately twice as large as the surface of the red blood cells. The authors concluded that the erythrocyte membrane consists of a

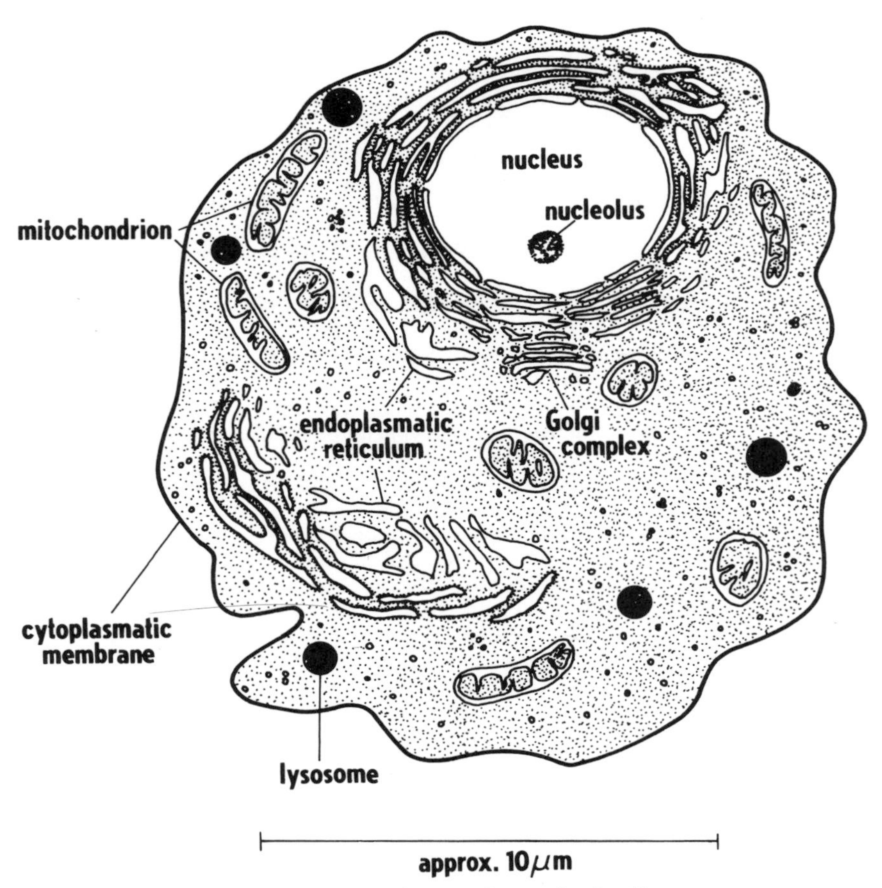

Fig. 6.7 A scheme of an animal cell

bimolecular layer (bilayer) of phospholipids, with their long alkyl chains oriented inwards (Fig. 6.9). Electron microscope studies have revealed that the central layer corresponds to this largely hydrocarbon region. In fact, biological membranes are still more complex as will be discussed in Section 6.4.2.

6.4.1 *Composition of biological membranes*

Biological membranes consist of lipids, proteins and also sugars, sometimes mutually bonded in the form of lipoproteins, glycolipids and glycoproteins. They are highly hydrated—water forms up to 25 per cent of the dry weight of the membrane. The content of the various protein and lipid components varies with the type of biological membrane. Thus, in

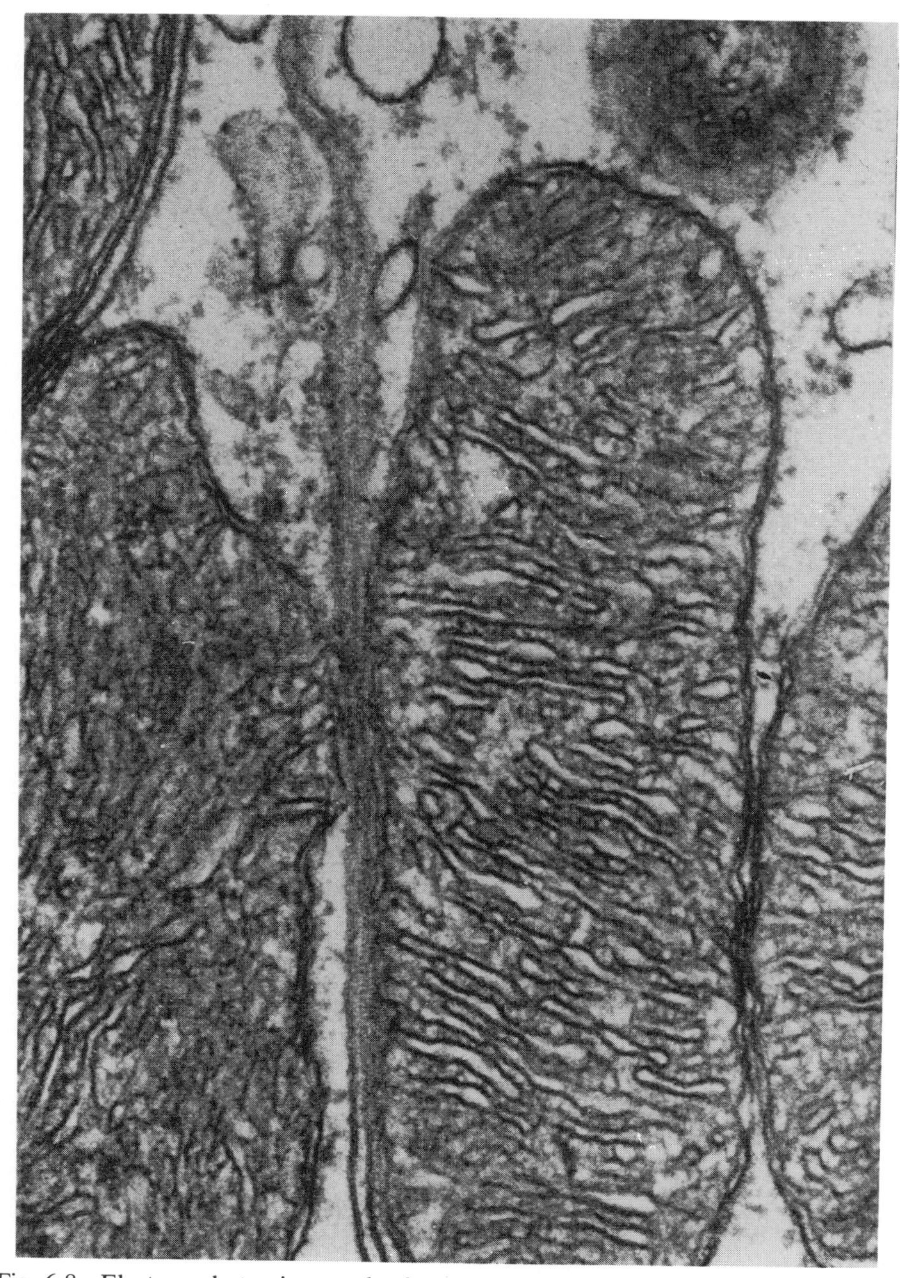

Fig. 6.8 Electron photomicrograph of mouse kidney mitochondria. The structure of both the cytoplasmatic membrane (centre) and the mitochondrial membranes is visible on the ultrathin section. Magnification 70,000×. (By courtesy of J. Ludvík)

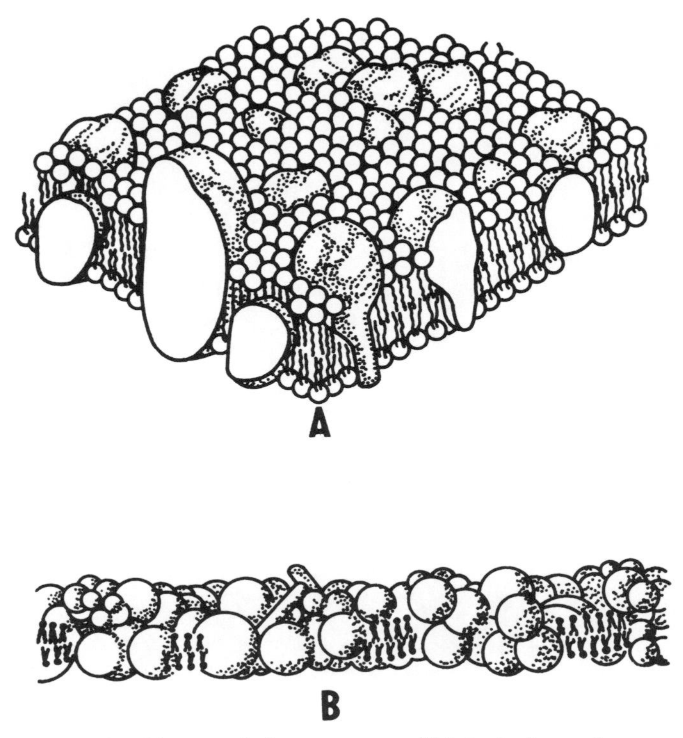

Fig. 6.9 Characteristic structures of biological membranes. (A) The fluid mosaic model (S. J. Singer and G. L. Nicholson) where the phospholipid component is predominant. (B) The mitochondrial membrane where the proteins prevail over the phospholipids

myelin cell membranes (myelin forms the sheath of nerve fibres) the ratio protein : lipid is 1 : 4 while in the inner membrane of a mitochondrion (cf. page 464) 10 : 3.

The lipid component consists primarily of phospholipids and cholesterol. The most important group of phospholipids are phosphoglycerides, based on phosphatidic acid (where X = H), with the formula

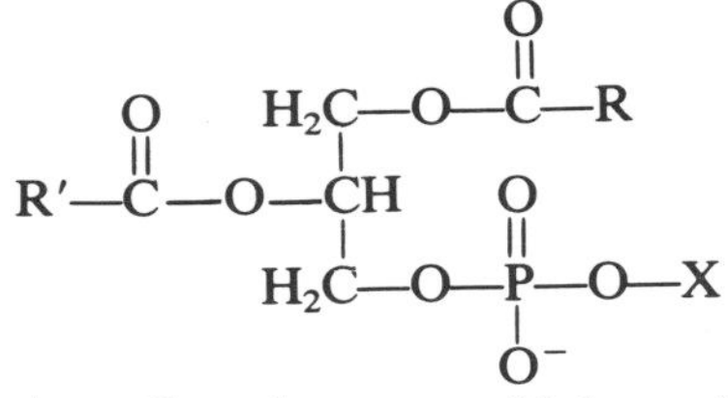

where R and R′ are alkyl or alkenyl groups with long chains. Thus, glycerol is esterified at two sites by higher fatty acids such as palmitic, stearic, oleic, linolic, etc., acids and phosphoric acid is bound to the remaining alcohol group. In phospholipids, the phosphoric acid is usually bound to nitrogen-substituted ethanolamines. For example, the phospholipid lecithin is formed by combination with choline ($X = —CH_2CH_2N^+(CH_3)_3$). Other molecules can also be bound, such as serin, inositol or glycerol. Some phospholipids contain ceramide in place of glycerol:

$$\begin{array}{l} CH(OH)CH{=}CH(CH_2)_{12}CH_3 \\ | \\ CHNHCOR \\ | \\ CH_2OH \end{array}$$

where R is a long-chain alkyl group. The ester of ceramide with phosphoric acid bound to choline is sphingomyelin. When the ceramide is bound to a sugar (such as glucose or galactose) through a β-glycosidic bond, cerebrosides are formed. The representation of various lipidic species strongly varies among biological membranes. Thus, the predominant component of the cytoplasmatic membrane of bacterium *Bacillus subtilis* is phosphatidyl glycerol (78 wt.%) while the main components of the inner membrane from rat liver mitochondrion are phosphatidylcholine = lecithin (40 wt.%) and phophatidylethanolamine = cephalin (35 wt.%).

Cholesterol contributes to greater ordering of the lipids and thus decreases the fluidity of the membrane.

Proteins either strengthen the membrane structure (building proteins) or fulfil various transport or catalytic functions (functional proteins). They are often only electrostatically bound to the membrane surface (extrinsic proteins) or are covalently bound to the lipoprotein complexes (intrinsic or integral proteins). They are usually present in the form of an α-helix or random coil. Some integral proteins penetrate through the membrane (see Section 6.4.2).

Saccharides constitute 1–8 per cent of the dry membrane weight in mammals, for example, while this content increases to up to 25 per cent in amoebae. They are arranged in heteropolysaccharide (glycoprotein or glycolipid) chains and are covalently bound to the proteins or lipids. The main sugar components are L-fucose, galactose, manose and sometimes glucose, *N*-acetylgalactosamine and *N*-acetylglucosamine. Sialic acid is an important membrane component:

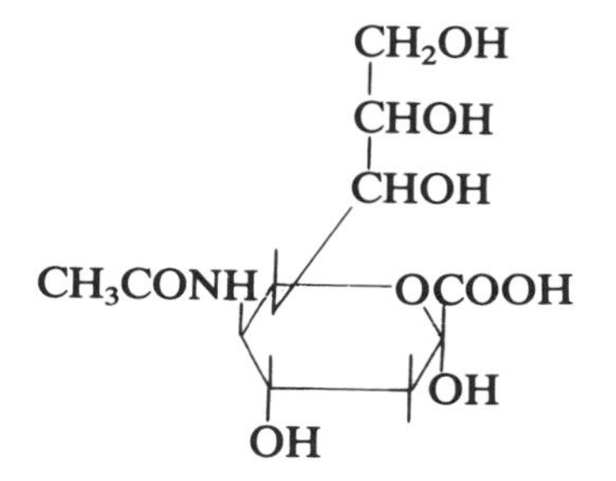

This substance often constitutes the terminal unit in heteropolysaccharide chains, and contributes greatly to the surface charge of the membrane.

6.4.2 *The structure of biological membranes*

Following the original simple concepts of Gorter and Grendel, a large number of membrane models have been developed over the subsequent half a century; the two most contrasting are shown in Fig. 6.9.

The basic characteristic of the membrane structure is its asymmetry, reflected not only in variously arranged proteins, but also in the fact that, for example, the outside of cytoplasmatic (cellular) membranes contains uncharged lecithin-type phospholipids, while the polar heads of strongly charged phospholipids are directed into the inside of the cell (into the cytosol).

Phospholipids, which are one of the main structural components of the membrane, are present primarily as bilayers, as shown by molecular spectroscopy, electron microscopy and membrane transport studies (see Section 6.4.4). Phospholipid mobility in the membrane is limited. Rotational and vibrational motion is very rapid (the amplitude of the vibration of the alkyl chains increases with increasing distance from the polar head). Lateral diffusion is also fast (in the direction parallel to the membrane surface). In contrast, transport of the phospholipid from one side of the membrane to the other (flip-flop) is very slow. These properties are typical for the liquid–crystal type of membranes, characterized chiefly by ordering along a single coordinate. When decreasing the temperature (passing the transition or Kraft point, characteristic for various phospholipids), the liquid–crystalline bilayer is converted into the crystalline (gel) structure, where movement in the plane is impossible.

The cells of the higher plants, algae, fungi and higher bacteria have a cell wall in addition to the cell membrane, protecting them from mechanical damage (e.g. membrane rupture if the cell is in a hypotonic solution). The cell wall of *green plants* is built from polysaccharides, such as cellulose or hemicelluloses (water-insoluble polysaccharides usually with branched structure), and occasionally from a small amount of glycoprotein (sugar-protein complex). Because of acid groups often present in the pores of the cell wall it can behave as an ion-exchanger membrane (Section 6.2).

Gram-positive bacteria (which stain blue in the procedure suggested by H. Ch. J. Gram in 1884) have their cell walls built of cross-linked polymers of amino acids and sugars (peptidoglycans). The actual surface of the bacterium is formed by teichoic acid (a polymer consisting of a glycerol-phosphate backbone with linked glucose molecules) which makes it hydrophilic and negatively charged. In this case the cell wall is a sort of giant macromolecule—a bag enclosing the whole cell.

The surface structure of *gram-negative bacteria* (these are not stained by Gram's method and must be stained red with carbol fuchsin) is more diversified. It consists of an outer membrane whose main building unit is a lipopolysaccharide together with phospholipids and proteins. The actual cell wall is made of a peptidoglycan. A loosely bounded polysaccharide layer termed a glycocalyx is formed on the surface of some animal cells.

6.4.3 *Experimental models of biological membranes*

Phospholipids are amphiphilic substances; i.e. their molecules contain both hydrophilic and hydrophobic groups. Above a certain concentration level, amphiphilic substances with one ionized or polar and one strongly hydrophobic group (e.g. the dodecylsulphate or cetyltrimethylammonium ions) form micelles in solution; these are, as a rule, spherical structures with hydrophilic groups on the surface and the inside filled with the hydrophobic parts of the molecules (usually long alkyl chains directed radially into the centre of the sphere). Amphiphilic substances with two hydrophobic groups have a tendency to form bilayer films under suitable conditions, with hydrophobic chains facing one another. Various methods of preparation of these *bilayer lipid membranes* (BLMs) are demonstrated in Fig. 6.10.

If a dilute electrolyte solution is divided by a Teflon foil with a small window to which a drop of a solution of a lipid in a suitable solvent (e.g. octane) is applied, the following phenomenon is observed (Fig. 6.10A). The layer of lipid solution gradually becomes thinner, interference rainbow bands appear on it, followed by black spots and finally the whole layer becomes black. This process involves conversion of the lipid layer from a multimolecular thickness to form a bilayer lipid membrane. Similar effects were observed on soap bubbles by R. Hooke in 1672 and I. Newton in 1702. Membrane thinning is a result of capillary forces and dissolving of components of the membrane in the aqueous solution with which it is in contact. The thick layer at the edge of the membrane is termed the Plateau–Gibbs boundary. Membrane blackening is a result of interference between incident and reflected light. If the membrane thickness is much less than one quarter of the wavelength of the incident light, then the waves of the incident and reflected light interfere and the membrane appears black against a dark background. The membrane thickness can be found from the reflectance of light with a low angle of incidence, from measurements of the membrane capacity and from electron micrographs (application of a metal coating to the membrane can, however, lead to artefact formation). The thickness of a BLM prepared from different materials lies in the range between 4 and 13 nm.

A BLM can even be prepared from phospholipid monolayers at the water–air interface (Fig. 6.10B) and often does not then contain unfavourable organic solvent impurities. An asymmetric BLM can even be prepared containing different phospholipids on the two sides of the membrane. A method used for preparation of tiny segments of biological membranes (patch-clamp) is also applied to BLM preparation (Fig. 6.10C).

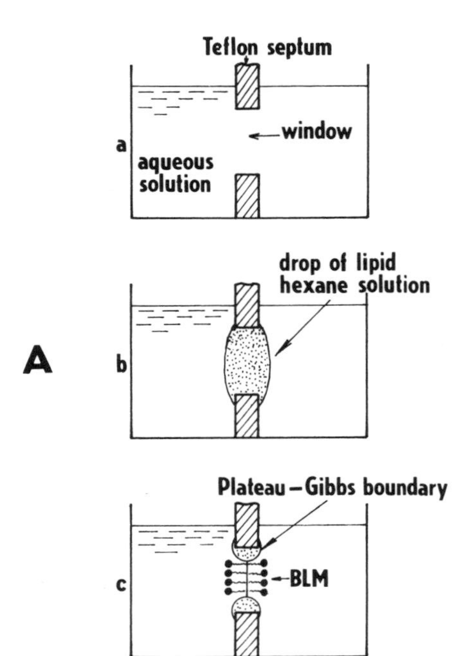

Fig. 6.10 Methods of preparation of bilayer lipid membranes. (A) A Teflon septum with a window of approximately 1 mm^2 area divides the solution into two compartments (a). A drop of a lipid–hexane solution is placed on the window (b). By capillary forces the lipid layer is thinned and a bilayer (black in appearance) is formed (c) (P. Mueller, D. O. Rudin, H. Ti Tien and W. D. Wescot). (B) The septum with a window is being immersed into the solution with a lipid monolayer on its surface (a). After immersion of the whole window a bilayer lipid membrane is formed (b) (M. Montal and P. Mueller). (C) A drop of lipid–hexane solution is placed at the orifice of a glass capillary (a). By slight sucking a bubble-formed BLM is shaped (b) (U. Wilmsen, C. Methfessel, W. Hanke and G. Boheim)

The conductivity of membranes that do not contain dissolved ionophores or lipophilic ions is often affected by cracking and impurities. The value for a completely compact membrane under reproducible conditions excluding these effects varies from 10^{-8} to $10^{-10}\,\Omega^{-1} \cdot cm^{-2}$. The conductivity of these simple 'unmodified' membranes is probably statistical in nature (as a result of thermal motion), due to stochastically formed pores filled with water for an instant and thus accessible for the electrolytes in the solution with which the membrane is in contact. Various active (natural or synthetic) substances

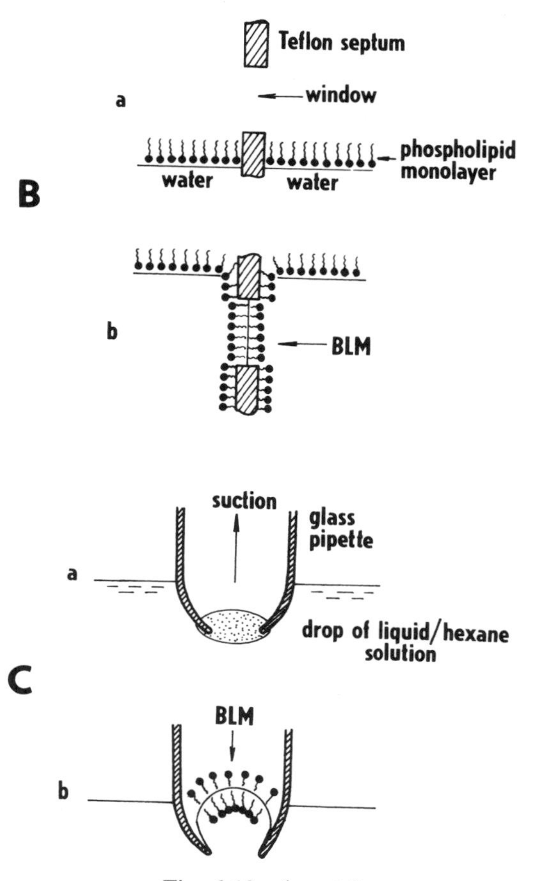

Fig. 6.10 (*cont'd*)

are often introduced into a BLM that preferably consists of synthetic phospholipids to limit irreproducible effects resulting from the use of natural materials. Sometimes definite segments of biological membranes (e.g. photosynthetic centres of thylakoids) are imbedded in a BLM. The described type of planar BLM can be used for electrochemical measurements (of the membrane potential or current–potential dependence) and radiometric measurements of the permeation of labelled molecules.

The second model of a biological membrane is the *liposome* (lipid vesicle), formed by dispersing a lipid in an aqueous solution by sonication. In this way, small liposomes with a single BLM are formed (Fig. 6.11), with a diameter of about 50 nm. Electrochemical measurements cannot be carried out directly on liposomes because of their small dimensions. After addition of a lipid-soluble ion (such as the tetraphenylphosphonium ion) to the bathing solution, however, its distribution between this solution and the liposome is measured, yielding the membrane potential according to Eq.

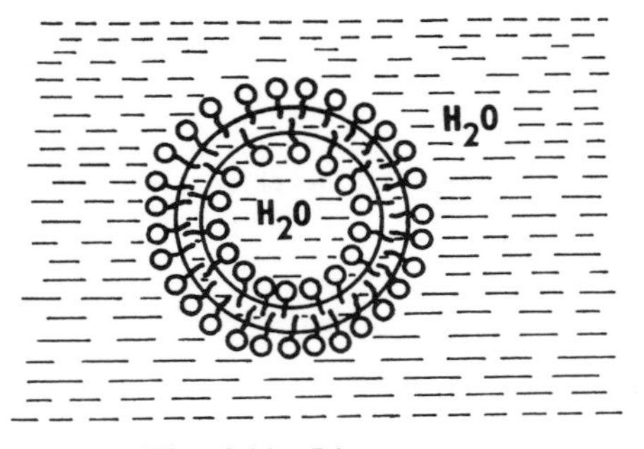

Fig. 6.11 Liposome

(6.1.3). A planar BLM cannot be investigated by means of the molecular spectroscopical methods because of the small amount of substance in an individual BLM. This disadvantage is removed for liposomes as they can form quite concentrated suspensions. For example, in the application of electron spin resonance (ESR) a 'spin-labelled' phospholipid is incorporated into the liposome membrane; this substance can be a phospholipid with, for example, a 2,2,6,6-tetramethylpiperidyl-*N*-oxide (TEMPO) group:

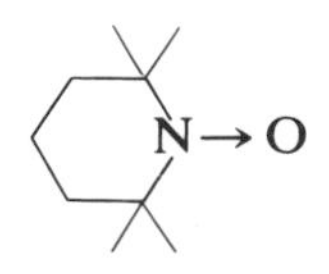

containing an unpaired electron. The properties of the ESR signal of this substance yield information on its motion and location in the membrane.

Liposomes also have a number of practical applications and can be used, for example, to introduce various pharmaceuticals into the organism and even for the transfer of plasmids from one cell to another.

6.4.4 *Membrane transport*

Passive transport. Similar to the lipid model, the non-polar interior of a biological membrane represents a barrier to the transport of hydrophilic substances; this is a very important function in preventing unwelcome substances from entering the cell and essential substances from leaving the cell. This property is demonstrated on a BLM which is almost impermeable for hydrophilic substances. Polar non-electrolytes that are hydrated in the aqueous solution can pass through the membrane, with a high activation energy for this process, 60–80 kJ · mol^{-1}. This activation is primarily a result of the large value of the Gibbs energy of transfer, i.e. the difference in solvation energy for the substance during transfer from the aqueous phase into the non-polar interior of the membrane.

In contrast, hydrophobic molecules and ions pass quite readily through the cell or model membrane. This phenomenon is typical for hydrophobic ions such as tetraphenylborate, dipicrylaminate, triiodide and tetraalkylammonium ions with long alkyl chains, etc. The transfer of hydrophobic ions through a BLM with applied external voltage consists of three steps:

(a) Adsorption of the ion on the BLM
(b) Transfer across the energy barrier within the BLM
(c) Desorption from the other side of the BLM

As the membrane has a surface charge leading to formation of a diffuse electrical layer, the adsorption of ions on the BLM is affected by the potential difference in the diffuse layer on both outer sides of the membrane ϕ_2 (the term *surface potential* is often used for this value in biophysics). Figure 6.12 depicts the distribution of the electric potential in the membrane and its vicinity. It will be assumed that the concentration c of the transferred univalent cation is identical on both sides of the membrane and that adsorption obeys a linear isotherm. Its velocity on the p side of the membrane (see scheme 6.1.1) is then

$$\frac{\mathrm{d}\Gamma(\mathrm{p})}{\mathrm{d}t} = k_\mathrm{a}c_\mathrm{s} - k_\mathrm{d}\Gamma(\mathrm{p}) - \vec{k}\Gamma(\mathrm{p}) + \overleftarrow{k}\Gamma(\mathrm{q}) \tag{6.4.1}$$

where $\Gamma(\mathrm{p})$ and $\Gamma(\mathrm{q})$ are the surface concentrations of the adsorbed ion on the membrane surfaces, k_a and k_d are the rates of adsorption and desorption, respectively, c_s is the volume concentration of the ion at the surface of the membrane given by Eq. (4.3.5) and $\vec{k}$ and $\overleftarrow{k}$ are the rate constants for ion transfer. This process can be described by using the relationship of electrochemical kinetics (5.2.24) for $\alpha = \frac{1}{2}$:

$$\begin{aligned} j &= Fk^{\ominus}\left[\Gamma(\mathrm{p})\exp\left(\frac{F\Delta\phi}{2RT}\right) - \Gamma(\mathrm{q})\exp\left(-\frac{F\Delta\phi}{2RT}\right)\right] \\ &= \vec{k}\Gamma(\mathrm{p}) - \overleftarrow{k}\Gamma(\mathrm{q}) \end{aligned} \tag{6.4.2}$$

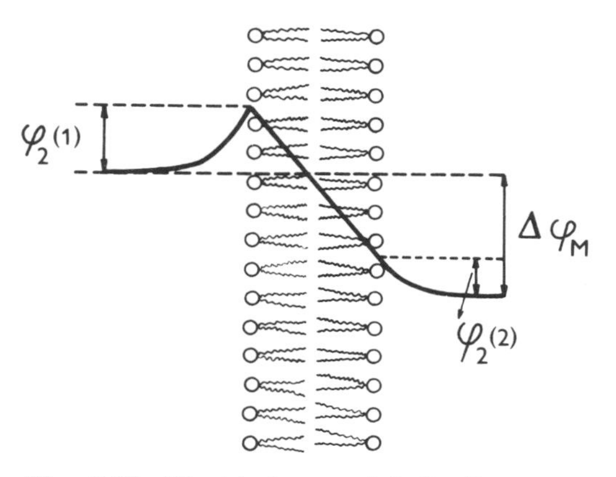

Fig. 6.12 Electrical potential distribution in the BLM and in its surroundings

where $\Delta\phi$ is the applied potential difference. At steady state $(\mathrm{d}\Gamma(\mathrm{p})/\mathrm{d}t = \mathrm{d}\Gamma(\mathrm{q})/\mathrm{d}t = 0)$,

$$\begin{aligned}\Gamma(\mathrm{p}) &= \frac{k_\mathrm{a} c_\mathrm{s}(k_\mathrm{d} + 2\overleftarrow{k})}{k_\mathrm{d}(k_\mathrm{d} + \overrightarrow{k} + \overleftarrow{k})} \\ \Gamma(\mathrm{q}) &= \frac{k_\mathrm{a} c_\mathrm{s}(k_\mathrm{d} + 2\overrightarrow{k})}{k_\mathrm{d}(k_\mathrm{d} + \overrightarrow{k} + \overleftarrow{k})}\end{aligned} \tag{6.4.3}$$

The conductivity of the membrane at the equilibrium potential $(\Delta\phi = 0)$, equal to the reciprocal of the polarization resistance value (Eq. 5.2.31), follows from Eqs (6.4.2), (6.4.3) and (4.3.5):

$$\begin{aligned}G_0 &= \frac{1}{R_\mathrm{p}} = \frac{z^2F^2k^\ominus}{RT} \exp\left(-\frac{F\phi_2}{RT}\right)\frac{k_\mathrm{a}}{k_\mathrm{d}}c \\ &= \frac{z^2F^2k^\ominus \beta c}{RT} \exp\left(-\frac{F\phi_2}{RT}\right)\end{aligned} \tag{6.4.4}$$

where $\beta = k_\mathrm{a}/k_\mathrm{d}$ is the adsorption coefficient and c the bulk concentration of the ion.

Equation (6.4.4) is valid when the coverage of the electrolyte–membrane interface is small. At higher concentrations of transferred ion, the ion transfer is retarded by adsorption on the opposite interface, so that the dependence of G_0 on c is characterized by a curve with a maximum, as has been demonstrated experimentally.

Under certain conditions, the transfer of various molecules across the membrane is relatively easy. The membrane must contain a suitable 'transport mediator', and the process is then termed 'facilitated membrane transport'. Transport mediators permit the transported hydrophilic substance to overcome the hydrophobic regions in the membrane. For example, the transport of glucose into the red blood cells has an activation energy of only $16\,\mathrm{kJ \cdot mol^{-1}}$—close to simple diffusion.

Either the transport mediators bind the transported substances into their interior in a manner preventing them from contact with the hydrophobic interior of the membrane or they modify the interior of the membrane so that it becomes accessible for the hydrophilic particles.

A number of transport mediators are *transport proteins*; in the absence of an external energy supply, thermal motion leads to their conformational change or rotation so that the transported substance, bound at one side of the membrane, is transferred to the other side of the membrane. This type of mediator has a limited number of sites for binding the transported substance, so that an increase in the concentration of the latter leads to saturation. Here, the transport process is characterized by specificity for a given substance and inhibition by other transportable substances competing for binding sites and also by various inhibitors. When the concentrations of the transported substance are identical on both sides of the membrane, *exchange transport* occurs (analogous to the exchange current—see Section 5.2.2). When, for example, an additional substance B is added to solution 1 under these conditions, it is also transported and competes with substance A for binding sites on the mediator; i.e. substance A is transported from solution 2 into solution 1 in the absence of any differences in its concentration.

A further type of mediator includes substances with a relatively low molecular weight that characteristically facilitate the transport of ions across biological membranes and their models. These transport mechanisms can be divided into four groups:

(a) Transport of a stable compound of the ion carrier (*ionophore*) with the transported substance.
(b) *Carrier relay*. The bond between the transported particle and the ionophore is weak so that it jumps from one associate with the ionophore to another during transport across the membrane.
(c) An *ion-selective channel*. The mediator is incorporated in a transversal position across the membrane and permits ion transport. Examples are substances with helical molecular structure, where the ion passes through the helix. The selectivity is connected with the ratio of the helix radius to that of the ion.
(d) A *membrane pore* that permits hydrodynamic flux through the membrane. No selectivity is involved here.

A number of substances have been discovered in the last thirty years with a macrocyclic structure (i.e. with ten or more ring members), polar ring interior and non-polar exterior. These substances form complexes with univalent (sometimes divalent) cations, especially with alkali metal ions, with a stability that is very dependent on the individual ionic sort. They mediate transport of ions through the lipid membranes of cells and cell organelles, whence the origin of the term ion-carrier (ionophore). They ion-specifically uncouple oxidative phosphorylation in mitochondria, which led to their discovery in the 1950s. This property is also connected with their antibiotic action. Furthermore, they produce a membrane potential on both thin lipid and thick membranes.

These substances include primarily depsipeptides (compounds whose structural units consist of alternating amino acid and α-hydroxy acid units). Their best-known representative is the cyclic antibiotic, valinomycin, with a 36-membered ring [L-Lac–L-Val–D-Hy-i-Valac–D-Val]$_3$, which was isolated from a culture of the microorganism, *Streptomyces fulvissimus*. Figure 6.13 depicts the structure of free valinomycin and its complex with a potassium ion, the most important of the coordination compounds of valinomycin.

Complex formation between a metal ion and a macrocyclic ligand involves interaction between the ion, freed of its solvation shell, and dipoles inside the ligand cavity. The standard Gibbs energy for the formation of the complex, ΔG^0_{JV}, is given by the difference between the standard Gibbs

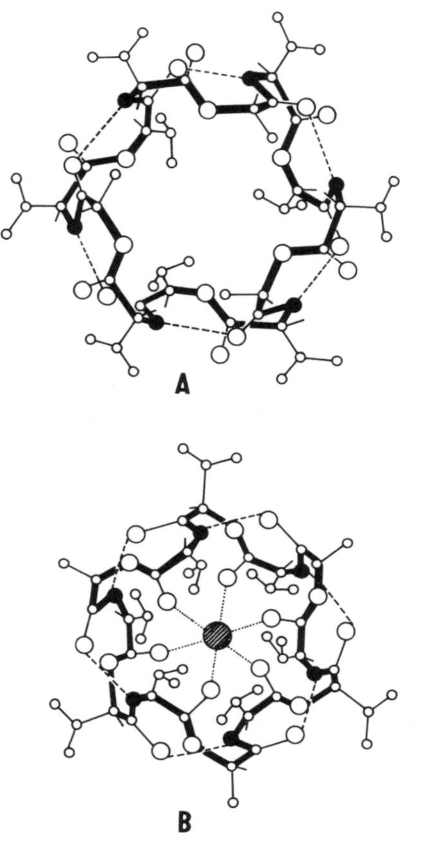

Fig. 6.13 Valinomycin structure in (A) a non-polar solvent (the 'bracelet' structure) and (B) its potassium complex. ○ = O, ○ = C, ● = N (According to Yu. A. Ovchinnikov, V. I. Ivanov and M. M. Shkrob)

energy for ion transfer from the vapour phase into the interior of the ligand in solution, $\Delta G^0_{J\to V}$, and the standard Gibbs energy of ion solvation, $\Delta G^0_{s,J}$:

$$\Delta G^0_{JV} = \Delta G^0_{J\to V} - \Delta G^0_{s,J} \tag{6.4.5}$$

As the quantities on the right-hand side are different functions of the ion radius, the Gibbs energy of complex formation also depends on this quantity, but not monotonously. The dependence of the Gibbs energy of solvation, e.g. for the alkali metal ions, is an increasing function of the ion radius, i.e. ions with a larger radius are more weakly solvated. In contrast, the interactions of the desolvated ions with the ligand cavity are approximately identical if the radius of the ion is less than that of the cavity. If, however, the ion radius is comparable to or even larger than the energetically most favourable cavity radius, then the ligand structure is strained or conformational changes must occur. Thus, the dependence of the stability constant of the complex (and also the rate constant for complex formation) is usually a curve with a maximum.

The following conditions must be fulfilled in the ion transport through an *ion-selective transmembrane channel*:

(a) The exterior of the channel in contact with the membrane must be lipophilic.
(b) The structure must have a low conformational energy so that conformational changes connected with the presence of the ion occur readily.
(c) The channel must be sufficiently long to connect both sides of the hydrophobic region of the membrane.
(d) The ion binding must be weaker than for ionophores so that the ion can rapidly change its coordination structure and pass readily through the channel.

The polarity of the interior of the channel, usually lower than in the case of ionophores, often prevents complete ion dehydration which results in a decrease in the ion selectivity of the channel and also in a more difficult permeation of strongly hydrated ions as a result of their large radii (for example Li^+).

The conditions a–d are fulfilled, for example, by the pentadecapeptide, valingramicidine A (Fig. 6.14):

```
HC=O
 |
 L-Val–Gly–L-Ala–D-Leu–L-Ala–D-Val–L-Val–D-Val
                                          |
 NH–L-Try–D-Leu–L-Try–D-Leu–L-Try–D-Leu–L-Try
 |
(CH2)2
 |
 OH
```

whose two helices joined tail-to-tail form an ion-selective channel. The selectivity of this channel for various ions is given by the series

$$H^+ > NH_4^+ \geqq Cs^+ > Rb^+ > K^+ > Na^+ > Li^+$$

If a substance that can form a transmembrane channel exists in several conformations with different dipole moments, and only one of these forms is permeable for ions, then this form can be 'favoured' by applying an electric potential difference across the membrane. The conductivity of the membrane then suddenly increases. Such a dependence of the conductivity of the membrane on the membrane potential is characteristic for the membranes of excitable cells.

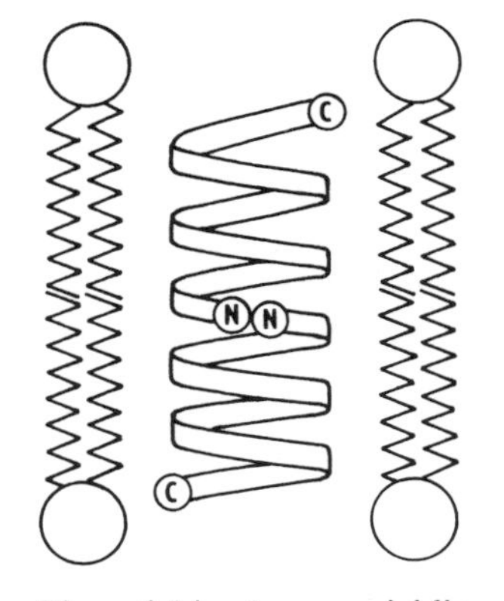

Fig. 6.14 A gramicidin channel consisting of two helical molecules in the head-to-head position. (According to V. I. Ivanov)

If a small amount of gramicidin A is dissolved in a BLM (this substance is completely insoluble in water) and the conductivity of the membrane is measured by a sensitive, fast instrument, the dependence depicted in Fig. 6.15 is obtained. The conductivity exhibits step-like fluctuations, with a roughly identical height of individual steps. Each step apparently corresponds to one channel in the BLM, open for only a short time interval (the opening and closing mechanism is not known) and permits transport of many ions across the membrane under the influence of the electric field; in the case of the experiment shown in Fig. 6.15 it is about 10^7 Na^+ per second at 0.1 V imposed on the BLM. Analysis of the power spectrum of these

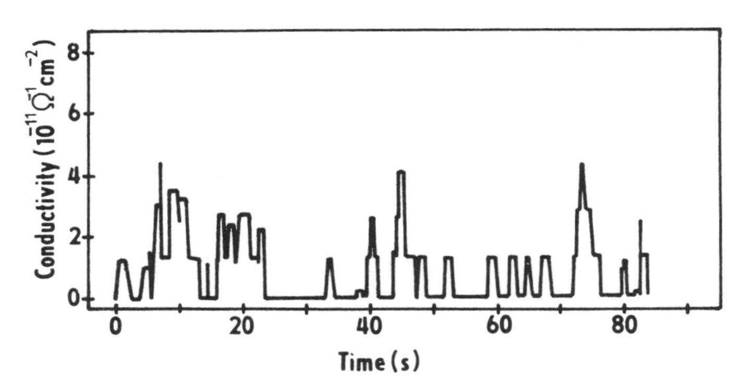

Fig. 6.15 Fluctuation of the conductivity of BLM in the presence of gramicidin A. (According to D. A. Haydon and B. S. Hladky)

stochastic events (cf. page 373) has shown that the rate-detemining reaction of channel formation is the bimolecular reaction of two gramicidin molecules.

A similar effect has been observed for alamethicin I and II, hemocyanin, antiamoebin I and other substances. Great interest in the behaviour of these substances was aroused by the fact that they represent simple models for ion channels in nerve cells.

Some substances form pores in the membrane that do not exhibit ion selectivity and permit flow of the solution through the membrane. These include the polyene antibiotics amphotericin B,

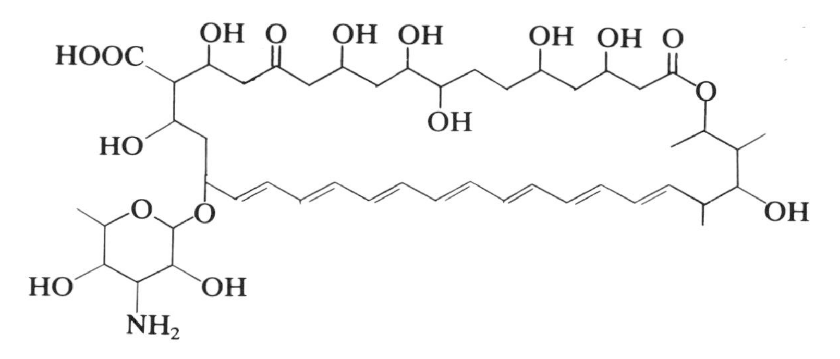

nystatin and mycoheptin, forming pores in the membrane with a diameter of 0.7–1.1 nm. Proteins also form pores, e.g. the protein from the sea anemone *Stoichactis helianthus* or colicin E1.

In all these systems, the energy source is an electrochemical potential gradient and transport occurs in the direction $-\text{grad}\,\tilde{\mu}_i$ (i.e. in the direction of decreasing electrochemical potential). It is often stated in the literature that this spontaneous type of transport occurs in the direction of the electrochemical potential gradient; this is an imprecise formulation.

Transport occurring in the absence of another source of transport energy is termed *passive transport.*

Active transport. The definition of active transport has been a subject of discussion for a number of years. Here, active transport is defined as a membrane transport process with a source of energy other than the electrochemical potential gradient of the transported substance. This source of energy can be either a metabolic reaction (primary active transport) or an electrochemical potential gradient of a substance different from that which is actively transported (secondary active transport).

A classical example of active transport is the transport of sodium ions in frog skin from the epithelium to the corium, i.e. into the body. The principal ionic component in the organism of a frog, sodium ions, is not washed out of its body during its life in water. That this phenomenon is a result of the active transport of sodium ions is demonstrated by an experiment in which the skin of the common green frog is fixed as a membrane between two separate compartments containing a single electrolyte solution (usually Ringer's solution, 0.115 M NaCl, 0.002 M KCl and 0.0018 M $CaCl_2$). In the absence of an electric current a potential difference is formed between the electrolyte at the outside and at the inside of the skin, equal to about -100 mV, even though the compositions of the two solutions are identical. This potential difference decreases to zero when electrodes on both sides of the membrane are short-circuited and a current flows between them. If the compartment on the outer side of the skin contains radioactively labelled ^{22}Na, then sodium ions are shown to be transported from the outer side to the inner side of the skin. Sodium ion transport in this direction occurs even when the electrolyte in contact with the inside of the skin contains a higher concentration of sodium ions than that at the outside.

When the temperature of the solution is increased, then the current as well as the sodium transport rate increase far more than would correspond to simple diffusion or migration. When substances inhibiting metabolic processes are added to the solution, e.g. cyanide or the glycoside, ouabain,

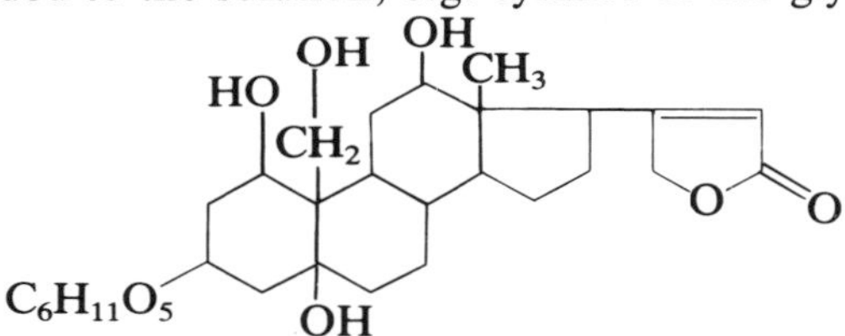

the current decreases. For example, in the presence of 10^{-4} M ouabain, it decreases to 5 per cent of its original value.

The rate of the active transport of sodium ion across frog skin depends both on the electrochemical potential difference between the two sides of this complex membrane (or, more exactly, membrane system) and also on the affinity of the chemical reaction occurring in the membrane. This combination of material flux, a vector, and 'chemical flux' (see Eq. 2.3.26), which is scalar in nature, is possible according to the Curie principle only when the medium in which the chemical reaction occurs is not homogeneous but anisotropic (i.e. has an oriented structure in the direction perpendicular to the surface of the membrane or, as is sometimes stated, has a vectorial character).

It is assumed that the chloride ion is transported passively across the membrane. Using an approach similar to the formulation of Eqs (2.1.2), (2.3.26) and (2.5.23), relationships can be written for the material fluxes of sodium and chloride ions, J_{Na^+} and J_{Cl^-} (the driving force is considered to be the electrochemical potential difference), and for the flux of the chemical reaction, J_{ch}:

$$\begin{aligned} J_{Na^+} &= L_{11}\Delta\tilde{\mu}_{Na^+} && + L_{1r}A \\ J_{Cl^-} &= \qquad L_{22}\Delta\tilde{\mu}_{Cl^-} && \\ J_{ch} &= L_{r1}\Delta\tilde{\mu}_{Na^+} && + L_{rr}A \end{aligned} \tag{6.4.6}$$

where $\Delta\tilde{\mu}_{Na^+}$ and $\Delta\tilde{\mu}_{Cl^-}$ designate the electrochemical potential difference between the inner and outer sides of the skin and A is the affinity of the chemical reaction. As the material flux in the direction from inside to outside is considered as positive, then the coefficients L_{11} and L_{22} will also be positive. On the other hand, the coefficients L_{1r} and L_{r1} will be negative.

When the electrolyte concentration is identical on both sides of the membrane and $\Delta\phi_M = 0$, then $\Delta\tilde{\mu}_{Na^+} = \Delta\tilde{\mu}_{Cl^-} = 0$ and Eqs (6.4.6) yield the equation for the current density,

$$j = FJ_{Na^+} = \frac{L_{1r}}{L_{rr}} J_{ch} = L_{1r}A \tag{6.4.7}$$

The resulting current is thus positive and proportional to the rate of the metabolic reaction.

Several types of active transport (also called a pump) can be distinguished. *Uniport* is a type of transport in which only one substance is transported across the membrane. An example is the transport of protons across the purple membrane of the halophilic bacterium *Halobacterium halobium,* where the energy for transport is provided by light activating the transport protein bacteriorhodopsin. The calcium pump in the membrane of the rod-shaped cells of the retina is also light-activated; the active substance is the related rhodopsin. On radiation, its prosthetic group, retinal, is transformed from the cis form to the trans form and the protein component is protonated by protons from the cytoplasm. After several milliseconds, retinal is converted back to the cis form and the protons are released into the surrounding medium. Another example of uniport is the functioning of H^+-ATPase, described in detail in Section 6.5.2.

Symport is the interrelated transport of two substances in the same direction, with opposite electrochemical potential gradients. Here the energy of the electrochemical potential gradient of one substance is used for active transport of the other substance. An example is the combined transport of sodium ions (in the direction of decreasing electrochemical potential) and glucose or amino acid (along its concentration gradient) across the membranes of the cell of the intestinal epithelium. In this process, the organism transfers the products of digestion from the intestine into the bloodstream.

Antiport is the coupling of two opposite transport processes. Na^+,K^+-ATPase activated by magnesium ions is a typical transport protein, enabling transport of sodium and potassium ions against their electrochemical potential gradients. This process employs the energy gained from the rupture of the 'macroergic' polyphosphate bond of adenosinetriphosphate (ATP) during its conversion to adenosinediphosphate (ADP) which is catalysed by the enzyme, ATPase. In the actual process, the asparagine unit of the enzyme is first phosphorylated with simultaneous conformational changes, connected with the transfer of sodium ions from the cell into the intercellular fluid. The second step involves dephosphorylation of the

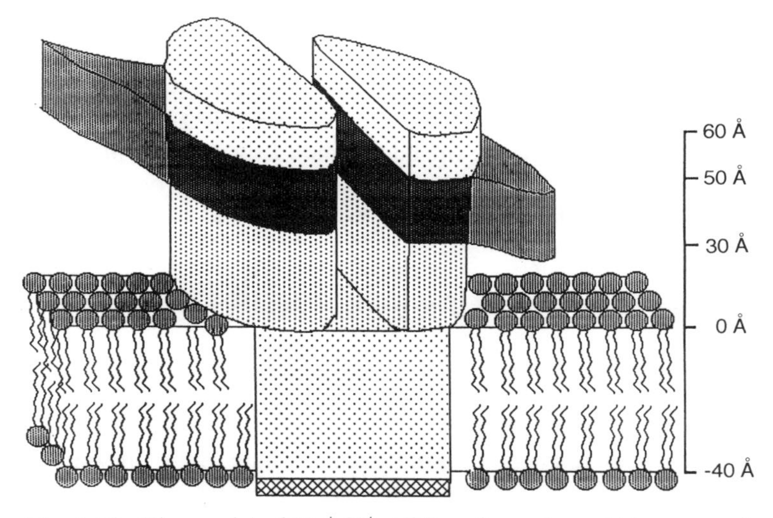

Fig. 6.16 The model of Na^+,K^+-ATPase from sheep kidney according to E. Amler, A. Abbott and W. J. Ball. The shading indicates various functional areas of the dimeric enzyme:

- top of yet unknown function
- binding site of the nucleotide (ATP or ADP)
- phosphorylation site
- ion transporting part (with the channel)
- ouabain binding site

protein with transfer of a potassium ion from the intercellular fluid into the cell. The overall process corresponds to the scheme

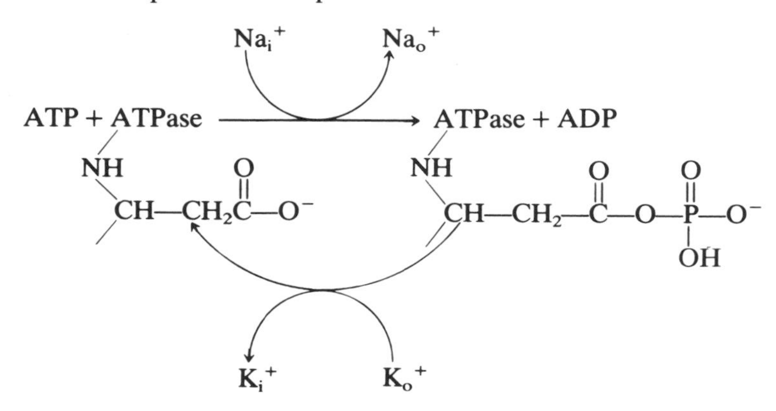

(where the subscripts designate: i, intracellular and o, extracellular).

The active form of the enzyme is, in fact, a dimer situated in the phospholipid region of the cytoplasmic membrane (Fig. 6.16) and the ATP molecule is bound between both the macromolecules.

This important enzyme is responsible, for example, for the active transport of sodium ions in frog skin (see page 450), maintains a concentration of sodium ions inside most cells lower than in the intercellular liquid (see page 454), etc. The function of Na^+,K^+-ATPase is reversible as it catalyses the synthesis of ATP under artificial conditions that do not occur in the organism (at high concentrations of Na^+ inside the cells, low concentrations of K^+ outside the cells and high ADP concentrations).

From the point of view of the stoichiometry of the transported ions during active transport, the electroneutral pump, where there is no net charge transfer or change in the membrane potential, must be distinguished from the electrogenic pump connected with charge transfer.

Active transport is of basic importance for life processes. For example, it consumes 30–40 per cent of the metabolic energy in the human body. The nervous system, which constitutes only 2 per cent of the weight of the organism, utilizes 20 per cent of the total amount of oxygen consumed in respiration to produce energy for active transport.

References

Abrahamson, S. A., and I. Pascher (Eds), *Structure of Biological Membrane,* Plenum Press, New York, 1977.

Amler, E., A. Abbott and W. J. Ball, Structural dynamics and oligometric interactions of Na^+,K^+-ATPase as monitored using fluorescence energy transfer, *Biophys. J.*, **61,** 553 (1992).

Bittar, E. E. (Ed.), *Membrane Structure and Function,* John Wiley & Sons, New York, Vol. 1, 1980; Vol. 2, 1981.

Bittar, E. E. (Ed.), *Membranes and Ion Transport,* John Wiley & Sons, New York, Vol. 1, 1970; Vol. 2, 1971; Vol. 3, 1971.

Bureš, J., M. Petráň, and J. Zachar, *Electrophysiological Methods in Biological Research,* Academic Press, New York, 1967.

Ceve, G., and D. Marsh, *Phospholipid Bilayers,* John Wiley & Sons, New York, 1987.

Chaplan, S. R., and A. Essig, *Bioenergetics and Linear Nonequilibrium Thermodynamics. The Steady State,* Harvard University Press, Cambridge, Mass., 1983.

Chapman, D. (Ed.), *Biological Membranes,* a series of advances published since 1968, Academic Press, London.

Chapman, D. (Ed.), *Biomembrane Structure and Function,* Vol. 4 of *Topics in Molecular and Structural Biology,* Verlag Chemie, Weinheim, 1984.

Eisenman, G., J. Sandblom, and E. Neher, Interactions in cation permeation through the gramicidin channel—Cs, Rb, K, Na, Li, Tl, H and effects of anion binding, *Biophys. J.*, **22,** 307 (1978).

Fineau, J. B., R. Coleman, and R. H. Michell, *Membranes and Their Cellular Function,* 2nd ed., Blackwell, Oxford, 1978.

Gómez-Puyon, A., and C. Gómez-Lojero, The use of ionophores and channel formers in the study of the function of biological membranes, in *Current Topics in Bioenergetics,* Vol. 6, p. 221, Academic Press, New York, 1977.

Gregoriadis, G. (Ed.), *Liposome Technology,* Vols 1–3, CRC Press, Boca Raton, 1984.

Haydon, D. A., and B. S. Hladky, Ion transport across thin lipid membranes: a critical discussion of mechanisms in selected systems, *Quart. Rev. Biophys.*, **5,** 187 (1972).

Henderson, R., The purple membrane from *Halobacterium halobium*, *Ann. Rev. Biophys. Bioeng.*, **6,** 87 (1977).
Katchalsky, A., and P. F. Curran, *Non-Equilibrium Thermodynamics in Biophysics,* Harvard University Press, Cambridge, Mass., 1967.
Knight, C. G. (Ed.), *Liposomes: From Physical Structure to Therapeutic Applications,* Elsevier, North-Holland, Amsterdam, 1981.
Kotyk, A., and K. Janáček, *Cell Membrane Transport, Principles and Techniques,* 2nd ed., Plenum Press, New York, 1975.
Läuger, P., Transport noise in membranes, *Biochim. Biophys. Acta,* **507,** 337 (1978).
Martonosi, A. N. (Ed.), *Membranes and Transport,* Vols. 1 and 2, Plenum Press, New York, 1982.
Montal, M., and P. Mueller, Formation of bimolecular membranes from lipid monolayers and a study of their electrical properties, *Proc. Natl. Acad. Sci. USA,* **69,** 3561 (1972).
Ovchinnikov, Yu. A., V. I. Ivanov, and M. M. Shkrob, *Membrane Active Complexones,* Elsevier, Amsterdam, 1974.
Papahadjopoulos, D. (Ed.), *Liposomes and Their Use in Biology and Medicine,* Annals of the New York Academy of Science, Vol. 308, New York, 1978.
Pullmann, A. (Ed.), *Transport Through Membranes: Carriers, Channels and Pumps,* Kluwer, Dordrecht, 1988.
Singer, S. I., and G. L. Nicholson, The fluid mosaic model of the structure of cell membranes, *Science,* **175,** 720 (1977).
Skou, J. C., J. G. Norby, A. B. Maunsbach and M. Esmann (Eds), *The Na^+, K^+-Pump,* Part A, A. R. Liss, New York, 1988.
Spack, G. (Ed.), *Physical Chemistry of Transmembrane Motions,* Elsevier, Amsterdam, 1983.
Stryer, L., *Biochemistry,* W. H. Freeman & Co., New York, 1984.
Ti Tien, H., *Bilayer Lipid Membrane (BLM),* M. Dekker, New York, 1974.
Urry, D. W., Chemical basis of ion transport specificity in biological membranes, in *Topics in Current Chemistry* (Ed. F. L. Boschke), Springer-Verlag, Berlin.
Wilschüt, J., and D. Hoekstra, Membrane fusion: from liposomes to biological membranes, *Trends Biochem. Sci.,* **9,** 479 (1984).

6.5 Examples of Biological Membrane Processes

6.5.1 *Processes in the cells of excitable tissues*

The transport of information from sensors to the central nervous system and of instructions from the central nervous system to the various organs occurs through electric impulses transported by nerve cells (see Fig. 6.17). These cells consist of a body with star-like projections and a long fibrous tail called an axon. While in some molluscs the whole membrane is in contact with the intercellular liquid, in other animals it is covered with a multiple myeline layer which is interrupted in definite segments (nodes of Ranvier). The Na^+,K^+-ATPase located in the membrane maintains marked ionic concentration differences in the nerve cell and in the intercellular liquid. For example, the squid axon contains 0.05 M Na^+, 0.4 M K^+, 0.04–0.1 M Cl^-, 0.27 M isethionate anion and 0.075 M aspartic acid anion, while the intercellular liquid contains 0.46 M Na^+, 0.01 M K^+ and 0.054 M Cl^-.

The relationship between the electrical excitation of the axon and the membrane potential was clarified by A. L. Hodgkin and A. P. Huxley in

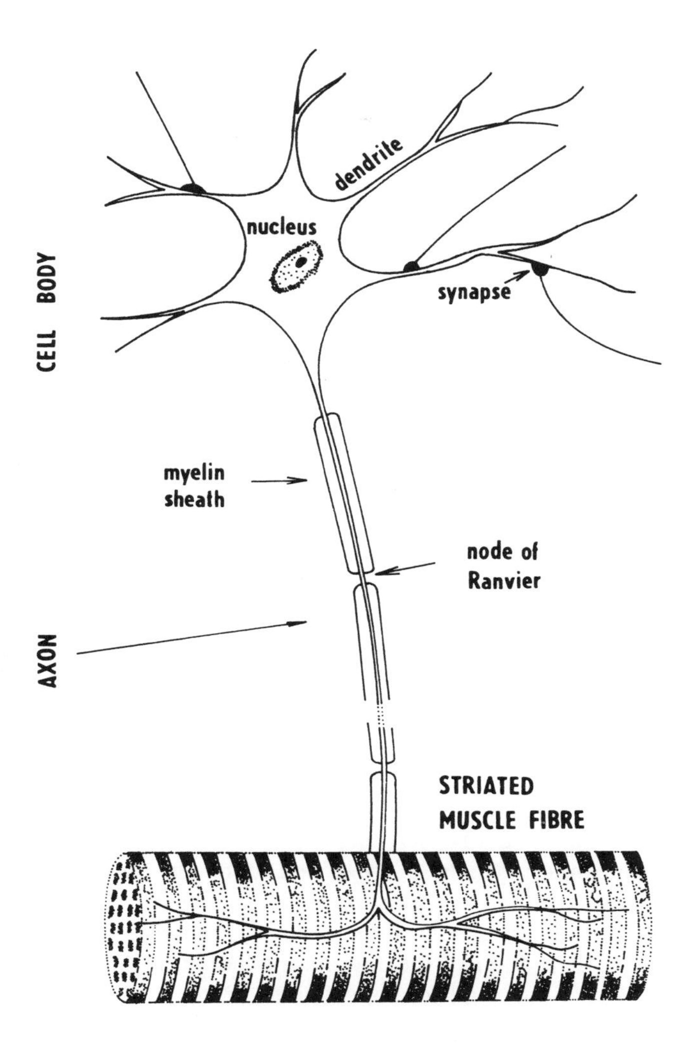

Fig. 6.17 A scheme of motoric nerve cell

experiments on the giant squid axon with a thickness of up to 1 mm. The experimental arrangement is depicted in Fig. 6.18. The membrane potential is measured by two identical reference electrodes, usually micropipettes shown in Fig. 3.8. If the axon is not excited, the membrane potential $\Delta\phi_M = \phi(\text{in}) - \phi(\text{out})$ has a rest value of about -90 mV. When the cell is excited by small square wave current impulses, a change occurs in the membrane potential roughly proportional to the magnitude of the excitation current impulses (see Fig. 6.19). If current flows from the interior of the cell to the exterior, then the absolute value of the membrane potential increases and the membrane is hyperpolarized. Current flowing in the opposite direction has a *depolarization* effect and the absolute potential value

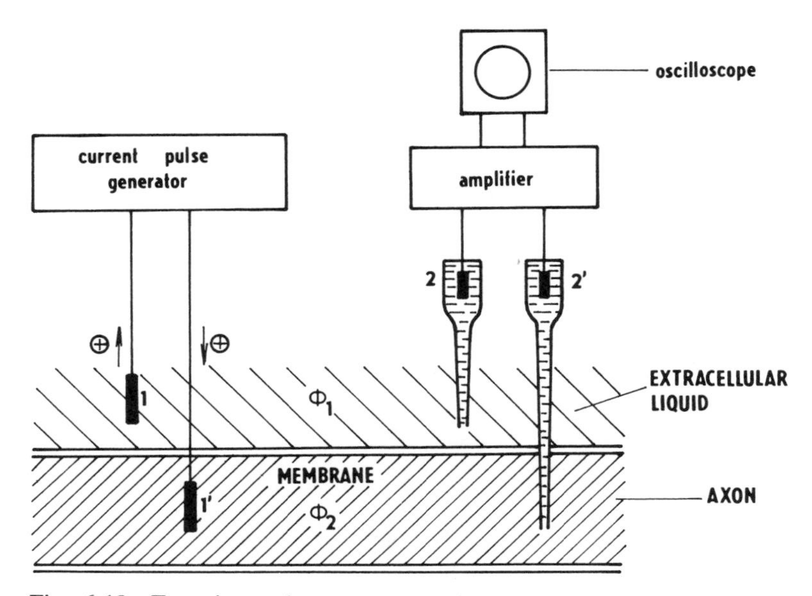

Fig. 6.18 Experimental arrangement for measurement of the membrane potential of a nerve fibre (axon) excited by means of current pulses: (1) excitation electrodes, (2) potential probes. (According to B. Katz)

decreases. When the depolarization impulse exceeds a certain 'threshold' value, the potential suddenly increases (Fig. 6.19, curve 2). The characteristic potential maximum is called a *spike* and its height no longer depends on a further increase in the excitation impulse. Sufficiently large excitation of the membrane results in a large increase in the membrane permeability for sodium ions so that, finally, the membrane potential almost acquires the value of the Nernst potential for sodium ions ($\Delta\phi_M = +50\,\text{mV}$).

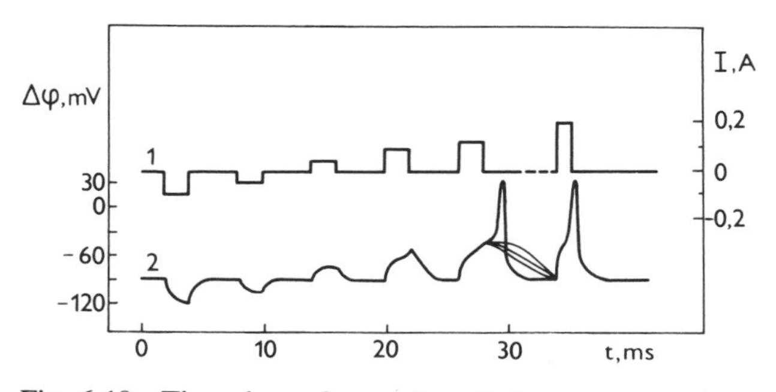

Fig. 6.19 Time dependence of excitation current pulses (1) and membrane potential $\Delta\phi$. (2) The abrupt peak is the spike. (According to B. Katz)

A potential drop to the rest value is accompanied by a temporary influx of sodium ions from the intercellular liquid into the axon.

If the nerve is excited by a 'subthreshold' current impulse, then a change in the membrane potential is produced that disappears at a small distance from the excitation site (at most 2 mm). A spike produced by a threshold or larger current impulse produces further excitation along the membrane, yielding further spikes that are propagated along the axon. As already pointed out, sodium ions are transferred from the intercellular liquid into the axon during the spike. This gradual formation and disappearance of positive charges corresponds to the flow of positive electric current along the axon. An adequate conductance of thick bare cephalopod axons allows the flow of sufficiently strong currents. In myelinized axons of vertebrates a much larger charge is formed (due to the much higher density of sodium channels in the nodes of Ranvier) which moves at high speed through much thinner axons than those of cephalopods. The myelin sheath then insulates the nerve fibre, impeding in this way the induction of an opposite current in the intercellular liquid which would hinder current flow inside the axon.

The electrophysiological method described above is, in fact, a galvanostatic method (page 300). A more effective method (the so-called *voltage-clamp* method) is based on polarizing the membrane by using a four-electrode potentiostatic arrangement (page 295). In this way, Cole, Hodgkin and Huxley showed that individual currents linked to selective ion transfer across the membrane are responsible for impulse generation and propagation. A typical current–time curve is shown in Fig. 6.20. Obviously, the membrane ion transfer is activated at the start, but after some time it becomes gradually inhibited. The ion transfer rate typically depends on both

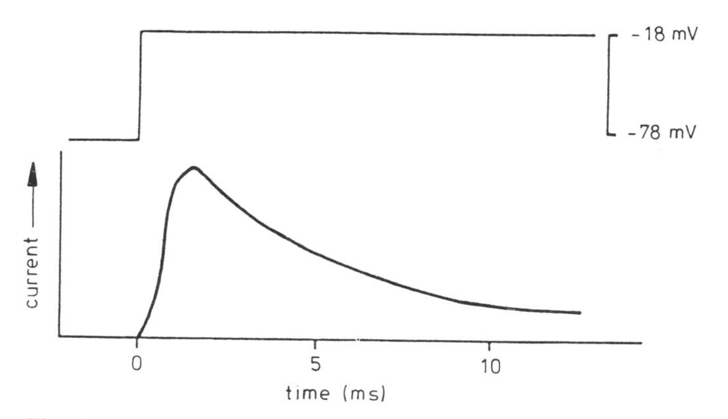

Fig. 6.20 Time dependence of the membrane current. Since the potassium channel is blocked the current corresponds to sodium transport. The upper line represents the time course of the imposed potential difference. (According to W. Ulbricht)

the outer (bathing) and inner solution (the inside of a cephalopod membrane as much as 1 mm thick can be rinsed with an electrolyte solution without affecting its activity). The assumption that the membrane currents are due to ion transfer through *ion-specific channels* was shown correct by means of experiments where the channels responsible for transfer of a certain ionic species were blocked by specific agents. Thus, the sodium transfer is inhibited by the toxin, tetrodotoxin:

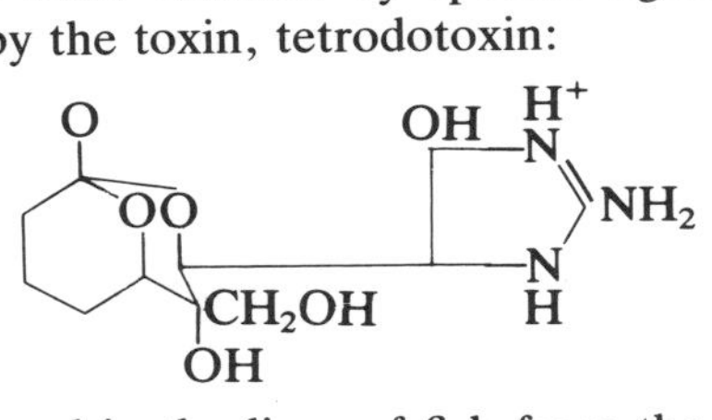

found in the gonads and in the liver of fish from the family *Tetraodontidae* (e.g. the Japanese delicacy, *fugu*) and by a number of similar substances. Their physiological effect is paralysis of the respiratory function. The transport of potassium ions is blocked by the tetraalkylammonium ion with three ethyl groups and one longer alkyl group, such as a nonyl.

The effect of toxins on ion transport across the axon membrane, which occurs at very low concentrations, has led to the conclusion that the membrane contains ion-selective channels responsible for ion transport. This assumption was confirmed by analysis of the noise level in ionic currents resulting from channel opening and closing (cf. page 448). Single-channel recording was a decisive experiment. Here, a glass capillary (a micropipette) with an opening of about 1 μm^2 is pressed onto the surface of the cell and a very tight seal between the phospholipids of the membrane and the glass of the micropipette is formed (cf. page 441). Because of the very low extraneous noise level, this *patch-clamp* method permits the measurement of picoampere currents in the millisecond range (Fig. 6.21). Obviously, this course of events is formally similar to that observed with the gramicidin channel (page 448) but the mechanism of opening and closing is different. In contrast with the gramicidin channel, the nerve cell channels are much larger and stable formations. Usually they are glycoproteins, consisting of several subunits. Their hydrophobic region is situated inside the membrane while the sugar units stretch out. Ion channels of excitable cells consist of a narrow pore, of a *gate* that opens and closes the access to the pore, and of a sensor that reacts to the stimuli from outside and issues instructions to the gate. The outer stimuli are either a potential change or binding of a specific compound to the sensor.

The nerve axon sodium channel was studied in detail (in fact, as shown by the power spectrum analysis, there are two sorts of this channel: one with fast opening and slow inactivation and the other with opposite properties). It is a glycoprotein consisting of three subunits (Fig. 6.22), the largest (mol. wt. 3.5×10^5) with a pore inside and two smaller ones (mol.wt 3.5×10^4 and 3.3×10^4). The attenuation in the orifice of the pore is a kind of a filter

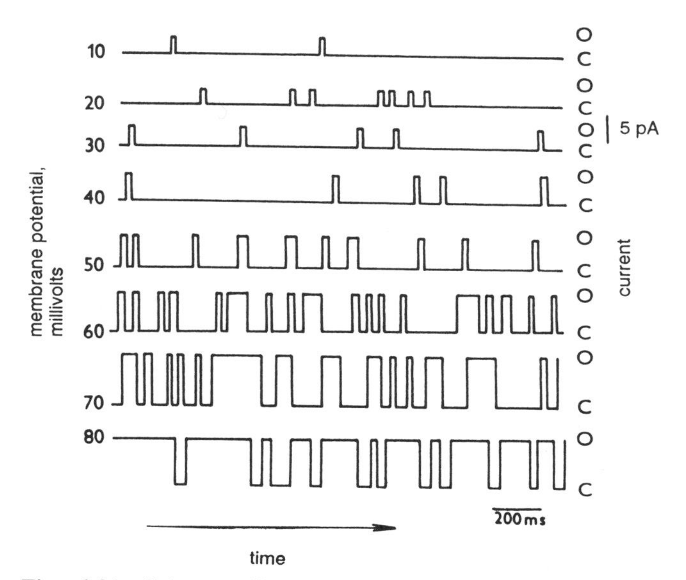

Fig. 6.21 Joint application of patch-clamp and voltage-clamp methods to the study of a single potassium channel present in the membrane of a spinal-cord neuron cultivated in the tissue culture. The values indicated before each curve are potential differences imposed on the membrane. The ion channel is either closed (C) or open (O). (A simplified drawing according to B. Hille)

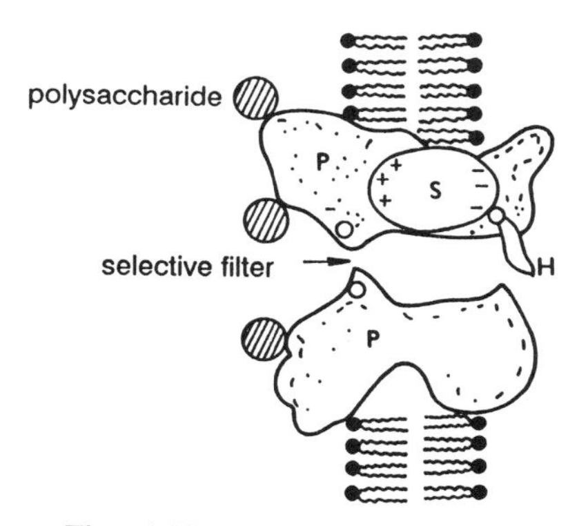

Fig. 6.22 A function model of the sodium channel. P denotes protein, S the potential sensitive sensor and H the gate. The negative sign marks the carboxylate group where the guanidine group of tetrodotoxin can be attached. (According to B. Hille)

(0.5 × 0.4 nm in size) controlling the entrance of ions with a definite radius. The rate of transport of sodium ions through the channel is considerable: when polarizing the membrane with a potential difference + 60 mV a current of approximately 1.5 pA flows through the channel which corresponds to 6×10^6 Na^+ ions per second—practically the same value as with the gramicidin A channel. The sodium channel is only selective but not specific for sodium transport. It shows approximately the same permeability to lithium ions, whereas it is roughly ten times lower than for potassium. The density of sodium channels varies among different animals, being only 30 μm^{-2} in the case of some marine animals and 330 μm^{-2} in the squid axon, reaching 1.2×10^4 μm^{-2} in the mammalian nodes of Ranvier (see Fig. 6.17).

The potassium channel mentioned above (there are many kinds) is more specific for K^+ than the sodium channel for Na^+ being almost impermeable to Na^+.

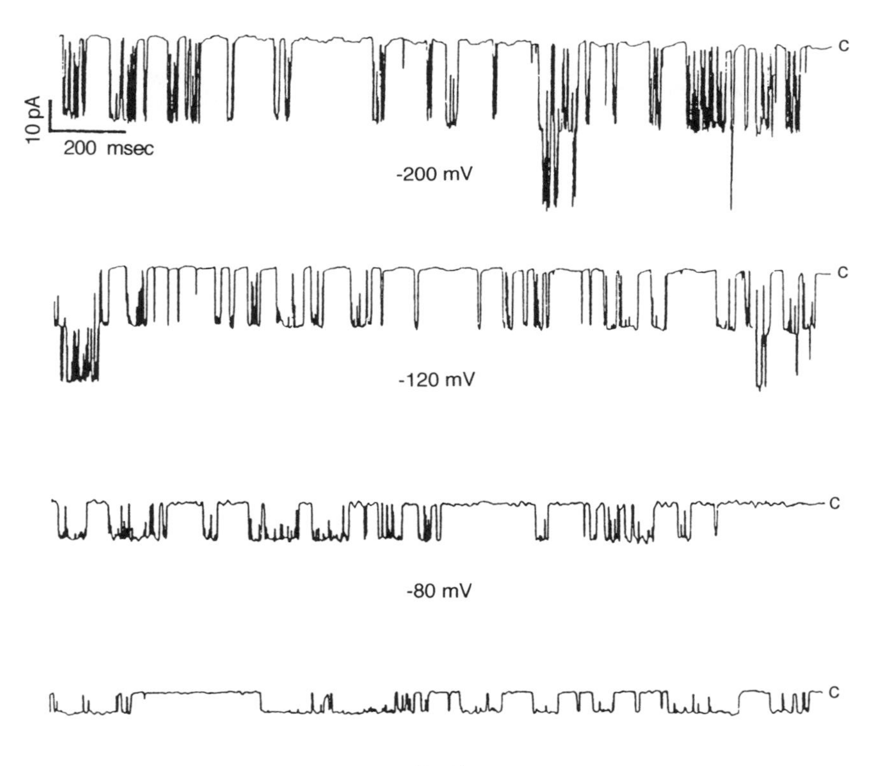

Fig. 6.23 Single-channel currents flowing across the membrane between the protoplast and vacuole of *Chara corallina.* Among several channels with different conductivity the recordings of the 130 pS channel are recorded here. The zero line is at the top of each curve. (By courtesy of F. Homblé)

The stochastic nature of membrane phenomena originating in channel opening and closing is not restricted to excitable cells. Figure 6.23 shows the time dependence of currents flowing through a patch of the membrane between the tonoplast (protoplasma) and the vacuole in the isolated part of a cell of the freshwater alga, *Chara corallina.* In this membrane there are three types of potassium channels with different conductivity and the behaviour of the 130 pS channel is displayed in the figure.

D. E. Goldman, A. L. Hodgkin, A. P. Huxley and B. Katz (A. L. H. and A. P. H., Nobel Prize for Physiology and Medicine, 1963, B.K., 1970) developed a theory of the resting potential of axon membranes, based on the assumption that the strength of the electric field in a thin membrane is constant and that ion transport in the membrane can be described by the Nernst–Planck equations. It would appear that this approach does not correspond to reality—it has been pointed out that ions are transported through the membrane in channels that are specific for a certain kind of ion. Thus, diffusion is not involved, but rather a jumping of the ions through the membrane, that must overcome a certain energy barrier.

In deriving a relationship for the resting potential of the axon membrane it will be assumed that, in the vicinity of the resting potential, the frequency of opening of a definite kind of ion channel is not markedly dependent on the membrane potential. The transport of ions through the membrane can be described by the same equations as the rate of an electrode reaction in Section 5.2.2. It will be assumed that the resting potential is determined by the transport of potassium, sodium and chloride ions alone. The constants $k_i^{\ominus}$ are functions of the frequency of opening and closing of the gates of the ion-selective channels. The solution to this problem will be based on analogous assumptions to those employed for the mixed potential (see Section 5.8.4).

The material fluxes of the individual ions are given by the equations

$$\begin{aligned} J_{K^+} &= k_{K^+}^{\ominus} c_{K^+}(1) \exp\left(\frac{-F\Delta\phi_M}{2RT}\right) - k_{K^+}^{\ominus} c_{K^+}(2) \exp\left(\frac{F\Delta\phi_M}{2RT}\right) \\ J_{Na^+} &= k_{Na^+}^{\ominus} c_{Na^+}(1) \exp\left(\frac{-F\Delta\phi_M}{2RT}\right) - k_{Na^+}^{\ominus} c_{Na^+}(2) \exp\left(\frac{F\Delta\phi_M}{2RT}\right) \\ J_{Cl^-} &= k_{Cl^-}^{\ominus} c_{Cl^-}(1) \exp\left(\frac{F\Delta\phi_M}{2RT}\right) - k_{Cl^-}^{\ominus} c_{Cl^-}(2) \exp\left(\frac{-F\Delta\phi_M}{2RT}\right) \end{aligned} \tag{6.5.1}$$

At rest, no current passes through the membrane and thus the material flux of chloride ions compensates the material flux of sodium and potassium ions:

$$J_{Na^+} + J_{K^+} = J_{Cl^-} \tag{6.5.2}$$

Equations (6.5.1) and (6.5.2) yield the membrane rest potential in the form

$$\Delta\phi_M = \frac{RT}{F} \ln \frac{k_{K^+}^{\ominus} c_{K^+}(1) + k_{Na^+}^{\ominus} c_{Na^+}(1) + k_{Cl^-}^{\ominus} c_{Cl^-}(2)}{k_{K^+}^{\ominus} c_{K^+}(2) + k_{Na^+}^{\ominus} c_{Na^+}(2) + k_{Cl^-}^{\ominus} c_{Cl^-}(1)} \tag{6.5.3}$$

Ion transport is characterized by conditional rate constants $k^{\ominus}_{K^+}$, $k^{\ominus}_{Na^+}$ and $k^{\ominus}_{Cl^-}$ which can be identified with the permeabilities of the membrane for these ions.

These relationships can be improved by including the effect of the electrical double layer on the ion concentration at the membrane surface (cf. page 444 and Section 5.3.2).

Equation (6.5.3) is identical with the relationship derived by Hodgkin, Huxley and Katz. It satisfactorily explains the experimental values of the membrane rest potential assuming that the permeability of the membrane for K^+ is greater than for Na^+ and Cl^-, so that the deviation of $\Delta\phi_M$ from the Nernst potential for K^+ is not very large. However, the permeabilities for the other ions are not negligible. In this way the axon at rest would lose potassium ions and gain a corresponding concentration of sodium ions. This does not occur because of the action of Na^+,K^+-ATPase, transferring potassium ions from the intercellular liquid into the axon and sodium ions in the opposite direction, through hydrolysis of ATP.

When the nerve cells are excited by an electric impulse (either 'natural', from another nerve cell or another site on the axon, or 'artificial', from an electrode), the membrane potential changes, causing an increase in the frequency of opening of the gates of the sodium channels. Thus, the flux of sodium ions increases and the membrane potential is shifted towards the Nernst potential value determined by sodium ions, which considerably differs from that determined by potassium ions, as the concentration of sodium ions in the extracellular space is much greater than in the intracellular space, while the concentration ratio of potassium ions is the opposite. The potential shift in this direction leads to a further opening of the sodium gates and thus to 'autocatalysis' of the sodium flux, resulting in a spike which is stopped only by inactivation of the gates.

In spite of the fact that the overall currents flowing across the cell membranes consist of tiny stochastic fluctuating components, the resulting dependences, as shown in Figs 6.19 and 6.20, are smooth curves and can be used for further analysis (the situation is, in fact, analogous to most of phenomena occurring in nature). Thus, the formation of a spike can be shown to be a result of gradual opening and closing of very many potassium and sodium channels (Fig. 6.24).

According to Fig. 6.17 the nerve cell is linked to other excitable, both nerve and muscle, cells by structures called, in the case of other nerve cells, as partners, *synapses,* and in the case of striated muscle cells, *motor end-plates* (*neuromuscular junctions*). The impulse, which is originally electric, is transformed into a chemical stimulus and again into an electrical impulse. The opening and closing of ion-selective channels present in these junctions depend on either electric or chemical actions. The substances that are active in the latter case are called *neurotransmitters.* A very important member of this family is acetylcholine which is transferred to the cell that receives the signal across the postsynaptic membrane or motor endplate through a

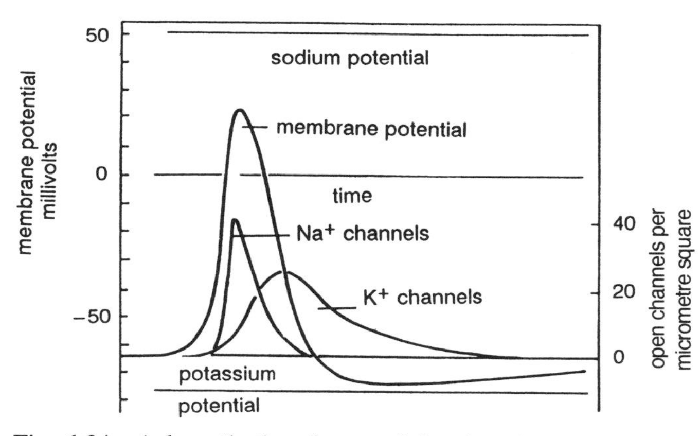

Fig. 6.24 A hypothetic scheme of the time behaviour of the spike linked to the opening and closing of sodium and potassium channels. After longer time intervals a temporary hyperpolarization of the membrane is induced by reversed transport of potassium ions inside the nerve cell. Nernst potentials for Na^+ and K^+ are also indicated in the figure. (According to A. L. Hodgkin and A. F. Huxley)

specific channel, the *nicotinic acetylcholine receptor.* This channel has been investigated in detail, because, among other reasons, it can be isolated in considerable quantities as the membranes of the cells forming the electric organ of electric fish are filled with this species.

Acetylcholine, which is set free from vesicles present in the neighbourhood of the presynaptic membrane, is transferred into the recipient cell through this channel (Fig. 6.25). Once transferred it stimulates generation of a spike at the membrane of the recipient cell. The action of acetylcholine is inhibited by the enzyme, acetylcholinesterase, which splits acetylcholine to choline and acetic acid.

Locomotion in higher organisms and other mechanical actions are made possible by the *striated skeletal muscles.* The basic structural unit of muscles is the muscle cell—muscle fibre which is enclosed by a sarcoplasmatic membrane. This membrane invaginates into the interior of the fibre through transversal tubules which are filled with the intercellular liquid. The inside of the fibre consists of the actual sarcoplasm with inserted mitochondria, sarcoplasmatic reticulum and minute fibres called myofibrils, which are the organs of muscle contraction and relaxation. The membrane of the sarcoplasmatic reticulum contains Ca^{2+}-ATPase, maintaining a concentration of calcium ions in the sarcoplasm of the relaxed muscle below $10^{-7}\,\text{mol}\cdot\text{dm}^{-3}$. Under these conditions, the proteins, actin and myosin, forming the myofibrils lie in relative positions such that the muscle is relaxed. When a spike is transferred from the nerve fibre to the sarcoplasmatic membrane, another spike is also formed there which continues

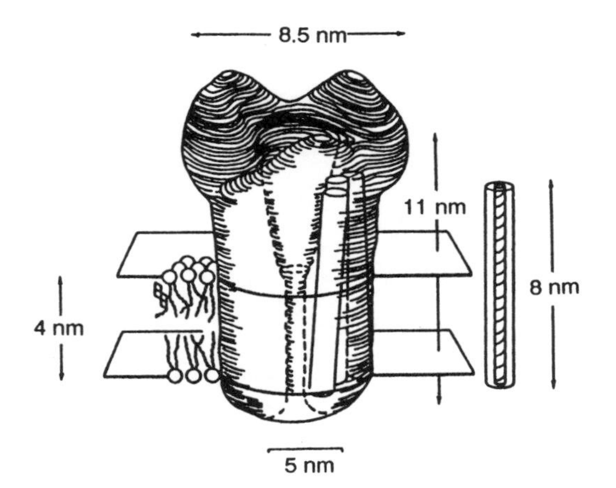

Fig. 6.25 The nicotinic acetylcholine receptor in a membrane. The deciphering of the structure is based on X-ray diffraction and electron microscopy. (According to Kistler and coworkers)

through the transverse tubules to the membrane of the sarcoplasmatic reticulum, increasing the permeability of the membrane for calcium ions by five orders within 1 millisecond, so that the concentration of Ca^{2+} ions in the sarcoplasm increases above $10^{-3}\,mol \cdot dm^{-3}$. This produces a relative shift of actin and myosin molecules and contraction of the muscle fibre. After disappearance of the spike, Ca^{2+}-ATPase renews the original situation and the muscle is relaxed.

6.5.2 *Membrane principles of bioenergetics*

The basic process by which living organisms obtain energy is *photosynthesis* in green plants, yielding carbonaceous substances (fuels) and *oxidative phosphorylation* in all organisms, in which the Gibbs energy obtained by oxidation of these carbonaceous substances with oxygen is converted into the energy of the polyphosphate bond in adenosinetriphosphate (ATP) through the reaction of adenosine diphosphate with dihydrogen phosphate ions (P_{inorg}).

This process occurs in the mitochondria, organelles present in the cells of all multicellular organisms (see Figs 6.8 and 6.26). Mitochondria have two membranes. The invaginations of the internal membrane into the inner space of the organelle (matrix space) are termed crests (from the Latin, *cristae*).

The oxidation of the carbonaceous substrate with oxygen cannot occur directly as the released energy would be converted into heat, but must proceed almost reversibly, through a number of poorly water-soluble redox

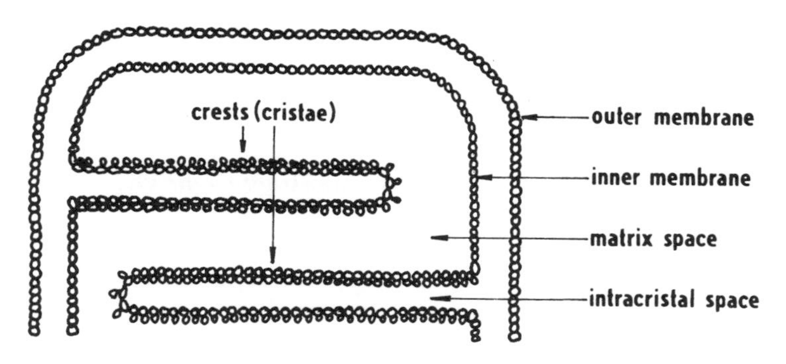

Fig. 6.26 Section of a mitochondrion

systems forming an electron transfer chain located in the internal membrane of the mitochondrion. They include the system of the oxidized and reduced forms of nicotinamide adenine dinucleotide (NAD^+/NADH), several flavoprotein systems, especially FMN (flavinmononucleotide), Fe–S proteins (where the iron ions are bound in a complex with inorganic sulphide, cystein and a protein), coenzyme Q (ubiquinone) and several cytochromes. These redox systems are bound to phospholipids and other molecules in several integrated systems depending on their redox potential (see Table 6.2). The redox potentials of these systems, measured in solution, can be quite different from the potentials in the mitochondrial membrane. The sequence of the redox components in this electron transfer permits the oxidation of the carbonaceous substrates to occur 'vectorially', i.e. the chemical change is oriented in space.

The theory of the accumulation of the Gibbs energy of oxidation of carbonaceous substrates in the form of the bond of the phosphate ion to

Table 6.2 Apparent formal redox potentials of systems present in the electron-transfer chain (pH = 7). It should be noted that the potential values were obtained in the homogeneous phase. Due to stabilization in a membrane, the oxidation–reduction properties vary so that the data listed below are of orientation character only

Oxidized form	Reduced form	$E^{0'}$ (pH = 7) (V)
O_2	H_2O	0.8
Cytochrome $\mathbf{a}_3$ (ox)	Cytochrome $\mathbf{a}_3$ (red)	0.5
Cytochrome **c** (ox)	Cytochrome **c** (red)	0.22
Cytochrome **b** (ox)	Cytochrome **b** (red)	0.07
Coenzyme Q	Coenzyme QH_2	0.10
Flavinadenin-dinucleotide (FAD)	$FADH_2$	−0.03
Fumarate	Succinate	−0.03
Nicotinamide adenin-dinucleotide (NAD^+)	NADH	−0.32

adenosine diphosphate (oxidative phosphorylation) must explain the following facts:

(a) A topologically closed vesicle (a cell or an organelle) which is not permeable to passively transported H^+ is always necessary for transformation of the chemical energy of the substrate to the chemical energy of ATP.
(b) This process is connected with a change in the pH difference between the intracristal space and the matrix space of the mitochondrion as well as with a change in the membrane potential.
(c) An artificially generated difference in the electrochemical potential of H_3O^+ between the intracristal space and the matrix space can result in ATP synthesis.
(d) Uncoupling of oxidative phosphorylation (i.e. the process in which the substrate is oxidized but no ATP is synthesized) occurs in the presence of certain proton and alkali metal ion carriers.

P. Mitchell (Nobel Prize for Chemistry, 1978) explained these facts by his chemiosmotic theory.† This theory is based on the ordering of successive oxidation processes into reaction sequences called loops. Each loop consists of two basic processes, one of which is oriented in the direction away from the matrix surface of the internal membrane into the intracristal space and connected with the transfer of electrons together with protons. The second process is oriented in the opposite direction and is connected with the transfer of electrons alone. Figure 6.27 depicts the first Mitchell loop, whose first step involves reduction of NAD^+ (the oxidized form of nicotinamide adenosine dinucleotide) by the carbonaceous substrate, SH_2. In this process, two electrons and two protons are transferred from the matrix space. The protons are accumulated in the intracristal space, while electrons are transferred in the opposite direction by the reduction of the oxidized form of the Fe–S protein. This reduces a further component of the electron transport chain on the matrix side of the membrane and the process is repeated. The final process is the reduction of molecular oxygen with the reduced form of cytochrome oxidase. It would appear that this reaction sequence includes not only loops but also a proton pump, i.e. an enzymatic system that can employ the energy of the redox step in the electron transfer chain for translocation of protons from the matrix space into the intracristal space.

The energy obtained by oxidation of the substrate with oxygen through the electron transport chain is thus accumulated as a difference in the electrochemical potential for H^+ between the intracristal and matrix spaces.

† The term 'chemiosmotic' theory is not very suitable; 'electrochemical membrane theory' would probably be better, but the former has become accepted. Similarly, Mitchell's term 'protonmotive' force is rather unfortunate for the potential difference produced by the transfer of a proton across the mitochondrion membrane. Even this term, imitating 'electromotive' force (retained only because of tradition), seems rather awkward.

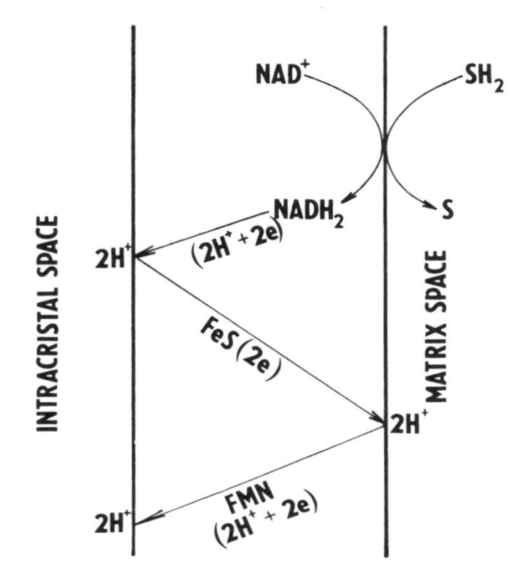

Fig. 6.27 An example of Mitchell's 'loops'. The substrate SH_2 is oxidized by NAD^+ (nicotinamideadeninedinucleotide) and the reduced form transports two protons and two electrons, of which two protons remain in the intracristal space and two electrons are transported back by the Fe–S protein to reduce FMN (flavinmononucleotide), the reduced form of which transports two protons and two electrons in the opposite direction

This can be used in several ways. H^+-ATPase plays a key role in oxidative phosphorylation (ATP synthesis) as it transfers protons back into the matrix space with simultaneous synthesis of ATP (with temporary enzyme phosphorylation, cf. page 451).

In the presence of certain substances, oxidative phosphorylation is uncoupled, i.e. while the carbonaceous substrate is oxidized, ATP is not synthesized, but can even be hydrolysed to ADP and a phosphate ion. The uncouplers of oxidative phosphorylation belong to the two groups of ionophores, one of which transfers protons across the mitochondrial membrane and the other alkali metal ions. The first group includes the semihydrophobic anions of weak acids such as 2,4-dinitrophenolate, dicumurate, azide, etc. By resonance effects resulting in spreading of the negative charge over the whole structure of these anions, they can pass through the lipid part of the inner membrane of the mitochondrion in the direction of increasing electric potential into the intracristal space, with higher concentration of hydrogen ions. They are then converted into the undissociated form of the corresponding acid, which is transported in the opposite direction into the matrix space, where protons obtained in the intracristal space are dissociated. The transport of the anions and of the undissociated acid is passive and decreases both the membrane potential and the activity of hydrogen ions in the intracristal space, so that the energy obtained in the oxidation of the carbonaceous substrate is dissipated as heat. The decrease in the hydrogen ion activity in the intracristal space leads to the opposite function of H^+-ATPase resulting in ATP hydrolysis.

The second group of uncouplers includes macrocyclic complexing agents of the valinomycin type (see page 445), which have the strongest effect in the presence of potassium ions. The hydrophobic (lipophilic) ionophore freely diffuses through the lipid part of the mitochondrial membrane. It binds potassium ions on the intracristal side, which are then passively transported into the matrix space in the complex, simulating proton transport. The explanation of the effect of uncouplers strongly supports the Mitchell theory of oxidative phosphorylation.

The difference in the hydrogen ion electrochemical potential, formed in bacteria similarly as in mitochondria, can be used not only for synthesis of ATP but also for the electrogenic (connected with net charge transfer) symport of sugars and amino acids, for the electroneutral symport of some anions and for the sodium ion/hydrogen ion antiport, which, for example, maintains a low Na^+ activity in the cells of the bacterium *Escherichia coli.*

It should be noted that, while the Mitchell theory explains some of the basic characteristics of oxidative phosphorylation, a number of phenomena connected with this process remain unclear from both the biochemical and the electrochemical points of view. Biochemical processes in the mitochondrial membrane are still far from being clarified completely. Although the Mitchell theory is called electrochemical, it is obvious that it was not formulated by an electrochemist. It employs only thermodynamic quantities and does not consider the kinetics of the individual steps. Charge balance is neglected in charge transfer across the mitochondrial membrane, although it is obvious that part of the transferred charge must be consumed in charging the electrical double layer at the membrane surface, which is very diversified; because of the presence of cristae the capacity of the double layer is large. The role of transferred protons in acid–base equilibria is also not clear. Nonetheless, the Mitchell theory has been a great step forward.

The second basic bioenergetic process consists of consumption of solar radiation as a source of energy in the organisms. One of the ways of employing solar radiation energy in the organism has already been mentioned in connection with the explanation of the function of the purple membrane of *Halobacterium halobium* (page 450); however, a far more important means of employing this energy for bioenergetic purposes is *photosynthesis*

$$6CO_2 + 6H_2O \xrightarrow[\text{chlorophyll}]{h\nu} C_6H_{12}O_2 + 6O_2 \qquad (6.5.4)$$

while the process in photosynthetic bacteria is

$$6CO_2 + 12H_2S \xrightarrow[\text{bacteriochlorophyll}]{h\nu} C_6H_{12}O_6 + 12S + 6H_2O \qquad (6.5.5)$$

However, process (6.5.5) cannot be a universal photosynthetic process because H_2S is unstable and is not available in sufficient quantities in nature. Water is the only substance that can be used in the reduction of carbon dioxide whose presence in nature is independent of biological processes.

Photosynthesis in green plants occurs in two basic processes. In the dark (the Calvin cycle) carbon dioxide is reduced by a strong reducing agent, the reduced form of nicotinamide adeninedinucleotide phosphate, $NADPH_2$, with the help of energy obtained from the conversion of ATP to ADP:

$$6CO_2 + 12NADPH_2 + 6H_2O \xrightarrow[\text{ATP} \;\rightarrow\; \text{ADP} + \text{P}_{\text{inorg}}]{} C_6H_{12}O_6 + 12NADP$$

The process occurring on the light is a typical membrane process. The membrane of the *thylakoid,* an organelle that constitutes the structural unit of the larger photosynthetic organelle, the chloroplast, contains chlorophyll (actually an integrated system of several pigments of the chlorophyll and carotene types) in two photosystems, I and II, and an enzymatic system permitting accumulation of light energy in the form of $NADPH_2$ and ATP (this case of formation of ATP from ADP and inorganic phosphate is termed *photophosphorylation*).

Figure 6.28 shows how the process occurring on the light proceeds in the two photosystems. Absorption of light energy involves charge separation in photosystem II, i.e. formation of an electron-hole pair. The hole passes to the enzyme containing manganese in the prosthetic group, which thus attains such a high redox potential that it can oxidize water to oxygen and hydrogen ions, set free on the internal side of the thylakoid membrane. The electron released from excited photosystem II reacts with plastoquinone (plastoquinones are derivatives of benzoquinone with a hydrophobic alkyl side chain), requiring transfer of hydrogen ions from the external side of the thylakoid membrane. The plastohydroquinones produced reduce the enzyme, plastocyanine, releasing hydrogen ions that are transferred across the internal side of the thylakoid membrane. On excitation by light, chlorophyll-containing photosystem I accepts an electron from plastocyanine by mediation of further enzymes. This high-energy electron then reduces the Fe–S protein, ferredoxin, which subsequently reduces NADP via further enzymes. Part of the energy is thus accumulated in the form of increased electrochemical potential of the hydrogen ions whose concentration—as has been seen—increases inside the thylakoid. The reverse transport of hydrogen ions provides H^+-ATPase with energy for synthesis of ATP. In contrast to cellular respiration, where the energy

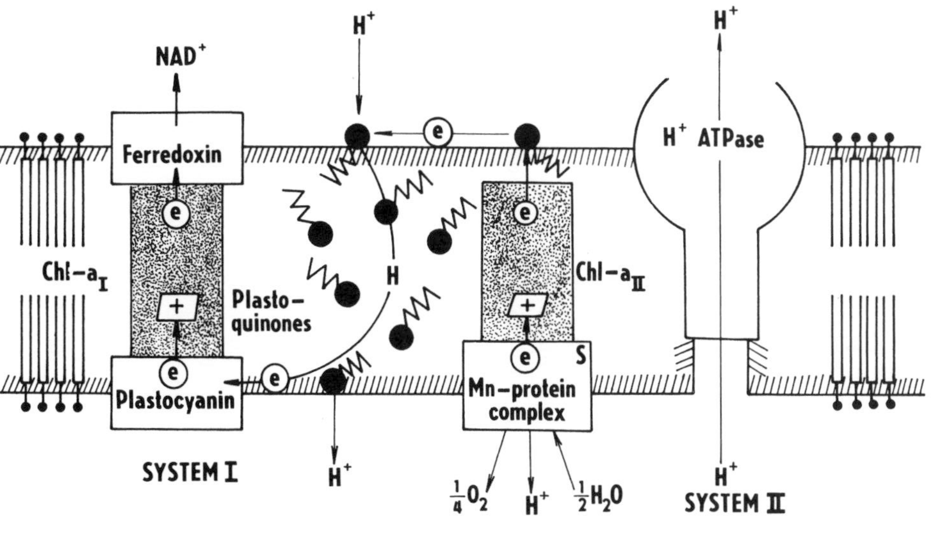

Fig. 6.28 The light process of photosynthesis. (According to H. T. Witt)

obtained in the oxidation of the carbonaceous substrate accumulates in the form of a single 'macroenergetic' compound of ATP, two energy-rich compounds are formed during photosynthesis, $NADPH_2$ and ATP, both of which are consumed in the synthesis of sugars. The subsequent steps appear to be purely biochemical in character and therefore will not be considered here.

In a partly biological, partly artificial model (page 397) reduced anthraquinone-2-sulphonate plays the role of NAD^+ and tetramethyl-*p*-phenylenediamine that of plastoquinones.

References

Baker, P. F. (Ed.), *The Squid Axon. Current Topics in Membranes and Transport,* Vol. 22, Academic Press, Orlando, 1984.

Barber, J. (Ed.), *Topics in Photosynthesis,* Vols. 1–3, Elsevier, Amsterdam, 1978.

Briggs, W. R. (Ed.), *Photosynthesis,* A. R. Liss, New York, 1990.

Calahan, M., Molecular properties of sodium channels in excitable membranes, in *The Cell Surface and Neuronal Function* (Eds C. W. Cotman, G. Poste and G. L. Nicholson), P. I, Elsevier, Amsterdam, 1980.

Catteral, W. A., The molecular basis of neuronal excitability, *Science,* **223,** 653 (1984).

Cole, K. S., *Membranes, Ions and Impulses,* University of California Press, Berkeley, 1968.

Conti, F., and E. Neher, Single channel recordings of K^+ currents in squid axons, *Nature,* **285,** 140 (1980).

Cramer, W. A., and D. B. Knaff, *Energy Transduction in Biological Membranes, A Textbook of Bioenergetics,* Springer-Verlag, Berlin, 1989.

French, R. J., and R. Horn, Sodium channel gating: models, mimics and modifiers, *Ann. Rev. Biophys. Bioeng.,* **12,** 319 (1983).

Govindjee (Ed.), *Photosynthesis,* Vols. 1 and 2, Academic Press, New York, 1982.

Hagiwara, S., and L. Byerly, Calcium channel, *Ann. Rev. Neurosci.,* **4,** 69 (1981).

Hille, B., *Ionic Channels of Excitable Membranes,* Sinauer, Sunderland, 1984.

Katz, B., *Nerve, Muscle and Synapse,* McGraw-Hill, New York, 1966.

Keynes, R. D., The generation of electricity by fishes, *Endeavour,* **15,** 215 (1956).

Keynes, R. D., Excitable membranes, *Nature,* **239,** 29 (1972).

Lee, C. P., G. Schatz, and L. Ernster (Eds), *Membrane Bioenergetics,* Addison-Wesley, Reading, Mass., 1979.

Mitchell, P., Davy's electrochemistry—Nature's protochemistry, *Chemistry in Britain,* **17,** 14 (1981).

Mitchell, P., Keilin's respiratory chain concept and its chemiosmotic consequences, *Science,* **206,** 1148 (1979).

Moore, J. W. (Ed.), *Membranes, Ions and Impulses,* Plenum Press, 1976.

Nicholls, D. G., *Bioenergetics, An Introduction to the Chemiosmotic Theory,* Academic Press, Orlando, 1982.

Sakmann, B., and E. Neher (Eds), *Single-Channel Recordings,* Plenum Press, New York, 1983.

Sauer, K., Photosynthetic membranes, *Acc. Chem. Res.,* **11,** 257 (1978).

Skulatchev, V. P., Membrane bioenergetics—Should we build the bridge across the river or alongside of it?, *Trends Biochem. Sci.,* **9,** 182 (1984).

Stein, W. D. (Ed.), *Current Topics in Membranes and Transport,* Vol. 21, *Ion Channels: Molecular and Physiological Aspects,* Academic Press, Orlando, 1984.

Stevens, C. F., Biophysical studies in channels, *Science,* **225,** 1346 (1984).
Stroud, R. M., and J. Tinner-Moore, Acetylcholine receptor, structure, formation and evolution, *Ann. Rev. Cell Biol.,* **1,** 317 (1985).
Tsien, R. W., Calcium channels in excitable cell membranes, *Ann. Rev. Physiol.,* **45,** 341 (1983).
Ulbricht, W., Ionic channels and gating currents in excitable membranes, *Ann. Rev. Biophys. Bioeng.,* **6,** 7 (1977).
Urry, D. W., A molecular theory of ion-conducting channels: A field dependent transition between conducting and non-conducting conformations, *Proc. Natl. Acad. Sci. USA,* **69,** 1610 (1972).
Van Dam, K., and B. F. Van Gelder (Eds.), *Structure and Function of Energy Transducing Membranes,* Elsevier, Amsterdam, 1977.
Witt, H. T., Charge separation in photosynthesis, *Light-Induced Charge Separation in Biology and Chemistry* (ed. H. Gerischer and J. J. Katz), Verlag Chemie, Weinheim, 1979.

Appendix A

Recalculation Formulae for Concentrations and Activity Coefficients

	Expressed in		
	x_1	m_1	c_1
(a) Concentration term			
Mole fraction	x_1	$\dfrac{m_1 M_0}{1 + \nu m_1 M_0}$	$\dfrac{c_1 M_0}{\rho + (\gamma M_0 + M_1)c_1}$
In dilute solution	$x_1 \approx \dfrac{n_1}{n_0}$	$m_1 M_0$	$\dfrac{c_1 M_0}{\rho_0}$
Molality	$\dfrac{x_1}{M_0 + \nu x_1 M_0}$	m_1	$\dfrac{c_1}{\rho + M_1 c_1}$
In dilute solution	$\dfrac{x_1}{M_0}$	m_1	$\dfrac{c_1}{\rho_0}$
Molar concentration	$\dfrac{x_1}{M_0 + (M_1 + \nu M_0)x_1}$	$\dfrac{\rho m_1}{1 + m_1 M_1}$	c_1
In dilute solution	$\dfrac{x_1 \rho_0}{M_0}$	$m_1 \rho_0$	c_1
(b) Activity coefficient			
$\gamma_{\pm,x}$	$\gamma_{\pm,x}$	$\gamma_{\pm,m}(1 + \nu m_1 M_0)$	$\gamma_{\pm,c} \dfrac{\rho + (M_0 - M_1)c_1}{\rho_0}$
$\gamma_{\pm,m}$	$\gamma_{\pm,x}(1 - \nu x_1)$	$\gamma_{\pm,m}$	$\gamma_{\pm,c} \dfrac{\rho - c_1 M_1}{\rho_0}$
$\gamma_{\pm,c}$	$\gamma_{\pm,x} \dfrac{\rho_0}{\rho}\left(1 + \nu x_1 + x_1 \dfrac{M_1}{M_0}\right)$	$\gamma_{\pm,m} \dfrac{\rho_0}{\rho}(1 + m_1 M_1)$	$\gamma_{\pm,c}$

M_0 and M_1 denote molar mass ($kg \cdot mol^{-1}$) of the solvent and of the solute, respectively, ν is the number of ions formed by an electrolyte molecule on dissolution, ρ and ρ_0 denote the density of the solution and of the pure solvent, respectively, and n_1 and n_0 are the amount of the solute and of the solvent (mol), respectively. The above relationships also hold for the ionic quantities m_+, m_-, c_+, etc., where m_1, c_1 and x_1 are replaced by $m_+ = \nu_+ m_1$, $m_- = \nu_- m_1$, $c_+ = \nu_+ c_1$, etc.

Appendix B
List of Symbols

A	affinity, surface
a	activity, effective ion radius, interaction coefficient
c	concentration
C	differential capacity, dimensionless concentration
D	diffusion coefficient, relative permittivity
E	electromotive force (EMF), electrode potential, electric field strength, energy; $E_{1/2}$ half-wave potential, $E^0_{\text{subscript}}$ standard electrode potential, E_{pzc} potential of zero charge, E_{p} potential of ideal polarized electrode, potential energy
e	proton charge
F	Faraday constant
G	Gibbs energy, conductance
g	gaseous phase
g	gravitation constant
H	enthalpy
H^0	acidity function
I	electric current, ionization potential
J, $\mathbf{J}$	flux of thermodynamic quantity
j, $\mathbf{j}$	current density
K	equilibrium constant, integral capacity
k	Boltzmann constant, rate constant, k_{a} anodic electrode reaction rate constant, adsorption rate constant, k_{c} cathodic electrode reaction rate constant, k_{d} desorption rate constant, $k^{\ominus}$ conditional electrode reaction rate constant
L	phenomenological coefficient
L_{D}	Debye length
l	length
l	liquid phase
M	relative molar mass ('molecular weight'), amount of thermodynamic quantity
m	molality, flowrate of mercury
m	metallic phase, membrane
N	number of particles; N_{A} Avogadro constant
n	amount of substance (mole number), number of electrons exchanged, charge number of cell reaction

P	solubility product
R	electric resistance, gas constant
r	radius
s	overall concentration, number of components
s	solid phase
S	entropy
T	absolute temperature
t	time, transport (transference) number; t_1 drop time
U	electrolytic mobility, internal energy, potential difference ('voltage')
u	mobility
V	volume, electric potential
v	molar volume, velocity
X, $\mathbf{X}$	thermodynamic force
x	molar fraction, coordinate
Z	impedance
z	charge number
α	dissociation degree, charge transfer coefficient, real potential
β	buffering capacity, adsorption coefficient
γ	activity coefficient, interfacial tension
Γ	surface concentration, surface excess
ΔG	reaction Gibbs energy change
$\Delta \tilde{H}$	activation enthalpy (energy) of electrode reaction
$\Delta \phi$	Galvani potential difference
$\Delta \psi$	Volta potential difference
δ	thickness of a layer, deviation
ε	permittivity, electron energy, ε_F Fermi level energy
ζ	electrokinetic potential
η	overpotential, viscosity coefficient
Θ	relative coverage
θ	wetting angle
κ	conductivity, mass transfer coefficient
Λ	molar conductivity
λ	ionic conductivity, fugacity
μ	chemical potential, reaction layer thickness; $\tilde{\mu}$ electrochemical potential
ν	kinematic viscosity, number of ions in a molecule, radiation frequency
π	osmotic pressure
ρ	density, space charge density
σ	surface charge density, rate of entropy production
τ	transition time, membrane transport number
ϕ	inner electrical potential, osmotic coefficient
χ	surface electrical potential

ψ electrical potential, outer electrical potential, density of a thermodynamic quantity

ω frequency, angular frequency

Φ dissipation function; Φ_e electron work function

Subscripts and superscripts of these symbols give reference to chemical species, charge, standard state, etc., of quantities concerned. Symbols that seldom occur have not been included in the above list, their meaning having been given in the text.

Index